T0135375

Studies in Computational Intelligence

Volume 502

Series Editor

J. Kacprzyk, Warsaw, Poland

For further volumes:
http://www.springer.com/series/7092

Peijun Guo · Witold Pedrycz
Editors

Human-Centric Decision-Making Models for Social Sciences

Editors
Peijun Guo
Faculty of Business Administration
Yokohama National University
Yokohama
Japan

Witold Pedrycz
Department of Electrical and Computer Engineering
University of Alberta
Edmonton
Canada

ISSN 1860-949X ISSN 1860-9503 (electronic)
ISBN 978-3-662-51069-8 ISBN 978-3-642-39307-5 (eBook)
DOI 10.1007/978-3-642-39307-5
Springer Heidelberg New York Dordrecht London

Softcover reprint of the hardcover 1st edition 2014

Printed on acid-free paper

Springer is part of Springer Science+Business Media (www.springer.com)

Preface

Decision making in business, economics, and social sciences is omnipresent yet at the same time it becomes highly difficult to comprehend and model human mental processes. In general, in spite of their diversity the decision problems exhibit a number of highly visible features:

- the objective of the decision problem is ambiguous;
- the problem structure describing the relationship among sub-problems might be loosely specified;
- preference relations are not explicitly stated;
- knowledge of the organizational environments is uncertain;
- available information is often imprecise, uncertain or there might be an acute lack of information; and
- one-time decision models are needed when dealing with unrepeated problems with partially available information.

The classical decision theories, such as the expected utility (EU) theory of von Neumann and Morgenstern, and the subjective expected utility (SEU) theory of Savage cannot fully address the complexity of the problems and still a number of open critical questions remain that need to be thoroughly addressed.

This edited volume aims to offer effective methods to deal with different types of uncertainty inherently existing in decision problems and deliver comprehensive decision frameworks to handle different decision scenarios under various facets of uncertainty. The objective is to bring forward diverse decision-making models, which help use effectively the explicit and tacit knowledge and intuition, model perceptions and preferences in a more human-oriented style, and form decisions which become more in rapport with a human line of thinking.

The volume presents original approaches and delivers new results in fundamentals and applications related to human-centered decision making approaches to business, economics, and social systems. It includes multi-criteria (multiattribute) decision making, decision making with prospect theory, decision making with incomplete probabilistic information, granular models of decision making and decision making realized with the use of non-additive measures. New emerging decision theories being presented as along with a wide spectrum of ongoing research make the book valuable to all interested in the field of advanced decision making.

An overall concise characterization of the objectives of this edited volume is captured by highlighting several focal points:

- Systematic exposure of the concepts, design methodologies, and detailed algorithms. This is a self-explanatory feature of the volume; the systematic, well-organized flow of the presentation of the ideas is directly supported by a way in which the material is structured.
- Individual chapters with clearly delineated agenda and well-defined focus and additional reading material available via carefully structured references.
- Self-containment. The intent is to provide a material, which is self-contained and provides the reader with all necessary prerequisites and, if necessary, augments some parts of the material with a step-by-step explanation. More advanced concepts are supported by a significant amount of illustrative numeric material. Furthermore several detailed application scenarios are offered to motivate the reader and make some abstract concepts more tangible and easy to follow.

This book is aimed at a broad audience of researchers and practitioners. The areas of particular interest include industrial engineering, informatics, business, economics, social systems. The material could be also of interest to those involved in operations research, management, and various branches of engineering. A prudently struck balance between the theoretical studies and applications makes the material suitable for researchers as well as graduate students especially in courses such as information, computer sciences, psychology, cognitive science, economics, system engineering, operation research and management science, risk management, public and social policy.

We would like to take this opportunity to express our sincere thanks to the authors for reporting on their innovative research and sharing their insights into the area. The reviewers deserve our thanks for their constructive input. We highly appreciate a continuous support and encouragement from the Editor-in-Chief, Prof. Janusz Kacprzyk whose leadership and vision makes this book series a unique vehicle to disseminate the most recent, highly relevant, and far-fetching publications in the domain of Computational Intelligence and intelligent systems.

We hope that the readers will find this volume of genuine interest and the research reported here will help foster further progress in research, education, and numerous practical endeavors.

Peijun Guo
Witold Pedrycz

Contents

Decision Making in the Environment of Heterogeneous Uncertainty 1
Phan H. Giang

One-Shot Decision Theory: A Fundamental Alternative for Decision Under Uncertainty 33
Peijun Guo

On the Influence of Emotion on Decision Making: The Case of Charitable Giving 57
Ryan Kandrack and Gustav Lundberg

Decision Theory and Rules of Thumb 75
Konstantinos V. Katsikopoulos

Aggregating Imprecise Linguistic Expressions 97
Edurne Falcó, José Luis García-Lapresta and Llorenç Roselló

Risk Perception and Ambiguity in a Quantile Cumulative Prospect Theory 115
Marcello Basili

Effective Decision Making in Changeable Spaces, Covering and Discovering Processes: A Habitual Domain Approach 131
Moussa Larbani and Po Lung Yu

Decision Making Under Interval Uncertainty (and Beyond) 163
Vladik Kreinovich

Dealing with Imprecision in Consumer Theory: A New Approach to Fuzzy Utility Theory 195
David Gálvez Ruiz and José Luís Pino Mejías

Decision Making Under Z-Information 233
R. A. Aliev and Lala M. Zeinalova

Approximations of One-dimensional Expected Utility Integral of Alternatives Described with Linearly-Interpolated *p*-Boxes 253
N. D. Nikolova, S. Ivanova and K. Tenekedjiev

Human-Centric Cognitive Decision Support System for Ill-Structured Problems 289
Tasneem Memon, Jie Lu and Farookh Khadeer Hussain

Decision-Making Under Conditions of Multiple Values and Variation in Conditions of Risk and Uncertainty 315
Ewa Roszkowska and Tom R. Burns

Supporting Ill-Structured Negotiation Problems 339
Ewa Roszkowska, Jakub Brzostowski and Tomasz Wachowicz

Personalised Property Investment Risk Analysis Model in the Real Estate Industry 369
Nur Atiqah Rochin Demong, Jie Lu and Farookh Khadeer Hussain

The Logic and Ontology of Assessment of Conditions in Older People 391
Patrik Eklund

Decision Making on Energy Options: A Case Study 401
V. Jain, D. Datta and A. Deshpande

Decision Making in the Environment of Heterogeneous Uncertainty

Phan H. Giang

Abstract The environment of heterogeneous uncertainty is characterized by the presence of variables in multiple uncertainty formalisms. This paper provides an overview of decision models under several uncertainty frameworks including probability theory, Dempster-Shafer belief function theory and possibility theory. It explores the challenges in pulling them together for decision making. We show that the information of sequence of variable resolution, which was often neglected, actually plays a key role in decision making under heterogeneous uncertainty. A novel approach, based on the well-known folding-back principle, to find the certainty equivalent of acts under heterogeneous uncertainty is proposed.

Keywords Decision making · Possibility theory · Dempster-Shafer belief function · Ignorance

1 Introduction

Most models of uncertainty used in science, engineering and medicine are probabilistic in nature. That is, all kinds of relevant uncertainty are forced to be quantified in terms of probability, chance or risk. While probabilistic modeling may apply naturally to many uncertain variables of interest, for others, it is simply a choice motivated by convenience rather than objective justification. When reliable data are scarce, a reliable estimation of probability would not be possible. If a concept is vaguely defined, its frequency count is inevitably imprecise. Even when the objective data are lacking, some would still argue that subjective/personal probability can be determined by observable preference via the subjective expected utility theory (SEU) according to the tradition established by de Finetti and Savage. However, the

P. H. Giang (✉)
George Mason University, 4400 University Dr., Fairfax, VA 22030, USA
e-mail: pgiang@gmu.edu

P. Guo and W. Pedrycz (eds.), *Human-Centric Decision-Making Models for Social Sciences*, Studies in Computational Intelligence 502,
DOI: 10.1007/978-3-642-39307-5_1,

results of research in last several decades show that SEU is not adequate for decision under uncertainty.

For example, if you ask an expert about how the risk of a nuclear disaster of the scale of Fukushima Daiichi is estimated. The expert will tell you that the probability of an accident at a nuclear power plant is often the result of elaborate modelings in which besides the laws of physics, also included are engineers' subjective assumptions. While the laws of physics are accurate, the subjective assumptions are much less reliable.

For another example, economists of liberal inclination rarely agree with their colleagues on conservative side about the consequences of a proposed policy. They disagree despite having the same academic training, using the same modeling methodology and accessing the same data. In particular, there is an on-going consequential debate among economists about the value of the *fiscal multiplier* which measures the change in national income relative to the change in government spending. Depending on that value, one can either justify a policy to increase government spending to fight an economic recession or recommend against the policy as an ineffective remedy.

In fact, in most practical situations where an accurate estimation of probability is not possible, practitioners have to supply subjective "educated" guess. As a consequence, the optimality of a decision derived from the models is a conditional property that depends on the validity of such guesses. In the quest for robustness, that dependency must be minimized.

The probability doctrine seems especially deficient to deal with situations of ignorance where reliable evidence is not available, data are noisy and expert opinions are contradictory.

Several formal theories have been proposed to capture the notions of non-additive uncertainty which elude the probability theory. In computer science and statistics, the possibility theory rooted in Zadeh's fuzzy set theory [8, 30], Dempster-Shafer belief function theory [5, 23] and imprecise probability theory [28] have been extensively investigated.

There is some misconception about the roles of those uncertainty theories. For example, we often hear the claim that non-probabilistic formalisms such as DS belief function theory can replace the role played by probability theory because the latter is just a special case of the former. This claim ignores an important fact that unlike probability which is updated according to Bayes' rule, belief functions can be updated in several ways that produce different updated belief [18].

The literature on uncertainty representation has been converging on the consensus that different uncertainty formalisms are designed to capture different aspects of the uncertainty that exists in the real world. In particular, the notion of chance is best represented by using probability theory, the vagueness inherent in linguistic variables by using possibility theory, the ambiguity associated with evidence by using DS belief function and the imprecision due to lack of data is conveyed by imprecise probability. This fact points naturally to the need to combine them in decision making situations. It is not difficult to find a real world decision problem that involves various kinds of uncertainty. For example, the choice in a national election depends on various considerations such as the chance that economy improves under candidate's plan, the vague perception that he is a good and honest person, the evidence of achievement or failure of his leadership in the past, and so on.

An important motivation of non-additive uncertainty formalisms, besides expressive power consideration, is to account for systematic violations of Savage's postulates which form the foundation of the subjective expected utility theory (SEU) [22]. The problem was first analyzed by scholars such as Allais [1] and Ellsberg [9] and subsequently has been the subject of many studies (see [16] and [26] for systematic discussions). Some of such violations can be explained by the cognitive perception bias, others such as Ellsberg's paradox is clearly due to the fact that the relevant uncertainty is not reducible to a probability measure.

It is well accepted common sense that people perceive and process risk/probability and uncertainty/ambiguity differently. But only recently, neuroscientists have accumulated scientific evidence to support the idea that the difference is truly fundamental and occurring at the physiological level. They have discovered that the regions of brain that deal with probability and the regions that handle ambiguous situations are distinct [17].

The advantage of using entire arsenal of uncertainty formalisms is that decision makers can capture, express and process information about the world in most intuitive and credible way, without invoking unsupported assumptions.

While the literature on individual non-additive uncertainty formalisms is extensive, the research that brings them together under one roof for decision making is almost nonexistent. Many scholars pursue an approach which places focus on decision making under more and more general uncertainty formalism. Underlying such "generalization approach" is the assumption, which has been questioned in the recent literature, that every relevant piece of uncertainty information can be expressed in a common uncertainty language.

In contrast, we want to aggregate different uncertainty frameworks to make them work together. In other words, our objective is to create a decision making framework that involves uncertain variables of different types including probabilistic, possibilistic, DS belief variables, variables with indeterminate probability and ignorant variables (those for which there is total lack of knowledge).

The plan of this chapter is as follows. In the next section we provide a systematic review of decision making models for probability theory, possibility theory and belief function theory. In the main section, we will pull those theories together to analyze decision situations in which uncertainty variables of different types are involved.

First, we list notation convention used in this chapter. Following a convention in literature, we make "you" the decision maker to help streamlining the discussion. For example, we would write: your knowledge about the world is encoded by some uncertainty function or you want to choose an action among available alternatives.

We use the capital letters to the end of the alphabet e.g., X, Y, Z. to denote variables. Their instances are denoted by lower case letters. A state is a tuple of instances of all variables. The set of states is denoted by Ω. Events or subsets of *states* are denoted by capital letters to the start of the alphabet e.g., A, B, C.

An act is a mapping from the set of states Ω to the set of outcomes $\mathcal{W} = [0, 1]$. Two acts f, g are equivalent if $\forall s \in \Omega$, $f(s) = g(s)$. Besides the mapping notation, an act is also recorded in the *rule* form: $\{A_i \hookrightarrow x_i\}_{i=1}^k$. The reading of rule $A_i \hookrightarrow x_i$ is "if event A_i occurs then you get outcome x_i". A value $c \in \mathcal{W}$ is called a constant act which in the rule form is $\Omega \hookrightarrow c$. A chain rule $E_1 \hookrightarrow E_2 \hookrightarrow \ldots E_k \hookrightarrow x$ means "if A_1 and A_2 and $\ldots$ A_k then x". Acts are denoted by lower case letters d, f, h, p, q, r etc. The set of acts is denoted by $\mathcal{D}$.

Given that your information is encoded as a measure of uncertainty over Ω, your choice behavior over acts is modeled by a preference $\succeq$ over acts. We assume that $\succeq$ is a weak order (reflexive, complete and transitive). The symmetric part $\sim$ and asymmetric part $\succ$ of $\succeq$ are defined as usual: $f \succ g$ iff $f \succeq g$ and $g \not\succeq f$. $f \sim g$ iff $f \succeq g$ and $g \succ f$. The restriction of $\succeq$ on the set of constant acts ($\mathcal{W}$) is denoted by the "greater than or equal to" symbol $\geq$. If f is an act and c is a constant act such that $f \sim c$, c is called the certainty equivalent of f.

2 Common Structure of Decision Making

This section, we start with the review of the classical expected utility theory for probability then proceed to describe decision theories developed for other non-additive uncertainty frameworks. The goal is to highlight both similarity and difference between them.

2.1 Expected Utility Theory

In this subsection, we assume that the risk relevant to your decision problem is described by a probability measure P on Ω. Your preference relation is denoted by $\succeq_{vnm}$. We say that your preference has the expected utility representation if

$$p \succeq_{vnm} q \quad \text{iff} \quad \mathbb{E}_P[U(p)] \geq \mathbb{E}_P[U(q)] \tag{1}$$

where U, utility function, is an increasing function from $\mathcal{W} \to \mathbb{R}$ where $\mathbb{R}$ is the set of real numbers.

The expected utility for an act q is calculated by

$$\mathbb{E}_P[U(q(s))] = \sum_{s \in \Omega} P(s)\ U(q(s)) \tag{2}$$

The certainty equivalent of q is calculated by

$$c = U^{-1}(\mathbb{E}_P[U(q(s))]) \tag{3}$$

where U^{-1} is the inverse function of U.

Note that the certainty equivalent remains unchanged if U is transformed by a positive linear transformation. To see that suppose $U_1 = aU + b$ with constants $a > 0, b$. Consider the inverse function U_1^{-1}. By the defining property definition $U_1(U_1^{-1}(x)) = x$ and Eq. $\mathbb{E}_P[U_1(q(s))] = \mathbb{E}_P[aU(q(s))+b] = a\mathbb{E}_P[U(q(s))]+b$, we have

$$\forall x,\ aU(U_1^{-1}(x)) + b = x \tag{4}$$

$$\forall x,\ U(U_1^{-1}(x)) = \frac{x-b}{a} \tag{5}$$

$$U(U_1^{-1}(\mathbb{E}_P[U_1(q(s))])) = \frac{\mathbb{E}_P[U_1(q(s))] - b}{a} = \mathbb{E}_P[U(q(s))] \tag{6}$$

$$U^{-1}(U(U_1^{-1}(\mathbb{E}_P[U_1(q(s))]))) = U^{-1}(\mathbb{E}_P[U(q(s))]) \tag{7}$$

$$U_1^{-1}(\mathbb{E}_P[U_1(q(s))]) = U^{-1}(\mathbb{E}_P[U(q(s))]) \tag{8}$$

In literature, the expected utility model has been characterized in several ways. A characterization is the necessary and sufficient condition that $\succeq_{vnm}$ must satisfy in order for representation equivalence $p \succeq_{vnm} q$ iff $\mathbb{E}_P[U(p)] \geq \mathbb{E}_P[U(q)]$ to hold. The following characterization is proposed by Jensen [21].

A key concept in this characterization is that of compound acts. Suppose p, q are acts, $\{\alpha/p, (1-\alpha)/q\}$ denotes an act r that pays p with probability α and q with probability $1-\alpha$. The implementation mechanism of the compound act is described in Fig. 1. A biased coin (chance of Head is α and the chance of Tail is $1-\alpha$) is tossed. If the coin lands Head, you will get act p, if Tail then you get q.

Technically, a compound act is a conditional act not a mapping of signature $\Omega \to \mathcal{W}$. Given the coin landed Head $r_{\text{Head}}(s) = p(s)$, and given Tail $r_{\text{Tail}}(s) = q(s)$. However, in characterization of $\succeq_{vnm}$, the compound act is identified with the linear combination as follows.

$$\{\alpha/p, (1-\alpha)/q\} \overset{\text{def}}{=} \{\alpha\ p + (1-\alpha)q\} \tag{9}$$

$$\forall s \in \Omega,\ \{\alpha\ p + (1-\alpha)q\}(s) \mapsto \alpha\ p(s) + (1-\alpha)q(s) \tag{10}$$

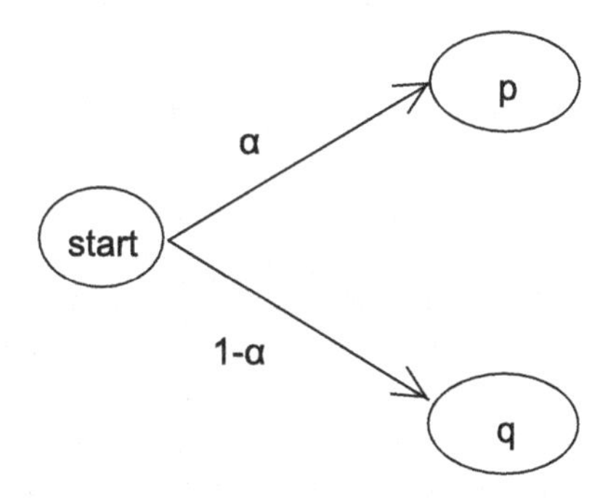

Fig. 1 Compound act

Notation $\{\cdot\}$ denotes acts. The reading of Eq. (9) is that compound act $\{\alpha/p, (1-\alpha)/q\}$ is taken to be equivalent to a linear combination of acts $\{\alpha\ p + (1-\alpha)q\}$. The Eq. (10) explains that the linear combination of p, q is a an act whose outcome is the linear combination of the outcomes of p and q. Note that the difference between the definition of compound act and that of linear combination is when the coin toss is used. In the former, coin is tossed before state realization. In the latter, the order is reversed. The need to distinguish between compound acts and linear combinations will be made clear later. The main reason is that this identity holds only for probabilistic acts. When the uncertainty is non-probabilistic such as belief function or possibility function the identity no longer hold.

Now let us look at the axioms more closely.

J1 Weak order. $\succeq_{vnm}$ is complete and transitive.
J2 Archimedean. For $p, q, r \in \mathcal{D}$ such that $p \succ_{vnm} q \succ_{vnm} r$ then there exist $\alpha, \beta \in [0, 1]$ such that $\{\alpha/p, (1-\alpha)/r\} \succ_{vnm} q$ and $q \succ_{vnm} \{\beta/p, (1-\beta)/r\}$.
J3 Independence. For all $p, q, r \in \mathcal{D}$ and $\alpha \in [0, 1]$, $p \succeq_{vnm} q$ iff $\{\alpha/p,\ (1-\alpha)/r\} \succeq_{vnm} \{\alpha/q,\ (1-\alpha)/r\}$

Some comments are in order. The weak order axiom conveys well accepted requirements for rationality. The completeness says that between two acts, you are always able to tell which one you prefer. The transitivity is meant to ensure that your preference is immune from the Dutch book or "money pump" attack against you. The last two axioms involve the compound acts. The Archimedean axiom implies that given two simple acts p, r, by choosing a suitable bias for a coin you can create a compound act that is indifferent to any act q which lies in between p and r. The independence axiom says that contingent branches of a compound act are evaluated separately from other branches.

A theorem proved by Jensen [21] states that the necessary and sufficient condition for preference $\succeq_{vnm}$ to have the representation $p \succeq_{vnm} q$ iff $\mathbb{E}_P[U(p)] \geq \mathbb{E}_P[U(q)]$ is the axioms $J1$, $J2$ and $J3$ together with the definition of compound acts and by linear combinations.

2.2 Decision Under Ignorance

An important reason to consider non-probabilistic uncertainty is the desire to express the notion of ignorance which probability theory has been shown to be unable to adequately capture. The examination of decision under ignorance is critical for two reasons. From practical point of view, ignorance is much more pervasive in real life situations than we are ready to admit. From a theoretical point view, ignorance is a test bed for any decision under uncertainty theory.

In literature on decision making under ignorance, an act can be identified with its set of possible outcomes because the uncertainty information associated with outcomes is absent. Hurwicz and Arrow [2] made the ground-breaking work on decision under ignorance in early 1950s. Their basic construct is a choice operator

($\hat{\cdot}$) that returns a subset of optimal acts $\hat{D}$ from the the set of available acts D. Arrow and Hurwicz examined the implication of imposing rationality postulates (properties A to D) that the choice operator must satisfy. Property A requires that if $D_1 \subset D_2$ and $\hat{D}_2 \cap D_1 \neq \emptyset$ then $\hat{D}_1 = \hat{D}_2 \cap D_1$. Property B requires that relabeling actions and states does not change the optimal status of actions. Property C says that deletion of a duplicate state does not change the optimality status of actions. A state s is a duplicate of another state s' with respect to set of acts D if $\forall f, f' \in D$ $f(s) = f(s')$. Property D states that for $f, f' \in D$ and f dominates f' if $f' \in \hat{D}$ then f is also in $\hat{D}$ and if $f \notin \hat{D}$ then $f' \notin \hat{D}$. An act is said to dominate another act if for every state the outcome of the former is as least as good as the outcome of the latter.

They have shown that under complete ignorance, only extreme (the best and the worst) outcomes matter. For example, $f(\Omega) = \{0.1, 0, 3, 0, 4, 0, 7\}$ is indifferent to $g(\Omega) = \{0.1, 0, 7\}$. A family of utility functions including max, min and linear combinations of minimal and maximal values is permissible by this criterion.

The *sequential consistency* is a property that essentially imposes the condition that starting from the epistemic stage of ignorance, one can not manipulate the value of an act by repackaging it with hypothetical reasoning. Specifically, suppose that you are ignorant about states in Ω and f is an act. Suppose that $\mathcal{H} = \{A, \bar{A}\}$ is a partition of Ω. You can reason as follows. If A occurs, then you will encounter a new act under ignorance denoted by f_A whose outcomes is $f(A)$. Alternatively if $\bar{A}$ occurs you will encounter another act under ignorance $f_{\bar{A}}$ with the set of outcome $f(\bar{A})$. If the certainty equivalent of f_A is x and that of $f_{\bar{A}}$ is y. So before the question which of A or $\bar{A}$ occurs is resolved, you have the set of potential outcomes $\{x, y\}$ but you are still ignorant about which will obtains. So, without any new information from outside, by purely hypothetical reasoning you can translate the original act under ignorance f to another act $\{x, y\}$ also under ignorance. The sequential consistency condition says that you cannot improve an act by such a trick. In other words, f and $\{x, y\}$ must be indifferent.

In [11] it has been shown that if the sequential consistency condition is imposed in addition to Arrow-Hurwicz four conditions then the utility function under ignorance must have the following form: There exists a value a which is called your *default* value, such that

$$U_I(D) = \begin{cases} \max(D) & \text{if } \max(D) \leq a \\ a & \text{if } \min(D) \leq a \leq \max(D) \\ \min(D) & \text{if } \min(D) \geq a \end{cases} \tag{11}$$

Utility function U_I has an appealing interpretation. Each decision maker has a pre-determined default value which she considers "good enough", "satisfactory" or "acceptable" in circumstance that she knows nothing about. If the potential outcome range of an act covers the default value, she is willing to exchange the act for that value. In case the outcome range does not include the default, she is willing to exchange the act for the value in the range which is closest to the default value.

2.3 Decision with Possibility Theory

The possibility theory has a root in Zadeh's works in fuzzy set theory [30] and substantially developed in theory and methodology by Dubois and Prade group in Toulouse [8]. This theory is used to represent partial ignorance and uncertainty on ordinal structure. Zadeh's original idea is to use possibility theory to formalize the semantics of statements in natural languages.

A *possibility* function is a mapping $\pi : 2^{\Omega} \rightarrow [0, 1]$ that satisfies constraints $\pi(\Omega) = 1, \pi(\emptyset) = 0$ and $\forall A, B \subseteq \Omega$

$$\pi(A \cup B) = \max(\pi(A), \pi(B)) \tag{12}$$

The non-additivity of possibility functions can be seen by the fact that the sum $\pi(A)+\pi(\bar{A})$ where $\bar{A}$ is the negation (complement) of A, is not invariant for different A. In fact, this sum is constrained by $1 \leq \pi(A) + \pi(\bar{A}) \leq 2$ because $1 = \pi(\Omega) = \pi(A \cup \bar{A}) = \max(\pi(A), \pi(\bar{A}))$, so $1 = \max(\pi(A), \pi(\bar{A})) \leq \pi(A) + \pi(\bar{A}) \leq 2 * \max(\pi(A), \pi(\bar{A})) = 2$.

From the possibility function, a dual construct can be derived. A *necessity* function ν on the 2^{Ω} is defined as $\nu(A) = 1 - \pi(\neg A)$. Clearly for

$$\nu(A \cap B) = \min(\nu(A), \nu(B)) \tag{13}$$

There are two regimes that a possibility function can be updated given a new information about realization of some event A. A *quantitative* conditional possibility of B given A is defined when $\pi(A) > 0$

$$\pi(B|A) = \frac{\pi(A \cap B)}{\pi(A)} \tag{14}$$

Under the qualitative regime, the *qualitative* conditional possibility

$$\pi(B|A) = \begin{cases} 1 & \text{if } \pi(A \cap B) = \pi(A) \\ \pi(A \cap B) & \text{otherwise} \end{cases} \tag{15}$$

Note that in the qualitative regime of possibility theory any ordinal set can be used as the uncertainty scale because the only operation needed is maximization.

For the rest of the chapter we will assume the possibility theory under quantitative regime. The discussion can be translated to the qualitative regime with minimal change.

The relationship between the possibility theory and belief function theory is not as simple it seems. One the one hand, the quantitative possibility can be seen as a special case of belief function theory. On the other hand, the qualitative version of possibility has no representation in belief function theory.

Besides the fuzzy-set interpretation, possibility theory has an interpretation in terms of statistical likelihood function. In this interpretation, the set of states Ω is the set of probabilistic models that you entertain for your stochastic problem. Specifically, a state $s \in \Omega$, is a probability measure over a sample space $\mathcal{X}$. Your information about the models is extracted from observed data. Suppose $\mathbf{x}$ is the data collected about the stochastic problem. According to the likelihood principle of statistics [3, 4], all information about the models is contained in the likelihood function calculated from the observation. The probability of observing $\mathbf{x}$ predicted by model s is $s(\mathbf{x})$. As a function of the models given observation $lik_{\mathbf{x}}(s) \mapsto s(\mathbf{x})$ is the likelihood function. According to the likelihood principle, proportional likelihood functions are equivalent. The likelihood function can be normalized by

$$Lik_{\mathbf{x}}(s) = \frac{lik_{\mathbf{x}}(s)}{\max_{s' \in \Omega} lik_{\mathbf{x}}(s')} \tag{16}$$

Although likelihood is defined only for individual models, it can be extended to sets of models in the tradition of the maximum likelihood procedure of statistics.

$$Lik_{\mathbf{x}}(A) \stackrel{\text{def}}{=} \max_{s \in A} Lik_{\mathbf{x}}(s) \tag{17}$$

It is easy to verify that function *Lik* defined in Eqs. (16) and (17) satisfies the definition of a possibility function.

We want to draw attention on how the view of likelihood information as a possibility function is different from the way Bayesian statistics uses it. Bayesian approach combines the likelihood with the prior probability on the set of model Ω to arrive at the posterior probability. The key difference between two approaches is the presence or lack of prior probability. As Bayesian approach would not be applicable without prior probability, its proponents insist on availability of prior one way or another. In many practical situations when such prior is not available, Bayesian approach would resort to the use of artificial non-informative prior probability.

We focus on the certainty equivalent operator for possibilistic acts which is given by the construct of binary utility [10, 13]. As it is much less familiar to readers, we provide a brief self-contained review of the construct. The following are the axioms that the preference relation on possibilistic acts, $\succeq_{pos}$, have to satisfy.

A1 Weak order. $\succeq_{pos}$ is complete and transitive.

A2 Archimedean axiom. If $f \succ_{pos} g \succ_{pos} h$ then there exist possibility vectors (α_1, α_2) and (β_1, β_2) such that $\{\alpha_1/f, \alpha_1/h\} \succ_{pos} g$ and $g \succ_{pos} \{\beta_1/f, \beta_1/h\}$.

A3 Independence. Suppose $f \succeq_{pos} g$ and let (α_1, α_2) is a possibility vector then for any h, $\{\alpha_1/f, \alpha_2/h\} \succeq_{pos} \{\alpha_1/g, \alpha_2/h\}$.

A4 Compound gamble. Suppose $f_1 = \{\beta_{1i}/f_{1i} | 1 \le i \le I\}$ and $f_2 = \{\beta_{2j}/f_{2j} | 1 \le j \le J\}$ then $\{\alpha_1/f_1, \alpha_2/f_2\} \sim_{pos} \{\alpha_1\beta_{1i}/f_{1i} | 1 \le i \le I\} \cup \{\alpha_2\beta_{2j}/f_{2j} | 1 \le j \le J\}$

A5 Idempotence. For any index set I $\{\alpha_i/f | 1 \le i \le I\} \sim_{pos} f$

Clearly, $A1 - A3$ are similar to $J1 - J3$. They differ on the account of possibility function being used in the former and probability function used in the latter. $A1$ says that even as your uncertainty is represented by possibility, your preference relation still complete and transitive as in the case of probability. Suppose act g is in between acts f and h, axiom $A2$ requires that one can construct a possibilistic compound act from f, h that is strictly better (worst) than g. Axiom $A3$ says that for compound possibilistic act, each branch evaluated independently of other branches. A less noticeable fact is that $A3$ is weaker than $J3$ because $A3$ is an "if" statement whereas $J3$ is an "iff" statement. The axiom $A4$ makes clear that the concept of linear combination of possibilistic acts is not the same as the compound act. It is because in the possibilistic world, you are not allowed to use probability generating devices such as dice to combine them. This restriction seems odd at first but actually reasonable in possibilistic world because there every thing you know or don't know is expressed by possibility functions.

If you insist on the ability to make linear combination of possibilistic acts, later in this chapter we present a method to evaluate such combination of possibilistic acts.

The possibilistic expected utility theory, a counterpart for the expected utility theory, starts with a new utility scale.

Definition 1. *Binary* or *polar* utility scale is the ordered set of pairs

$$\daleth = \{\langle\alpha, \beta\rangle \mid 0 \leq \alpha \leq 1,\ 0 \leq \beta \leq 1 \text{ and } \max(\alpha, \beta) = 1\} \tag{18}$$

equipped with an order $\gtrsim\!\!\!\gtrsim$ defined by

$$\langle\alpha, \beta\rangle \ggcurly \langle\alpha', \beta'\rangle \quad \text{iff} \quad \alpha \geq \alpha' \text{and } \beta \leq \beta' \tag{19}$$

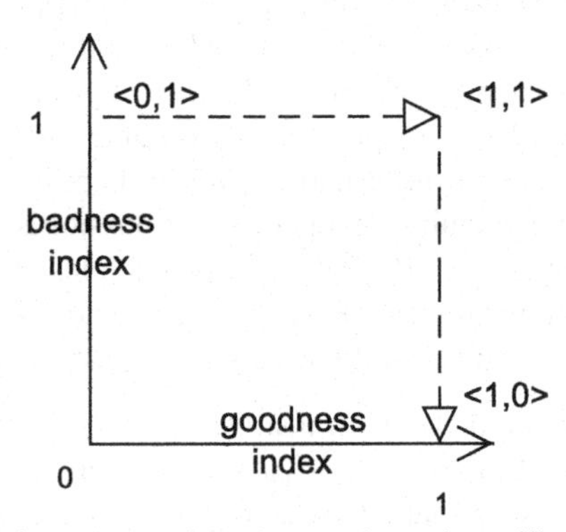

Hebrew letter $\daleth$ (daleth) is used to denote the polar utility set to remind the fact that it is the set of points on the top and the right sides of the unit square. Intuitively, the left number of a pair (the x-coordinate) is the index of "goodness" and the right number (the y-coordinate) is the index of "badness". This explains the intuition of the order $\ggcurly$. A pair is "better" than another pair if the goodness index of the former is higher than the goodness index of the later *and* the badness index of the former is lower than that of the latter. The relation $\gg$ ("strictly better") if at least one of two inequalities is strict.

Definition 2. Three operations *scalar multiplication* ($*$), *component-wise maximization* (max) and $\underline{\gg}$ *-maximization* ($\Uparrow$) are defined on polar pairs as follows:

$$c * \langle \alpha, \beta \rangle \stackrel{\text{def}}{=} \langle c * \alpha, c * \beta \rangle \tag{20}$$

$$\max(\langle \alpha, \beta \rangle, \langle \alpha', \beta' \rangle) \stackrel{\text{def}}{=} \langle \max(\alpha, \alpha'), \max(\beta, \beta') \rangle \tag{21}$$

$$\Uparrow (\langle \alpha, \beta \rangle, \langle \alpha', \beta' \rangle) \stackrel{\text{def}}{=} \begin{cases} \langle \alpha, \beta \rangle \ \text{ if } \langle \alpha, \beta \rangle \underline{\gg} \langle \alpha', \beta' \rangle \\ \langle \alpha', \beta' \rangle \text{ if } \langle \alpha', \beta' \rangle \underline{\gg} \langle \alpha, \beta \rangle \end{cases} \tag{22}$$

Note that the result of component-wise maximization is not the greatest element according to the order $\underline{\gg}$. We reserve the symbol $\Uparrow$ for the taking the maximum element in the set according to $\underline{\gg}$.

Definition 3. A strictly increasing mapping $V : \mathcal{W} \to \urcorner$ is called *scalar to polar utility transform* if

$$V(0) = \langle 0, 1 \rangle, \ V(1) = \langle 1, 0 \rangle, \ V(c) \gg V(c') \ \text{ iff } \ c > c'. \tag{23}$$

Definition 4. Suppose $f = \{A_i \hookrightarrow w_i \mid i = 1, m\}$ is a possibilistic act i.e., $\{A_i\}_{i=1}^{m}$ is a partition of Ω, and π is a possibility measure on 2^{Ω}, the possibilistic expectation of f with respect to π is defined by

$$\mathbb{Q}_{\pi}[V(f)] = \max_{1 \le i \le m} \{\pi(A_i)V(f(A_i))\} \tag{24}$$

Let us take a moment to explain the construct $\mathbb{Q}_{\pi}[\cdot]$. It is instructive to compare it with the more familiar expected utility construct for a probabilistic act f: $\mathbb{E}_P[f] = \sum_{i=1}^{m} P(A_i)U(f(A_i))$. The similarity is obvious. The differences are: (1) between functions U and V, while U maps to scalar utility, V maps to polar utility; (2) P is a additive uncertainty while π is a possibility measure and (3) summation is used in the expected utility wrt additive uncertainty while max is used in the possibilistic expected utility wrt non-additive uncertainty. In the same way as the utility function U is said to reflect the *risk attitude* of a decision maker, we say that function V reflects her *ambiguity attitude* (for more details see [10]).

A theorem proved in [10] states that the necessary and sufficient condition for preference $\succeq_{pos}$ to have the representation $f \succeq_{pos} g$ iff $\mathbb{Q}_{\pi}[V(f)] \underline{\gg} \mathbb{Q}_{\pi}[V(g)]$ is the axioms $A1 - A5$. We note that this model subsumes two regimes of decision making with possibility proposed in Dubois et al. [7].

Because V is a strictly increasing mapping, its inverse function $V^{-1} : \urcorner \to \mathcal{W}$ exists and defined uniquely by $\forall w \in \mathcal{W}$, $V^{-1}(V(w)) = w$. Using V^{-1}, the *certainty equivalent* of a possibilistic act can be defined by $V^{-1}(\mathbb{Q}[V(f)])$.

The ambiguity attitude conveyed by function V can be quantified by an index ρ in the range $(0, 1)$ with $\rho = 0.5$ means ambiguity neutral, the smaller ρ you have, the more ambiguity averse you are. A handy class of polar utility function V_ρ, *constant ambiguity attitude* utility function, is characterized by the following equation for $0 < x < 1$

$$V_\rho(x) = \langle \alpha, \beta \rangle \quad \text{iff} \quad \log\left(\frac{x}{1-x}\right) = \log\left(\frac{\alpha}{\beta}\right) + \log\left(\frac{\rho}{1-\rho}\right) \tag{25}$$

It follows that

$$V_\rho(x) = \langle \alpha, \beta \rangle \quad \text{where} \quad \begin{cases} \alpha = \min(1, \exp(\text{logit}(x) - \text{logit}(\rho))) \\ \beta = \min(1, (\exp(\text{logit}(x) - \text{logit}(\rho)))^{-1}) \end{cases} \tag{26}$$

where $\text{logit}(x) \stackrel{\text{def}}{=} \log(x)/\log(1-x)$.

Example 1. (Fig. 2) Suppose $f = \{A_1 \hookrightarrow 0.1;\ A_2 \hookrightarrow 0.4; A_3 \hookrightarrow 0.9\}$, $\pi(A_1) = 1; \pi(A_2) = 0.7; \pi(A_3) = 0.3$ and $V_{0.4}(0.1) = \langle 0.17, 1 \rangle$; $V_{0.4}(0.4) = \langle 1, 1 \rangle$; $V_{0.4}(0.9) = \langle 1, 0.07 \rangle$.

$$\mathbb{Q}[V_{0.4}(f)] = \max\{\pi(A_1)V_{0.4}(0.1), \pi(A_2)V_{0.4}(0.4), \pi(A_3)V_{0.4}(0.9)\} \tag{27}$$

$$= \max\{1\,\langle 0.17, 1 \rangle, 0.7\,\langle 1, 1 \rangle, 0.3\,\langle 1, 0.07 \rangle\} = \langle 0.7, 1 \rangle \tag{28}$$

So, we have $V_{0.4}^{-1}(\langle 0.7, 1 \rangle) = 0.32$ i.e., the certainty equivalent of f is 0.32. ■

Using possibility theory, the state of ignorance can be represented by possibility function $\forall s \in \Omega, \pi(s) = 1$. It can be shown that possibilistic expected utility (24) reduces to (11) for decision under ignorance.

Example 2. For act f as in the previous example. Suppose we have an ignorant possibility function $\pi(A_1) = \pi(A_2) = \pi(A_3) = 1$.

$$\mathbb{Q}[V_{0.4}(f)] = \max\{V_{0.4}(0.1), V_{0.4}(0.4), V_{0.4}(0.9)\} \tag{29}$$

$$= \max\{\langle 0.17, 1 \rangle, \langle 1, 1 \rangle, \langle 1, 0.07 \rangle\} = \langle 1, 1 \rangle \tag{30}$$

$V_{0.4}^{-1}(\langle 1, 1 \rangle) = 0.4$. ■

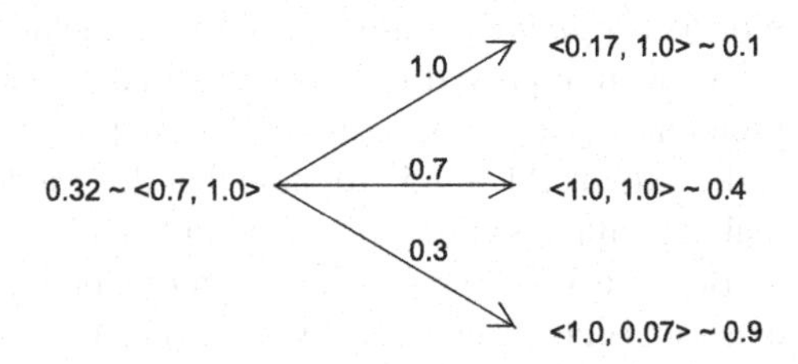

Fig. 2 Calculation of possibilistic expected utility

2.4 Decision Models for Dempster-Shafer Belief Function Theory

The belief function theory was originally developed by Dempster in 1960s to generalize Bayesian statistics [6]. Later, Shafer extended this proposal for evidential reasoning [23]. A major advantage belief function theory is its ability to express, in a more faithful manner, information availability: from complete ignorance to partial ignorance to full information.

The classic exposition of the DS belief function theory is Shafer's book "A Mathematical Theory of Evidence". We present a brief introduction to the concepts, notations and important results. A basic probability assignment (bpa) function m is defined by

$$m : 2^{\Omega} \to [0, 1] \text{such that} \sum_{A \subseteq \Omega} m(A) = 1. \tag{31}$$

A subset with positive mass is a *focus*. Other forms of a belief function are *belief* (*Bel*), *plausibility* (*Pl*) and *commonality* (Com) that are defined from m as follows: $\forall B \subseteq \Omega$

$$Bel(B) = \sum_{A \subseteq B} m(A); \; Pl(B) = \sum_{A \cap B \neq \emptyset} m(A); \; Com(B) = \sum_{A \supseteq B} m(A) \tag{32}$$

All the forms m, *Bel*, *Pl* and Com are equivalent in the sense that given any form the others are completely determined. For example, given a belief function in *Bel* form, its bpa form is completely determined by Möbius inverse transform

$$m(A) = \sum_{B \subseteq A} (-1)^{|A \setminus B|} Bel(B) \tag{33}$$

Although in literature *Bel* is often referred to as belief function. However, in our usage, "belief function" is reserved for the body of information that has many incarnations m, *Bel*, *Pl* and Com. The choice of a form to work with depends on manipulation convenience. In our case, we use plausibility form largely because of its conditional expression.

If you hold belief function m_1 and learn about new evidence represented by a second belief function m_2, you should update your belief by using Dempster's rule. The result is a new belief function denoted by $m_1 \oplus m_2$ defined as follows

$$\forall A \neq \emptyset, \; m_1 \oplus m_2(A) = K \cdot \left(\sum_{A_i \cap B_j = A} m_1(A_i)\, m_2(B_j) \right) \tag{34}$$

where K is the normalization constant defined by

$$K = \left(1 - \sum_{A_i \cap B_j = \emptyset} m_1(A_i)\, m_2(B_j)\right)^{-1} \tag{35}$$

In a case of special interest, m_2 is a belief function with a single focus i.e., $m_2(B) = 1$ for some B, belief combination becomes conditionalization. It is easy to verify that in the plausibility form, conditional has a familiar form of probability conditioning

$$Pl(A|B) = \frac{Pl(A \cap B)}{Pl(B)} \tag{36}$$

Dempster [5] views of belief function as multiple value mapping from a sample space to another space. Suppose there is probability function P on sample space Θ and a set-valued mapping h from Θ to the power set of another space Ω. For example Θ can be observation space and Ω —a hypothesis space. Each observation points to a set of hypotheses. A belief function m on Ω is induced by P and h by equation

$$\forall C \in 2^{\Omega},\ m(C) = \sum_{h(w)=C} P(w) \tag{37}$$

Note that $h : \Theta \to 2^{\Omega}$ is not necessarily one-to-one. Therefore, the probability of more than one points can be transferred to one subset (see Fig. 3). Dempster's view is the basis of many decision models.

There are several important subclasses of belief functions depending on the topology of the foci. In Shafer's book, the belief functions that have nested foci are singled out. This class is called *consonant* belief function. Suppose the foci of a belief function are $B_1, B_2, \ldots B_n$ and $B_1 \subset B_2 \subset \ldots \subset B_n$. It is easy to verify for such a belief function $Pl(.)$ is a possibility function, *Bel* is a necessity function. In particular $Pl(A \cup B) = \max(Pl(A), Pl(B))$ and Dempster's combination rule is reduced to the quantitative conditioning formula. Conversely, given a possibility function π (and its dual form ν), the bpa and foci of a consonant belief function can be determined by the inverse Mobius transform (33).

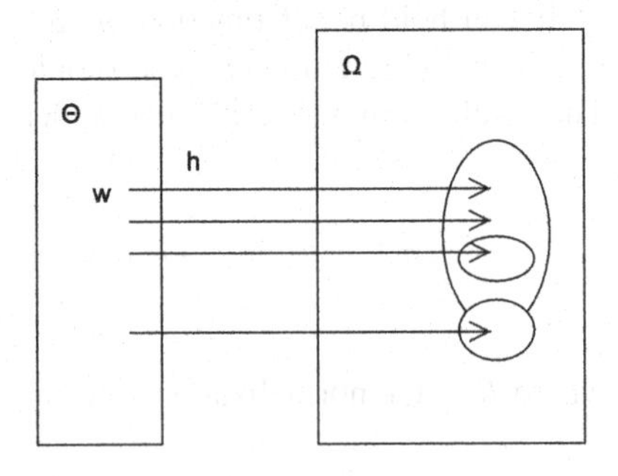

Fig. 3 Belief function as a multiple valued mapping from a probability function

Starting from the premise that belief functions can be used to represent statistical evidence, Walley [27] asks what condition that belief functions must satisfy in order to be consistent with a fundamental principle in statistics —the likelihood principle. He proves the condition that the set of foci must be partitioned into non-intersecting groups and within each group the foci are nested. Such belief functions are called *partially consonant*. In other words, the foci of a partially consonant belief function can be rearranged $B_1, B_2, \dots B_n$ with the following property. There exist k numbers $0 = i_0 < i_1 < i_2 \dots < i_k = n$ that serve as the boundaries for the groups (see Fig. 4). Group j has foci with index i in between i_{j-1} and i_j i.e., $G_j = \{B_i | i_{j-1} < i \leq i_j\}$

$$B_{i_{j-1}+1} \subset B_{i_{j-1}+2} \dots \subset B_{i_j} \text{for } 1 \leq j \leq k \tag{38}$$

$$B_{i_j} \cap B_{i_{j'}} = \emptyset \quad \text{for} \quad j \neq j' \tag{39}$$

Earlier we noted that in the possibilistic world, there is no such thing as linear combination of possibilistic functions or acts. If we insist on that ability we have to go beyond the world of possibility functions. In particular, a partially consonant belief function can be construed as a linear combination of several possibility function.

Abusing notation slightly, we use symbol G_j also to denote the set of elements of Ω that belong to group j, clearly $\{G_j | 1 \leq j \leq k\}$ is a the collection of disjoint subsets. We have $Pl(G_j) = Bel(G_j)$ because every focus that intersects with G_j is completely enclosed in it. Also $\sum_{j=1}^{k} Pl(G_j) = 1$ because that sum takes into account the mass of every focus and no focus is double accounted. So, vector $(Pl(G_j))_{j=1}^{k}$ can be viewed as a probability function on a partition $\{G_1, G_2, \dots G_k, \Omega_{-G}\}$ where Ω_{-G} consists of elements that are not in any G_j. Also note that the conditional belief function given G_i is consonant. So $Pl(\cdot|G_j)$ is a possibility function.

The advantage of using belief function theory is the power to express the state of partial lack of information. Specifically, the partiality of information is expressed by assigning the probability mass not to singleton-states as in probability but to sets of states. The ignorance is pertinent to question how that mass is distributed within the focus.

There is a convenient convention in discussion of decision making. Given a belief function $m : 2^{\Omega} \rightarrow [0, 1]$ with foci $B_1, B_2, \dots B_K$ an act $f : \Omega \rightarrow \mathcal{W}$ induces a

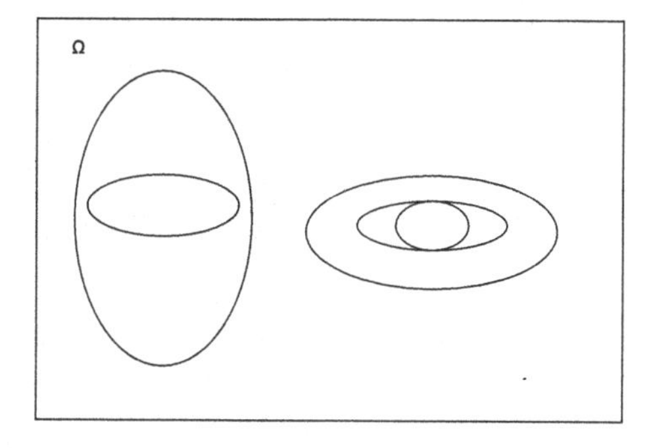

Fig. 4 Foci of a partially consonant belief function

belief function $m_f : 2^{\mathcal{W}} \rightarrow [0, 1]$ (on the space of outcomes) by transferring the probability mass from B_i to $f(B_i)$:

$$\forall i, m_f(f(B_i)) = m(B_i) \quad \text{and} \quad m_f(C) = 0 \text{ otherwise.} \tag{40}$$

With $f \Leftrightarrow m_f$ identity, one can safely discuss the choice of belief functions (on the outcome space) instead of the choice of acts.

There are several proposals in literature on decision making with belief function such as Jaffray [19], Yager [29], Smets [25] and Giang and Shenoy [14]. Most of them relies on the view that a belief function is a linear combination of elementary belief functions. A belief function is *elementary* if it has only one focus. Suppose belief function f has foci $B_1, B_2, \ldots B_K$, f can be written as

$$f = \sum_{i=1}^{K} m_f(B_i)\, e_{B_i} \tag{41}$$

where e_{B_i} is an elementary belief function with single focus B_i.

2.4.1 Jaffray's Model

Suppose f is belief function on the outcome space. The main argument for this model is that since f is a linear combination of elementary belief functions, the utility of f is a linear combination of the utilities of those elementary belief functions. Suppose f has foci $\{B_i\}_{i=1}^{K}$

$$v_J(f) = \sum_{i=1}^{K} m_f(B_i) v_J(e_{B_i}) \tag{42}$$

where $v_J(f)$ is utility of f. Since the elementary e_B represents ignorance conditional on B, Jaffray used the Hurwicz's solution for decision under ignorance

$$v_J(e_B) = \alpha_B \bot_B + (1 - \alpha_B)\top_B \tag{43}$$

where $\bot_B$ and $\top_B$ are the smallest (bottom) and largest (top) elements in B and $0 \leq \alpha_B \leq 1$ is a constant that depends on the values $\bot_B$ and $\top_B$ only. Hence the utility expression of a general belief function is

$$v_J(f) = \sum_{i=1}^{K} m_f(B_i) \left(\alpha_{B_i} \bot_{B_i} + (1 - \alpha_{B_i})\top_{B_i}\right) \tag{44}$$

The axiomatic justification of Jaffray's model is given in [20].

2.4.2 Yager's Method

The only difference between Jaffray's method and Yager's method is in the treatment of elementary belief functions.

$$v_Y(f) = \sum_{i=1}^{K} m_f(B_i) v_Y(e_{B_i}) \tag{45}$$

Here $v_Y(e_B)$ is computed according to the *ordered weighted average* (OWA) method. Suppose $B = \{w_1, \ldots w_k\}$ and its elements are arranged in a decreasing order i.e., $w_1 > w_2 > \ldots > w_k$. Yager assumes that for each DM, there are non-negative *weights* $c_1, c_2, \ldots c_k$ that sum to 1. $v_Y(e_B) = \sum_{i=1}^{k} c_i.w_i$.

OWA subsumes familiar criteria such as max ($c_1 = 1$), min ($c_k = 1$), Hurwicz's ($c_1 = 1 - \alpha, c_k = \alpha$). It is not clear how to justify Yager's model from axiomatic perspective. Without going into details, note that, except three special cases mentioned above, $v_Y(e_B)$ accounts for non-extreme values in B, hence, OWA does not satisfy Arrow-Hurwicz's postulates for decision under ignorance [2].

2.4.3 Smets' Transferable Belief Model

Smets and Kennes [24, 25] describe an approach called the *transferable belief model* (TBM) in which "beliefs can be held at two levels: (1) a credal level where beliefs are entertained and quantified by belief functions, (2) a pignistic level where belief can be used to make decisions and are quantified by probability functions." Given a belief function with foci $\{B_i\}_{i=1}^{K}$, probability function Pr_b defined by dividing the mass of each focus evenly on its elements:

$$\forall s \in \Omega,\ Pr_b(s) = \sum_{B_i \supseteq s} \frac{m(B_i)}{|B_i|} \quad \text{and} \quad \forall A \subseteq \Omega,\ Pr_b(A) = \sum_{s \in A} Pr_b(s) \tag{46}$$

The expected utility of act d wrt Pr_b

$$v_T(d) = \sum_{s \in \Omega} Pr_b(s)\, d(s) = \sum_{i=1}^{K} m(B_i) \frac{\sum_{s \in B_i} d(s)}{|B_i|} \tag{47}$$

It can be seen that (47) is a special case of (45) when all the weights c_i are equal. As such, TBM model also does not satisfy the requirements of decision under ignorance and consistency. It can be argued that in this approach, the problem of decision under ignorance does not even exist because the notions of ignorance and ambiguity are meaningful in the "credal level" only. They cease to exist when it comes to decision.

2.4.4 Giang-Shenoy's Model

Unlike the previous models, Giang and Shenoy [15] propose an axiomatic decision model applied not for belief functions in general but for the class of partially consonant belief function. For a consonant belief function f with foci $B_1, B_2, \ldots B_n$ which are divided into k nested groups with group boundary indices $0 = i_0 < i_1 < i_2 \ldots < i_k = n$. Denote by G_j the largest element of group j.

$$v_{GS}(f) = \sum_{j=1}^{k} Pl(G_j)V^{-1}(\mathbb{Q}_{Pl_{G_j}}(V(f_{G_j}))) \quad (48)$$

where $Pl(G_j)$ is the plausibility of G_j, Pl_{G_j} is the conditional plausibility given G_j. Because the belief function conditional on G_i is consonant, Pl_{G_j} is a possibility function. f_{G_j} is the restriction of act f on G_j i.e., $f_{G_j}(w) = f(w)$ if $w \in G_j$, it is undefined outside G_j.

The meaning of (48) can be explained in Fig. 5. It is a two stage folding-back evaluation. In the first stage (right), act f_{G_j} is evaluated to a binary utility value $\langle \alpha_j, \beta_j \rangle$ under possibility function Pl_{G_j} by the possibilistic expectation operator (24). The certainty equivalent x_j of the conditional act is found by applying V^{-1}. In the second stage (left), the probabilistic act $\{Pl(G_j)/x_j | 1 \leq j \leq k\}$ is evaluated by expected utility.

2.4.5 Examples

To conclude this section, we illustrate how different methods work on two examples.

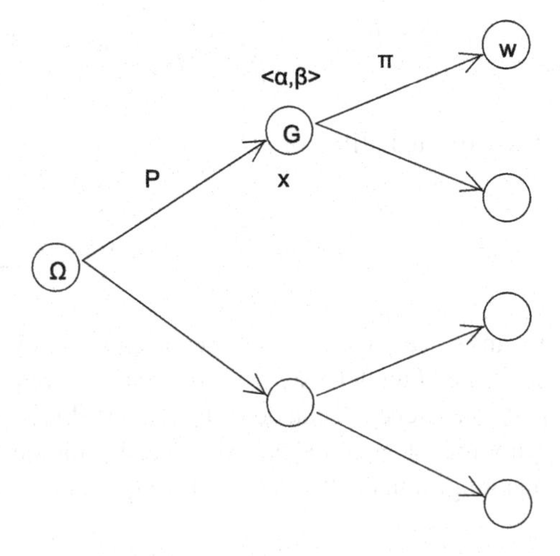

Fig. 5 Evaluation of act in Giang-Shenoy's model

Example 3. The state of ignorance can be represented by a belief function of a single focus which is the entire set of states Ω i.e., $m(\Omega) = 1$. Let us consider the act $f = \{A_1 \hookrightarrow 0.1;\ A_2 \hookrightarrow 0.4; A_3 \hookrightarrow 0.9\}$.

The certainty equivalent of f in the Giang-Shenoy's model with ambiguity attitude $\rho = 0.4$ is 0.4 as calculated in the example 2.

Using Jaffray's model with $\alpha = 0.6$ (ambiguity averse), we have the certainty equivalent calculated by $\alpha * 0.1 + (1 - \alpha) * 0.9 = 0.42$.

Using Yager's model with weight vector $(0.15, 0.45, 0.40)$, the certainty equivalent is calculated by $0.15 * 0.1 + 0.45 * 0.4 + 0.40 * 0.9 = 0.555$

Under Smets' TBM model the certainty equivalent is calculated by $(0.1 + 0.4 + 0.9)/3 = 0.47$. ■

Example 4. Ellsberg's paradox [9]

Ellsberg's paradox is one in a series of experiments used to demonstrate that rational behavior under ambiguity violates Savage's sure-thing principle. In an urn, there are 90 balls of the same size. The balls are painted one of three colors: red, yellow and white. It is known that 30 balls are red. The proportions of yellow and white are not known.

Ellsberg considers four gambles. $IA = \{\text{red} \hookrightarrow 1, \{\text{white, yellow}\} \hookrightarrow 0\}$ that offers \$1 if a randomly drawn ball is red, nothing otherwise. $IB = \{\text{yellow} \hookrightarrow 1, \{\text{white, red}\} \hookrightarrow 0\}$ offers \$1 if the ball is yellow, nothing otherwise. $IIA = \{\{\text{red, white}\} \hookrightarrow 1, \text{yellow} \hookrightarrow 0\}$ offers \$1 if the ball is red or white, nothing if the ball is yellow. $IIB = \{\{\text{yellow, white}\} \hookrightarrow 1, \text{red} \hookrightarrow 0\}$ offers \$1 if the ball is yellow or white and nothing if it is red.

Ellsberg discussed findings that a sizable proportion of respondents preferred IA to IB and, at the same time, preferred IIB to IIA. This observed preference is not consistent with the sure-thing principle because the pair (IIA, IIB) is different from the pair (IA, IB) only by the level of prize for white balls.

The uncertainty in the problem is nicely described by a pcb with 2 foci (Fig. 6). $m(\{\text{red}\}) = \frac{1}{3}$ and $m(\{\text{yellow, white}\}) = \frac{2}{3}$. This pcb decomposes into $P(\{\text{red}\}) = \frac{1}{3}$ and $P(\{\text{yellow, white}\}) = \frac{2}{3}$ and $\Pi(\text{yellow}|\{\text{yellow, white}\}) = \Pi(\text{white}|\{\text{yellow, white}\}) = 1$.

We show how Giang-Shenoy model works under three different scenarios of ambiguity aversion, ambiguity neutrality and ambiguity seeking.

Case 1 (*ambiguity aversion*). We assume binary utility function $V_a(\$1) = \langle 1, 0\rangle$, $V_a(\$0) = \langle 0, 1\rangle$ and $V_a(\$0.4) = \langle 1, 1\rangle$. The first two equalities are natural since \$1 is the best outcome and \$0 is the worst outcome. The default value under ignorance is

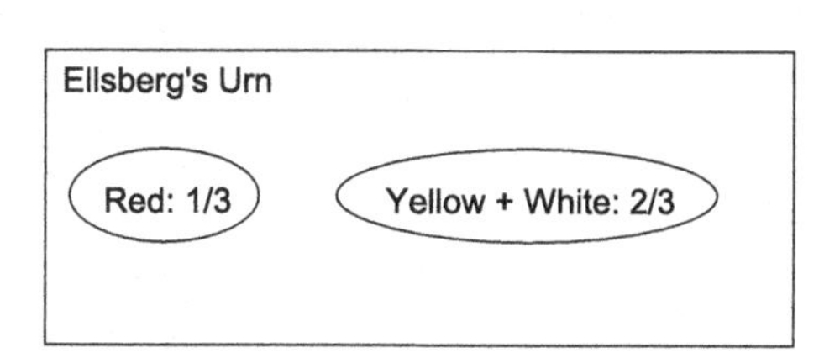

Fig. 6 Belief function for Ellsberg's Urn

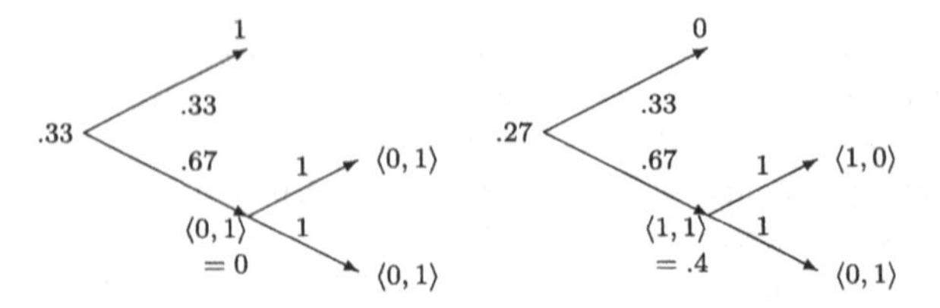

Fig. 7 Utility calculation for IA and IB under ambiguity aversion $\rho = 0.4$

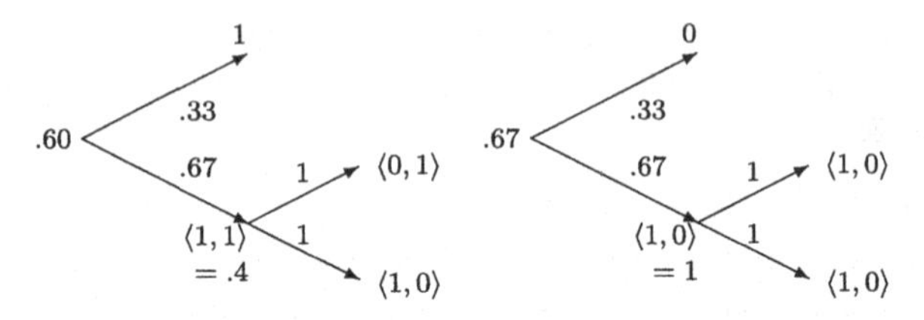

Fig. 8 Utility calculation for IIA and IIB under ambiguity aversion $\gamma = 0.4$

0.4 indicating somewhat ambiguity aversion. In Figs. 7 and 8 we show the calculation of utility for the gambles. We have $v_{GS}(IA) = 0.33$, $v_{GS}(IB) = 0.27$, $v_{GS}(IIA) = 0.60$ and $v_{GS}(IIB) = 0.67$. This means $IIB \succ IIA \succ IA \succ IB$. These preferences are consistent with the observed behavior.

Case 2 (*ambiguity neutrality*). We have $V_n(\$1) = \langle 1, 0 \rangle$, $V_n(\$0) = \langle 0, 1 \rangle$ and $V_n(\$0.5) = \langle 1, 1 \rangle$ and $v_{GS}(IA) = \frac{1}{3}$, $v_{GS}(IB) = \frac{1}{3}$, $v_{GS}(IIA) = \frac{2}{3}$, $v_{GS}(IIB) = \frac{2}{3}$. This means $IIA \sim IIB \succ IA \sim IB$.

Case 3 (*ambiguity seeking*). We have $V_s(\$1) = \langle 1, 0 \rangle$, $V_s(\$0) = \langle 0, 1 \rangle$ and $V_s(\$.6) = \langle 1, 1 \rangle$ and $v_{GS}(IA) = 0.333$, $v_{GS}(IB) = 0.4$, $v_{GS}(IIA) = 0.733$, $v_{GS}(IIB) = 0.667$. This means $IIA \succ IIB \succ IB \succ IA$.

Jaffray's model. In this model the ambiguity attitude is controlled by the Hurwicz's coefficient α. We also consider three scenarios for $\alpha = 0.6$, $\alpha = 0.5$ and $\alpha = 0.4$ corresponding to ambiguity averse, neutral and seeking. In case $\alpha = 0.6$, $v_J(IA) = 0.33$, $v_J(IB) = 0.33 * 0 + 0.67 * (0.6 * 0 + 0.4 * 1) = 0.27$.

The result of calculation is shown in Table 1. The second row has the color proportion data and the ambiguity attitude parameter for Jaffray's and Giang-Shenoy's models. Smets' TMB model does not have ambiguity attitude parameter. In this example, the set of outcomes has only two elements (0, 1), Yager's model is identical to Jaffray's. Also because of that, the parameters in Jaffray's and Giang-Shenoy's models can be set so that they produce the same result. In general, however, they have different behavior. For more details on how these two models are different see [12]. ■

Table 1 The certainty equivalent under different models and ambiguity assumptions

	Red	Yellow	White	Jaffray	Jaffray	Jaffray	GS	GS	GS	TBM
Prob/α/ρ	0.333	0.667		0.6	0.5	0.4	0.4	0.5	0.6	
IA	1	0	0	0.33	0.33	0.33	0.33	0.33	0.33	0.33
IB	0	1	0	0.27	0.33	0.40	0.27	0.33	0.40	0.33
IIA	1	0	1	0.60	0.67	0.73	0.60	0.67	0.73	0.67
IIB	0	1	1	0.67	0.67	0.67	0.67	0.67	0.67	0.67

2.5 Relationship Between Uncertainty Models

We have considered several models of uncertainty that extend the traditional representation by probability. The main motivation for such models is to capture different aspects of uncertainty existed in the real world, to faithfully represent the extent of knowledge and ignorance that you have about the real world.

It is useful at this point to summarize their relationship (see Fig. 9). The oval nodes are the models and their special cases. The arrow signifies "is a" relationship.

For example, there is an arrow from "ignorance" to "consonant belief function" because the epistemic state of ignorance can be represented by a belief function with a single focus which is exactly the set of states Ω i.e., $m(\Omega) = 1$ and $m(A) = 0$ for any $A \subset \Omega$. That ignorant belief function is consonant. Also the arrow from the "ignorance" node to "possibility function" node signifies that in the language of possibility theory, the state of ignorance is represented by possibility function in which the possibility of any element of the power set is 1 i.e., $\forall A \subseteq \Omega, Pl(A) = 1$.

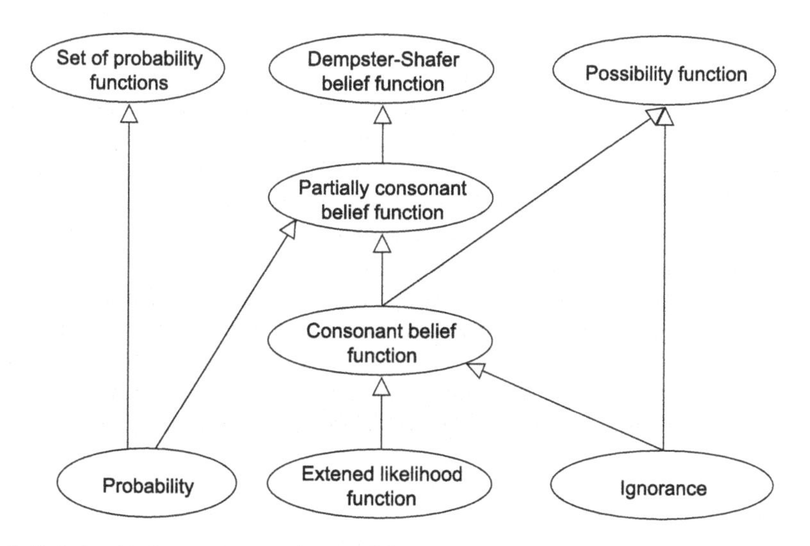

Fig. 9 Relationship between uncertainty models

The arrow from "probability" to "partially consonant belief function" is because a belief function with all singleton foci is a probability function. The arrow from "consonant belief function" to "possibility function" is justified by the fact that the plausibility form of a consonant belief function is a possibility function with quantitative conditioning. The lack of a reverse arrow from "possibility function" to "consonant belief function" is justified by the fact that possibility theory also accepts qualitative conditioning (15) which does not have counterpart in DS belief function theory.

Another model which should be mentioned here is the representation of uncertainty by sets of probability functions. This is a natural extension of the representation by single probability functions. The clear advantage is keeping the Bayes rule for updating. Given a set $S = \{P_i\}$ of probability measures and the arrival of evidence A, the update belief is the set of probability measure $S' = \{P_i(\cdot|A)\}$. The arrow from "probability" to "set of probability functions" is obvious. However, the lack of arrow between "set of probability functions" and "DS belief functions" needs some clarification. It is well known [18] that *Pl* and *Bel* forms of a DS belief function can be interpreted as the upper and the lower envelops of some set of probability functions. In this sense, we can say that the belief function and the set of probabilities are matched. However, upon arrival of a new evidence, the DS belief function is updated according to Dempster's rule while the set of probabilities is updated via Bayes rule for each member of the set. The resulting belief function and the set of updated probabilities may no longer match.

The relationship between uncertainty models is useful in comparative analysis of the decision proposals for different uncertainty models. For example, as probability is a special case of partially consonant belief function (hence also a special case of belief function), we want to make sure that the decision model for (partially consonant) belief function applied for probability reduces to EU model. Also the state of ignorance, as a special case of both belief function and possibility function, can be used to compare their decision models.

All the models for decision making with belief function including those proposed by Jaffray, Yager, Smets and Giang-Shenoy are reduced to expected utility model when the foci of belief function are singletons. When belief function reduces to ignorance the models by Jaffray and Giang-Shenoy satisfy four rationality postulates by Arrow and Hurwicz. The models by Yager and Smets do not. Only Giang-Shenoy model satisfies, in addition, the sequential consistency postulate. For the consonant belief function—the possibility function, Giang-Shenoy model for partial consonant belief and that for possibility theory have identical solutions which subsume the two models proposed by Dubois et al. [7]. Applied for consonant belief function, the models by Jaffray, Yager and Smets have different behaviors than the models proposed for possibility theory—by Giang-Shenoy and Dubois et al.

3 Decision Making Under Heterogeneous Uncertainty

The environment of heterogeneous uncertainty is characterized by the presence of variables in multiple uncertainty formalisms. In previous section, we have reviewed various uncertainty formalisms together with decision models proposed. Clearly, the real world does not fit comfortably within any individual formalism. For example, probability poorly severs the need to express the notion and degree of ignorance. On the other hand, within the possibility model, you are not allowed to express the notion of linear combination of acts even though such acts can arise naturally. For example, given two possibilistic acts and a coin, a new act can be created by choosing one alternative act depending coin toss. The focus of this section is a decision model that can deal with different formalisms of uncertainty.

3.1 A Motivating Example

Let us consider a simple example in which the outcome of an act depends on both risk and uncertainty.

Example 5. An investor considers an investment instrument d whose return, measured in risk-adjusted utility unit (util), depends on two binary variables. X, the weather in US, is a chance variable with $Pr(X = 1) = Pr(X = 0) = 0.5$. $X = 1$ means favorable weather and $X = 0$ means unfavorable. Variable Y represents the outcome of a political event of an obscure tribe in South Asia. The information about Y is unreliable and mostly contradictory. Y is an *ignorant* variable. The contingent outcome of d is as follows. $d(s_{00}) = 0$, $d(s_{01}) = 1$, $d(s_{10}) = 1$ and $d(s_{11}) = 0$ where s_{ij} stands for proposition $(X = i \;\&\; Y = j)$. The information is summarized in Fig. 10.

Suppose that you consider to buy the instrument and wonder what is its worth. You can reason as follows.

Scenario 1 Hypothetically assume $X = 0$ (unfavorable weather), the choice then reduces to a decision problem under ignorance (political outcome) $d_{X=0}(Y = 0) = 0$ and $d_{X=0}(Y = 1) = 1$ since no information about Y is available. If you are uncertainty averse (e.g. your preference pattern is highlighted in Ellsberg's paradox), the certainty equivalent you attach to $d_{X=0}$ is c which is strictly less than 0.5. Note that 0.5 is the value of $d_{X=0}$ if Y were a fair coin toss. A symmetrical argument

Fig. 10 A chance variable and an ignorant variable of an investment

d(X,Y)		*X*		*I(Y)*
		1	0	
Y	1	0	1	*ignorance*
	0	1	0	*ignorance*
Pr(X)		0.5	0.5	

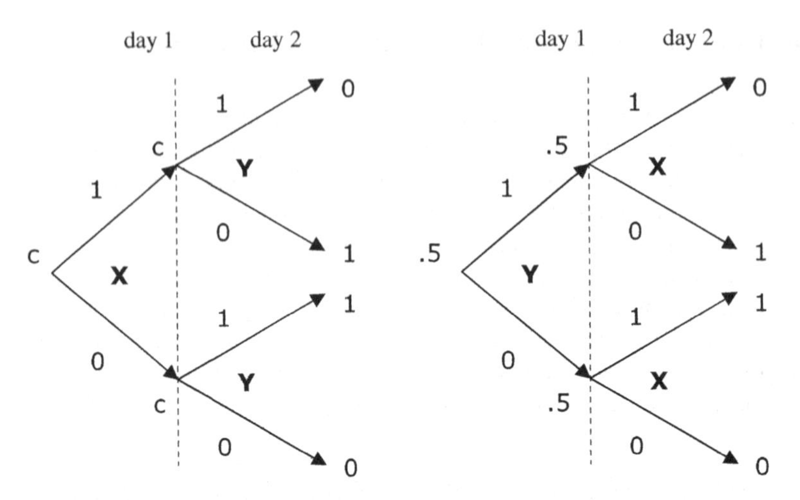

Fig. 11 Evaluation of uncertain investment under two orders of variable resolution

shows that the value of $d_{X=1}$ is also c. So, no matter what the actual value of X will be, the investor gets value c, she concludes that the value of investment is c (see in Fig. 11).

Scenario 2 This scenario mirrors Scenario 1, but instead of conditioning on X, you choose to condition the investment on the values of Y. $d_{Y=0}(X = 0) = 0$, $d_{Y=0}(X = 1) = 1$, since X is a chance variable with two equi-probable states and the investment return is risk-adjusted, the value of $d_{Y=0}$ is 0.5 (write $d_{Y=0} = 0.5$). Similarly, $d_{Y=1} = 0.5$. Thus, no matter what Y turns out to be, the value that you get is 0.5, hence the value of the investment is also 0.5.

The apparent conflict between two perfectly reasonable evaluation scenarios begs for a resolution. ■

First of all, we need to figure out the answer for the following question. "Can and how to represent the uncertainty pertaining to the instrument by one of the uncertainty formalisms?"

On the one hand, probability theory is excluded because it is not able to handle the ignorant variable (Y). On the other hand, possibility theory also does not fit the bill because it can not describe the chance variable (X).

This leaves the belief function as the only candidate. The state space is naturally formed by Cartesian product of variables' domains. $\Omega = \{s_{00}, s_{01}, s_{10}, s_{11}\}$ as s_{ij} stands for $(X = i \;\&\; Y = j)$. In this space, the basic probability assignment of the belief function is $m(\{s_{00}, s_{01}\}) = 0.5$ and $m(\{s_{10}, s_{11}\}) = 0.5$ (see Fig. 12).

Given that belief function, most models (Jaffray's, Yager's and Giang-Shenoy's) evaluate the instrument to c —the solution under Scenario 1. Smets' TBM model evaluates the instrument to 0.5 which in this case coincides with the solution under Scenario 2.

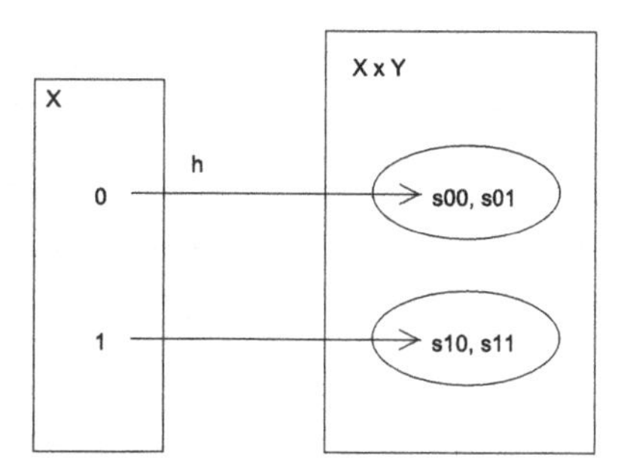

Fig. 12 Belief function for the investment instrument

The reason that TBM model can not deliver the solution under Scenario 1 is clear. TBM is actually an application of expected utility model for a probability that is derived from belief function. The uncertainty or partial ignorance contained in belief function is assumed away in the derivation.

3.2 Variable Resolution Order

We consider it a serious defect of a model if it cannot compute the solution in a perfectly reasonable scenario. In this sense, the inability of well-regarded decision models for belief functions to deliver the solution in Scenario 2 is even more problematic.

A careful examination of the problem reveals that the problem starts with representation of uncertainty. Maybe, it is not correct to insist on belief function representation of the uncertainty in the motivating example. This representation lumps together the information about two separate variables of different nature.

First, it is not an information-preserving transformation. It is enough to realize that there are infinite number of configurations (Θ, P, h) of a sample space Θ, a probability measure P and a set-valued mapping h that induce the belief function $m(\{s_{00}, s_{01}\}) = 0.5$ and $m(\{s_{10}, s_{11}\}) = 0.5$. For example, suppose Θ is the real line, P is a probability density function and a is the median point then the mapping h defined as follows creates the belief function

$$h(x) \mapsto \begin{cases} \{s_{00}, s_{01}\} \text{ if } & x < a \\ \{s_{10}, s_{11}\} \text{ if } & x > a \end{cases} \tag{49}$$

In fact, the configuration that consists of a binary chance variable of equi-probable probability and an binary ignorant variable is only one among infinite many. Any decision that is made on the basis of the belief function fails to account for the information that can be found in the original problem setting.

Unfortunately, even if you know that the use of belief function incurs loss of information, it still remains the only available choice if you insist on bring probability and ignorance variables into a single formal uncertainty formalism.

The novel approach we are proposing will not insist on representing the uncertainty about different variables in the common product state space. For example, we do not represent a chance variable X and an ignorant variable Y by an uncertainty measure on the product space $X \times Y$. Instead we connect those variables in a sequence.

Variable resolution order, the temporal order in which the variables involved in the decision problem are resolved (their realizations become known) is a novel concept that will play a key role in our approach.

Before presenting our approach, we speculate about the reason why the literature on non-expected utility never seriously considered this concept. It is well-known that decision models under uncertainty are heavily influenced by the knowledge about the expected utility theory. An important fact about the EU model is that the result does not depend on the order of variable resolution. This is an implication of the dynamic consistency property. As discussed in the Sect. 2.1, in the axiomatization of the expected utility theory, the compound act is identified with the linear combination of acts. This identification is considered self-evident even though they are two distinct concepts.

Let's calculate the expected utility in the case if both variables X and Y in the example are chance variables. There are three ways to compute the expected utility of the instrument.

$$\mathbb{E}_{X,Y}[d(X,Y)] = \sum_{x,y} Pr(x,y)d(x,y) \tag{50}$$

$$\mathbb{E}_X[\mathbb{E}_{Y|X}[d(X,Y)|X]] = \sum_{x} Pr(x) \sum_{y} Pr(y|x)d(x,y) \tag{51}$$

$$\mathbb{E}_Y[\mathbb{E}_{X|Y}[d(X,Y)|Y]] = \sum_{y} Pr(y) \sum_{x} Pr(x|y)d(x,y) \tag{52}$$

Because $Pr(x,y) = Pr(x)Pr(y|x) = Pr(y)Pr(x|y)$, the expected value calculated is the same no matter which method is used. This is a special case of the law of iterated expectation. You can interpret each calculation method in terms of order of variable resolution. The first calculation corresponds to the case when you don't know which variable realizes first or you know that they are realized at the same time. The second calculation applies when X realizes before Y. When the realization of X is known, you can condition the instrument and probability of Y on that event and calculate (conditional) expected utility. Before knowing X realization you can calculate the expected utility by accounting for the probability of X realizations. The third calculation applies when Y realizes before X. The invariance between three calculations tells us the value of the instrument does not depend on the order of variable resolution.

The situation becomes quite different when a non-probabilistic variable is involved. Consider the motivating example. The return of investment instrument

d depends on two variables, a chance variable X and an ignorant variable Y. For the sake of simplicity, assume that X and Y are *non-influential* (term *independent* is reserved for the relationship between probabilistic variables). The problem description, notably, does not say anything about when each variable is revealed to the investor. The only implicit assumption is that both variables are resolved before the instrument's maturity. The omission could be due to the fact that the resolution order is not known. But the order information, even known, could be left out because it is judged to be irrelevant to the value of the investment, especially when variables are supposed to be independent.

It turns out that the omission is an error and the *information about the resolution order is essential for evaluation* of acts under heterogeneous uncertainty. If the order is not given, one can list all the possibilities. There are two possibilities: (1) X is revealed before Y denoted by $X \rightharpoonup Y$ and (2) Y is revealed before X denoted by $Y \rightharpoonup X$. If the order of resolution is unknown, notation $X \rightleftharpoons Y$ is used.

Suppose that X is resolved at day 1 and Y a day later. Also assumed is that the variable realizations become public information instantly. In this case, the Scenario 1 where the investor conditions Y, the variable that remain unresolved, on the realization of X, seems more credible than Scenario 2. In fact, the conditional of X on Y makes little sense because given a realization of Y at day 2, X is no longer a chance variable. Once the realization of X is known, the instrument becomes a function of Y only – $d_{X=1}(Y = 1) = 0$ and $d_{X=1}(Y = 0) = 1$ or $d_{X=0}(Y = 1) = 1$ and $d_{X=0}(Y = 0) = 0$. The value (certainty equivalent) of such an instrument can be found by putting it on the market. Suppose the market value for the instrument is c (c is less than 0.5 if market is averse to uncertainty). So, the instrument that the decision maker holds today (day 0) can be sold in day 1 for c no matter what the realization of X will be. Therefore, in order to avoid arbitrage the value of the instrument today must be exactly c.

On the other hand, if $Y \rightharpoonup X$ meaning Y is resolved at day 1 and X is resolved at day 2, the Scenario 2 is more credible than Scenario 1. In this case, the value of the instrument at day 1, either $d_{Y=0}$ or $d_{Y=1}$, is 0.5 because they are instruments under risk (depend on X only) with equal probabilities of getting 0 and 1 util. Therefore, the value (fair price) of d today must be 0.5.

Thus, *the apparent conflict between two scenarios can be reconciled if the information about the order of variable resolution is considered.*

3.3 Decision Making Under Heterogeneous Uncertainty

Formally, we consider a decision situation described by a tuple $(\mathcal{V}, \Delta, \mathcal{W}, \mathcal{D})$. $\mathcal{V}$ is a finite collection of variables $\{X_1, X_2, \ldots X_n\}$. Each variable has a domain $\text{dom}(\cdot)$ —the set of values it can take. The Cartesian product of the domains of variables in $\mathcal{V}$ is the state-space denoted by $\text{dom}(\mathcal{V}) = \Omega = \times_{X \in \mathcal{V}} \text{dom}(X)$. $s \in \Omega$ is a tuple of values in the domains of X_i. Δ is a functional mapping from $\mathcal{V}$ to the set of functions of the form $2^{\text{dom}(.)} \to [0, 1]$. $\Delta(X)$ represents the uncertainty about X. $\mathcal{W} = [0, 1]$ is the *outcome space*. $\mathcal{D}$ is the *act space*. An *act* is a mapping $f : \Omega \to \mathcal{W}$.

Furthermore, we assume that each variable in $\mathcal{V}$ has a *type*. The type of a variable is determined by the properties that its uncertainty function Δ must satisfy. For example, X is of probabilistic type if $\Delta(X)$ is a probability measure. For each type of variable, there is a *certainty equivalent* operator ($\mathcal{CE}$). This operator maps a uncertainty measure $\Delta(X)$ and act f_X on the domain of X to a point in $\mathcal{W}$ i.e., $\mathcal{CE}(\Delta(X), f_X) \in \mathcal{W}$. For example, if X is probabilistic then $\mathcal{CE}_{Pr}$ is the expected utility operator.

For the sake of simplicity, we initially make an assumption that variables are mutually non-influential. That means $\Delta(X)$ does not depend on the occurrence of other variables.

Suppose $\mathcal{O}$ is a subset of the collection of variables $\mathcal{V}$ and $\bar{\mathcal{O}}$ is the complement of $\mathcal{O}$. Let s be a state or element in dom($\mathcal{V}$), denote by $s_{\mathcal{O}}$ the projection of s on $\mathcal{O}$. Conversely, if s is an element of dom($\mathcal{O}$) and t —an element of dom($\bar{\mathcal{O}}$) then $s.t$ denotes the element of dom($\mathcal{V}$) that is constructed by joining s and t.

For act f on Ω (i.e., dom($\mathcal{V}$)) and $s \in$ dom($\mathcal{O}$), denote by f_s the act on dom($\bar{\mathcal{O}}$) defined as follows

$$\forall t \in \text{dom}(\bar{\mathcal{O}}), \ f_s(t) = f(s.t) \tag{53}$$

We say that f_s is f conditional on s.

In the example in Sect. 3.1, $\mathcal{V} = \{X, Y\}$, $\Delta(X) = \{Pr(X = 0) = 0.5, Pr(X = 1) = 0.5\}$, $\Delta(Y) = \{\pi(Y = 0) = 1, \pi(Y = 1) = 1\}$, $\mathcal{W} = [0, 1]$ and $\mathcal{D}$ is the set of functions from dom(X) × dom(Y) → $\mathcal{W}$.

3.4 Folding Back Principle

For a given sequence of variable realization $X_1 \rightharpoonup X_1 \rightharpoonup \ldots X_n$, an act $f^{(0)}$ on Ω (i.e., dom($X_1, X_2, \ldots X_n$)) is transformed by folding back to an act $f^{(1)}$ on dom($X_1, X_2, \ldots X_{n-1}$). In the rule form $f^{(1)}$ is

$$f^{(1)} = \{s \hookrightarrow \mathcal{CE}(\Delta X_n, f_s^{(0)}) \mid s \in \text{dom}(X_1, X_2, \ldots X_{n-1})\} \tag{54}$$

In words, $f^{(1)}$ maps each state of the remaining domain dom($X_1, X_2, \ldots X_{n-1}$) into the certainty equivalent obtained by folding back the original act conditional on that state $f_s^{(0)}$. The superscript $^{(i)}$ tells the folding back step. Recursively, $f^{(1)}$ can be transformed to $f^{(2)}$ on dom($X_1, X_2, \ldots X_{n-2}$) defined by

$$f^{(2)} = \{s \hookrightarrow \mathcal{CE}(\Delta X_{n-1}, f_s^{(1)}) \mid s \in \text{dom}(X_1, X_2, \ldots X_{n-2})\} \tag{55}$$

Finally, the certainty equivalent of act $f^{(0)}$ is taken to be.

$$\mathcal{CE}(f^{(0)}) = f^{(n)} = \mathcal{CE}(\Delta(X_1), f^{(n-1)}) \tag{56}$$

The fold-back procedure is well known. We just propose to apply it in situations involving uncertain variables of different types. The major advantage of this approach is that it makes use of uncertainty that is local for each variable and eschews the need to have a *global* uncertainty on $\mathrm{dom}(X_1, X_2, \ldots X_n)$. This property is important because it could be impossible to have a global measure without loss of information. More importantly, this procedure allows you to have more flexibility in using different formal frameworks to model uncertainty about the world.

For an example, let us use this procedure to analyze the concept of linear combination of possibilistic acts which is impossible to handle within a purely possibilistic framework.

3.5 Linear Combination of Possibilistic Acts

The linear combination of possibilistic acts f, g can be constructed with the help of any probability generating device such as coin with bias p.

There are two schemes to construct such an act. In the first scheme, you toss the coin and depending on which side it lands you get either f if Head occurs or g if Tail occurs. You get an act h in the sense that if state s of the Ω realizes then you will be rewarded with the outcome of this act $h(s)$.

Alternatively, you can construct a new act by postponing the coin toss until after realization of s and depends on which way it lands you get either $f(s)$ if Head occurs or $g(s)$ if Tail occurs.

We can describe the situation with two variables X —a coin toss and Y —a possibilistic variable. The first way of construction corresponds to $X \rightharpoonup Y$ while the second to $Y \rightharpoonup X$.

Example 6. Figure 13 describes two schemes for the case where $\mathrm{dom}(X) = \{H, T\}$ and $\mathrm{dom}(Y) = \{M, L\}$ (M stands for "more" and L stands for "less"). The following configuration is used. $Pr(H) = 0.7$, $Pr(T) = 0.3$ and $\pi(M) = 1, \pi(L) = 0.4$. $f(M) = 1$, $f(L) = 0.2$ and $g(M) = 0.4$, $g(L) = 0.8$.

For risk attitude, we consider a utility function in the Constant Relative Risk Averse (CRRA) family $u = x^r$ with $r = 0.5$ (risk averse). For ambiguity attitude, we use ambiguity averse polar utility function $V_{0.4}(x)$. Attached to each node is the certainty equivalent of the sub-tree rooting at the node. For example, on the tree on the right, 0.79 is the certainty equivalent obtained by folding back the probabilistic act $\{H \hookrightarrow 1, T \hookrightarrow 0.4\}$ under utility function $u(x) = x^{0.5}$. ■

The calculation shows that the certainty equivalent if the coin is tossed before realization of Y (on the left) is 0.55. If the order of resolution is reversed, the certainty equivalent is 0.63. It shows that the two schemes of combination are not equivalent.

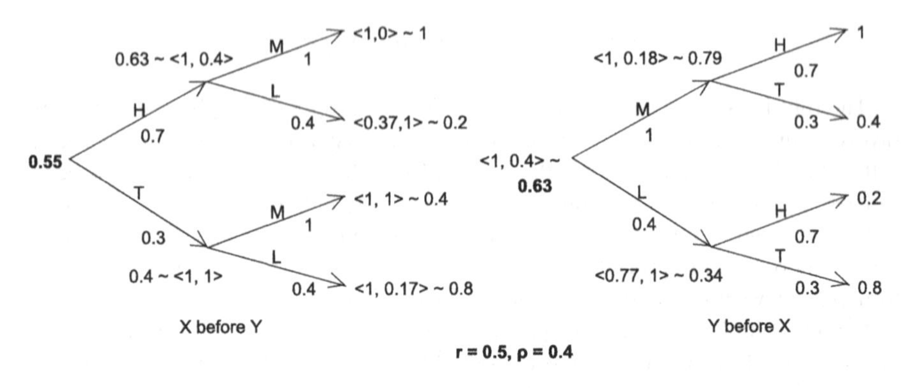

Fig. 13 Ambiguity of the linear combination of possibilistic acts

If that conclusion seems counter-intuitive it could be because of the habits of reasoning with probability.

Let consider some of the arguments. You could argue that if you are not allowed to exchange what you have after the first variable resolved (coin toss in the first scheme) for its certainty equivalent then the certainty equivalents of two scheme must be identical because the reward for each contingency (HM, HL, TM, TL) is the same.

The response to such argument is that first, the imposed ban can alter the value of an act or in the language of business, distorts the market. There is a more subtle explanation. If you are not in position to distinguish individual variables and all you know is the final four options (HM, HL, TM, TL) then the formulation of problem in terms of two variables X and Y is not correct. For you actually there is only one variable Z whose domain has four elements. Under this scenario, the trees in Fig. 13 do not represent your information. Note that our proposed method still applied to this one-variable scenario but of course it would give a different answer.

We have shown that the notion of "linear combination of possibilistic acts" is not well-defined. The ambiguity embedded explains the problems in developing an evaluation method for such a combination.

4 Summary

The environment of heterogeneous uncertainty is the situations in which a decision maker has to choose among actions that depend on variables of multiple uncertainty formalisms. This paper provides a systematic overview of decision models for individual uncertainty frameworks including probability theory, Dempster-Shafer belief function theory and possibility theory. We then propose a solution to an open problem of decision making in the environment of heterogeneous uncertainty. This is an issue of significant practical impact and many theoretically interesting questions. The key idea is to introduce the order of variable resolution to the analysis of decision making under heterogeneous uncertainty. Many important questions in this approach remain unsolved and are the subjects of future study.

References

1. Allais, M.: Le comportement de l'homme rationel devant le risque: Critique des postulats et axioms de l'ecole americaine. Econometrica **21**, 503–546 (1953)
2. Arrow, K.J., Hurwicz, L.: An optimality criterion for decision making under ignorance. In: Arrow, K.J., Hurwicz, L. (eds.) Studies in Resource Allocation Processes. Cambridge University Press, Cambridge (1977)
3. Birnbaum, A.: On the foundation of statistical inference. J. Am. Stat. Assoc. **57**(298), 269–306 (1962)
4. Cox, D.R., Hinkley, D.V.: Theoretical Statistics. Chapman and Hall, London (1974)
5. Dempster, A.: Upper and lower probability induced by multivalued mapping. Ann. Probab. **38**, 325–339 (1967)
6. Dempster, A.: A generalization of Bayesian inference. J. Roy. Stat. Soc. B **30**, 205–247 (1968) with discussion
7. Dubois, D., Godo, L., Prade, H., Zapico, A.: Making decision in a qualitative setting: from decision under uncertaity to case-based decision. In: Cohn, A., Schubert, L., Shapiro, S. (eds.) Sixth International Conference on Principles of Knowledge Representation and Reasoning (KR98), pp. 594–605. Morgan Kaufmann, California (CA) (6 1998)
8. Dubois, D., Nguyen, T.H., Prade, H.: Possibility theory, probability and fuzzy sets. In: Dubois, D., Prade, H. (eds.) Fundamentals of Fuzzy Sets, pp. 344–438. Kluwer Academic, Boston (2000)
9. Ellsberg, D.: Risk, ambiguity and the Savage's axioms. Q. J. Econ. **75**(4), 643–669 (1961)
10. Giang, P.H.: A new axiomatization for likelihood gambles. In: Dechter, R., Richardson, T.S. (eds.) Uncertainty in Artificial Intelligence: Proceedings of the 22nd Conference, pp. 192–199. AUAI Press, (2006)
11. Giang, P.H.: Dynamic consistency and decision making under vacuous belief. In: Cozman, F.G., Pfeffer, A. (eds.) Uncertainty in Artificial Intelligence: Proceedings of the 27nd Conference (UAI-2011), pp. 230–237. AUAI Press, (2011)
12. Giang, P.H.: Decision with Dempster-Shafer belief functions: decision under ignorance and sequential consistency. Int. J. Approximate Reasoning **53**(1), 38–53 (2012)
13. Giang, P.H., Shenoy, P.P.: Two axiomatic approaches to decision making using possibility theory. Eur. J. Oper. Res. **162**(2), 450–467 (2005)
14. Giang, P.H., Shenoy, P.P.: A decision theory for partially consonant belief functions. Int. J. Approximate Reasoning **52**(3), 375–394 (2011)
15. Giang, P.H., Shenoy, P.P.: A decision theory for partially consonant belief functions. Int. J. Approximate Reasoning **52**, 375–394 (2011)
16. Gilboa, I.: Theory of Decision Under Uncertainty. Cambridge University Press, New York (2009)
17. Glimcher, P.W.: Foundations of Neuroeconomic Analysis. Oxford University Press, New York (2011)
18. Halpern, J.Y.: Reasoning About Uncertainty. MIT Press, Cambridge (2003)
19. Jaffray, J.-Y.: Linear utility theory for belief functions. Oper. Res. Lett. **8**, 107–112 (1989)
20. Jaffray, J.-Y., Wakker, P.: Decision making with belief functions: compatibility and incompatibility with the sure-thing principle. J. Risk Uncertainty **8**(3), 255–271 (1994)
21. Jensen, N.E.: An introduction to Bernoullian utility theory I: utility functions. Swed. J. Econ. **69**, 163–183 (1967)

22. Savage, L.J.: The Foundations of Statistics, 2nd edn. Dover, New York (1972)
23. Shafer, G.: A Mathematical Theory of Evidence. Princeton University Press, Princeton (1976)
24. Smets, P.: The transferable belief model for quantified belief representation. In: Gabbay, D.M., Smets, P. (eds.) Handbook of Defeasible Reasoning and Uncetainty Management System, vol. 1, pp. 267–301. Kluwer, Doordrecht (1998)
25. Smets, P., Kennes, R.: The transferable belief model. Artif. Intell. **66**(2), 191–234 (1994)
26. Wakker, P.P.: Prospect Theory for Risk and Ambiguity. Cambridge University Press, Cambridge (2010)
27. Walley, P.: Belief function representation of statistical evidence. Ann. Stat. **15**(4), 1439–1465 (1987)
28. Walley, P.: Statistical Reasoning with Imprecise Probabilities. Chapman and Hall, London (1991)
29. Yager, R.: Decision making under Dempster-Shafer uncertainties. Int. J. Gen Syst **20**, 223–245 (1992)
30. Zadeh, L.: Fuzzy set as a basis for a theory of possibility. Fuzzy Sets Syst. **1**, 3–28 (1978)

One-Shot Decision Theory: A Fundamental Alternative for Decision Under Uncertainty

Peijun Guo

Abstract The attempts of this paper are as follows: clarifying the fundamental differences between the one-shot decision theory which was initially proposed in the paper [16] and other decision theories under uncertainty to highlight that the one-shot decision theory is a scenario-based decision theory instead of a lottery-based one; pointing out the instinct problems in other decision theories to show that the one-shot decision theory is necessary to solve one-shot decision problems; manifesting the relation between the one-shot decision theory and the probabilistic decision methods. As regret is a common psychological experience in one-shot decision making, we propose the one-shot decision methods with regret in this paper.

Keywords Decision making · One-shot decision · Regret · Regret focus points · Scenario-based decision theory · Human-centric decision-making · Behavioral operations research

1 Introduction

In many decision problems encountered in practice, a decision maker has one and only chance to make a decision under uncertainty. Such decision problems are called one-shot decision problems. Let us begin with several real examples to show the features of one-shot decision problems. An article in NIKKAN SPORTS (10-28-2005) stated that Hanshin Electric Railway Co., Ltd., which owns Hansin Tiger baseball team, lost nearly 500 thousand dollars because Hansin Tiger was beaten by Chiba Lotte Marines in Japanese National Baseball Championship in 2005. The huge loss resulted from the production cost of commemorative goods. The Hanshin

P. Guo (✉)
Faculty of Business Administration, Yokohama National University,
79-4 Tokiwadai, 240-8501 Hodogaya-ku, Yokohama, Japan
e-mail: guo@ynu.ac.jp

P. Guo and W. Pedrycz (eds.), *Human-Centric Decision-Making Models for Social Sciences*, Studies in Computational Intelligence 502,
DOI: 10.1007/978-3-642-39307-5_2,

Electric Railway Co., Ltd. had one and only one chance to make a decision whether to prepare the commemorative goods and decide how many goods to be produced before the final result of the game was known. Another example is the Great Sichuan Earthquake that occurred at 14:28:01 CST on May 12,2008. Official figures stated that 69,197 people were confirmed dead. Amongst many serious problems caused by the earthquake, Tangjiashan Lake particularly drew the attention of the world because it was seriously threatening the lives of 1300,000 people, Lanchengyu Oil Pipeline and one of the arterial railways in China, Chengbao Railway. To prevent damage to the dam, the water in the lake needed to be drained away as soon as possible by building a sluice channel. There were only two alternatives for building a sluice channel, using explosives or digging by excavators. It was a one-shot decision to decide which method should be utilized in the face of the uncertainties from rain, aftershock, dam stability, land slip and time.

Quoting from King ([23], p. 102) "There is a strong basis for the belief that the decision to outsource-particularly offshore-is a "one-time and-never-return" decision because the loss of capability by the client in activities that are outsourced is well known and the cost of re-creating those capabilities may be prohibitive." Clemen and Kwit ([7], p. 74) stated that "Because of the one-time nature of typical decision-analysis projects, organizations often have difficulty identifying and documenting their value. Based on Eastman Kodak Company's records for 1990 to 1999, we estimated that decision analysis contributed around a billion dollars to the organization over this time." Fine [11] emphasized that technological innovation and competitive intensity have been acting as two major drivers to speed up the rates of evolution i.e. "industry clock speeds", with regard to the product, the process, and the organization of each industry. Accelerated industry clock speed makes one-shot decision problem highly relevant. Lastly, the growing dominance of service industries makes one-shot decision problems especially applicable.

It can be seen that one-shot decision is a kind of irreversible action for problems with partially known information. Such decision problems are commonly encountered in business, social systems and economics.

Guo [16] proposed the one-shot decision theory (OSDT) for solving one-shot decision problems. In OSDT, we argue that a person makes a one-shot decision based on some particular scenario which is regarded as the most appropriate one for him/her while considering the satisfaction level incurred by this scenario and its possibility degree. The one-shot decision process involves two steps. The first step is to identify which state of nature should be taken into account for each alternative. The identified state of nature is called focus point. The second step is to evaluate the alternatives based on the outcomes brought by the focus points to obtain the optimal alternative. As an application, a duopoly market of a new product with a short life cycle is analyzed where three kinds of firms, i.e. normal, active and passive firms are considered. Possibilistic Cournot equilibriums are obtained for different kinds of pairs of firms in a duopoly market. The results of analysis are quite in agreement with the situations encountered in the real business world [14]. Private real estate investment is a typical one-shot decision problem for personal investors due to the huge investment expense and the fear of substantial loss.

In Guo [15], private real estate investment problem is analyzed using one-shot decision framework. The analysis demonstrates the relation between the amount of uncertainty and the investment scale for different types of personal investors. The proposed model provides insights into personal real estate investment decisions and important policy implications in regulating urban land development.

In this research, we attempt to clarify the fundamental differences between OSDT and other decision theories under uncertainty, types of instinct problems in other decision theories that make OSDT necessary to solve one-shot decision problems, and the kind of relation that OSDT holds with the probabilistic decision methods. Realizing that regret is a common psychological experience in one-shot decision making, we propose the one-shot decision methods with regret in this paper.

The remainder of the paper is organized as follows. In Sect. 2, we address the fundamental differences between OSDT and other decision theories under uncertainty and why OSDT is necessary to solve certain types of problems. In Sect. 3, the one-shot decision methods with regret are proposed. In Sect. 4, a numerical example of a newsvendor problem is addressed. Finally, the relationship between OSDT and other decision theories under uncertainty is clarified and the future research directions are provided in Sect. 5.

2 The Need for the One-Shot Decision Theory

2.1 The Same Framework of Weighting Average for the Existing Decision Theories Under Uncertainty

In general, before taking an action a decision maker cannot know which outcome will occur. Such unknown situations can be divided into three categories: risk, uncertainty and ignorance. According to Knight [24], risk involves situations where the probabilities of all possible outcomes can be exactly calculated whereas uncertainty is related to the status when exact probabilities cannot be obtained due to inadequate information. Ignorance occurs when no information is available to distinguish which outcome is more likely to occur.

Different unknown situations require different decision theories. Decision rules for situations involving ignorance include maximin, maximax, minmax regret and Hurwicz criterion. The expected utility (EU) theory of Von Neumann and Morgenstern is appropriate for decision making under risk and the subjective expected utility (SEU) theory of Savage is appropriate for decision making under uncertainty where subjective probabilities are used to reflect an individual's belief. There is evidence that people systematically violate EU theory while making decisions [21, 25]. Most criticism of the Von Neumann-Morgenstern's and Savage's axioms mainly focus on independence axiom or sure thing principle [1, 10], transitivity axiom [26] and completeness axiom [5, 30].

Let us discuss the completeness axiom. Quoting Von Neumann and Morgenstern ([33], p. 17) "Let us for the moment accept the picture of an individual whose system of preferences is all-embracing and complete, i.e. who, for any two objects or rather for any two imagined events, possesses a clear intuition of preference. More precisely we expect him, for any two alternative events which are put before him as possibilities, to be able to tell which of the two he prefers." In fact, in the real world the decision maker does not have the capability to distinguish which alternative is better so that he/she asks a decision analyst to help solving the problem. Nevertheless, with the assumption that the completeness axiom holds for the decision maker the decision analyst builds decision models based on (subjective) expected utility theory. Obviously, it is logically inconsistent. It is natural to raise the questions: who is the protagonist? Is it the decision maker or the decision analyst?

Many theories have been proposed to react to such empirical evidence that human behavior often contradicts expected utility theory. One such theory, i.e. prospect theory developed by Kahneman and Tversky [20] is a non-additive probability model. In prospect theory, value is assigned to gains and losses based on a reference point rather than to the final asset as in EU and SEU. Also, probabilities are replaced by decision weights which do not satisfy the probability additivity. The value function is defined on deviations from a reference point. Value functions are normally concave for gains (implying risk aversion), and convex for losses (implying risk seeking). Regret theory [26] uses modified utility of choosing one alternative instead of another which consists of a choiceless utility and a regret-rejoice function.

Other models such as, second-order probabilities models [19, 29] and non-additive probability models [13, 28] have also been proposed in this empirical challenge. It should be noted that these decision theories follow the same framework of weighting average of all outcomes no matter how they revise their models. In the context of fuzzy decision making, Yager [35] proposed the optimistic utility and Whalen ([34]) gave the pessimistic utility. These two utilities were axiomatized in the style of Savage by Dubois et al. [9]. Giang and Shenoy [12] generalized them by introducing an order on a class of canonical lotteries. In fact, the optimistic utility is a sort of a weighted average where multiplication and addition is replaced by T-norm, min and Co-norm, max, respectively. The pessimistic utility is a counterpart of the optimistic utility in the sense of possibility and necessity measures. Brandstatter et al. [6] proposed the priority heuristic where the lotteries are chosen by lexicographic rules for the four reasons, i.e. minimum gain, maximum gain and their respective probabilities.Katsikopoulos and Gigerenzer [22] showed that the priority heuristic can predict human decision-making better than the most popular modifications of utility theory, such as cumulative prospect theory, and is, in this sense, close to human psychology.

It can be concluded that decision theories under uncertainty are theories of choice under uncertainty where the objects of choice are lotteries. In light of the features of the one-shot decision problem, this raises two problems: Is the probability distribution suitable for characterizing the uncertainty? Is the expected utility a reasonable index for evaluating the performance of a one-shot decision? The answers are given in the following two subsections.

2.2 Is the Probability Distribution Suitable for Characterizing the Uncertainty in the One-Shot Decision Problem?

In general, the one-shot decision problem involves the situation that has seldom or never happened so far so that the decision maker can not obtain the objective probability distribution. Subjective probability enters as a means of describing the belief about how likely a particular event is to occur. Mainly, there are two kinds of approaches for obtaining subjective probabilities, i.e. the lottery method and the exchangeability method. The lottery method determines the subjective probability of an event in terms of simple betting odds [27]. The exchangeability method consists in the subsequent splitting of the state space into equally likely events via binary choices between binary prospects. Baillon [2] argued that subjective probabilities elicited by the exchangeability method might violate the additivity. The lottery method also was examined by the following experiment.

Subjects:

Fifty subjects participated in the experiment conducted on May 25, 2011. All the participants were undergraduate students who took the course of Decision Sciences, at Faculty of Business Administration, Yokohama National University. The experiment started at the beginning of the lesson. None of them were aware of the true goal of the experiment. The experiment was conducted before teaching them what the subjective probability and the additivity of probability measure are.

Procedure:

The following questions are independently asked:

Q1: If it rains next Wednesday, you will get 10,000Yen. However, if it does not rain, you will get nothing. How much would you be willing to pay for this proposition?
Q2: If it does not rain next Wednesday, you will get 10,000Yen. However, if it rains, you will get nothing. How much would you be willing to pay for this proposition?

Results:

The prices for Q1 and Q2 are denoted as X_1 and X_2, respectively. The mean of $X_1 + X_2$ is 4773.98 and the standard deviation of $X_1 + X_2$ is 2858.73. If the additivity property holds, then the mean of $X_1 + X_2$ should be 10,000. We set up the null hypothesis: the mean of $X_1 + X_2$ is 10,000. The value of the test statistic is calculated as -12.93 so that this null hypothesis is rejected at the 0.01 level of significance by two-tail test. It means that the additivity can not always be guaranteed while using the lottery method.

Possibility is an alternative for characterizing the uncertain situation. It can be explained from three semantic aspects, i.e. ease of achievement, plausibility referring to the propensity of events to occur (which relates to the concept "potential surprise") and logical consistency of available information. Possibility distribution is a function whose value shows the degree to which an element is to occur, as defined as follows.

Definition 1 Given a function $\pi : S \rightarrow [0, 1]$ if $\max_{x \in S} \pi(x) = 1$, then $\pi(x)$ is called a possibility distribution where S is the sample space. $\pi(x)$ is the possibility degree of x.

$\pi(x) = 1$ means that it is normal that x occurs and $\pi(x) = 0$ means that it is abnormal that x occurs. The smaller the possibility degree of x, the more surprising the occurrence of x. Obtaining the possibility distribution always poses a fundamental problem for decision with possibilistic information. Guo and Tanaka [17] proposed the method for identifying the possibility distribution of the stock returns with the idea of similarity. Guo et al. [18] obtained the possibility distribution of the demand for a new product with the idea of potential surprise. Guo [16] presented a general method for identifying the possibility distribution by voting described as follows:

Suppose $S = \{x_1, x_2, \ldots, x_n\}$. We ask multiple experts to select the most possible events from S. In other words, if an expert selects the event x_i, the expert will not be surprised by its occurrence. The number of experts who select x_i is denoted as k_i. Setting $K = \max_{i=1,\ldots,n} k_i$, the possibility degree of x_i is obtained as k_i/K in the sense that each expert has equal reliability for judging which event will occur.

It is a valid question to ask which is better, probability or possibility. To answer this question, let us take a look at the following example.

Example 1 [15] Who is guilty?

A car has been destroyed by somebody in a parking lot. After careful investigation, it is sure that one and only one of three suspects A, B and C must be guilty of the crime. However, who is guilty of the crime is still unknown. Suppose, based on the currently obtained evidence subjective probabilities are used to characterize the belief about who is guilty amongst the three suspects and given as e.g. $P(A) = 0.4$, $P(B) = 0.4$ and $P(C) = 0.2$. Considering the relation $P(A) = 1 - P(\overline{A})$ where $\overline{A}$ is the complement of A, it can be concluded that none of these three suspects is guilty in the context of probability ($P(A) < P(\overline{A})$, $P(B) < P(\overline{B})$, $P(C) < P(\overline{C})$). This conclusion is in conflict with the antecedent one, i.e. one and only one of three suspects A, B and C must be guilty. This conflict originates from the existence of incomplete information. In this example, the possibility distributions showing the degrees to which a person is guilty might be given as e.g. $\pi(A) = 1$, $\pi(B) = 1$ and $\pi(C) = 0.7$. $\pi(A) = \pi(B) = 1$ means that based on the obtained evidence, A or B is most possible to be guilty. The relation $\pi(A) \neq 1 - \pi(\overline{A})$ implies that the possibility degree of A being guilty does not provide any information on A not being guilty.

It follows from this example that the possibility distribution is a less restricted framework than single probability measures and hence can be used for encoding ill-known subjective probability information. The answer to the question which is better, probability or possibility is that the possibility distribution might be effective for representing the rough knowledge or judgment of human being when the information is not rich enough.

2.3 Is the Expected Value a Reasonable Index for Evaluating the Performance of a One-Shot Decision?

To answer this question, let us consider the following example.

Example 2 Is Mr. Smith is taller than Mr. Tanaka?

Let us consider two populations:

Population A: The heights of male undergraduate students in Yokohama National University (YNU)

Population B: The heights of male undergraduate students in University of Alberta (UA)

For instance, we take 100 samples from the populations A and B, respectively. The sample mean of A, say 175cm is less than the sample mean of B, say 180cm. You randomly select one male undergraduate student from UA, say Mr. Smith and select one from YNU, say Mr. Tanaka. Can you say Mr. Smith is taller than Mr. Tanaka? The answer will be "no" because the statistical property by itself does not imply anything about what might happen in just one sample. Next, let us take into account two other populations as follows:

Population I: The outcomes generated by an alternative C

Population II: The outcomes generated by an alternative D

Suppose that the mean of the population I is larger than the one of II. Randomly select one outcome from I, that is, x, and one outcome from II, that is, y. Can you say x is larger than y? Can you say C is better than D? Both of answers will be "no". From the above examples, it is easy to understand that for the one-shot decision problem the expected value might not be a suitable index for evaluating the performance of an alternative.

In conclusion, a new decision theory is needed to solve one-shot decision problems featured by partially known information and the occurrence of only one outcome. Guo [16] initially proposed the one-shot decision theory (OSDT) which is scenarios-based instead of lotteries-based as in other decision theories under uncertainty. In OSDT, we argue that a person makes a one-shot decision based on some particular scenario which is regarded as the most appropriate one for him/her while considering the satisfaction level incurred by this scenario and its possibility degree. Because regret is a common emotion in one-shot decision problems, we propose one-shot decision methods with regret in the following section.

3 One-Shot Decision Methods with Regret

Some people find decision making under uncertainty difficult because they fear making the "wrong decision", wrong in the sense that the outcome of their chosen alternative proves to be worse than could have been achieved with another alternative ([3], p. 1156). This kind of situation can be described by the word "regret" which is "the painful sensation of recognizing that 'what is' compares unfavorably with 'what might have been"' ([32], p. 77). Shimanoff pointed out that regret was the most frequently named negative emotion in a study of verbal expressions of emotions in everyday conversation [31]. Decision with regret has been researched by Savage [27], Loomes and Sugden [26], Bell [3], Sugden [32] and so on. In one-shot decision problems, the decision maker has one and only opportunity to make a decision so that there is no chance to correct his/her decisions once the decision has been made. Hence, regret emotion is an especially important factor that affects the decision maker's behavior.

3.1 Regret Function

Denote the set of an alternative a as A and the set of a state of nature x as S. The degree to which a state of nature is to occur in the future is characterized by a possibility distribution $\pi(x)$ defined by the definition 1. The consequence resulting from the combination of an alternative a and a state of nature x is refereed to as a payoff, denoted as $v(x, a)$. Suppose that after a decision maker chooses an alternative a, a state of nature x appears. The decision maker might regret his/her choice. The regret value is $p(x, a) = \max_{b \in A} v(x, b) - v(x, a)$. Then the regret quantile denoted as $w(x, a)$, is calculated as follows:

$$w(x, a) = p(x, a) / \max_{d \in A} p(x, d). \tag{1}$$

The regret level of a decision maker for a regret quantile can be expressed by a regret function, as defined below.

Definition 2 Denote the set of a regret quantile $w(x, a)$ as W. The following function

$$r : W \to [0, 1] \tag{2}$$

with

$$r(w_1) > r(w_2) \quad \text{for} \quad w_1 > w_2, \tag{3}$$

is called a regret function. Because the regret quantile is the function of x and a, we can rewrite the regret function as $r(w(x, a))$. For the sake of simplification, we write $r(w(x, a))$ as $r(x, a)$ in this paper. Regret function is a nonlinear transformation of the regret quantile and represents the relative position of the regret.

The information for one-shot decision with regret can be summarized as a quadruple (A, S, π, r). One-shot decision is to choose one alternative based on (A, S, π, r) when only one decision chance is given.

It is well recognized that when you ask some person why he/she makes such a one-shot decision with little information, he/she always tells you just one scenario which is crucial to him/her and is the basis for achieving some conclusion. For instance, empirical evidence suggests that insurance buyers focus on the potential large loss even at the low probabilities; lottery ticket buyers focus on the big gains even at small probabilities [8]. Interestingly, Bertrand and Schoar [4] found out that financial decision depended not just on the nature of the firm and its economic environment, but also the personalities of the firm's top management. For instance, while older CEOs tended to be more conservative and pushed their firms towards lower debt, CEOs with MBA degrees tended to be more aggressive.

For the one-shot decision methods with regret, we think that a person make a one-shot decision based on some particular scenario while considering the possibility degree and the regret level. Selecting the scenario depends on the personalities of the decision maker for example one person may be active whereas another may be passive. The one-shot decision making procedure consists of the following three steps. In Step 1, a decision maker identifies some state of nature (particular scenario), called regret focus point for each alternative according to his/her own characteristic. In Step 2, the validity of the regret focus points is checked. In Step 3, the decision maker evaluates the alternatives based on the regret level brought by regret focus point to obtain the best alternative. These three steps are addressed in detail in the following subsections.

3.2 Identifying Regret Focus Points

Since one and only one state of nature will come up for a one-shot decision problem, a decision maker needs to decide which state of nature ought to be considered for making a one-shot decision. Each state of nature is equipped with a pair of possibility and regret so that how to determine the states of nature depends on his/her attitudes about possibility and regret. The selected state of nature is call regret focus point. Twelve types of regret focus points are provided to help a decision maker in finding out his/her own appropriate one. The characteristics of these focus points are depicted below (shown in Tables 1, 2, 3). Type I and II regret focus points are the states of nature that have the highest and the lowest regret levels, respectively, amongst the ones that have high possibility degrees. Type III and IV regret focus points are the states of nature that have the highest and the lowest regret levels, respectively, amongst the ones that have low possibility degrees. Type V and VI regret focus points are the states of nature that have the highest and the lowest possibility degrees, respectively, amongst the ones that have high regret levels. Type VII and VIII regret focus points are the states of nature that have the highest and lowest possibility degrees, respectively, amongst the ones that have low regret levels. Type IX regret focus point is the state of nature with the higher possibility degree and the higher regret level. Type X regret focus point is the state of nature that has the lower possibility degree and the lower regret level. Type XI regret focus point is the state of nature with the higher possibility degree but the lower regret level. Type XII regret focus point is the state of nature that has the lower possibility degree but the higher regret level.

Table 1 The characteristics of regret focus points (types I–IV)

	High possibility	Low possibility
The highest regret	Type I regret focus point	Type III regret focus point
The lowest regret	Type II regret focus point	Type IV regret focus point

Table 2 The characteristics of regret focus points (types V–VIII)

	High regret	Low regret
The highest possibility	Type V regret focus point	Type VII regret focus point
The lowest possibility	Type VI regret focus point	Type VIII regret focus point

Table 3 The characteristics of regret focus points (types IX–XII)

	Higher regret	Lower regret
Higher possibility	Type IX regret focus point	Type XI regret focus point
Lower possibility	Type XII regret focus point	Type X regret focus point

In the following we will provide mathematical formulas to find out the above mentioned twelve types of regret focus points. For establishing the focus points, we use the operators

$$\min[b_1, b_2, \cdots, b_n] = [\bigwedge_{i=1,\ldots,n} b_i, \bigwedge_{i=1,\ldots,n} b_i, \cdots, \bigwedge_{i=1,\ldots,n} b_i], \tag{4}$$

and

$$\max[b_1, b_2, \cdots, b_n] = [\bigvee_{i=1,\ldots,n} b_i, \bigvee_{i=1,\ldots,n} b_i, \cdots, \bigvee_{i=1,\ldots,n} b_i]. \tag{5}$$

$\min[b_1, b_2, \cdots, b_n]$ and $\max[b_1, b_2, \cdots, b_n]$ are lower and upper bounds of $[b_1, b_2, \cdots, b_n]$, respectively. For example, $\min[0.3, 0.8] = [0.3, 0.3]$ and $\max[0.3, 0.8] = [0.8, 0.8]$. Twelve kinds of regret focus points are as follows:

Type I: $x_\alpha^{1*}(a) = \arg\max_{x \in X^{\geq \alpha}} r(x, a)$ where $X^{\geq \alpha} = \{x | \pi(x) \geq \alpha\}$.

The given parameter α is a level used to distinguish whether the possibility degree is evaluated as 'high' by a decision maker. If $\alpha = 1$ then only the normal case ($\pi(x) = 1$) is considered. The states of nature belonging to $X^{\geq \alpha} = \{x | \pi(x) \geq \alpha\}$ are regarded as having the equivalent possibility to occur. $x_\alpha^{1*}(a)$ is a state of nature with high occurrence possibility. Once it occurs, the decision maker will most regret his/her choice of the alternative a. $x_\alpha^{1*}(a)$ is Type I regret focus point.

Type II: $x_\alpha^{2*}(a) = \arg\min_{x \in X^{\geq \alpha}} r(x, a)$ where $X^{\geq \alpha} = \{x | \pi(x) \geq \alpha\}$.

$x_{\alpha}^{2*}(a)$ is a state of nature with high occurrence possibility. Its occurrence will lead to the lowest regret level of the decision maker for choosing the alternative a. $x_{\alpha}^{2*}(a)$ is Type II regret focus point.

Type III: $x_{\alpha}^{3*}(a) = \arg \max_{x \in X^{\leq \alpha}} r(x, a)$ where $X^{\leq \alpha} = \{x | \pi(x) \leq \alpha\}$.

The occurrence of $x_{\alpha}^{3*}(a)$ will make the decision maker most regret his/her choice of the alternative a. However, the possibility of its occurrence is low. $x_{\alpha}^{3*}(a)$ is Type III regret focus point.

Type IV: $x_{\alpha}^{4*}(a) = \arg \min_{x \in X^{\leq \alpha}} r(x, a)$ where $X^{\leq \alpha} = \{x | \pi(x) \leq \alpha\}$.

The occurrence of $x_{\alpha}^{4*}(a)$ will make the decision maker have the lowest regret level for choosing the alternative a. However, the possibility of its occurrence is low. $x_{\alpha}^{4*}(a)$ is Type IV regret focus point.

Type V: $x_{\beta}^{5*}(a) = \arg \max_{x \in X^{\geq \beta}(a)} \pi(x)$ where $X^{\geq \beta}(a) = \{x | r(x, a) \geq \beta\}$.

The given parameter β is the level to distinguish whether the regret level is evaluated as 'high' by a decision maker. The states of nature belonging to $X^{\geq \beta}(a) = \{x | r(x, a) \geq \beta\}$ are regarded as having the same regret level generated by the alternative a. $x_{\beta}^{5*}(a)$ is an undesirable (the regret level is high) state of nature that has the highest possibility to occur. $x_{\beta}^{5*}(a)$ is Type V regret focus point.

Type VI: $x_{\beta}^{6*}(a) = \arg \min_{x \in X^{\geq \beta}(a)} \pi(x)$ where $X^{\geq \beta}(a) = \{x | r(x, a) \geq \beta\}$, which called Type VI regret focus point, is an undesirable state of nature that has the smallest possibility to occur.

Type VII: $x_{\beta}^{7*}(a) = \arg \max_{x \in X^{\leq \beta}(a)} \pi(x)$ where $X^{\leq \beta}(a) = \{x | r(x, a) \leq \beta\}$, which called Type VII regret focus point, is a desirable (the regret level is low) state of nature that has the highest possibility to occur.

Type VIII: $x_{\beta}^{8*}(a) = \arg \min_{x \in X^{\leq \beta}(a)} \pi(x)$ where $X^{\leq \beta}(a) = \{x | r(x, a) \leq \beta\}$,

which called Type VIII regret focus point, is a desirable state of nature that has the smallest possibility to occur.

Type IX:

$$x^{9*}(a) = \arg \max_{x \in S} \min[\pi(x), r(x, a)]. \tag{6}$$

It follows from (6) that $x = x^{9*}(a)$ maximizes $g(x, a) = \min[\pi(x), r(x, a)]$. In consideration of (4), we know that $\min[\pi(x), r(x, a)]$ represents the lower bound of the vector $[\pi(x), r(x, a)]$. Increasing $\min[\pi(x), r(x, a)]$ ($\max_{x \in S} \min[\pi(x), r(x, a)]$) will increase the possibility degree and the regret level simultaneously. Therefore, $\arg \max_{x \in S} \min[\pi(x), r(x, a)]$ is for seeking a state of nature that has the higher possibility degree and brings the higher regret level due to the choice of the alternative

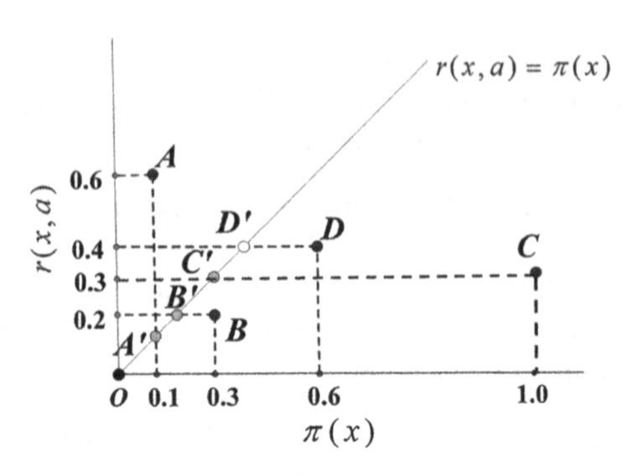

Fig. 1 The explanation of the formula (6)

a. $x^{9*}(a)$ is Type IX regret focus point. For easily understanding (6), let us have a look at Fig. 1. There are four states of nature x_1, x_2, x_3 and x_4 whose $[\pi(x), r(x, a)]$ are respectively $[0.1, 0.6]$, $[0.3, 0.2]$, $[1.0, 0.3]$ and $[0.6, 0.4]$ represented by A, B, C and D. $\min[\pi(x), r(x, a)]$ transfers A, B, C and D into A', B', C' and D', which are $[0.1, 0.1]$, $[0.2, 0.2]$, $[0.3, 0.3]$ and $[0.4, 0.4]$, respectively. $\max_{x \in S} \min[\pi(x), r(x, a)]$, that is,$\max([0.1, 0.1], [0.2, 0.2], [0.3, 0.3], [0.4, 0.4]) = [0.4, 0.4]$ corresponds to D'. $\arg\max_{x \in S} \min[\pi(x), r(x, a)]$ chooses x_4. It follows from Fig. 1 that x_4 is a state of nature with a higher possibility degree and a higher regret level.

Type X:

$$x^{10*}(a) = \arg\min_{x \in S} \max[\pi(x), r(x, a)]. \tag{7}$$

(7) shows that $x = x^{10*}(a)$ minimizes $h(x, a) = \max[\pi(x), r(x, a)]$. In consideration of (5), we know that $\max[\pi(x), r(x, a)]$ represents the upper bound of the vector $[\pi(x), r(x, a)]$. Decreasing $\max[\pi(x), r(x, a)](\min_{x \in S} \max[\pi(x), r(x, a)])$ will decrease the possibility degree and the regret level simultaneously. Therefore, $\arg\min_{x \in S} \max[\pi(x), r(x, a)]$ is for seeking a state of nature that has the lower possibility degree and generates the lower regret level due to the choice of the alternative a. $x^{10*}(a)$ is Type X regret focus point.

Type XI:

$$x^{11*}(a) = \arg\min_{x \in S} \max[1 - \pi(x), r(x, a)]. \tag{8}$$

Likewise, we understand that $x^{11*}(a)$ is the state of nature that has the higher possibility degree and causes the lower regret level when choosing the alternative a. $x^{11*}(a)$ is Type XI regret focus point.

Type XII:

$$x^{12*}(a) = \arg\min_{x \in S} \max[\pi(x), 1 - r(x, a)]. \tag{9}$$

Following (9), we know that $x^{12*}(a)$ is the state of nature that has the lower possibility degree and incurs the higher regret level when choosing the alternative a. $x^{12*}(a)$ is Type XII regret focus point.

For one alternative, more than one state of nature might exist as one type of regret focus point. We denote the sets of twelve types of regret focus points of the alternative a as $X_\alpha^1(a)$, $X_\alpha^2(a)$, $X_\alpha^3(a)$, $X_\alpha^4(a)$, $X_\beta^5(a)$, $X_\beta^6(a)$, $X_\beta^7(a)$, $X_\beta^8(a)$, $X^9(a)$, $X^{10}(a)$, $X^{11}(a)$, and $X^{12}(a)$, respectively. It should be noted that $X_\alpha^3(a)$ and $X_\alpha^4(a)$ are empty sets when $X^{\leq\alpha} = \oslash$; $X_\beta^5(a)$ and $X_\beta^6(a)$ are empty sets when $X^{\geq\beta}(a) = \oslash$; $X_\beta^7(a)$ and $X_\beta^8(a)$ are empty sets when $X^{\leq\beta}(a) = \oslash$. The relationships between different focus points are shown in the following theorem.

Theorem 1

$$(I)\quad X_\alpha^1(a) \cup X_\beta^5(a) \subseteq X^9(a), \tag{10}$$

where

$$\alpha = \beta = \max_{x\in S} \min(\pi(x), r(x,a)). \tag{11}$$

$$(II)\quad X_\alpha^4(a) \cup X_\beta^8(a) \subseteq X^{10}(a), \tag{12}$$

where

$$\alpha = \beta = \min_{x\in S} \max(\pi(x), r(x,a)). \tag{13}$$

$$(III)\quad X_\alpha^2(a) \cup X_\beta^7(a) \subseteq X^{11}(a), \tag{14}$$

where

$$\alpha = 1 - \beta = \max_{x\in S} \min(\pi(x), 1 - r(x,a)). \tag{15}$$

$$(IV)\quad X_\alpha^3(a) \cup X_\beta^6(a) \subseteq X^{12}(a), \tag{16}$$

where

$$1 - \alpha = \beta = \max_{x\in S} \min(1 - \pi(x), r(x,a)). \tag{17}$$

Proof The proof is similar to the proof of Theorem 1 in the paper [16].

Theorem 1 shows the relationships between the different types of regret focus points. The inclusion relations (10), (12), (14) and (16) hold by choosing the suitable values of parameters α and β shown in (11), (13), (15) and (17). Expressed in detail, the set of regret focus points with the higher regret and the higher possibility ($X^9(a)$) includes the set of regret focus points with the highest regret and the high possibility ($X_\alpha^1(a)$) and the set of regret focus points with the highest possibility and the high regret ($X_\beta^5(a)$). The set of regret focus points with the lower regret and the lower possibility ($X^{10}(a)$) includes the set of regret focus points with the lowest regret and the low possibility ($X_\alpha^4(a)$) and the set of regret focus points with the lowest possibility and the low regret ($X_{\beta}^{8}(a)$). The set of focus points with the lower regret and the higher possibility ($X^{11}(a)$) includes the set of regret focus points with the lowest regret and the high possibility ($X_{\alpha}^{2}(a)$) and the set of regret focus points with the highest possibility and the low regret ($X_{\beta}^{7}(a)$). The set of regret focus points with the higher regret and the lower possibility ($X^{12}(a)$) include the set of regret focus points with the highest regret and the low possibility ($X_{\alpha}^{3}(a)$) and the set of regret focus points with the lowest possibility and the high regret ($X_{\beta}^{6}(a)$).

Comments: It raises one question how a decision maker would choose among the twelve focus points. The answer is choosing which type focus point completely depends on which kind of the combination of possibility and regret, for example, the higher possibility and the higher regret, is most worth taking into account for his/her making a one-shot decision. It should be decided by the decision maker himself/herself instead of a decision analyst. Sometimes, a decision maker may consider several types or all types of focus points to make a final decision.

3.3 Checking the Validity of Regret Focus Points (Type IX, X, XI and XII)

In Step 1, twelve types of regret focus points are identified. These regret focus points will be used for determining the optimal alternative. Before that, the validity of Type IX, X, XI and XII regret focus points needs to be checked.

Definition 3 Given the thresholds of the possibility degree α and the regret level β, we say that $x^{9*}(a)$, $x^{10*}(a)$, $x^{11*}(a)$ and $x^{12*}(a)$ are acceptable for α and β if $x^{9*}(a) \in X^{\geq\alpha} \cap X^{\geq\beta}(a)$, $x^{10*}(a) \in X^{\leq\alpha} \cap X^{\leq\beta}(a)$, $x^{11*}(a) \in X^{\geq\alpha} \cap X^{\leq\beta}(a)$ and $x^{12*}(a) \in X^{\leq\alpha} \cap X^{\geq\beta}(a)$ hold, respectively.

We denote the sets of Type IX, X, XI and XII acceptable regret focus points for α and β as $X_{\alpha,\beta}^{9}(a)$, $X_{\alpha,\beta}^{10}(a)$, $X_{\alpha,\beta}^{11}(a)$ and $X_{\alpha,\beta}^{12}(a)$, respectively. For easily understanding the definitions 3, let us consider the following example.

Example 3 The sets of alternatives and states of nature are $A = \{a_1, a_2\}$ and $S = \{x_1, x_2\}$, respectively. For illustrative purposes, let us assume that the estimated possibility degrees of states of nature and the regret levels for two alternatives on each state of nature are shown in Table 4. We set α and β, e.g. as 0.5 and 0.5, respectively. $x^{9*}(a_2)$, $x^{10*}(a_2)$, $x^{11*}(a_1)$, and $x^{12*}(a_1)$ are not acceptable because $x^{9*}(a_2) = x_1 \notin X^{\geq\alpha} \cap X^{\geq\beta}(a_2) = \oslash$, $x^{10*}(a_2) = x_1, x_2 \notin X^{\leq\alpha} \cap X^{\leq\beta}(a_2) = \oslash$, $x^{11*}(a_1) = x_1 \notin X^{\geq\alpha} \cap X^{\leq\beta}(a_1) = \oslash$ and $x^{12*}(a_1) = x_1 \notin X^{\leq\alpha} \cap X^{\geq\beta}(a_1) = \oslash$ hold. We can always obtain Type IX, X, XI and XII regret focus points by (6), (7), (8) and (9). However, in some cases, they are not intuitively accepted as the states of nature with the higher possibility and the higher regret, the lower possibility and the lower regret, the higher possibility and the lower regret, the lower possibility and the higher regret as shown in this example.

Table 4 Data of the example 3

	x_1	x_2
$\pi(x_i)$	0.2	1
$r(x_i,a_1)$	0.3	0.85
$r(x_i,a_2)$	1	0.1

3.4 Obtaining Optimal Alternatives

A decision maker identifies the valid regret focus points of each alternative according to his/her own attitude about possibility and regret as shown in Sects. 3.2 and 3.3. He/she contemplates that the regret focus points are the most appropriate states of nature (scenarios) for him/her and then chooses the alternative which can bring about the best consequence (the lowest regret level) once the regret focus point (scenario) comes true. The procedure for choosing the optimal alternative with regret focus points are given below. Since there are twelve types of regret focus points, there are twelve types of optimal alternatives.

Type I optimal alternative $a^{1*}(\alpha)$: $a^{1*}(\alpha) = \arg\min_{a\in A} r(x_\alpha^{1*}(a), a)$.

Type II optimal alternative $a^{2*}(\alpha)$: $a^{2*}(\alpha) = \arg\min_{a\in A} r(x_\alpha^{2*}(a), a)$.

Type III optimal alternative $a^{3*}(\alpha)$: If $X^{\leq\alpha} = \oslash$, then $a^{3*}(\alpha) \in \oslash$; else $a^{3*}(\alpha) = \arg\min_{a\in A} r(x_\alpha^{3*}(a), a)$.

Type IV optimal alternative $a^{4*}(\alpha)$: If $X^{\leq\alpha} = \oslash$, then $a^{4*}(\alpha) \in \oslash$; else $a^{4*}(\alpha) = \arg\min_{a\in A} r(x_\alpha^{4*}(a), a)$.

Type V optimal alternative $a^{5*}(\beta)$: If $\forall a X_\beta^5(a) \neq \oslash$, then $a^{5*}(\beta) = \arg\min_{a\in A} \max_{x_\beta^{5*}(a)\in X_\beta^5(a)} r(x_\beta^{5*}(a), a)$; if $\forall a \; X_\beta^5(a) = \oslash$, then $a^{5*}(\beta) \in \oslash$; else $a^{5*}(\beta) \in \{a | X_\beta^5(a) = \oslash\}$. The minmax operator is needed for the cases where multiple focus points of an alternative a exist. It reflects the conservative attitude of a decision maker.

Type VI optimal alternative $a^{6*}(\beta)$: If $\forall a X_\beta^6(a) \neq \oslash$, then $a^{6*}(\beta) = \arg\min_{a\in A} \max_{x_\beta^{6*}(a)\in X_\beta^6(a)} r(x_\beta^{6*}(a), a)$; if $\forall a \; X_\beta^6(a) = \oslash$, then $a^{6*}(\beta) \in \oslash$; else $a^{6*}(\beta) \in \{a | X_\beta^6(a) = \oslash\}$.

Type VII optimal alternative $a^{7*}(\beta)$: If $\forall a \; X_\beta^7(a) = \oslash$, then $a^{7*}(\beta) \in \oslash$; else $a^{7*}(\beta) = \arg\min_{a\in A^-} \max_{x_\beta^{7*}(a)\in X_\beta^7(a)} r(x_\beta^{7*}(a), a)$ where $A^- = \{a | X_\beta^7(a) \neq \oslash\}$.

Type VIII optimal alternative $a^{8*}(\beta)$: If $\forall a \; X_\beta^8(a) = \oslash$, then $a^{8*}(\beta) \in \oslash$; else $a^{8*}(\beta) = \arg\min_{a\in A^-} \max_{x_\beta^{8*}(a)\in X_\beta^8(a)} r(x_\beta^{8*}(a), a)$ where $A^- = \{a | X_\beta^8(a) \neq \oslash\}$.

Type IX optimal alternative $a^{9*}(\alpha,\beta)$: If $\forall a X^{9}_{\alpha,\beta}(a) \neq \oslash$, then $a^{9*}(\alpha,\beta) = \arg\min_{a\in A} \max_{x^{9*}(a)\in X^{9}_{\alpha,\beta}(a)} r(x^{9*}(a),a)$; if $\forall a \;\; X^{9}_{\alpha,\beta}(a) = \oslash$, then $a^{9*}(\alpha,\beta) \in \oslash$; else $a^{9*}(\alpha,\beta) \in \{a|X^{9}_{\alpha,\beta}(a) = \oslash\}$.
Type X optimal alternative $a^{10*}(\alpha,\beta)$: If $\forall a \;\; X^{10}_{\alpha,\beta}(a) = \oslash$, then $a^{10*}(\alpha,\beta) \in \oslash$; else $a^{10*}(\alpha,\beta) = \arg\min_{a\in A^-} \min_{x^{10*}(a)\in X^{10}_{\alpha,\beta}(a)} r(x^{10*}(a),a)$ where $A^- = \{a|X^{10}_{\alpha,\beta}(a) \neq \oslash\}$. The minmin operator is used for the cases where multiple focus points of an alternative a exist. It reflects the aggressive attitude of a decision maker.
Type XI optimal alternative $a^{11*}(\alpha,\beta)$: If $\forall a \;\; X^{11}_{\alpha,\beta}(a) = \oslash$, then $a^{11*}(\alpha,\beta) \in \oslash$; else $a^{11*}(\alpha,\beta) = \arg\min_{a\in A^-} \min_{x^{11*}(a)\in X^{11}_{\alpha,\beta}(a)} r(x^{11*}(a),a)$ where $A^- = \{a|X^{11}_{\alpha,\beta}(a) \neq \oslash\}$.
Type XII optimal alternative $a^{12*}(\alpha,\beta)$: If $\forall a X^{12}_{\alpha,\beta}(a) \neq \oslash$, then $a^{12*}(\alpha,\beta) = \arg\min_{a\in A} \max_{x^{12*}(a)\in X^{12}_{\alpha,\beta}(a)} r(x^{12*}(a),a)$; if $\forall a \;\; X^{12}_{\alpha,\beta}(a) = \oslash$, then $a^{12*}(\alpha,\beta) \in \oslash$; else $a^{12*}(\alpha,\beta) \in \{a|X^{12}_{\alpha,\beta}(a) = \oslash\}$.

Comments:

This research extends the results of the paper [16] in two aspects. The first aspect is that instead of the satisfaction level we utilize the regret level to seek focus points because regret is a common emotion in one-shot decision problems. The second aspect is introducing the step for checking the validity of Type IX, X, XI and XII regret focus points. It should be noted that such a step is also applicable to the focus points with satisfaction levels. We also can define dissatisfaction function and use possibility and dissatisfaction to find out focus points. It is especially appropriate for emergency management problems where the upper and lower bounds of losses correspond to the dissatisfaction levels 1 and 0, respectively.

4 Numerical Example: The Newsvendor Problem

In this study, we consider the newsvendor problem for a new product with a short life cycle. As the product is new, there is no data available for forecasting the upcoming demand via statistical analysis. As the life cycle of the product is short, determining optimal order quantity is a typical one-shot decision problem.

The newsvendor problem is described as follows. The retailer orders q units before the season at the unit wholesale price W. When the demand x is observed, the retailer sells goods (limited by the supply q and the demand x) at the unit revenue R with $R > W$. Any excess units can be salvaged at the unit salvage price S_o with $W > S_o$. If there is a shortage, the lost chance price is S_u. The profit function of the retailer is

$$r(x,q) = \begin{cases} Rx + S_o(q-x) - Wq; if x < q \\ (R-W)q - S_u(x-q); if x \geq q. \end{cases} \tag{18}$$

Table 5 Profits obtained for each order quantity

		Demand					
		5	6	7	8	9	10
	5	200	180	160	140	120	100
	6	150	240	220	200	180	160
Orders	7	100	190	280	260	240	220
	8	50	140	230	320	300	280
	9	0	90	180	270	360	340
	10	−50	40	130	220	310	400

Table 6 Regret levels for each order quantity

		Demands					
		5	6	7	8	9	10
	5	0	0.3	0.8	1	1	1
	6	0.2	0	0.4	0.667	0.75	0.8
Orders	7	0.4	0.25	0	0.333	0.5	0.6
	8	0.6	0.5	0.333	0	0.25	0.4
	9	0.8	0.75	0.667	0.278	0	0.2
	10	1	1	1	0.556	0.208	0

Table 7 Possibility degrees of demands

Demands	5	6	7	8	9	10
Possibility degrees	0.2	0.5	0.7	1	0.8	0.6

The unit wholesale price W, the unit revenue R, the unit salvage price S_o, and the lost chance price S_u are set, e.g. as 60 \$, 100 \$, 10 \$ and 20 \$, respectively. Following (18), we calculate the profits (see Table 5). Using (1), we obtain the regret quantile for each order and demand. In this example we set $r(w) = w$, that is, the regret quantile is the same as the regret level. The regret levels for each order and demand are listed in Table 6.

Let us analyze this one-shot decision problem in the form of (A, S, π, u). The set of alternatives is the set of order quantities $A = \{5, 6, 7, 8, 9, 10\}$. The set of states of nature is the set of demands $S = \{5, 6, 7, 8, 9, 10\}$. The regret levels are shown in Table 6. We assume that the possibility degrees of the demands 8, 9, 7, 10, 6, and 5 are 1, 0.8, 0.7, 0.6, 0.5, and 0.2, respectively (shown in Table 7).

The thresholds of possibility degrees and satisfaction levels, α and β, are set, e.g. as 0.55 and 0.52, respectively. In Step 1, all regret focus points are obtained and listed in Table 8. For avoiding unnecessary repetition, only some results are explained below. Amongst the high possible demands $\{7, 8, 9, 10\}$, 8, 9 or 10 makes order 5 most regretful. In other words, any other order will be better than them if demand 5 comes true. As a result, demands 8, 9 and 10 are Type I regret focus point. Demand

Table 8 Regret focus points of order quantities

	Order quantities					
	5	6	7	8	9	10
Type I	8, 9, 10	10	10	10	7	7
Type II	7	7	7	8	9	10
Type III	6	5	5	5	5	5, 6
Type IV	5	6	6	6	6	5, 6
Type V	8	8	10	5	7	8
Type VI	10	10	10	5	5	5
Type VII	6	7	8	8	8	9
Type VIII	5	5	5	6	10	10
Type IX	8	9	10	6	7	7
Type X	5	5	5	6	10	10
Type XI	6	7	7	8	9	9
Type XII	10	10	5, 10	5	5	5

7 can lead to the least regret for order 5 amongst high possible demands so that it is Type II regret focus point. A decision maker might think about the scenarios which have the low possibility to occur. They correspond to Type III and IV regret focus points. Amongst the demands with low possibilities, that is $\{5, 6\}$, demand 6 makes order 5 more regrettable than demand 5. Thus, demands 6 and 5 are regarded as Type III and IV regret focus points, respectively. Amongst the demands $\{7, 8, 9, 10\}$ which can generate the high regret for order 5, demand 8 is Type V regret focus point due to its highest possibility whereas demand 10 is Type VI regret focus point due to its lowest possibility. Amongst the demands $\{6, 7, 8, 9, 10\}$ which can bring about low regret for order 8, demand 8 is identified as Type VII regret focus point because of its highest possibility whereas demand 6 is chosen as Type VIII regret focus points because of its lowest possibility. Type IX, X, XI and XII regret focus points are obtained according to (6), (7), (8) and (9). The regret levels brought by twelve types of regret focus points for each order quantity are listed in Table 9. In Step 2, let us examine the validity of the obtained Type IX, X, XI and XII regret focus points. Since x^{9*} (8) $\notin X^{\geq\alpha} \cap X^{\geq\beta}$ (8), x^{10*} (9) $\notin X^{\leq\alpha} \cap X^{\leq\beta}(9)$, x^{10*} (10) $\notin X^{\leq\alpha} \cap X^{\leq\beta}$ (10), x^{11*} (5) $\notin X^{\geq\alpha} \cap X^{\leq\beta}$ (5), x^{12*} (5) $\notin X^{\leq\alpha} \cap X^{\geq\beta}$ (5), x^{12*} (6) $\notin X^{\leq\alpha} \cap X^{\geq\beta}$ (6) and x^{12*} (7) $\notin X^{\leq\alpha} \cap X^{\geq\beta}$ (7) hold, x^{9*} (8), x^{10*} (9), x^{10*} (10), x^{11*} (5), x^{12*} (5), x^{12*} (6) and x^{12*} (7) are not acceptable for $\alpha = 0.55$ and $\beta = 0.52$. The regret levels brought by twelve types of valid regret focus points for each order quantity are listed in Table 10.

In Step 3, the optimal order quantities are selected based on the regret levels of valid regret focus points. The optimal orders are 8, $\{7, 8, 9, 10\}$, 6, $\{5, 6\}$, 10, $\{7, 8\}$, 8, $\{5, 10\}$, 8, 5, $\{7, 8, 9\}$ and $\{5, 6, 7\}$ which corresponds to Types I to XII regret focus points, respectively. As the retailer sells seasonal goods, there is one and only one chance for him/her to decide how many should be ordered. Hence, considering a reasonable level of demand before determining how many products should be ordered is appropriate for such one-shot decision problems.

Table 9 Regret levels for regret focus points

	Order quantities					
	5	6	7	8	9	10
Type I	1,1,1	0.8	0.6	0.4	0.667	1
Type II	0.8	0.4	0	0	0	0
Type III	0.3	0.2	0.4	0.6	0.8	1,1
Type IV	0	0	0.25	0.5	0.75	1,1
Type V	1	0.667	0.6	0.6	0.667	0.556
Type VI	1	0.8	0.6	0.6	0.8	1
Type VII	0.3	0.4	0.334	0	0.278	0.208
Type VII	0	0.2	0.4	0.5	0.2	0
Type IX	1	0.75	0.6	0.5	0.667	1
Type X	0	0.2	0.4	0.5	0.2	0
Type XI	0.3	0.4	0	0	0	0.208
Type XII	1	0.8	0.4,0.6	0.6	0.8	1

Table 10 Regret levels for valid regret focus points

	Order quantities					
	5	6	7	8	9	10
TypeI	1,1,1	0.8	0.6	0.4	0.667	1
TypeII	0.8	0.4	0	0	0	0
TypeIII	0.3	0.2	0.4	0..6	0.8	1,1
TypeIV	0	0	0.25	0.5	0.75	1,1
TypeV	1	0.667	0.6	0.6	0.667	0.556
TypeVI	1	0.8	0.6	0.6	0.8	1
TypeVII	0.3	0.4	0.334	0	0.278	0.208
TypeVII	0	0.2	0.4	0.5	0.2	0
TypeIX	1	0.75	0.6	*	0.667	1
TypeX	0	0.2	0.4	0.5	*	*
TypeXI	*	0.4	0	0	0	0.208
TypeXII	*	*	*	0.6	0.8	1

5 Conclusions

The difference between OSDT and the decision based on optimistic and pessimistic utilities have been comprehensively addressed in the paper [14]. It is especially worthy making a detailed comparison between OSDT and SEU as follows:
Comparison 1: In SEU, there are two steps:
Step 1: Evaluating each alternative by using the weighted average utility of all outcomes;
Step2: Selecting the alternative with the maximum average.
In OSDT, there are two steps:
Step 1: Scenario (focus point) seeking for each alternative;

Step 2: Choosing the alternative with the maximal satisfaction level or minimal regret level of the focus point.

Comparison 2: In SEU the utility function is used whereas in OSDT the satisfaction function or regret function is used. Utility function is associated with risky situations. If a person is a risk avoider, the utility function is concave. If a person is a risk taker, the utility function is convex. If a person is risk neutral, the utility function is linear. Satisfaction function or regret function has no relation with risk situations, which just represents the relative position of payoff or regret. In OSDT, taking into account which kind of focus point reflects the attitude of the individual to uncertainty.

Comparison 3: SEU uses subjective probability to characterize uncertainty whereas OSDT applies possibility distribution.

Comparison 4: SEU amounts to the expected payoff based on the distorted probabilities as follows:

$$EU = \sum p_i u(x_i) = k \sum p_i' x_i,$$

where k is a positive constant and p_i' is a distorted probability. The conventional explanation of the optimal decision with SEU is that it can lead to the maximal average utility when the decision is repeated infinite time in the sense of strong law of large numbers. Hence, it is lack of consistency for the one-shot decision cases because the expected value will never appear. On the other hand, OSDT give a clear answer to why the decision maker makes such a decision in the face of uncertainty and why the decision might not generate a satisfactory result after the uncertainty resolving.

In conclusion, OSDT provides a scenario-based choice instead of the lottery-based choices as in other decision theories under uncertainty. Therefore, it is a scenario-based decision theory. OSDT is a fundamental alternative theory for decision under uncertainty with greater appeal to intuition, simplicity of application and explicability. Because it is very close to the human way of thinking, the decision with OSDT is of human-centric decision making. OSDT also provides one of the basic theories for behavioral operations research.

It is pointless to dispute which decision theory is better. There is no simple theory which is appropriate for any decision situation and in this respect the one-shot decision theory is no exception. It is true that different theories play different roles for different decision situations.

The one-shot decision theory is mainly utilized in the situation where a decision is experienced only once and the probability distribution is unavailable due to lack of enough information. However it might play an indispensable role of a bridge in linking decision under ignorance and decision with probabilities (shown in Fig. 2). For a repeatable decision problem, at the beginning, a decision maker has to make a decision under ignorance because the decision situation is completely new for him/her and therefore he/she has no ability to tell the difference between the states of the nature. After the first decision is made based on maximin or maximax or minmax regret or Hurwicz criterion, he/she would has some knowledge about the state of nature so that it is possible to construct an initial possibility distribution of states

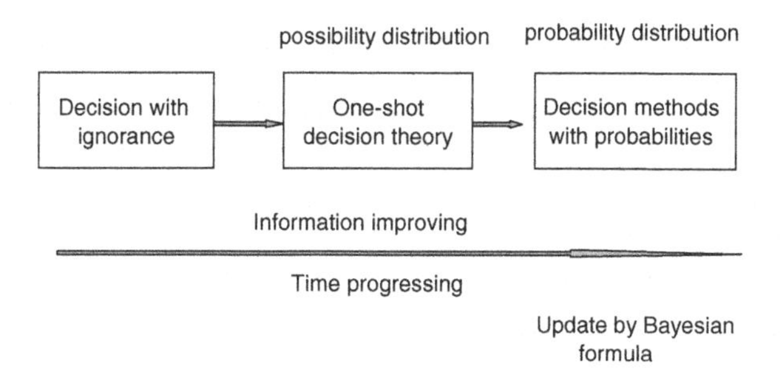

Fig. 2 The role of bridge between decision with ignorance and decision with probabilities

of nature. He/she could make a one-shot decision and repeat such decision with the updated possibility distributions. As time progresses, the information improves. The possibility distribution will switch into a probability distribution when the data is rich enough. The switching criterion is the hypothesis test for the probability distribution. After that decision methods with probability distributions should be utilized with the probabilities updated using Bayesian formula.

Finally, let us give some comments on the case of one-shot decision under risk. In such a case, for example, a game of tossing an ordinary coin, the objective probabilities are exactly known. When making a one-shot decision under risk, we can obtained the possibility distribution by normalizing a probability mass function (for a discrete random variable) or a probability density function (for a continuous random variable) and make a decision with OSDT.

The research on one-shot decision under uncertainty is in its early stages. There is potential for research on theoretical and applied aspects. As a direct extension of this research, multistage one-shot decision problems can be studied. One-shot game theory can be developed and the case studies of international conflict resolutions can be done. Newsvendor problems and supply chain management for innovative products are other interesting and important applications of OSDT. Use of OSDT in behavioral finance problems is another interesting research area. Other decision problems, such as mergers and acquisitions (M&A), emergency management for irregular events such as earthquakes, or nuclear power plant accidents, social policy decision making for environment, energy, social insurance and infrastructure can also be analyzed using OSDT. It may be especially interesting to test the hypotheses—the aggregation result of individual decision making with OSDT can be approximated by the decision result with SEU by empirical studies.

References

1. Allais, M.: Le comportement de l'homme rationnel devant le risque: Critique des postulats et axiomes de l'école américaine. Econometrica **21**(4), 503–546 (1953)

2. Baillon, A.: Eliciting subjective probabilities through exchangeable events: an advantage and a limit. Decis. Anal. **5**(2), 76–87 (2008)
3. Bell, D.E.: Risk premiums for decision regret. Manag. Sci. **29**(10), 1156–1166 (1983)
4. Bertrand, M., Schoar, A.: Managing with style: the effect of managers on firm policies. Q. J. Econ. **118**(4), 1169–1208 (2003)
5. Bewley, T.: Knightian decision theory, part I. Decis. Econ. Finan. **25**(2), 79–110 (2002)
6. Brandstatter, E., Gigerenzer, G., Hertwig, R.: The priority heuristic: making choices without trade-offs. Psychol. Rev. **113**(2), 409–432 (2006)
7. Clemen, R.T., Kwit, R.C.: The value of decision analysis at Eastman Kodak company, 1990–1999. Interfaces **31**(5), 74–92 (2001)
8. Daston, L.J.: Classical probability in the enlightenment. Princeton University Press, Princeton, NJ (1988)
9. Dubois, D., Prade, H., Sabbadin, R.: Decision-theoretic foundations of possibility theory. Eur. J. Oper. Res. **128**(3), 459–478 (2001)
10. Ellsberg, D.: Risk, ambiguity and savage axioms. Q. J Econ. **75**(4), 643–669 (1961)
11. Fine, C.H.: Clock Speed: Winning Industry Control in the Age of Temporary Advantage. Perseus Books, New York (1998)
12. Giang, P.H., Shenoy, P.: Two axiomatic approaches to decision making using possibility theory. Eur. J. Oper. Res. **162**(2), 450–467 (2005)
13. Gilboa, I.: Expected utility with purely subjective nonadditive probabilities. J. Math. Econ. **16**(1), 65–88 (1987)
14. Guo, P.: One-shot decision approach and its application to Duopoly market. Int. J. Inf. Decis. Sci. **2**(3), 213–232 (2010a)
15. Guo, P.: Private real estate investment analysis within one-shot decision framework. Int. Real Estate Rev. **13**(3), 238–260 (2010b)
16. Guo, P.: One-shot decision theory. IEEE Trans. Syst. Man Cybern. Part A Syst. Hum. **41**(5), 917–926 (2011)
17. Guo, P., Tanaka, H.: Decision analysis based on fused double exponential possibility distributions. Eur. J. Oper. Res. **148**(3), 467–479 (2003)
18. Guo, P., Yan, R., Wang, J.: Duopoly market analysis within one-shot decision framework with asymmetric possibilistic information. J. Comput. Intell. Syst. **3**(6), 786–796 (2010)
19. Kaha, B.E., Sarin, R.K.: Modelling ambiguity in decision under uncertainty. J. Consum. Res. **15**(2), 265–272 (1988)
20. Kahneman, D., Tversky, A.: Prospect theory: an analysis of decision under risk. Econometrica **47**(2), 263–291 (1979)
21. Kahneman, D., Tversky, A. (eds.): Choices, Values, and Frames. Cambridge University Press, Cambridge (2000)
22. Katsikopoulos, K.V., Gigerenzer, G.: One-reason decision-making: modeling violations of expected utility theory. J. Risk Uncertainty **37**(1), 35–56 (2008)
23. King, W.R.: Offshoring decision time is at hand. Inf. Syst. Manag. **23**(1), 102–103 (2006)
24. Knight, F.H.: Risk. Uncertainty and Profit. Houghton Mifflin, Boston (1921)
25. List, J.A., Mason, C.F.: Are CEOs expected utility maximizers? J. Econometrics **162**(1), 114–123 (2011)
26. Loomes, G., Sugden, R.: Regret theory: an alternative theory of rational choice under uncertainty. Econ. J. **92**(368), 805–824 (1982)
27. Savage, L.J.: The Foundations of Statistics. Wiley, New York (1954)
28. Schmeidler, D.: Subjective probability and expected utility without additivity. Econometrica **57**(3), 571–587 (1989)
29. Segal, U.: The Ellsberg paradox and risk aversion: an anticipated utility approach. Int. Econ. Rev. **28**(1), 175–202 (1987)
30. Shafer, G.: Savage revisited. Stat. Sci. **1**(4), 463–501 (1986)
31. Shimanoff, S.B.: Commonly named emotions in everyday conversations. Percept. Mot. Skills **58**(2), 514 (1984)
32. Sugden, R.: Regret, recrimination and rationality. Theor. Decis. **19**(1), 77–99 (1985)

33. Von Neumann, J., Morgenstern, O.: Theory of Games and Economic Behavior. Princeton University Press, Princeton (1944)
34. Whalen, T.: Decision making under uncertainty with various assumptions about available information. IEEE Trans. Syst. Man Cybern. **14**(6), 888–900 (1984)
35. Yager, R.R.: Possibilistic decision making. IEEE Trans. Syst. Man Cybern. **9**, 388–392 (1979)

On the Influence of Emotion on Decision Making: The Case of Charitable Giving

Ryan Kandrack and Gustav Lundberg

Abstract This chapter summarizes and discusses methodologies and findings of recent research focused on the influence of emotion on decision-making in general and charitable giving in particular. Exploring how appraisal theory findings carry over to the decision of charitable giving, we experimentally examine the influence of incidental sadness and anger on charitable donations to an identified or a statistical victim. First, subjects viewed a previously validated film clip and provided a written response to how they would feel in the situation in the clip. Subjects then viewed a charity letter and had the opportunity to make a donation. Overall, participants in both the sad and angry conditions donated more than participants in the control condition. Sad individuals donated more money to a statistical victim relative to individuals in a neutral condition. This finding is consistent with appraisal-tendency theories. Angry individuals, however, did not donate significantly more to either an identified or statistical victim relative to individuals in a neutral condition. Self-reported emotions reveal discrete levels of sadness elicited in the sad condition, but elevated levels of additional negative emotions in the anger conditions.

Keywords Identifiable victim effect · Charitable giving · Incidental emotion · Appraisal tendency framework

R. Kandrack · G. Lundberg (✉)
Palumbo-Donahue School of Business Administration, Duquesne University, Pittsburgh, PA, USA
e-mail: lundberg@duq.edu

R. Kandrack
McAnulty College of Liberal Arts,Duquesne University, Pittsburgh, PA, USA

G. Lundberg
Swedish School of Economics and Business Administration, Helsinki, Finland

P. Guo and W. Pedrycz (eds.), *Human-Centric Decision-Making Models for Social Sciences*, Studies in Computational Intelligence 502,
DOI: 10.1007/978-3-642-39307-5_3,

1 Introduction

Charitable giving represents a substantial economic transaction in the United States and around the world. According to the American Association of Fundraising Counsel, Americans donated over $290 billion to charities in 2010, over $211 billion of which was donated by individuals. The amount of money donated to different causes has led researchers in various fields, from psychology to economics, to investigate the influences of altruistic behavior. In the words of Harbaugh et al. ([19], p. 1622), "[t]o economists, charitable giving is a puzzle: Money is good, so why are people willing to give it away?"

It is clear that people give for many reasons, and equally clear that much effort is focused on how to get people to give. The reasons why people give include guilt [5, 30, 50], sympathy and empathy [11, 40, 48], happiness [12, 35], self-therapy [5], and donor (e.g. moral) identity [1]. Cialdini et al. [9] claim that since altruism has reinforcing properties, it is employed by people who wish to make themselves feel better. Increased self-gratification following negative mood priming (e.g. sadness) is mediated by an attempt to comfort oneself, to engage in self-therapy. With regard to donor identity, Aaker and Akutsu [1] argue that there are contexts in which a person thinks of her/himself as a giver (cf. [39, 45]). Referring back to the above list of reasons why people give, there is mounting evidence that spending money (or time through volunteering) on other people has a more positive impact on happiness than spending on oneself [1, 22, 35]. Interestingly, however, Dunn and colleagues [12] show that a significant majority of participants in their study thought that personal spending would make them happier than pro-social spending. In three early studies, Cialdini et al. [9], Cialdini and Kenrick [10] and Baumann et al. [5] explore altruism as hedonism, finding support for a view of adult benevolence as self-gratification. Cialdini and Kenrick primed subjects to think of either depressing or neutral events and subsequently gave them the opportunity to be privately generous. They found that subjects in the most socialized (oldest) group in the negative-mood condition were significantly more generous than subjects in the neutral-mood control group. Thus Cialdini and Kenrick showed the influencing of an action by an idea, a process that has become known as the ideomotor effect. In the same vein of research, Vohs et al. [52], show that study participants primed with money donated significantly less money to a student fund than participants not primed with money. For further insights on ideomotor processes and priming see Vohs et al. [53] and Kahneman [21].

Harbaugh and colleagues [19] discuss two possible motives for charitable contributions: "pure altruism" and "warm glow." The first motive is satisfied by increases in the public good no matter the source or intent. The second motive is only fulfilled by an individual's own voluntary donations. The fMRI studies of Harbaugh and his colleagues show that neural activation in very similar areas of the brain increased with the monetary payoff to both the subject and to the charity. They demonstrate that mandatory taxation for a good cause can produce activation in specific areas of the brain associated with concrete, individualistic rewards; that transfers to others are associated with neural activation akin to that of receiving money (rewards) for oneself. This finding was anticipated by Cialdini and his research associates in the 1970s and 1980s who argued that "... individuals often behave charitably in order to provide themselves with reward" ([5], p. 1039).

Finally, Dickert et al. [11] explored the role of affective versus deliberate information processing in decisions to provide financial aid to people in need. They found that different mechanisms influence the decision to donate money compared to subsequent decisions on how much money to donate. Whereas motivations for mood management were predictive of donation decision, empathic feelings were predictive of the amount.

A key distinction in studies of charitable giving has been made between donations to an identifiable victim versus a statistical victim. Substantial research has focused on how and why donors are affected by the two forms of presenting need, as well as on in what conditions donors may be swayed in either direction. Thomas Schelling first commented on this social phenomenon when he made the distinction between an identified individual and a statistical life. For example, when the media reports on a young girl's need of funds for a life-saving operation, many individuals quickly respond with donations. However, when an announcement is made about a need to fund a hospital, few would act with equal generosity [42]. Within this framework, an identified victim is one whose fate is seemingly certain in the mind of a potential donor in the absence of action. A statistical victim is one whose fate is uncertain as increased funding could represent only a possibility of saving more lives, not a guarantee. Researchers have since expanded on this notion. Small and Loewenstein [47] find support for the identifiable victim effect in the first explicit lab experiment structured as a dictator game with a weak form of identification. Continuing this research, Small and colleagues [48] find that priming a "feeling" mode of thought, one driven first by emotion, as opposed to a deliberative mode of thought, increases giving.

Psychologists have long been concerned with emotion and its influence on decision-making. Though at first concerned with examining emotions in terms of pleasantness and arousal, a more recent strand of research has shown that not all positive or negative emotions are equal. According to cognitive appraisal theory people extract emotions from evaluations (appraisals) of events in their environment. Smith and Ellsworth [49] experimentally study emotions on eight dimensions (pleasantness, attention, control, certainty, perceived obstacle, legitimacy, responsibility, and anticipated effort), finding that emotions are closely linked to specific cognitive evaluations. For example, if an individual thinks that a negative event is caused by another individual, she will feel anger. In contrast, an individual who sees a negative event as controlled by situational factors will feel sadness. Building on cognitive appraisal theory, the appraisal-tendency framework [8, 18, 32] serves as a framework for distinguishing and predicting the influence of specific emotions on judgment and decision making. The appraisal-tendency framework posits that specific emotions trigger specific cognitive and emotional processes, which delineate the effects of each emotion on decision making [18]. For example, the individual who feels sadness from some negative event will then make a subsequent decision formed by the appraisals which characterize sadness.

This chapter summarizes and discusses methodologies and findings of recent research focused on the influence of emotion on decision-making in general and charitable giving in particular. After reviewing the relevant research, we present the results of an experiment designed to examine the influence of incidental sadness and anger on charitable donations to an identified or a statistical victim. That is, we explore how appraisal theories of incidental sadness and anger carry over to the decision of charitable giving.

2 Review of Key Literature

2.1 Emotion and Decision-Making

The early study of decision-making paid rather little attention to the role of emotions. Instead, researchers focused on cognitive errors/biases and heuristics in judgment. More recently, social scientists have turned their attention to the study of emotions, arriving at a granular perspective on emotion and its influence on decision-making. Before reviewing recent research on emotion, it is useful to present the conceptual distinction between emotion, affect, and mood, three terms sometimes used interchangeably for emotional states. Affect refers to a general emotional state without deliberation on cause. It has traditionally been studied in terms of positive and negative valence. Emotion is characterized by a specific cause or behavior, a short duration, and a physiological manifestation. For example, when coming into contact with a grotesque image an individual might feel disgust. When looking away or leaning backwards (physiological manipulation), an individual immediately wishes to reverse the feeling of disgust and thus the emotion does not last. In addition, emotion can be *incidental* or *integral*. Incidental emotions are caused by dispositional factors and are unrelated to the decision faced by an individual. Integral emotions occur at the time of making a decision and are derived from considering the consequence of a decision. Mood, however, is distinguished by its long duration and diffuse cause. For example, an individual might be in an irritable mood for no particular reason, simply feeling vexed by the world in general.[1]

The study of incidental emotion and its influence on subsequent decisions has blossomed recently, and results suggest that the carry-over effects of incidental emotions are robust to a variety of judgment scenarios and economic decisions. A number of methodologies have been used to elicit emotion. A frequently-used method involves reading an emotionally-charged scenario and then performing a writing task where participants imagine themselves in the scenario and write about how they might feel. Keltner et al. [24] examined the influence of incidental sadness and anger on causal judgments. In several experiments, subjects were first presented with ambiguous scenarios in order to induce emotion (e.g. the death of a family member to elicit sadness)

[1] On the mapping of the distinction between emotion, affect, and mood, Ryan Kandrack has benefitted from personal communication with Dr. Nicole Verrochi Coleman.

and subsequently were instructed to imagine how they would feel or what they might think in the given situations. Subjects then judged the likelihood of future life events associated with an individual or a situational cause. Keltner and colleagues find angry individuals likely to blame someone else while sad individuals are likely to find fault with situational factors, results which are consistent with cognitive appraisal theories of emotion [18, 28, 49].

This methodology has been extended to the study of incidental emotion's influence on risk-taking. There is evidence that sad individuals are risk-taking and anxious individuals are risk averse [27, 37]. In addition, fear has been shown to be associated with risk-aversion, while anger has been associated with risk-taking [32, 33]. The results of these studies were instrumental in creating the appraisal-tendency framework (ATF) through which researchers have been able to differentiate specific emotions regardless of valence [18, 32, 33]. The ATF creates an emotion-to-cognition pathway that relies on appraisal dimensions which fuel motivation to appraise, or evaluate, future decisions by using the appraisal dimensions of the specific emotion. Small and Lerner [46] provide an extension of the ATF by examining the effects of incidental sadness and anger on the judgment and justification of a welfare recipient's amount of assistance. Participants in this study wrote about the cause of the person's need and selected a recommendation to increase or decrease poverty assistance. The researchers find that incidental anger decreases recommended assistance while sadness increased assistance.

Expanding the range of decision contexts influenced by emotion, as well as the methodologies to induce emotion, Lerner and colleagues [34] examine the impact of incidental sadness and disgust on the endowment effect, a notion that individuals value things they own more than things they do not own. Their experiment crossed an emotion manipulation (disgust, sadness, neutral) with an ownership condition in which half of the subjects were given an object and presented with the opportunity to sell it, while the other half were shown the object and asked if they would like to receive cash or the object. To induce emotion, subjects viewed one of three film clips: *The Champ* in the sadness condition, *Trainspotting* in the disgust condition, and a *National Geographic* depiction of fish to induce neutrality. Subjects then wrote a self-reflective response on how they might feel had they been in the situation viewed in the film clip. The results suggest that disgust reduces buying and selling prices, while sadness increases buying but decreases selling prices. The endowment effect is eliminated in the disgust condition and reversed in the sadness condition.

In daily activities individuals frequently encounter events that trigger emotional responses, many of which occur in succession. Winterich and colleagues [55], following cognitive appraisal theories, utilize film clips to induce different emotions of the same valence in succession to examine the blunting effects of subsequent emotion elicitation. In one study, subjects watch a film clip to induce sadness (*The Champ*) or to induce a neutral state (*National Geographic*). A second study induces anger by assigning subjects as the recipient of an unfair offer ($8 dictator/$2 receiver) in a dictator game, and then giving them the choice to accept or reject. Following the dictator game, subjects recorded emotional responses to the allocation and completed the Life Events Questionnaire adapted from Lerner and Keltner [33]. The results suggest that sadness mediates the effect of subsequent anger, and that the reverse also holds.

2.2 Charitable Giving and the Identified Victim Effect

The identified victim effect refers to the propensity of donors to give more assistance to a single, specific, and vivid victim. On the other hand, a statistical victim refers to a large, ambiguously-defined entity (e.g. starving children in Africa). This phenomenon has been attributed to an individual's judgment of the relative size of the reference group given aid [20]. That is, the identified victim is one of one, whereas a statistical victim represents a vaguely defined set. Small and Loewenstein [47] provide the first explicit test of the identifiable victim effect, (1) in a dictator game lab experiment, and (2) in a field experiment where people in an airport terminal were given a chance to donate all or any part of $5 given to them by the experimenters. The studies employed a weak form of identifiability—determining the victim without providing any personalized information—focusing on determined versus not-yet-determined victims. In both experiments the contributions were larger when the recipients had already been determined than when they were yet to be determined.

Kogut and Ritov [25, 26] study the identifiable/statistical victim phenomenon to examine its boundary conditions and find that a single, identified victim (in this case a child identified by age, name and picture) gains greater contributions than one which is non-identified, but that fully identified groups of children do not gain more than non-identified groups. The researchers argue that in the donors' information processing the singularity of the individual victim represents coherency. The expectation of coherency leads to greater information processing and generates a higher level of empathy for the single victim [17, 51].

Small et al. [48] test the effect of educating people about the inconsistent valuation of lives when considering an identified or a statistical victim. The researchers provide a written explanation of the differences between the two and then present experiment participants with the choice to give. The authors find that providing education on the identifiable victim bias decreased donations to the identified victim, but did not increase donations to the statistical victim. While priming with education was not successful to counter the predispositional bias, there is evidence that priming with an emotional task increases the amount donated [11, 48].

3 Experiment Overview

The goal of this experiment is to investigate the influence of incidental sadness and anger on an individual's propensity to donate to a victim. The experiment follows a 3×2 between-subject design, crossing an emotion manipulation (sadness, anger, neutral) with the decision to give to a victim (identified, statistical). The experiment was presented as two short studies to reduce demand effects (cf. [34, 46]). The first study follows cognitive appraisal theories of emotion and the appraisal-tendency framework, eliciting sadness and anger to examine the influence on subsequent decisions [28, 32–34, 37, 46, 49]. Incidental emotion elicitation has been shown to influence subsequent, unrelated decisions [6, 13, 43, 44]. Those who participated received $5 compensation. In the second study a short charity letter was presented to subjects along with two envelopes in which they had the opportunity to make a donation using the $5 compensation, or retain any amount of that money. Study 2 follows the identified victim effect literature and adapts the procedure used in inducing an affective mode of thinking prior to a donating decision [11, 48].

4 Propositions

Proposition. 1 Anger is associated with appraisals of increased certainty and human agency. The identifiable victim effect has been shown to be a dispositional bias in decision-making, yielding increased giving to the victim. An individual primed to feel anger will feel more certain of his/her decision, and will also find the plight of the identified victim more likely, which will intensify the identifiable victim effect. That is, individuals in the anger condition are predicted to give more money to an identifiable victim relative to individuals in the neutral condition.

Proposition. 2 Sadness is associated with cognitive appraisals of decreased certainty and situational agency. An individual primed with sadness will therefore require more cognitive processing to make a decision and will find the plight of the statistical victim more likely. Therefore, individuals in the sadness condition are predicted to give more to the statistical victim relative to individuals in the neutral condition.

Proposition. 3 Drawing on earlier research relating altruism and spending money on others to happiness, we expect that participants who donate more to charity will report greater happiness than participants who keep more of the money for themselves. We expect this relationship to hold in all three conditions, and to be most clearly evident in the neutral condition.

5 Participants

Two hundred and thirty five undergraduate students in the school of business at Duquesne University participated in the experiment. The mean age of the subjects was 20 years. About 52 % of subjects were male, and 57 % of subjects reported having a part-time job. About 95 % of the subjects reported that they enjoyed the experiment or were indifferent, and 4 % reported they did not.

6 Methodology

Following the completion of consent forms, subjects received their $5 compensation which had been placed in a blank envelope beneath their survey packet. Due to budget restrictions on the project, the money was allocated through a randomized lottery at the end of the study such that roughly 40 % of subjects had the opportunity to leave with the share of the $5 which they did not donate. The five dollar compensation consisted of four one-dollar bills and four quarters. Next, subjects completed a baseline survey of affect (PANAS, adapted from Watson et al. [54]). The baseline affect survey has been used in past research to simply ease participants into the emotion elicitation task by instructing them to begin thinking about and feeling emotions (cf. [34, 46]). The survey consists of twenty emotions, both positive and negative, which the subjects rate on a scale of one (very slightly/not at all) to five (extremely) based on how they felt at that time.

Following this initial survey, subjects began the "imagination study" in which they watched one of three film clips (sad, angry, neutral) and were asked to imagine themselves in the situations in the clip. For the neutral conditions, subjects were asked to simply watch the clip, a documentary on the Great Barrier Reef from *National Geographic*. In the sadness condition, a scene from *The Champ* showed a young boy grieving over the death of a boxer. In the anger condition, a scene from *My Bodyguard* portrayed a bully scene (the film clips were adapted from [16, 34, 55]). After viewing the clip, subjects wrote about how they would feel if they were in the situation in the clip in order to create a deeper personal connection. Subjects in the neutral condition wrote about what they had done that day (cf. [33]). The use of film clips and a writing response has been shown to be a reliable method of eliciting target emotions [31, 33].

Study 2 consisted of the charity letter and the exit survey. Subjects were given two envelopes (labeled "me" and "charity") along with a charity letter in which they read about a single identified child (name, age, picture) or factual information on poverty in the United States. The child's picture and poverty information was obtained from *Save the Children.org*. Subjects were then asked if they would like to donate any amount of their $5 compensation by placing a donation into the envelope labeled *charity*; otherwise they could retain any share of the five dollars by placing that amount into the envelope labeled *me*. The exit survey, adapted from Rottenberg et al. [41], asked subjects to rate how they felt during the film clip anchored on 0 ("not at all/none") to 8 ("extremely/a great deal"). The survey consisted of eighteen emotions, of which only three were of primary interest (sad, angry, and happy). This scale has been used extensively in past research (see [33, 34, 46]). These survey questions were asked toward the end of each session to prevent subjects from thinking about or labeling their emotions felt as a result of watching the film clip (cf. [33, 34]). Subjects also answered simple demographic questions such as age and gender, and answered yes/no to "do you have a part-time job" and "did you enjoy this study".

Once subjects completed the exit survey, those with randomly chosen participant IDs were able to keep the envelope labeled *me*. Forty percent of subjects were randomly chosen to keep the money they chose not to donate.

7 Results

The subjects' donations ranged from $0 to $5 and 97 % of all subjects donated some amount of the $5 compensation. About 70 % of all subjects donated the entire $5. Descriptive statistics on donations across conditions are presented in Tables 1 and 2. Overall, participants in both the sad and angry conditions (identified and statistical victims), donated marginally more than participants in the neutral condition; $t(95) = 1.704$ ($p = 0.092$) and $t(95) = 1.770$ ($p = 0.080$), respectively. Note that the degrees of freedom in each case reflect unequal variances. Given the relatively high mean donations, Table 2 summarizes the proportions of participants in the various categories who donated the full $5 amount, along with the proportions of participants who donated half or less ($\leq$\$2.50) of the received payment. The highest proportion of full-amount-donors is associated with the angry-identified (81.40 %) and sad-statistical (76.60 %) conditions. This donating behavior provides directional (but not statistically significant, $\chi^2 = 1.315$, d.f. = 1, p = 0.251) support for our expectation

Table 1 Descriptive statistics of overall donations

Emotion, Victim	n	Mean Donation	Standard Deviation	Standard Error	Coefficient of Variation
Sad, Identified	44	4.22	1.319	0.199	0.312
Sad, Statistical	47	4.24	1.448	0.211	0.341
Angry, Identified	43	4.44	1.259	0.192	0.284
Angry, Statistical	44	4.07	1.433	0.216	0.352
Neutral, Identified	29	3.89	1.674	0.311	0.431
Neutral, Statistical	28	3.61	2.025	0.383	0.561
Total	235	4.12	1.510	0.099	0.367

Table 2 Proportion of participants donating all versus half or less

Emotion, Victim	n	Donated Full Amount (%)	Donated Half or Less (%)
Sad, Identified	44	68.18	15.91
Sad, Statistical	47	76.60	17.02
Angry, Identified	43	81.40	11.63
Angry, Statistical	44	63.64	18.18
Neutral, Identified	29	65.52	31.03
Neutral, Statistical	28	64.29	32.14
Total	235	70.64	19.57

that, relative to the neutral condition, angry individuals will contribute more to identified victims and sad individuals more to statistical victims. The highest proportions of participants who donated \$2.50 or less are associated with the neutral-statistical (32.14 %) and neutral-identified (31.03 %) conditions. These two proportions clearly differ from the proportions in the other four conditions, as reflected in the noticeably higher coefficients of variation in Table 1.

Subjects felt significantly more sad than angry in the sad conditions ($t(89) = 20.47$, $p < 0.001$), but did not feel significantly more angry than sad in the angry conditions ($t(86) = -0.779$, $p = 0.438$; see Fig. 1). The effect of gender on donations was not significant ($t(225) = -1.599$, $p = 0.112$). Having a part-time job also did not have a significant effect on donations ($t(227) = 0.746$, $p = 0.457$).

Examining the influence of emotion on donations, one-way analysis of variance (ANOVA) revealed a marginally significant effect ($p = 0.098$). Post-hoc LSD tests revealed the difference in mean donations between sad and neutral conditions to be marginally significant at 5 % ($p = 0.057$), and the difference in mean donations between angry and neutral conditions to be significant ($p = 0.05$).

Mean donations in the sad and angry conditions were not significantly different ($p = 0.931$). One way ANOVA between all six conditions revealed an overall insignificant difference in mean donations, $F(5,229) = 1.305$ ($p = 0.263$), but post-hoc LSD tests revealed a marginally significant difference between the sad and neutral statistical conditions ($p = 0.076$). The difference in donations between the angry and neutral identified conditions was not significant ($p = 0.126$; see Fig. 2 below).

A further dissection of how cleanly the various emotions were elicited helps us understand why our results were not as strong as expected. Sadness was cleanly elicited such that the self-reported levels of sadness were significantly higher than the anger level, but the same does not hold for anger. Subjects felt high levels of sadness and low levels of anger in the sad conditions while subjects felt high levels of both anger and sadness in the angry conditions. Figure 3 shows self-reported sadness and

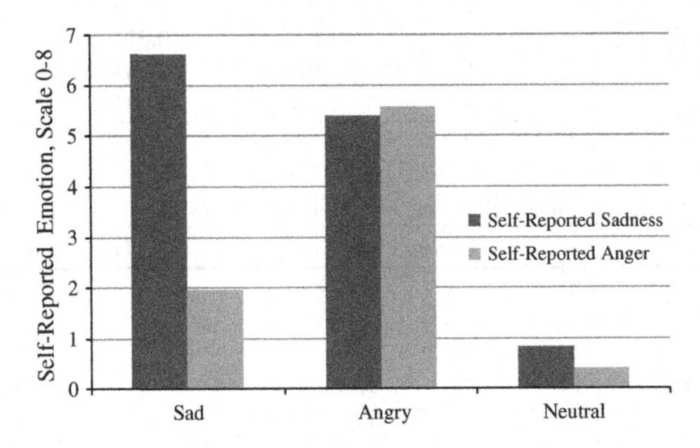

Fig. 1 Mean self-reported emotion across the three emotion manipulations

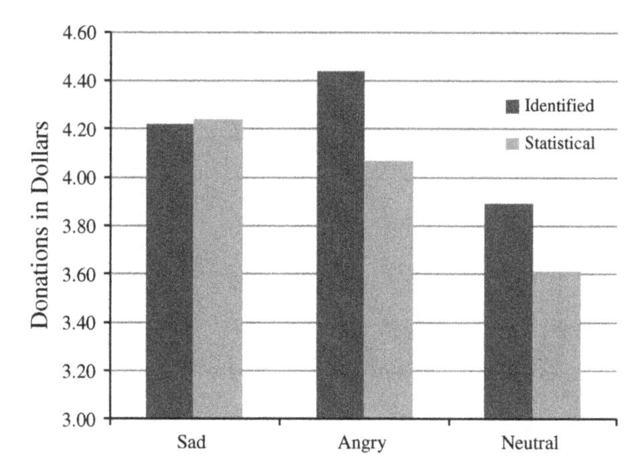

Fig. 2 Mean donations across all six conditions

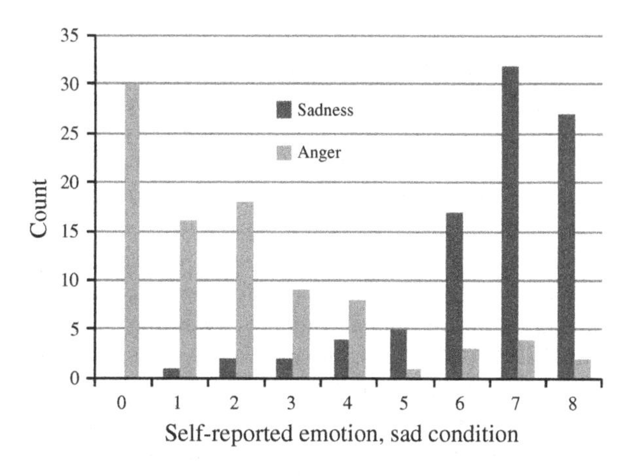

Fig. 3 Self-reported anger and sadness; sad condition

anger within the sad condition, whereas Fig. 4 shows self-reported anger and sadness in the angry condition.

In the exit survey, subjects responded to eighteen different emotions of which only three were of primary interest to the current study (i.e. sad, angry, and happy). However, it is interesting to note some of the additional negative emotions felt by the subjects, all of which have been studied in similar research. In addition to sadness and anger, we examined disgust and fear (see Table 3 for mean self-reported levels of emotion). Taking into account the additional negative emotions, the sad manipulation elicited a more discrete emotion while the anger manipulation appears to have generated an overall negativity, with elevated levels of disgust, anger, and sadness.

As proposed, sadness increased giving to a statistical victim relative to the neutral condition. Surprisingly, anger did not significantly increase donations to an identified

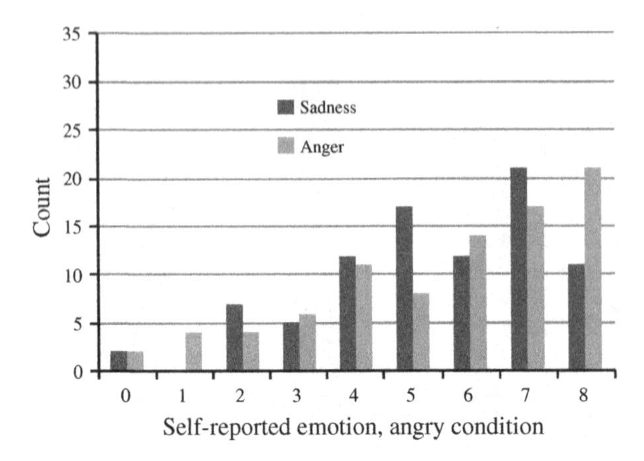

Fig. 4 Self-reported anger and sadness; angry condition

victim relative to the neutral condition. This could be due to a general negativity, characterized by multiple high-scoring components. Although no statistical significance was attained in the anger/identified victim condition, a case could be made that there is "practical" significance in that individuals in this condition averaged $0.20 more in donations than donors in any other condition (see Table 1).

With respect to the relationship between giving to others and happiness, we expected that participants who donate more to charity will report greater happiness than participants who keep more of the money for themselves. Our proposition—largely an extension of the above general tendency—was that this relationship would hold in all three conditions (sad, angry, and neutral), and be most clearly evident in the neutral condition. As illustrated in Table 4, we observe an unexpectedly complex pattern. In accordance with our expectations, the correlation between amount given to charity and happiness was indeed positive and marginally significant in the neutral condition. This is consistent with earlier findings. However, the correlation coefficients in the angry and sad conditions are near zero and strongly negative, respectively. The induced sadness seems to have trumped any happiness stemming from giving-to-others, whereas the induced anger largely seems to have mitigated that happiness (cf. [55]). Lerner and colleagues [34] report a similar finding with

Table 3 Mean self-reported emotion in sadness and anger manipulations

Emotion	Sadness Manipulation Mean	Anger Manipulation Mean
Anger	1.96	5.57
Disgust	1.88	6.06
Fear	2.40	2.95
Sadness	6.60	5.39

Table 4 Correlation coefficients relating amount given to charity and happiness for three induced conditions: sad, angry, and neutral

	Happiness	n	t	p
S_i and S_s	−0.3323	89	−3.286	0.0015
A_i and A_s	0.0340	87	0.314	0.7545
N_i and N_s	0.2452	57	1.876	0.0660

regard to the endowment effect, as discussed earlier, as in the case of disgust[2] the endowment effect is eliminated, whereas in the sadness case the endowment effect is reversed. A similar phenomenon is reported by Baumann et al. [5] who observed that for participants in a sad mood, altruistic activity canceled the enhanced tendency for self-gratification. All in all, these reversals illustrate that emotions of the same valence can have dissimilar effect (Table 4).

8 Discussion

We find support for our claim that incidental emotions can influence the decision to donate to a charity, that despite the fact that the two emotions elicited in our experiment were not equally unambiguous. The emotion elicited in the sad condition was clean in that self-reported sadness far exceeded any of the other emotions felt by the subjects. In contrast, disgust was the highest self-reported emotion felt by subjects in the anger manipulation, and there was no significant difference between self-reported anger and sadness. However, it is interesting to note the high levels of disgust in relation to its associated appraisal characteristics. Disgust, associated with an appraisal of being in close proximity to a disagreeable idea or object, has been shown to be further associated with an appraisal tendency to avert from accepting a new object or idea [34]—in the present case the $5 compensation for themselves. However, due to increased levels of anger and sadness in addition to disgust, this is a difficult assumption to tease out. Clearly, the development of methods used to induce specific emotions is in its infancy, albeit a promising one, and much additional work is necessary.

While many economists have been concerned with policies and tax implications relating to charities (cf. [36, 38]), relatively few have examined the determinants of charitable giving (e.g. [3, 23]). One such contribution in economics has been the "warm glow" theory, which states that individuals may simply gain positive util-

[2] Interestingly, disgust and anger—used in our study—are located in close proximity to each other in Smith and Ellsworth's [49] plot of 15 emotions where the vertical axis ranges from Situational to Human Control and the horizontal axis ranges from Other-Responsibility/Control to Self-Responsibility/Control. Both emotions are located in the Other-Responsibility/Control-Human Control quadrant.

ity from the act of giving [2]. The current research contributes to our knowledge by finding a determinant of increased donations founded on psychological theory. The current research also contributes an additional application of appraisal tendency theories to an economic decision. Recent behavioral economics research has sought to incorporate psychological insights into models and experiments to further understand decision-making (see [7]). This strand of research seeks to bridge the gap between social sciences to create stronger theories and expand the boundaries for decision-making. The current research supports those goals by providing experimental evidence of the influence of sadness and anger on charitable donations.

The present research is perhaps most limited by the sample size of subjects. With between 28 and 47 subjects in each of the six conditions, statistical significance could not be attained for all differences. However, it could be argued that the difference in mean donations between subjects in the angry and neutral manipulations shows directional support ($p = 0.126$), and with a larger sample could attain some level of statistical significance. Similarly, the difference in mean donations between males and females was also nearing significance ($p = 0.112$). Importantly, the sample consisted only of college students with an average age of 20 years. This homogeneous sample of business school students is not representative of the general population in terms of demographics. The subjects' age may indeed matter. Baumann and colleagues [5] note that with increasing age, helping becomes a progressively greater response of subjects. In contrast, saddened young children engage in a higher degree of helping (compared to neutral mood controls) only when it leads to external (social) reinforcement. Also, it has been suggested that a person's field of study in itself may influence social behaviors like cooperation and views on altruism [14, 15, 29]. Also on these fronts there is ample room for further research.

Given the elevated levels of emotion in the anger condition, the film clip used to elicit anger comes under question. Though past research has used this film clip without reports of elevated levels of other negative emotions, the anger condition in the present research is polluted with emotions such as disgust, sadness and fear. Gross and Levenson [16] found sixteen film clips which were moderately successful in eliciting discrete emotions. One such film clip was *My Bodyguard*, the clip used in the present research to elicit anger. Gross and Levenson note that anger is a complex emotion and difficult to elicit using a film clip, and also found that subjects reported high levels of both disgust and sadness. Instead of a film clip, Gross and Levenson suggest that eliciting anger may require a more personal involvement for subjects. Future research could try to elicit anger using unfair (rigged) offers to unknowing subjects in a dictator game, a procedure employed by Winterich et al. [55]. This method of eliciting anger has been successfully employed, resulting in a purer emotion compare to that/those induced by a film clip.

Future research should further examine the role of disgust and possibly moral outrage [4] in the context of charitable donations. In addition, future research should consider the effects of systematic processing and uncertainty associated with sadness related to the decision to give to an identified or statistical victim, possibly by inducing cognitive load prior to making the decision. We sense research opportunities in today's emotionally-charged political environment. For example, what would happen if an organization like Save the Children were to air a solicitation message after a polarizing political advertisement? Such a laboratory or field experiment could serve as a managerial or practical extension of the current research.

9 Conclusion

The present research indicates that incidental emotions carry over and influence the decision to donate to a charity, a finding that resonates with the appraisal tendency framework and extends the applications of the framework to a new decision environment. Sad individuals donated more to statistical victims—Americans in poverty—relative to individuals in the neutral condition. This result is supported by an appraisal tendency framework which suggests that sad individuals find events caused by situational factors more likely. Interestingly, angry individuals did not donate significantly more to an identified victim than did those in the neutral condition, although their contributions were larger than those of any other group in the experiment. Moreover, the identified victim effect was eliminated in the sad manipulation. This could be due to increased systematic processing associated with sadness. That is, individuals may read the description of the identified victim and think more about the plight instead of immediately making a donation. Likewise, sad individuals may see the description of the statistical victim and, instead of being distracted by the vague statistics, consider that this is indeed a victim which deserves aid. The elimination of the identified victim effect could also be due to the associated uncertainty. These issues merit future consideration.

While we found an expected positive relationship between the amount of money given to others and happiness, we also found no relationship between giving-to-others and happiness in the angry condition, and a strong negative relationship in the sad condition. While other researchers also have found reversals of established effects, e.g. the endowment effect, among subjects primed to be sad, collectively these reversals reveal how complex the impact of emotions are in diverse decision making and judgment contexts.

References

1. Aaker, J.L., Akutsu, S.: Why do people give? The role of identity in giving. J. Consum Psychol **19**(3), 267–270 (2009)
2. Andreoni, J.: Impure altruism and donations to public goods: a theory of warm-glow giving. Econ. J. **100**(401), 464–477 (1990)
3. Andreoni, J.: Philanthropy. In: Kolm, S.-C., Ythier, J.M. (eds.) Handbook of the Economics of Giving, Altruism and Reciprocity, pp. 1201–1269. North Holland, Amsterdam (2006)
4. Batson, C.D., Kennedy, C.L., Nord, L.-A., Stocks, E.L., Fleming, D.A., Marzette, C.M., Lishner, D.A., Hayes, R.E., Kolchinsky, L.M., Zerger, T.: Anger at unfairness: is it moral outrage? Eur. J. Soc. Psychol **37**(6), 1272–1285 (2007)

5. Baumann, D.J., Cialdini, R.B., Kenrick, D.T.: Altruism as hedonism: helping and self-gratification as equivalent responses. J. Pers. Soc. Psychol. **40**(6), 1039–1046 (1981)
6. Bodenhausen, G.V.: Emotions, arousal, and stereotypic judgments: a heuristic model of affect and stereotyping. In: Mackie, D.M., Hamilton, D.L. (eds.) Affect, Cognition, and Stereotyping: Interactive Processes in Group Perception, pp. 13–37. Academic Press, San Diego (1993)
7. Camerer, C., Loewenstein, G., Rabin, M.: Advances in Behavioral Economics. Russell Sage Foundation, New York (2004)
8. Cavanaugh, L.A., Bettman, J.R., Luce, M.F., Payne, J.W.: Appraising the appraisal-tendency framework. J. Consum. Psychol. **17**(3), 169–173 (2007)
9. Cialdini, R.B., Darby, B.L., Vincent, J.E.: Transgression and altruism: a case for hedonism. J. Exp. Soc. Psychol. **9**(6), 502–516 (1973)
10. Cialdini, R.B., Kenrick, D.T.: Altruism as hedonism: a social development perspective on the relationship of negative mood state and helping. J. Per. Soc. Psychol. **34**(5), 907–914 (1976)
11. Dickert, S., Sagara, N., Slovic, P.: Affective motivations to help others: a two-stage model of donation decisions. J. Behav. Decis. Making **24**(4), 361–376 (2011)
12. Dunn, E.W., Aknin, L.B., Morton, M.I.: Spending money on others promotes happiness. Science **319**(5870), 1687–1688 (2008)
13. Forgas, J.P.: Mood and judgment: the affect infusion model (AIM). Psychol. Bull. **117**(1), 39–66 (1995)
14. Frank, R.H., Gilovich, T., Regan, D.T.: Does studying economics inhibit cooperation? J. Econ. Persp. **7**(2), 159–171 (1993)
15. Frank, R.H., Gilovich, T., Regan, D.T.: Do economists make bad citizens? J. Econ. Persp. **10**(1), 187–192 (1996)
16. Gross, J., Levenson, R.: Emotion elicitation using films. Cogn. Emot. **9**(1), 87–108 (1995)
17. Hamilton, D., Sherman, S.: Perceiving persons and groups. Psychol. Rev. **103**(2), 336–355 (1996)
18. Han, S., Lerner, J.S., Keltner, D.: Feelings and consumer decision making: the appraisal-tendency framework. J. Consum. Psychol. **17**(3), 158–168 (2007)
19. Harbaugh, W.T., Mayr, U., Burghart, D.R.: Neural responses to taxation and voluntary giving reveal motives for charitable donations. Science **316**(5831), 1622–1625 (2007)
20. Jenni, K.E., Loewenstein, G.: Explaining the "identifiable victim effect". J. Risk and Uncertainty **14**(3), 235–257 (1997)
21. Kahneman, D.: Thinking Fast and Slow. Farrar, Straus, and Giroux, New York (2011)
22. Kahneman, D., Krueger, A.B., Schkade, D., Schwarz, N., Stone, A.A.: Would you be happier if you were richer? A focusing illusion. Science **312**(5782), 1908–1910 (2006)
23. Karlan, D., List, J.: Does price matter in charitable giving? Evidence from a large-scale natural field experiment. Am. Econ. Rev. **97**(5), 1774–1793 (2007)
24. Keltner, D., Ellsworth, P., Edwards, K.: Beyond simple pessimism: effects of sadness and anger on social perception. J. Pers. Soc. Psychol. **64**(5), 740–752 (1993)
25. Kogut, T., Ritov, I.: The "identified victim" effect: an identified group, or just a single individual? J. Behav. Decis. Making **18**(3), 157–167 (2005a)
26. Kogut, T., Ritov, I.: The singularity effect of identified victims in separate and joint evaluations. Organ. Behav. Hum. Decis. Processes **97**(2), 106–116 (2005b)
27. Kuhnen, C.M., Knutson, B.: The influence of affect on beliefs, preferences, and financial decisions. J. Financ. Quant. Anal. **46**(3), 605–626 (2011)
28. Lazarus, R.S.: Emotion and Adaptation. Oxford University Press, New York (1991)
29. Lea, S.E.G., Webley, P.: Money as tool, money as drug: the biological psychology of a strong incentive. Behav. Brain Sci. **29**(2), 161–209 (2006)
30. Lee-Wingate, S.N., Corfman, K.P.: A little something for me and maybe for you, too: promotions that relieve guilt. Mark. Lett. **21**(4), 385–395 (2010)
31. Lerner, J.S., Goldberg, J.H., Tetlock, P.E.: Sober second thought: the effects of accountability, anger, and authoritarianism on attributions of responsibility. Pers. Soc. Psychol. Bull. **24**(6), 563–574 (1998)

32. Lerner, J.S., Keltner, D.: Beyond valence: toward a model of emotion-specific influences on judgment and choice. Cogn. Emot. **14**(4), 473–493 (2000)
33. Lerner, J.S., Keltner, D.: Fear, anger, and risk. J. Pers. Soc. Psychol. **81**(1), 146–159 (2001)
34. Lerner, J.S., Small, D.A., Loewenstein, G.: Heart strings and purse strings: carryover effects of emotions on economic decisions. Psychol. Sci. **15**(5), 337–341 (2004)
35. Liu, W., Aaker, J.: The happiness of giving: the time-ask effect. J. Consum. Res. **35**(3), 543–557 (2008)
36. Peloza, J., Steel, P.: The price elasticities of charitable contributions: a meta- analysis. J. Public Policy Mark. **24**(2), 260–272 (2005)
37. Raghunathan, R., Pham, M.T.: All negative moods are not equal: motivational influences of anxiety and sadness on decision making. Organ. Behav. Hum. Decis. Processes **79**(1), 56–77 (1999)
38. Randolph, W.: Dynamic income, progressive taxes, and the timing of charitable contributions. J. Polit. Econ. **103**(4), 709–738 (1995)
39. Reed II, A., Aquina, K., Levy, E.: Moral identity and judgments of charitable behaviors. J. Mark. **71**(1), 178–193 (2007)
40. Rick, S.I., Cryder, C.E., Loewenstein, G.: Tightwads and spendthrifts. J. Consum. Res. **34**(6), 767–782 (2008)
41. Rottenberg, J., Ray, R.R., Gross, J.J.: Emotion elicitation using films. In: Coan, J.A., Allen, J.J.B. (eds.). The Handbook of Emotion Elicitation and Assessment. Oxford University Press, New York (2007)
42. Schelling, T.C.: The life you save may be your own. In: Chase, S. (ed.) Problems in Public Expenditure Analysis. The Brookings Institute, Washington, DC (1968)
43. Schwarz, N.: Feelings as information: informational and motivational functions of affective states. In: Higgins, E.T., Sorrentino, R.M. (eds.) Handbook of Motivation and Cognition: Foundations of Social Behavior, vol. 2, pp. 527–561. Guilford Press, New York (1990)
44. Schwarz, N., Clore, G.L.: Feelings and phenomenal experiences. In: Higgins, E.T., Kruglanski, A.W. (eds.) Social Psychology: Handbook of Basic Principles, pp. 433–465. Guilford Press, New York (1996)
45. Shang, J., Reed II, A., Croson, R.: Identity congruency effects on donations. J. Mark. Res. **45**(3), 351–361 (2008)
46. Small, D.A., Lerner, J.S.: Emotional policy: personal sadness and anger shape judgments about a welfare case. Political Psychol. **29**(2), 149–168 (2008)
47. Small, D.A., Loewenstein, G.: Helping *a* victim or helping *the* victim: altruism and identifiability. J. Risk Uncertainty **26**(1), 5–16 (2003)
48. Small, D.A., Loewenstein, G., Slovic, P.: Sympathy and callousness: the impact of deliberative thought on donations to identifiable and statistical victims. Organ. Behav. Hum. Decis. Processes **102**(2), 143–153 (2007)
49. Smith, C.A., Ellsworth, P.C.: Patterns of cognitive appraisal in emotion. J. Pers. Soc. Psychol. **48**(4), 813–838 (1985)
50. Strahilevitz, M., Meyers, J.G.: Donations to charity as purchase incentives: how well they work may depend on what you are trying to sell. J. Consum. Res. **24**(4), 434–446 (1998)
51. Susskind, J., Mauer, K., Thakkar, V., Hamilton, D., Sherman, J.: Perceiving individuals and groups: expectancies, dispositional inferences, and causal attributions. J. Pers. Soc. Psychol. **76**(2), 181–191 (1999)
52. Vohs, K.D., Mead, N.L., Goode, M.R.: The psychological consequences of money. Science **314**(5802), 1154–1156 (2006)
53. Vohs, K.D., Mead, N.L., Goode, M.R.: Merely activating the concept of money changes personal and interpersonal behavior. Curr. Dir. Psychol. Sci. **17**(3), 208–212 (2008)
54. Watson, D., Clark, L.A., Tellegen, A.: Development and validation of brief measures of positive and negative affect: the PANAS scales. J. Pers. Soc. Psychol. **54**(6), 1063–1070 (1988)
55. Winterich, K.P., Han, S., Lerner, J.S.: Now that I'm sad, it's hard to be mad: The role of cognitive appraisals in emotional blunting. Pers. Soc. Psychol. Bull. **36**(11), 1467–1483 (2010)

Decision Theory and Rules of Thumb

Konstantinos V. Katsikopoulos

Abstract This chapter presents a relatively new and rapidly developing interdisciplinary theory of decision making, the theory of fast and frugal heuristics. It is first shown how the theory complements most of the standard theories of decision making in the social sciences such as Bayesian expected utility theory and its variants: Fast and frugal heuristics are not derived from normatively compelling axioms but are inspired by the simple rules of thumb that people and animals have been empirically found to use. The theory is illustrated by presenting the basic concepts and mathematics of some fast and frugal heuristics such as the recognition heuristic, the take-the-best heuristic, and fast and frugal trees. Then, applications of fast and frugal heuristics in a number of problems are described (how do laypeople make investment decisions? how do military staff identify unexploded ordnance buried in the ground? how do doctors decide whether to admit a patient to the emergency care or not?) It is emphasized that there are no good or bad decision models per se but that all models can work well in some situations and not in others, and thus the goal is to find the right model for each situation. Accordingly, in all applications, the performance of fast and frugal heuristics is compared, by computer simulations and mathematical analyses, to the performance of standard models such as Bayesian networks, classification-and-regression trees and support-vector machines. Finally, ways of combining standard decision theory and rules of thumb are discussed.

Keywords Mathematical modeling · Utility · Rules of thumb · Heuristics · Psychology

---

Chapter for the book Human-Centric Decision-Making Models for Social Sciences, Edited by Peijun Guo and Witold Pedrycz.

K. V. Katsikopoulos (✉)
Max Planck Institute for Human Development, Center for Adaptive Behavior and Cognition (ABC), Lentzeallee 94, 14195 Berlin, Germany
e-mail: katsikop@mpib-berlin.mpg.de

P. Guo and W. Pedrycz (eds.), *Human-Centric Decision-Making Models for Social Sciences*, Studies in Computational Intelligence 502,
DOI: 10.1007/978-3-642-39307-5_4,

1 Introduction

Decision theory today is dominated by models that are exhaustive and integrative. Exhaustive means that as many decision options as possible are evaluated. For each option, the evaluation is integrative, meaning that tradeoffs are made among the option's good and bad features. For example, if a medical treatment has a low probability that the patient will live for at least ten more years, in the Bayesian expected utility model, the utility of this outcome is multiplied by its probability. When the situation is deterministic, the features of an option are its attributes and their importance to the decision-maker. For instance, if an apartment has a guest room, the multi-attribute utility model multiplies the utility of this attribute by a weight that expresses the attribute's importance. In this chapter, the theory that consists of exhaustive and integrative models of decision making is called *standard decision theory*.

Standard decision theory is taught widely at schools of management, engineering colleges and at departments of economics and psychology, among others. Although a flurry of standard decision models exists, it is possible to (over)simplify the picture and trace some core ideas of the theory to just a few pieces of work.

Savage [60] proved that if a decision maker accepts a set of apparently self-evident axioms—such as transitivity, which says that if options *a*, *b* and *c* are such that *a* is preferred to *b* and *b* is preferred to *c*, then *a* is also preferred to *c*—then, she will decide for the option that has the highest expected utility where the probabilities will be her subjective beliefs and, when new information is obtained, she will update those probabilities according to Bayes rule.

Such work on the mathematical foundations of the Bayesian expected utility model makes standard decision theory seem normatively compelling. That is, it can be argued that it is what we should do in an ideal world. Nevertheless, Savage himself did not appear to see his contribution as much more than a theoretical exercise. He pointed out that it is applicable to what he called *small worlds*, that is, situations where the decision maker can obtain the relevant information on decision options and attributes, and has the time and other resources necessary to computationally process this information. If these conditions are not met, Savage made no claim that decision makers should use standard decision theory.

Savage's hesitations were not taken that seriously, at least not by those eager to solve real decision problems. The perceived success of other mathematical models applied to decision making, such as linear- and dynamic programming, provided reasons for optimism around the world.

In Cambridge, Massachusetts, economist Howard Raiffa and his colleagues [44, 58] advocated the application of Bayesian and multi-attribute versions of utility theory to all kinds of decision problems such as choosing an apartment or designing an airport. In Ann Arbor, Michigan, where Savage also worked for a bit, psychologist Ward Edwards became so enamored with standard decision theory that he devoted more than five decades to its further development and application [17]. Edwards shaped decision theory and practice as few, if any, others [37]. His statement "no principle other than maximizing subjective expected utility deserves a moment of consideration" (quoted in [65]) may appear too single-minded but it is in fact exactly the dominant attitude in decision theory and practice today.

On the other hand, problems with standard decision models are well known. For example, it is challenging for decision makers to provide reliable and correct estimates of attribute weights, probabilities of possible outcomes or their utilities; suitable approximations to intractable Bayesian computations must be discovered; and decision makers find the models difficult to understand and resist them [41, 46]. Proponents of standard decision theory, of course, do not see these challenges as formidable [66]. Rather, whatever problems are acknowledged are seen as relating to practical application and not as undermining the accuracy of standard decision theory. Consequently, research effort is invested in improving the application of the standard theory rather than in developing alternative theories.

This chapter makes a more radical point—that there exists an alternative to standard decision theory which is simpler to understand, neither exhaustive nor integrative, and also, under some conditions, more accurate. The chapter reviews the basics of the theory and some applications—including yet unpublished studies—and also synthesizes the current state of knowledge and presents open problems.

Of course, this theory is not presented as a substitute for standard decision theory, but rather as a challenge and as complementary. It is a mathematical theory, which is not derived analytically from normatively compelling axioms but is synthesized from empirical knowledge from biology and psychology.

2 Rules of Thumb

Behavioral biologists use the term *rules of thumb* to describe how animals solve their basic problems, such as finding a home, foraging for food, avoiding a predator and choosing a mate. For example, the ant *Leptothorax albipennis* estimates the size of a candidate nest cavity as follows [50]. It first explores the cavity for a fixed time interval on an irregular path that covers the area fairly evenly; while doing this the ant lays down an individually distinct pheromone trail. Then, the ant leaves. When it returns, *Leptothorax albipennis* explores the nest again but now on a different irregular path. The rule of thumb is that cavity size is inversely proportionate to the frequency of encountering the old trail.

The ant constructs an attribute that it can use to make decisions. When honeybees have to identify the species of a flower, they use attributes which already exist and are easy for them to observe. They use a rule of thumb that relies on odor, color and shape, in that order [27]. That is, to choose one of two alternatives species, honeybees first attempt to decide based on odor only; then, if the odors of the two alternatives are the same, they use color; and finally, they use shape if both odors and colors of the two alternatives species are the same.

For a collection of rules of thumb that animals use, see Hutchinson and Gigerenzer [35]. These rules are neither exhaustive nor integrative. In the above examples, the first nest cavity with an acceptable size can be chosen without inspecting any other candidates; and, a flower species is evaluated without making trade-offs between the attributes of odor, color and shape.

Cognitive scientists, such as Gerd Gigerenzer and his colleagues, call rules of thumb *heuristics* [8, 25]. Heuristics have been proposed for describing people's judgments of salaries or values of assets, their choices of consumer goods, and so on. Like rules of thumb, heuristics are neither exhaustive nor integrative. For example, laypeople may invest on the stock of these companies they recognize, without even considering other companies [54]. Or, when customers choose between cameras varying on a number of attributes, in the majority of times, they use just one or two attributes [21]. For a collection of the heuristics that laypeople and professional decision-makers use, see Gigerenzer et al. [24]. Typically, heuristics are just assumed to be second-best to standard decision models. This chapter, however, makes the case that, when the accuracy of the two is actually compared empirically, heuristics are sometimes found to outperform the standard models.

It should be noted here that the heuristics discussed in this chapter are distinct from other conceptions of heuristics in the psychological literature. For example, Daniel Kahneman and his colleagues [36, 63] developed verbal models of heuristics, using labels such as "availability" and "representativeness", which do not lead to precise quantitative predictions. On the other hand, the heuristics developed in the program of Gigerenzer and his colleagues lead to precise quantitative predictions. Kahneman et al.'s research program on heuristics is known as the "heuristics-and-biases" program. Gigerenzer et al.'s research program is called the "fast-and-frugal-heuristics" program. Interestingly, both programs claim to continue the work of Herbert Simon [61, 62], a polymath who often presented himself as a cognitive psychologist but was also awarded the Nobel prize in economics. For details on the similarities and differences between the two programs, see [45].

Biologists emphasize that rules of thumb are adapted through natural selection while psychologists and other social scientists point out that heuristics can also be learned individually and socially or are formally taught, but this difference is ignored here. I focus on a similarity between rules of thumb and heuristics: that both can, and have been, defined by simple mathematical models.

Formally, a number of decision problems reduce to the identification of one out of many alternative options A, B, and so on, so that this option has a maximum value on a numerical criterion of interest Cr. The value of the criterion can be objectively determined as in the size of a nest cavity, or it can be determined subjectively by the decision-maker as in the satisfaction derived by using a camera. The important thing about the criterion values of the options is that they are unknown to the decision-maker at the time the decision has to be made.

The next four sections present simple models of rules of thumb for solving this problem. The goal of this presentation is to give a flavor of the kinds of models of rules of thumb that have been developed; the research that has compared the accuracy of these models with the accuracy of standard decision models is presented afterwards.

3 The Recognition Heuristic

Imagine that you are a contestant in a TV game show and face the $1 million question: Which city has more inhabitants: Detroit or Milwaukee? You cannot use an Internet connection to find the answer, but you have to infer it based on whatever you know about these two cities.

What is your answer? If you are American, then your chances of finding the right answer, Detroit, are not bad. 60 % of undergraduates at the University of Chicago did. If, however, you are German, your prospects look dismal because most Germans know little about Detroit, and many have not even heard of Milwaukee.

So, how many correct inferences did the less knowledgeable German group achieve? 90 % of the Germans answered the question correctly [26]! How can people who know less about a subject make more correct inferences? The answer seems to be that the Germans used the following heuristic: If you recognize the name of one city but not the other, then infer that the recognized city has the larger population. This heuristic is reasonable in the sense that one may expect to have heard of heavily populated cities because they generate a lot of news. Note that someone who happens to know many cities, as the American participants in this experiment, can not use the heuristic. She would have too much knowledge.

For simplicity, I assume here that the correlation between recognition and criterion is positive. For problems where the goal is to infer which one of two options (e.g., cities) has the higher value on a numerical criterion (e.g., population), the heuristic is stated as follows.

Recognition heuristic: If one of two options is recognized and the other is not, then infer that the recognized option has the higher value on the criterion.

The recognition heuristic builds on people's core capacity for recognition, of faces, voices and names. No computer program exists today that can perform face recognition as well as a human child does (with the possible exception of new anti-terrorist technologies). Note that the heuristic is not derived from any logical axioms, but is suggested by the empirical knowledge that people are excellent at recognizing things they have been experienced. It has been claimed that animals also use recognition to make decisions (for examples, see [35]).

Intuitively, one may expect the recognition heuristic to be successful when ignorance is systematic rather than random, that is, when recognition is strongly correlated with the criterion. Substantial correlations exist in competitive situations, that is, between name recognition and the excellence of colleges, the value of the products of companies or the quality of sports teams [26].

A strong prediction of the recognition heuristic is that no other pieces of information can change the decision to which recognition points. For example, suppose that a person (*i*) recognizes Detroit and not Milwaukee and (*ii*) recalls that the automobile industry in Detroit has been hit for long time by a recession. The prediction of the recognition heuristic is that she will infer that Detroit is more populous despite Detroit's recession. In other words, recognition is predicted to be used in a *noncompensatory* fashion. This is a strong prediction in the sense that it does not follow from other theories of decision making. Pachur et al. [55] reported that 50% of the participants in their study chose the recognized object consistently, that is, in every single trial, even when they had knowledge of three attributes indicating that the recognized object should have a low criterion value.

3.1 The Less-is-More Effect

Beyond the noncompensatory use of recognition, the recognition heuristic leads to another strong prediction, one that has to do with accuracy. This is the *less-is-more* effect, where less information leads to more accuracy [26]. The effect can be viewed as a violation of the celebrated effort-accuracy tradeoff where effort is measured by the amount of information used to make a decision. This tradeoff, touted as one of the most general laws of human cognition, holds that it is not possible to increase accuracy without increasing effort [26]. Below, I briefly present a theoretical analysis of the less-is-more effect.

Assume that there exist N options (e.g., cities) and the person performs all $N(N - 1)/2$ paired comparisons according to a numerical criterion involving two of these options, (e.g., compare two city populations). The amount of information a person uses is measured by the number of options the person recognizes, n. The question is if, and under what conditions, can a smaller n lead to higher accuracy than a larger n.

The probability of being able to use the recognition heuristic for a paired comparison equals the probability of exactly one option in the pair being recognized, or

$$r(n) = 2n(N - n)/[N(N - 1)]. \quad (1)$$

Similarly, the probability that both options are recognized, and thus other knowledge beyond recognition must be used equals

$$k(n) = n(n - 1)/[N(N - 1)]. \quad (2)$$

Finally, the probability that neither option is recognized, which means that the decision maker has to guess, equals

$$g(n) = (N - n)(N - n - 1)/[N(N - 1)]. \quad (3)$$

Let α be the accuracy of the recognition heuristic and β the accuracy when both options are recognized and other knowledge is used (where $\alpha, \beta > 1/2$ and both are constant across n). I also assume that accuracy equals 1/2 when none of the options is recognized. Based on these assumptions and (1–3), the overall accuracy of a person who recognizes n options equals

$$f(n) = r(n)\alpha + k(n)\beta + g(n)(1/2). \quad (4)$$

From (4), it can be shown analytically that $f(n)$ is an inverted-U-shaped function of n whenever $\alpha > \beta$ [26]. In other words, whenever $\alpha > \beta$, a less-is-more effect is predicted.

As an illustration, assume that there are three sisters who study for a geography quiz with $N = 100$ cities. The three sisters have the same $\alpha = 0.8$ and $\beta = 0.6$, but differ in the number of cities n they recognize. The little sister who recognizes zero cities has an accuracy of $f(0) = 0.50$. The eldest sister who recognizes all 100 cities has an accuracy of $f(100) = 0.60$. The middle sister, who recognizes 50 cities is the most accurate of the three sisters with $f(50) = 0.68$.

At a first glance, the less-is-more effect might appear paradoxical. But it is not because less recognition information may simply enable more accurate cognitive processing via the use of the recognition heuristic. This idea is expressed formally by the condition $\alpha > \beta$. Additionally, it is a mathematical fact that, whenever $\alpha > \beta$, a less-is-more effect is also predicted for groups who use a variety of majority rules [59].

The above analyses assumed that decision makers have a perfect recognition memory, in the sense that all options that have been experienced are recognized (and all options which have not been experienced are not recognized). Of course, this is a simplification. More realistically, it can be assumed that a decision maker falsely recognizes a city which she has not experienced with a probability of a false alarm f (and fails to recognize a city she has experienced with some other probability). As in the case of perfect recognition memory, it has been proven that, under some conditions, a less-is-more effect is predicted. The conditions are relatively cumbersome—for details, see Katsikopoulos [38]—but, informally, less experience leads to more accuracy when the probability of a false alarm f is either relatively low or relatively high, but not when f has a medium value.

Finally, how frequently is the less-is-more effect observed in practice and what is its magnitude? Katsikopoulos [38] reviewed four studies and found that a less-is-more effect was observed in two of them; additionally, the magnitude of the effect varied a lot, from 0.30 down to 0.01. Of course, it should be noted that even a tiny effect could be very important in actual decision-making as, for example, in large business contexts. Thus, decision makers and the analysts who support them should not assume that more information always leads to better decisions.

The recognition heuristic is not an exhaustive decision model, as it does not even consider unrecognized options. The next family of models presented also does not consider all available options.

4 Social Heuristics

When recognition is not strongly correlated with the criterion or the decision maker recognizes all options, decision making may involve a search for the possible outcomes of each option. A few years after his voyage on the Beagle, the 29-year-old Charles Darwin divided a scrap of paper (titled "This is the Question") into two columns with the headings "Marry" and "Not Marry" and listed favorable outcomes for each of the two options, such as "nice soft wife on a sofa with good fire" and "conversation of clever men at clubs." Darwin concluded that he should marry, writing "Marry–Marry–Marry Q. E. D" decisively beneath the first column. The following year, Darwin married his cousin, Emma Wedgwood, with whom he eventually had ten children.

How did Darwin decide to marry, based on the possible outcomes he envisioned, such as children, loss of time or having a constant companion? He did not tell us. One possibility is that he used a version of multi-attribute utility theory or some heuristic simplifications of it which I will present in the next sections. Another possibility is that he used one of the social heuristics which exploit the social core capacities of people, such as imitation which is unmatched among animal species. For instance, consider the following [47].

Do-what-the-majority-does heuristic: If the majority of your peers display a behavior, engage in it as well.

For the marriage problem, this heuristic makes a man start thinking of marriage at a time when most other men in one's social group do, say, around age 30. It is a most frugal heuristic, for one does not even have to think of pros and cons. Do-what-the-majority-does tends to perform well when (*i*) the observer and the demonstrators of the behavior are exposed to similar environments that (*ii*) are stable rather than changing and (*iii*) noisy, that is, where it is hard to see what the immediate consequence of one's action is [3].

Social heuristics appear to guide many of our decisions, and do-what-the-majority-does is only one such heuristic in the adaptive toolbox of decision makers. But there are other social heuristics as well.

Consider deciding about green versus gray energy. Assume you have moved into a new apartment, and you need to choose providers for the basic utilities. In the United States, the United Kingdom and many countries in Europe, 50–90 % of the people asked say that they would favor a green electricity carrier and are even willing to pay a small premium for it. But, unfortunately, these statements do not reflect behavior. The percentage of people who consume green electricity is marginal; for example, 2 % in Germany and 0.5 % in the United Kingdom. This discrepancy between what people say and what they do can be explained by the use of a social heuristic [56]. When one moves into their new apartment, there is typically an electricity carrier that provides a default (the carrier that was used by the previous tenant or the carrier that the landlord has chosen). The new tenants typically take no action and the default is used.

Default heuristic: If a decision is set as the default, do not change it.

The default heuristic can explain a flurry of phenomena such as peoples' retirement plans and whether they are organ donors or not. It is not an exhaustive model as it does not consider all available options and their attributes.

It is important to note that the default heuristic, as also the recognition- and do-what-the-majority-does heuristics, arrive at a decision without evaluating options. For problems where it is difficult to obtain high-quality input for evaluating options or to perform the computations necessary for the evaluation, sidestepping option evaluation can be a great relief. It is possible, however, that a decision maker feels uncomfortable to not perform any evaluations. In the next two sections, I describe two families of attribute-based heuristics, which evaluate options, albeit in a simpler way than standard decision theory does.

5 Lexicographic Heuristics

In this family of models, a decision is based on a subset of the attribute values of the oprions. I use the term "attributes" broadly here to include any piece of information that can be used as, for example, the probability of a particular outcome for a given option, and so on.

5.1 *Take-the-Best*

Let us say that you want to predict which one of Lufthansa and Southwest Airlines will have a higher stock price five years from now. You have heard of both Lufthansa and Southwest Airlines, and thus you cannot use the recognition heuristic, and you also hesitate to use a social heuristic. Relevant company attributes may be the number of years that the company has been operating, whether the country of origin is a G-8 country or not, and so on. I symbolize attributes by $a_1, a_2, \ldots$, and the values of option A on the attributes by $a_1(A), a_2(A), \ldots$ (attributes are coded so that their values are nonnegative and the correlation between each attribute and the criterion is positive).

A family of simple attribute-based models is *lexicographic heuristics* [20]:

$$\begin{aligned}&\text{Infer } Cr(A) > Cr(B) \text{if and only if}\\&a_i(A) > a_i(B), \text{ where } a_j(A) = a_j(B) \text{ for all } j < i.\end{aligned} \tag{5}$$

What does (5) say? Attributes are inspected one at a time until an attribute is found that has different values on the two objects; then, the object with the higher value on this attribute is inferred to have the higher criterion value. For example, suppose that a decision-maker orders the country-of-origin-in-G-8 attribute first and the number-of-years attribute second. The country-of-origin-in-G-8 attribute has the same value on Lufthansa and Southwest Airlines ("yes" that would be coded as 1), and Lufthansa has a higher value on the number-of-years attribute, so the decision-maker would infer that Lufthansa has a higher stock price.

The family of lexicographic heuristics is parameterized by the rule used to order attributes. For instance, in the *take-the-best* heuristic [23], attributes are ordered in descending order of their validity, v_i:

$$v_i = Pr[a_i(A) > a_i(B) \mid a_i(A) \neq a_i(B)], \text{ where } Cr(A) > Cr(B). \quad (6)$$

According to (6), the validity of an attribute is the probability that the attribute has a higher value on the option that has the higher criterion value (given that the attribute has different values on the two options).

Gigerenzer and Goldstein [23] postulated that people are able to calculate attribute validities based on their core capacity of monitoring frequencies of events, but this claim has been challenged [16]. Rules for ordering attributes, simpler than using validity, have also been proposed, as ordering attributes randomly [23]. In any case, when the decision-maker makes an inference by using a lexicographic heuristic such as take-the-best, she needs to retrieve attribute values from memory, one by one. Thus, lexicographic heuristics rely on peoples' core capacity for what psychologists call recall.

Note that, like the recognition heuristic, take-the-best is noncompensatory. Furthermore, take-the-best specifies the processes by which people make inferences. More specifically, it is specified how people search for information (by ordering attributes by validity), how they decide to stop the search (as soon as one attribute discriminates between the objects and allows making a decision) and how they decide based on the available information (by using the first discriminating attribute). There have been a number of laboratory tests of these processes (as well as of the decision outcomes predicted by take-the-best) and this research is summarized in Broeder and Newell ([6] see also the other articles in this journal's special issue). Overall, if people use heuristics such as take-the-best depends on the characteristics of the decision environment, as, for example, whether there is time pressure or not, and how skewed is the distribution of attribute validities. Animals have also been argued to use lexicographic heuristics (for examples, see [35]).

The standard decision-theoretic way of comparing two options is the family of *linear models*, in which a weighted sum of attribute values for each option is computed and the option with the higher sum is inferred to have the higher criterion value (if the sums are equal, one object is picked randomly). More formally,

$$\begin{aligned} &\text{Infer } Cr(A) > Cr(B) \text{ if and only if} \\ &\Sigma_i w_i a_i(A) > \Sigma_i w_i a_i(B), \text{ where } w_i \geq 0. \end{aligned} \quad (7)$$

Unlike lexicographic heuristics, linear models are compensatory. The weight w_i for attribute c_i can be computed in a number of ways as, for example, in ordinary linear regression, by minimizing the sum of squared differences between the real criterion values in the ecology and the criterion values estimated by the linear model.

Another family of standard decision-theoretic models for making paired comparisons is that of *Bayesian* models [14].

$$\begin{aligned} &\text{Infer } Cr(A) > Cr(B) \text{if and only if} \\ &Pr[Cr(A) > Cr(B) | a_i(A), a_i(B)] > 1/2. \end{aligned} \quad (8)$$

That is, the option that has, given all available information, the higher probability of having the higher criterion value, is inferred to have the higher criterion value. The probability in (8) is difficult to compute if the number of attributes is large or their interrelations are complicated. In practice, Bayesian models make simplifying assumptions about the interrelations among attributes. For example, *naïve Bayes* [14] assumes that attributes are conditionally independent given the criterion. It is easy to see [42] that, if attributes take binary values, naïve Bayes reduces to a linear model (7) with $w_i = \log[v_i/(1 - v_i)]$.

For making inferences in problems in which there is an objectively determined answer, as in comparing two companies' stock prices, the linear model (7) is the analogue of the additive multi-attribute utility model for making choices in problems where the criterion is subjectively determined, as in choosing an apartment: The relevant pieces of information (attribute values) are weighted and added. On the other hand, lexicographic heuristics such as take-the-best dispense with adding, and instead just use a simple form of weighing, ordering attributes.

Next, I present another kind of lexicographic heuristics, that are used in a decision problem different from the paired-comparison problem discussed so far.

5.2 *Fast and Frugal Trees*

A middle-aged man is taken to the hospital with complaints of intense chest pain. The doctors have to decide quickly whether he is at a low risk of having ischemic heart disease and just needs a regular nursing bed, or he is at a high risk and should be rushed to the emergency room. This decision problem is called a categorization problem. In the particular situation of categorizing a heart-disease patient, the available resources—such as time, information, and computation—are limited, there is pressure to be accurate and the stakes are big. The fast-and-frugal-heuristics research program has provided some answers to how professionals and laypeople make, or should make, accurate categorizations with limited resources, by using simple trees.

I first introduce some elements from the general theory of trees for categorization. In a categorization problem, the decision-maker has to assign objects to one of mutually exclusive categories, based on the values of the objects on some attributes. In the example above, the objects are the patients, there are two categories—having a low and a high risk of ischemic heart disease—and the attributes are the available pieces of medical information such as readings from an electrocardiogram.

A categorization tree can be graphically represented by the root node, on the tree's first level, and subsequent levels with one attribute processed at each level (see Fig. 1). There are two types of nodes. First, a node may specify a question about the value of the object on an attribute; the answer then leads to another node at the next level, and the process continues in this way. The root node is of that type. For nodes of the other type there is an exit; the object is categorized and the process stops. In sum, starting from the root node and answering a sequence of questions about attributes, an exit is reached and the object is categorized. For trees to be easy for

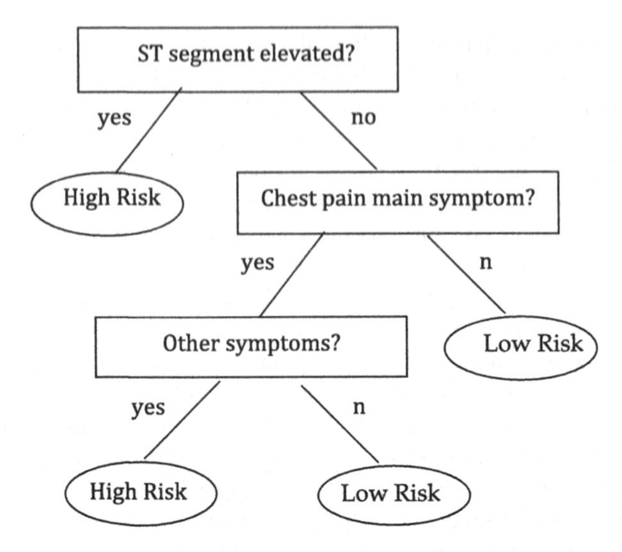

Fig. 1 A fast frugal tree for categorizing patients as having a high or low risk of ischemic heart disease (for more details, see [28])

people to understand and apply they should not have too many levels or attributes. For example, Fig. 1 shows such a tree, developed by two practicing doctors for the ischemic-heart-disease problem [28].

A categorization tree is called *fast and frugal* if and only if it has at least one exit at each level [53]. According to this definition, the tree of Fig. 1 is fast and frugal. If a second question were asked for all patients with elevated ST segment, the tree would not have been fast and frugal.

Fast and frugal trees are noncompensatory. They also specify a number of cognitive processes—how information is searched for, how search is stopped, and how a decision is taken based on the obtained information. For example, a physician using the tree of Fig. 1 first looks up the ST segment, then the chest pain, and finally other symptoms. There are a number of simple ways of ordering attributes, including straightforward extensions of the validity rule (7) for take-the-best (for details, see [53]).

Standard decision theory has produced a number of categorization models such as logistic regression [48] Vapnik's [64] support vector machines (SVM) and Breiman et al.'s [4] classification and regression trees (CART). These families of models are more mathematically sophisticated than fast and frugal trees. For example, CART use information theory for ordering attributes; and the resulting trees are not, in general, fast and frugal. Also, the standard models are compensatory.

Fast and frugal trees appear to be used by practitioners in a number of fields such as law and medicine [9, 10]. Louis Cook and his team at the Emergency Medical Services Division of the New York City Fire Department used a fast and frugal tree for deciding which of the victims of the September 11 terrorist attack needed urgent care [7].

Next, I mention a family of models that lies between standard decision theory and heuristics.

6 Tallying

Recall that lexicographic heuristics are a simplification of linear models where attributes are weighed but not added. *Tallying* heuristics are another simplification of linear models where attributes are added but not weighed. In other words, by setting $w_i = 1$ in (7), the following is obtained.

$$\text{Infer } Cr(A) > Cr(B) \text{ if and only if } \Sigma_i a_i(A) > \Sigma_i a_i(B). \qquad (9)$$

Tallying as described in (9) refers to paired comparisons. It is easy to see that it can also be applied to categorization, as well as to other problems such as deciding how to construct one's financial portfolio. Lay investors often use tallying in the sense that they allocate an equal amount of wealth to each asset in the portfolio (this is also known as naïve diversification or as the 1/*N* heuristic where *N* is the number of assets or alternatives; [2]).

Of course, tallying is exhaustive and integrative and, in this sense, it is a standard model. It can, however, also be seen as a heuristic in that it amounts to simply adding attribute values and "is not demanding from a cognitive viewpoint" [29]. Tallying is based on peoples' capacity for simple arithmetic, which even if not necessarily innate, it is, in most cases, easily learned by children.

In the previous four sections, I introduced some models of heuristics and discussed their basic properties (e.g., less-is-more effect, noncompensatoriness, etc.) This kind of theory is necessary from the perspective of cognitive science. But the crucial question for an engineer is how these models perform compared to standard decision models. This is an active topic of research. Even though not always framed as such, answers are being provided since the 1970s. In the last 15 years, many studies have been carried out and today there exists a relatively large repository of results (for a review, see [39]). The next section presents some basic findings, including some yet unpublished studies, and attempts to synthesize the current state of knowledge.

7 Decision Theory and Rules of Thumb: Comparisons

I first define the measure by which the performance of models is evaluated. The *accuracy* of a model is the proportion of problems in which it made the correct decision; for example, a correct inference is that Lufthansa has a higher stock price than Southwest Airlines, and the categorization that a patient is at a high risk of having heart disease is correct if the patient subsequently suffered a heart attack. Most of the studies in the heuristics literature investigate accuracy (in a few studies, a second performance measure is investigated, the financial gain from a model's inferences, but I will not present such studies).

Furthermore, there are two types of accuracy. Fitting accuracy refers to the situation where the parameters of a model (e.g., attribute weights, order of attributes) are estimated using all data available. Predictive accuracy is often measured by cross-validation where the model parameters are estimated by using a subset of all data—called the *training set*—and the same parameters are applied to make decisions for the rest of the data—the *test set*; this process is repeated many times to average out random variation. Often, the training set comprises of the attribute and criterion values, or categories, of 50 % of all options. Predictive accuracy is a relevant measure of performance because it refers to decisions not yet made, and this is what most studies have focused on.

The empirical evidence I review comes from computer simulation studies. By simulations I do not mean that the datasets discussed are fictitious—most of them are in fact real—but rather that the performance of models is not calculated by using closed-form equations, but by simulating how an ideal agent would apply the models. I focus on simulations because the goal is to first evaluate the performance of models per se, excluding the human factor in applying the models. Presumably, taking into account human errors would favor heuristics because they are simpler.

To keep things simple, I do not discuss research on the accuracy of the recognition heuristic; a complicating factor with this research is that each decision-maker has potentially different attribute values, so the performance of one decision maker does not say much about the performance of other decision makers (for a review of this work, see [24]). Also, there are no studies that I am aware of that evaluate the accuracy of social heuristics in real-world problems (for theoretical analyses, see [3]). As will be acknowledged in the concluding section of the chapter, I see the lack of research on these heuristics as a weakness of the heuristics program. In any case, below I discuss the accuracy of lexicographic heuristics (take-the-best and fast and frugal trees) and tallying, compared to standard decision models.

7.1 Empirical Findings

I first survey some results of comparisons among lexicographic heuristics, tallying and linear models. In the seventies, Robyn Dawes and his colleagues [11, 12] found that tallying had higher predictive accuracy than linear regression in two of three forecasting problems (e.g., one problem was to predict success in graduate school). Dorans and Drasgow [15] generated a number of artificial datasets so that they reflected characteristics of real forecasting problems and concluded that tallying overall outperformed a number of versions of regression.

It has been and should continue to be emphasized, however, that there are conditions under which regression has higher predictive accuracy than tallying as, for example, when the size of the training set is large [18] further highlighted such conditions.

Czerlinski et al. [8] performed a simulation study with 20 datasets from the fields of biology, environmental science, economics, demography, health, psychology, sociology, and transportation. The criterion varied widely from men's and women's attractiveness, to cities' populations and homelessness rates, to obesity rates and mammals' sleep amounts, and so on. First, continuous attributes were dichotomized by using the median. In fitting, regression was most accurate (77 %), tallying scored 73 % and take-the-best 75 %. In prediction, where the size of the training set was 50 % of the whole dataset, take-the-best was most accurate (71 %), and even tallying outperformed regression by 69–68 %. When continuous attributes were not dichotomized, the predictive accuracy of take-the-best and regression was equal, 76 %. More recently, in a series of papers, Hogarth and Karelaia [29–33] used mostly, though not exclusively, artificial datasets, and confirmed and extended these results: Take-the-best, tallying, and linear regression all sometimes had superior- and sometimes inferior performance.

I now discuss comparisons of lexicographic heuristics and tallying with Bayesian models. Martignon and Hoffrage [52] compared the predictive accuracy of take-the-best and tallying with two Bayesian models in the 20 datasets of [8] when the size of the training set equaled 50 % of the whole dataset. The first model was naïve Bayes where attributes were assumed to be conditionally independent given the criterion, and the second one was a Bayesian network where attributes were assumed dependent in a relatively simple Markov sense. Recall that the predictive accuracy of take-the-best with continuous attributes was 76 %, of take-the-best with binary attributes 71 %, and of tallying (with binary attributes) 69 %. The predictive accuracy of Naïve Bayes was 73 % and of the Bayesian network 75 % (both models used binary attributes).

Katsikopoulos et al. [43] also compared the predictive accuracy of take-the-best with continuous attributes and tallying with that of naïve Bayes with binary attributes. This study tested very small training set sizes, from 2 to 10 objects, that is, from 3 % to 15 % of all objects across 19 of the Czerlinski et al. [8] datasets. It was found that, for 2 objects, tallying had the highest predictive accuracy and take-the-best was more accurate than naïve Bayes; for 3–10 objects, take-the-best had the highest accuracy, with naïve Bayes being more accurate than tallying. For 5–10 objects, the predictive accuracy of take-the-best exceeded that of naïve Bayes by more than 5 %.

DeMiguel et al. [13] run a simulation study of models for deciding how to allocate one's wealth across assets in a financial portfolio. They tested tallying (here meaning the allocation of an equal amount of wealth to each asset), against Harry Markowitz's [51] mean-variance model (for details, see [13], pp. 1921–1922), and 13 variants of this model, some of them Bayesian, designed to deal with issues of statistical estimation. Tallying ignores the data on returns, whereas the mean-variance models use past returns to reallocate wealth. The authors used seven real portfolios (with data on the returns of the assets spanning form twenty to forty years) and one artificial portfolio. The performance of the models was evaluated according to three measures (Sharpe ratio, which is a risk-adjusted return; certainty-equivalent return; and turnover) in a test set, over many repetitions. The main result is that tallying was not consistently outperformed by any of the optimization models, in any of the three measures. On the other hand, the same authors have also developed more sophisticated Bayesian models that outperformed tallying [13].

Finally, I discuss comparisons of fast and frugal trees with standard models. Brighton [5] compared the predictive accuracy of fast and frugal trees with CART [4] and another popular family of decision trees, C4.5 [57]. He used eight of the problems of Czerlinski et al. [8], and, in each problem, varied the size of the training set. In four of the problems, the fast and frugal tree outperformed the other decision trees for all training set sizes. In the other four problems, the highest predictive accuracy was achieved by different models for different training set sizes: When the size of the training set was relatively small, the fast and frugal tree tended to do best, whereas when the size of the training set was larger, CART and C4.5 tended to do best.

Martignon et al. [53] compared two fast and frugal trees—that differed on the rules used for ordering attributes and for assigning, at each tree level, the exit to one category—with CART and logistic regression. They used 30 categorization problems from the UC Irvine Machine Learning Repository, of which 11 were medical decision problems. For each problem, three sizes of the training set were tested: 90, 50, and 15 % of all objects. The results were similar to Brighton's: When the training set size was large, one of CART or logistic regression outperform both fast and frugal trees, and when the training set size was small, a fast and frugal tree outperformed both CART and logistic regression. For example, in the 11 medical problems, when the training set included 90 % of the objects, logistic regression outperformed both fast and frugal trees (79 % vs. 76 % and 74 %); and when the training set included 15 % of the objects, a fast and frugal tree scored 74 %, whereas the other fast and frugal tree scored 72 %, which was equal to the accuracy of CART and logistic regression.

In a yet unpublished study, Fernandez, Katsikopoulos, and Shubitizde [19] applied fast and frugal trees, CART and SVM to the problem of detecting unexploded ordnance (UXO; this is munitions used in war or military practice). This is a very relevant problem as, for example, around 11 million acres contain UXO in the United States and in Afghanistan more people have lost their lives due to UXO than due to landmines between 2002 and 2006. In cross-validation, when the training set had up to 10, out of a total of 216, objects found in the military ground Camp Sibert in Alabama, fast and frugal trees had higher accuracy than CART and equal accuracy with SVM (when the training set had more than 10 objects, the three models had essentially equal accuracy).

In sum, there are three main findings of computer simulation studies comparing the accuracy of lexicographic heuristics and tallying with that of standard decision models such as linear and Bayesian models, classification and regression trees and support vector machines. First, when all evidence is taken into account, the accuracy of all models is not that different (with tallying possibly lagging a bit behind). It should be emphasized, however, that even a small difference of, say, 1 % in accuracy could translate to large differences in situations with high stakes. Second, all models can achieve relatively superior and inferior accuracy. Third, the accuracy of heuristics is surprisingly competitive to that of standard models, especially in prediction.

Thus, the picture is quite complicated. Nevertheless, we do have some understanding of the mathematical and conceptual reasons for the three findings. In the next discussion, I sketch this understanding (for more details, see [39]), while also trying to show how open the problem still really is.

7.2 Theoretical Analyses

I first sketch why one may expect that, to a first approximation, the accuracy of heuristics and standard models is not that different. This claim follows from the combination of two facts: (*i*) many of the models can be viewed as linear models, which differ just in their attribute weights, and (*ii*) the performance of linear models does not, informally stated, change "much" when the attribute weights change.

The second fact is well known in the statistics and forecasting literature [49]. The first fact can be seen if one considers various models. Tallying is by definition a linear model with attribute weights $w_i = 1$; naïve Bayes, as said earlier, is a linear model with $w_i = \log[v_i/(1 - v_i)]$ if attributes are binary-valued; and it turns out that, again if attributes are binary valued, lexicographic heuristics can also be seen as linear models where the attribute weights satisfy the condition $w_i \geq \Sigma_{k>i} w_k$ for all i ([52]; e.g., it is easy to verify that a lexicographic heuristic that first inspects a_1, then a_2 and finally a_3, makes identical paired comparisons with the linear model $4a_1 + 2a_2 + a_3$).

Explaining the second finding amounts to uncovering conditions under which heuristics are more accurate than standard models, and vice versa. There is actually a host of such conditions. I give two examples.

First, Katsikopoulos and Martignon [41] provided a necessary and sufficient condition for a lexicographic heuristic to achieve maximum accuracy among all possible models in paired comparisons (this accuracy equals that of naïve Bayes). Assuming conditional independence and that attributes are binary, the condition is that attributes have *noncompensatory validities*:

$$o_i \geq \Pi_{k>i} o_k, \text{ where } o_i = v_i/(1 - v_i); \text{ for all } i. \tag{10}$$

For example, if there are three attributes with $v_1 = 0.8$, $v_2 = 2/3$, and $v_3 = 0.6$, then (10) holds.

In a series of papers, Robin Hogarth and Natalia Karelaia [29–33], analyzed further the relative accuracy of linear models and lexicographic heuristics. There are three main differences between these studies and the studies by Laura Martignon and her colleagues [41, 42, 52]. First, Hogarth and Karelaia also considered continuous attributes [30, 33]. Second, they looked into issues such as the correlations among attributes, or errors in the application of the models [29].

Third, and most importantly, unlike Martignon and her colleagues, Hogarth and Karelaia modeled the decision environment, that is, the relationship between the criterion value of an option and the attribute values of the option. A simple version of the *environment model* is linear:

$$Cr(A) = \Sigma_i \beta_i a_i(A), \text{ where } \beta_i \geq 0. \tag{11}$$

In some of their papers, Hogarth and Karelaia made additional mathematical assumptions to (11), as, for example, that attributes are normally distributed random variables [30, 33]. They were then able to derive conditions for patterns of relative accuracy involving heuristics and linear models.

For example, Hogarth and Karelaia [30] showed that a lexicographic heuristic is at least as accurate as linear regression whenever the following condition holds (where a_1 is the attribute inspected first in the lexicographic heuristic, $\rho_{C,a1}$ is the correlation between a_1 and the criterion value C, and R^2_{adj} is an adjusted version of the correlation coefficient of linear regression; for details see [30], p. 118):

$$\rho^2_{C,a1} \geq R^2_{\text{adj}}. \tag{12}$$

An informal interpretation of (12) is that the single attribute used by the lexicographic heuristic (because attributes are continuous, the first attribute inspected, a_1, allows making a decision almost always) has a higher correlation with the criterion than does the sum of all attributes (weighed by the regression coefficients). In a sense, the attribute structure specified by (12) is noncompensatory, as is the attribute structure specified by (10).

In sum, even though it is an oversimplification, it can be said that the results of Hogarth and Karelaia converge with the results of Martignon and her colleagues on a condition for competitive accuracy of lexicographic heuristics. This condition is a noncompensatory attribute structure. We do not have analytical results on how inferior is the accuracy of lexicographic heuristics when this condition is not satisfied; all that is known is that there exist other models that outperform the heuristics.

There is a second condition that guarantees competitive accuracy for lexicographic heuristics. Baucells, Carasco, and Hogarth [1] showed that, assuming a linear environment model (11), a lexicographic heuristic achieves maximum accuracy across all possible models in paired comparisons, if the condition of *cumulative dominance* holds:

$$\text{There exists } A \text{ so that for all } B: \Sigma_{k\leq i}\, a_k(A) \geq \Sigma_{k\leq i}\, a_k(B), \text{ for all } i \text{ (inequality holds strictly for at least one } i). \tag{13}$$

For example, for two options, A and B, such that $a_1(A) = 1, a_2(A) = 0, a_3(A) = 1$, and $a_1(B) = 0, a_2(B) = 1, a_3(B) = 1$, (13) shows that$A$ cumulatively dominates B. The lexicographic heuristic that inspects attributes in the order a_1, a_2, and a_3, would infer A as having the highest criterion value, and this is correct for the linear environment model $C(A) = 5a_1(A) + 4a_2(A) + 3a_3(A)$. In a yet unpublished study, Katsikopoulos [40] showed that this result holds for a linear environment model that includes multiplicative interactions among attributes if and only if attributes are binary.

Do the two conditions of noncompensatoriness and cumulative dominance explain why lexicographic heuristics often achieve superior accuracy? Not fully. Noncompensatoriness seems to be infrequently satisfied. For example, Hogarth and Karelaia [29] pointed out that, in principle, attribute weights are seldom noncompensatory, and Katsikopoulos and Martignon [42] empirically found that attribute validities were noncompensatory in three of the 20 datasets of Czerlinski et al. [8]. Cumulative dominance is actually relatively common [1] for example, given two objects with three attributes each, one object cumulatively dominates the other in 97 % of all possible distributions of binary attributes across objects. But, it is not clear how often is the environment model linear.

In sum, there is still a lot to understand about why lexicographic heuristics often have superior accuracy. In fact, this is even truer when one tries to understand the third finding, the success of heuristics in prediction. It does seem that part of the answer has to do with heuristics needing less, and perhaps also lower-quality, information to get calibrated than the standard models, because they have fewer parameters and simpler functional forms. But how exactly does this affect predictive accuracy? Gigerenzer and Brighton [22] used an insight from the machine-learning literature [34] and conjectured that heuristics have lower variance in their decisions than the standard models but a comprehensive study of this conjecture is still lacking.

The next section concludes the chapter by briefly summarizing its message and speculating on how decision theory, as we know it today, and the new ideas of modeling rules of thumb, as presented in this chapter, can be combined in order to construct an adaptive and effective repertoire for decision making.

7.3 Decision Theory and Rules of Thumb, Together?

Until recently, it was basically taken for granted that the models of standard decision theory, such as linear models, Bayesian networks, classification and regression trees, perform better than their heuristic counterparts, such as lexicographic heuristics and tallying. The heuristics were viewed as simplifications, perhaps dictated by the constraints of the real world, but simplifications nevertheless, doomed to be second best.

As I hope to have shown in this chapter, this is not true. The standard decision models and the heuristics all have their regions of being best, and, in fact, this does not seem to be a strange accident but also fits with theoretical analyses. In a broader sense, this result could have been anticipated because people, as well as animals, have, for a long time, been using rules of thumb that are very intimately related to the models of heuristics presented here such as take-the-best and fast and frugal trees, and one may expect that these rules have some value, at least under some conditions.

In sum, it seems necessary to combine decision theory and rules of thumb, so that decision makers are supported in adaptively switching between the two, depending on the problem. But how?

I think it is fair to say that this question has not really been considered. There is some relevant work, as in Hogarth and Karelaia (2006) and [39] who have suggested "maps" that delineate how to decide which decision model to use depending on the characteristics of the problem at hand. These maps include standard models, lexicographic heuristics and tallying. The maps concentrate on option evaluation. What they are missing is a role for the more radical rules of thumb such as the recognition heuristic and social heuristics. These heuristics are radical because they are not exhaustive and thus do not require that all options are known. Given that not knowing all options is what makes real decision-making under uncertainty so challenging, studying social heuristics may well be the way in which it makes most sense to infuse today's decision theory with rules of thumb.

References

1. Baucells, M., Carrasco, J.A., Hogarth, R.M.: Cumulative dominance and heuristic performance in binary multi-attribute choice. Oper. Res. **56**, 1289–1304 (2008)
2. Benartzi, S., Thaler, R.H.: Naïve diversification strategies in defined contribution saving plans. Am. Econ. Rev. **91**(1), 79–98 (2001)
3. Boyd, R., Richerson, P.J.: The Origin and Evolution of Cultures. Oxford University Press, Oxford (2005)
4. Breiman, L., Friedman, J., Stone, C.J., Olshen, R.A.: Classification and Regression Trees. Chapman and Hall (1984)
5. Brighton, H.: Robust inference with simple cognitive models. In: Lebiere, C., Wray, R. (eds.) AAAI Spring Symposium: Cognitive Science Principles Meet AI-Hard Problems, pp. 17–22. AAAI Press, Menlo Park, CA (2006)
6. Broeder, A., Newell, B.R.: Challenging some common beliefs: Empirical work within the adaptive toolbox metaphor. Judgment Decis. Mak. **3**(3), 205–214 (2008)
7. Cook, L.: The world trade center attack–the paramedic response: an insider's view. Critical Care **5**, 301–303 (2001)
8. Czerlinski, J., Gigerenzer, G., Goldstein, D.G.: How good are simple heuristics? In: Gigerenzer, G., Todd, P.M. (eds.) & the ABC Research Group, Simple Heuristics that Make us Smart, pp. 97–118. Oxford University Press, New York (1999)
9. Dhami, M.K.: Psychological models of professional decision-making. Psychol. Sci. **14**, 175–180 (2003)
10. Dhami, M.K., Harries, C.: Fast and frugal versus regression models in human judgement. Think. Reason. **7**, 5–27 (2001)
11. Dawes, R.M.: The robust beauty of improper linear models. Am. Psychol. **34**, 571–582 (1979)
12. Dawes, R.M., Corrigan, B.: Linear models in decision making. Psychol. Bull. **81**(2), 95–106 (1974)
13. DeMiguel, V., Garlappi, L., Uppal, R.: Optimal versus naïve diversification: How inefficient is the 1/N portfolio strategy? Rev. Financial Studies **22**, 1915–1953 (2007)
14. Domingos, P., Pazzani, M.: On the optimality of the simple Bayesian classifier under zero-one loss. Machine Learning **29**, 103–130 (1997)
15. Dorans, N., Drasgow, F.: Alternative weighting schemes for linear prediction. Organ. Behav. Hum. Perfor. **21**, 316–345 (1978)

16. Dougherty, M.R., Franco-Watkins, A.M., Thomas, R.: Psychological plausibility of the theory of probabilistic mental models and the fast and frugal heuristics. Psychol. Rev. **115**, 199–213 (2008)
17. Edwards, W., Fasolo, B.: Decision technology. Ann. Rev. Psychol. **52**(1), 581–606 (2001)
18. Einhorn, H.J., Hogarth, R.M.: Unit weighting schemes for decision making. Organ. Behav. Hum. Perform. **13**, 171–192 (1975)
19. Fernandez, J.P., Katsikopoulos, K.V., Shubitizde, F.: Detecting unexploded ordnance by fast and frugal trees. (Unpublished manuscript), Dartmouth College,Hanover (2012)
20. Fishburn, P.C.: Lexicographic orders, decisions, and utilities: A survey. Manag. Sci. **20**, 1442–1471 (1974)
21. Ford, J., Schmitt, N., Schechtman, S.L., Hults, B.H., Dogherty, M.L.: Process tracing methods: Contributions, problems, and neglected research questions. Organ. Behav.Hum. Decis.Process. **43**(1), 75–117 (1989)
22. Gigerenzer, G., Brighton, H.: Homo heuristicus: Why biased minds make better inferences. Top. Cogn. Sci. **1**, 107–143 (2009)
23. Gigerenzer, G., Goldstein, D.G.: Reasoning the fast and frugal way: Models of bounded rationality. Psychol. Rev. **103**(4), 650–669 (1996)
24. Gigerenzer, G., Goldstein, D.G.: The recognition heuristic: A decade of research. Judgment Decis. Mak. **6**(1), 100–121 (2011)
25. Gigerenzer, G., Hertwig, R., Pachur, T. (eds.): Heuristics: The Foundations of Adaptive Behavior. Oxford University Press, New York (2011)
26. Goldstein, D.G., Gigerenzer, G.: Models of ecological rationality: The recognition heuristic. Psychol. Rev. **109**, 75–90 (2002)
27. Gould, J.L., Gould, C.G.: The Honey Bee. Scientific American Library, New York (1988)
28. Green, L., Mehr, D.R.: What alters physicians' decisions to admit to the coronary care unit? The J. Family Pract. **45**, 219–226 (1997)
29. Hogarth, R.M., Karelaia, N.: Simple models for multiattribute choice with many alternatives: When it does and does not pay to face tradeoffs with binary attributes? Manag. Sci. **51**(12), 1860–1872 (2005a)
30. Hogarth, R.M., Karelaia, N.: Ignoring information in binary choice with continuous variables: When is less "more"? J. Math. Psychol. **49**(2), 115–124 (2005b)
31. Hogarth, R.M., Karelaia, N.: Regions of rationality: Maps for bounded agents. Decis. Anal. **3**, 124–144 (2006a)
32. Hogarth, R.M., Karelaia, N.: "Take-the-best" and other simple strategies: Why and when they work "well" with binary cues. Theory and Decis. **61**, 205–249 (2006b)
33. Hogarth, R.M., Karelaia, N.: Heuristic and linear models of judgment: Matching rules and environments. Psychol. Rev. **114**(3), 733–758 (2007)
34. Holte, R.C.: Very simple classification rules perform well on most commonly used datasets. Machine Learning **3**(11), 63–91 (1993)
35. Hutchinson, J.M.C., Gigerenzer, G.: Simple heuristics and rules of thumb: Where psychologists and behavioural biologists might meet. Behav. Process. **69**, 97–124 (2005)
36. Kahneman, D., Slovic, P., Tversky, A. (eds.): Judgment Under Uncertainty: Heuristics and Biases. Cambridge University Press, Cambridge (1982)
37. Katsikopoulos, K.V.: Don't take gurus too seriously: Review of "A Science of Decision Making: The Legacy of Ward Edwards" (Eds. J. W. Weiss and D. J. Weiss). J. Math. Psychol. **54**, 401–403 (2010)
38. Katsikopoulos, K.V.: The less-is-more effect: Predictions and tests. Judgment Decis. Mak. **5**(4), 244–257 (2010)
39. Katsikopoulos, K.V.: Psychological heuristics for making inferences: Definition, performance, and the emerging theory and practice. Decis. Anal. **8**(1), 10–29 (2011)
40. Katsikopoulos, K.V.: Multi-attribute choice: An analysis of when to use heuristics. (Unpublished manuscript), Max Planck Institute for Human Development, Berlin (2012)
41. Katsikopoulos, K.V., Fasolo, B.: New tools for decision analysts. IEEE Transact. Syst. Man, Cybernetics: Syst. Hum. **36**(5), 960–967 (2006)

42. Katsikopoulos, K.V., Martignon, L.: Naïve heuristics for paired comparison: Some results on their relative accuracy. J. Math Psychol. **50**, 488–494 (2006)
43. Katsikopoulos, K.V., Schooler, L.J., Hertwig, R.: The robust beauty of mediocre information. Psychol. Rev. **117**(4), 1259–1266 (2010)
44. Keeney, R.L., Raiffa, H.: Decision-making with Multiple Objectives: Preferences and Value Tradeoffs. Wiley, New York (1976)
45. Kelman, M.G.: The Heuristics Debate. Oxford University Press, Oxford (2011)
46. Klein, G.A., Calderwood, R.: Decision models: Some lessons from the field. IEEE Transact. Syst. Man, Cybernetics **21**(5), 1018–1026 (1991)
47. Laland, K.N.: Imitation, social learning, and preparedness as mechanisms of bounded rationality. In: Gigerenzer, G., Selten, R. (eds.) Bounded Rationality: The Adaptive Toolbox. MIT Press, Cambridge, MA (2001)
48. Long, W.J., Griffith, J.L., Selker, H.P., D'Agostino, R.B.: A comparison of logistic regression to decision-tree induction in a medical domain. Comput. Biomedical Res. **26**, 74–97 (1993)
49. Lovie, A.D., Lovie, P.: The flat maximum effect and linear scoring models for prediction. J. Forecasting **5**, 159–168 (1986)
50. Mallon, E.B., Franks, N.R.: Ants estimate area using Buffon's needle. Proc. Royal Soc. Lond. Series B **267**, 765–770 (2000)
51. Markowitz, H.M.: Portfolio selection. J. Finance **7**, 77–91 (1952)
52. Martignon, L., Hoffrage, U.: Fast, frugal, and fit: Simple heuristics for paired comparison. Theory Decis. **52**, 29–71 (2002)
53. Martignon, L., Katsikopoulos, K.V., Woike, J.: Categorization with limited resources: A family of simple heuristics. J. Math. Psychol. **52**(6), 352–361 (2008)
54. Ortmann, A., Gigerenzer, G., Borges, B., and Goldstein, D. G.: The recognition heuristic: A fast and frugal way to investment choice? In: Plott, C. R. and Smith, V. L. (eds.) Handbook of Experimental Economics Results: Vol. 1, North-Holland, Amsterdam (2008)
55. Pachur, T., Broeder, A., Marewski, J.N.: The recognition heuristic in memory-based inference: Is recognition a non-compensatory cue? J. Behav. Decis. Mak. **21**, 183–210 (2008)
56. Pichert, D., Katsikopoulos, K.V.: Green defaults: Information presentation and pro-environmental behavior. J. Environ. Psychol. **28**, 63–73 (2008)
57. Quinlan, J.R.: Decision trees and decision-making. IEEE Transact. Syst. Man, Cybernetics **20**, 339–346 (1990)
58. Raiffa, H., Schleifer, R.: Applied Statistical Decision Theory, Harvard University Press, Boston(1961)
59. Reimer, T., Katsikopoulos, K.V.: The use of recognition in group decision-making. Cogn. Sci. **28**, 1009–1029 (2004)
60. Savage, L.J.: The Foundations of Statistics. Yale University Press, New Haven (1954)
61. Simon, H.A.: A behavioral model of rational choice. Q. J. Econ. **69**, 99–118 (1955)
62. Simon, H.A.: Rational choice and the structure of environments. Psychol. Rev. **63**, 129–138 (1956)
63. Tversky, A., Kahneman, D.: Heuristics and biases: Judgment under uncertainty. Science **185**, 1124–1130 (1974)
64. Vapnik, V.: The Nature of Statistical Learning Theory. Springer, Newyork (1995)
65. Vlek, C.: What constitutes a good decision?A panel discussion among Ward Edwards, Istvan Kiss, Giandomenico Majone, and Masanao Toda. Acta Psychol. **56**, 5–27 (1984)
66. von Winterfeldt, D. Edwards, W.: Decision Analysis and Behavioral Research,Cambridge University Press, Cambridge(1986)

Aggregating Imprecise Linguistic Expressions

Edurne Falcó, José Luis García-Lapresta and Llorenç Roselló

Abstract In this chapter, we propose a multi-person decision making procedure where agents judge the alternatives through linguistic expressions generated by an ordered finite scale of linguistic terms (for instance, 'very good', 'good', 'acceptable', 'bad', 'very bad'). If the agents are not confident about their opinions, they might use linguistic expressions composed by several consecutive linguistic terms (for instance, 'between acceptable and good'). The procedure we propose is based on distances and it ranks order the alternatives taking into account the linguistic information provided by the agents. The main features and properties of the proposal are analyzed.

Keywords Group decision-making · linguistic assessments · Imprecision · Distances

1 Introduction

People face a lot of decision-making problems in their everyday life. Some of these problems can be easily managed by means of numbers (How many tablespoons of sugar should I add to my coffee? How much is this computer?), but other problems are more complex and a numerical representation is more difficult to be implemented

E. Falcó
PRESAD Research Group, IMUVA, Universidad de Valladolid, Valladolid, Spain
e-mail: edurne@eco.uva.es

J. L. García-Lapresta (✉)
PRESAD Research Group, IMUVA, Dept. de Economía Aplicada, Universidad de Valladolid, Valladolid, Spain
e-mail: lapresta@eco.uva.es

L. Roselló
Dept. de Matemàtica Aplicada II, Universitat Politècnica de Catalunya, Barcelona, Spain
e-mail: llorenc.rosello@upc.edu

P. Guo and W. Pedrycz (eds.), *Human-Centric Decision-Making Models for Social Sciences*, Studies in Computational Intelligence 502,
DOI: 10.1007/978-3-642-39307-5_5, © Springer-Verlag Berlin Heidelberg 2014

(Which mean of transportation should I choose? How much is this brand preferred to this other?). Trying to assign a number to an opinion that could be imprecise makes it even harder. Human beings usually have difficulties representing uncertainty through numbers. As Zimmer [31] suggested, people generally prefer to handle the imprecision with linguistic terms rather than with numbers, because verbal expressions and their associated rules of conversation are more naturally included in people's thoughts.

Wallsten et al. [26] conducted an experimental research where they showed that people are more comfortable expressing the meanings of probability through words rather than through numbers. Following this line of thought, the program *Computing with Words* arises (see [15, 29], among others). In it, the objects of computation are words drawn from the natural language and agents express themselves through linguistic terms.

Among all possible kinds of decisions, this chapter focuses on the ones concerning voting systems. In voting, agents (or voters) have to show their preferences over multiple options (candidates or alternatives). Next, the individual preferences are somehow aggregated to yield a final result.

There are several voting systems where the agents assess linguistic terms to show their preferences. One of the most simple is *Approval Voting* [5, 6], where agents can either "approve of" or "not-approve of" the candidates. As an extension of Approval Voting, recently the voting system *Majority Judgment* [1–3] appears. In Majority Judgment, agents can assess to each candidate a linguistic term as 'excellent', 'very good', 'good', etc., from a fixed linguistic scale, to each candidate.

Majority Judgment is a controversial method and some authors have shown several paradoxes and inconsistences (see [9, 12, 17, 22], among others).

In order to solve some of these inconsistences, extensions of Majority Judgment have been developed. For instance, García-Lapresta and Martínez-Panero [12] proposed an extension for small committees where the linguistic information is aggregated by means of *centered OWA operators* [28] and the *2-tuple fuzzy linguistic representation* [14]. In Falcó and García-Lapresta [7, 8], an extension based on the distances between linguistic terms is introduced. Finally, Zahid [30] proposed a combination between Majority Judgment and the *Borda Count* [4].

There are other examples of voting systems using linguistic terms, such as García-Lapresta [10], who extended *simple majority* through linguistic preferences, or García-Lapresta et al. [11, 13] who generalized Borda rule assessing linguistic terms to the alternatives.

The introduction of linguistic terms partially captures agent's complexity. Nevertheless, this treatment does not necessarily include all the uncertainty that agents may feel. An agent might have some doubts about which linguistic term to assess. In this regard, allowing agents to assess several consecutive linguistic terms comes out as a possible solution (see Tang and Zheng [23], Ma et al. [16], Rodríguez et al. [18]). In this sense, our proposal deals with the matter by means of the *absolute order of magnitude spaces* introduced by Travé-Massuyés and Piera [25] and Travé-Massuyés and Dague [24]. More specifically, in the extension developed in Roselló et al. [19–21] as a starting point.

In this chapter we introduce a decision process where agents show their assessments over the feasible alternatives either through linguistic terms or through linguistic expressions. These expressions are generated by consecutive linguistic terms, and allow individuals to express imprecise assessments when they are not confident about their opinions.

The process aggregates the individual assessments by providing a weak order on the set of alternatives, satisfying some desirable properties. This weak order ranks the alternatives according to the distance between the corresponding individual assessments and the maximal linguistic term. These distances are defined through parameterized metrics in such a way that the values of the parameters allow to consider different ways of penalization on the agents' imprecision.

The chapter is organized as follows. Section 2 includes some notation and basic notions. Section 3 is devoted to analyze how to penalize the imprecision through appropriate parameterized metrics. Section 4 introduces the canonical linear order on the set of linguistic expressions and shows how this order can be reached through distances to the maximal linguistic term. Section 5 describes the decision process and some properties. Section 6 includes some illustrative examples. Finally, Sect. 7 includes some concluding remarks.

2 Preliminaries

Let $V = \{1, \ldots, m\}$, with $m \geq 2$, be a set of agents or voters and let $X = \{x_1, \ldots, x_n\}$, with $n \geq 2$, be the set of alternatives or candidates that have to be evaluated.

Let $L = \{l_1, \ldots, l_g\}$ be a linguistic ordered scale, where $l_1 < l_2 < \cdots < l_g$. The *granularity* of L is its cardinal, $\# L = g \geq 2$. The elements of L are linguistic terms as 'excellent', 'very good', 'good', etc.

A binary relation $\succcurlyeq$ on a set $A \neq \emptyset$ is a *weak order* (or *complete preorder*) if it is complete ($a \succcurlyeq b$ or $b \succcurlyeq a$, for all $a, b \in A$) and transitive (if $a \succcurlyeq b$ and $b \succcurlyeq c$, then $a \succcurlyeq c$, for all $a, b, c \in A$). On the other hand, a *linear order* on $A \neq \emptyset$ is an antisymmetric[1] weak order on A. Given a weak or linear order $\succcurlyeq$ on $A \neq \emptyset$, the asymmetric and symmetric parts of $\succcurlyeq$ are denoted by $\succ$ and $\sim$, respectively; in other words, $a \succ b$ if not $b \succcurlyeq a$, and $a \sim b$ if $a \succcurlyeq b$ and $b \succcurlyeq a$.

The set of weak orders on A is denoted by $W(A)$.

Based on the *absolute order of magnitude spaces* following Travé-Massuyès and Piera [25], we define the *set of linguistic expressions* as

$$\mathbb{L} = \{[l_h, l_k] \mid l_h, l_k \in L\,,\ 1 \leq h \leq k \leq g\}\,,$$

where $[l_h, l_k] = \{l_h, l_{h+1}, \ldots, l_k\}$. Since $[l_h, l_h] = \{l_h\}$, this linguistic expression can be replaced by the linguistic term l_h. In this way, $L \subset \mathbb{L}$.

[1] $\succcurlyeq$ is antisymmetric if for all $a, b \in A$ such that $a \neq b$ it holds $a \succ b$ or $b \succ a$.

Table 1 Meaning of the linguistic terms in Example 1

l_1	l_2	l_3	l_4	l_5
Very bad	Bad	Acceptable	Good	Very good

Given $\mathscr{E} = [l_h, l_k] \in \mathbb{L}$, with $\#\mathscr{E}$ we denote the number of linguistic terms of $\mathscr{E}$, i.e., $\#\mathscr{E} = k - h + 1$.

Example 1 Consider the set of linguistic terms $L = \{l_1, l_2, l_3, l_4, l_5\}$ with the meanings given in Table 1.

Each linguistic expression has a meaning on its own. For instance, $[l_2, l_4]$ means 'between bad and good', $[l_4, l_5]$ means between 'good and very good', or 'at least good', etc.

The set of all the linguistic expressions can be represented by a graph where the lowest layer represents the linguistic terms $l_h \in L \subset \mathbb{L}$, the second layer represents the linguistic expressions formed by two consecutive linguistic terms $[l_h, l_{h+1}]$, the third layer represents the linguistic expressions formed by three consecutive linguistic terms $[l_h, l_{h+2}]$, and so on up to the last layer where the linguistic expression $[l_1, l_g]$ is located. Consequently, the higher layer a linguistic expression is located, the more imprecise is.

Notice that $\#\mathbb{L} = g + (g-1) + \cdots + 1 = \dfrac{g(g+1)}{2}$.

Sometimes the computations in $\mathbb{L}$ will be done in $\mathbb{Z}^2$ by means of the injection $\psi : \mathbb{L} \longrightarrow \mathbb{Z}^2$ defined as $\psi([l_h, l_k]) = (k-1, h-1)$. Trough the function ψ we can represent a linguistic expression as a point in the plane. For instance, the linguistic expression $[l_2, l_5]$ can be represented as the point $(4, 1)$ in $\mathbb{Z}^2$. This function allows us to work in an easier computational setting.

The *Manhattan metric* on $\mathbb{R}^q$ is the function $d_M : \mathbb{R}^q \times \mathbb{R}^q \longrightarrow \mathbb{R}$ defined as

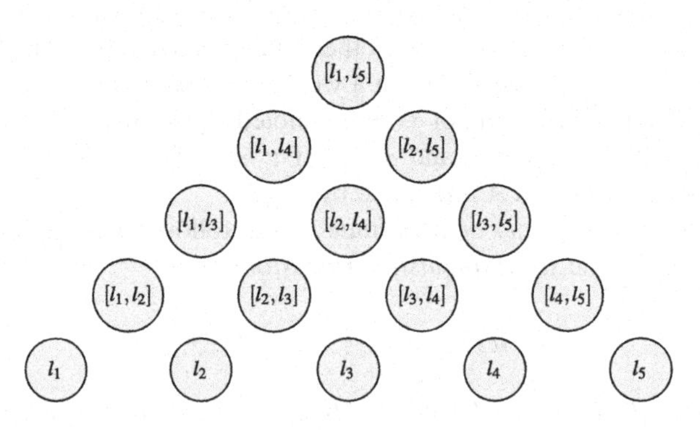

Fig. 1 Layers in the set of linguistic expressions for $g = 5$

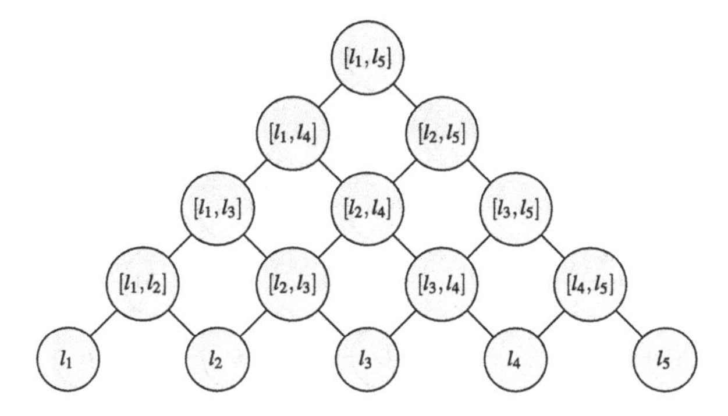

Fig. 2 Graph representation of the linguistic expressions for $g = 5$

$$d_M((a_1, \ldots, a_q), (b_1, \ldots, b_q)) = \sum_{r=1}^{q} |a_r - b_r|.$$

For $q = 1$ the Manhattan and the Euclidean metrics coincide: $d_M(a, b) = |a - b|$.

To define a first metric on the set of linguistic expressions $\mathbb{L}$, we adopt the treatment introduced by Roselló et al. [21] in the associated graph $G_{\mathbb{L}}$. The vertices in $G_{\mathbb{L}}$ are the elements of $\mathbb{L}$ and the edges $\mathscr{E} \sim \mathscr{F}$, where $\mathscr{E} = [l_h, l_k]$ and $\mathscr{F} = [l_h, l_{k+1}]$, or $\mathscr{E} = [l_h, l_k]$ and $\mathscr{F} = [l_{h+1}, l_k]$.

The graph representation for $g = 5$ is included in Fig. 2.

Definition 1 The *geodesic metric* on $\mathbb{L}$ is the function $d_G : \mathbb{L} \times \mathbb{L} \longrightarrow \mathbb{R}$ defined as

$$d_G(\mathscr{E}, \mathscr{F}) = d_M(\psi(\mathscr{E}), \psi(\mathscr{F})).$$

Notice that $d_G(\mathscr{E}, \mathscr{F})$ is the number of edges in one of the shortest paths connecting $\mathscr{E}$ and $\mathscr{F}$ in the graph associated with $\mathbb{L}$.

Example 2 The geodesic distance between the linguistic expressions $[l_1, l_4]$ and $[l_3, l_5]$ in Example 1 is

$$d_G([l_1, l_4], [l_3, l_5]) = d_M(\psi([l_1, l_4]), \psi([l_3, l_5])) = d_M((3, 0), (4, 2)) = 3,$$

just the length of one of the shortest paths from $[l_1, l_4]$ to $[l_3, l_5]$, for instance from vertex $[l_1, l_4]$ to vertex $[l_2, l_4]$, from vertex $[l_2, l_4]$ to vertex $[l_2, l_5]$ and, finally, from vertex $[l_2, l_5]$ to vertex $[l_3, l_5]$. This path is not unique, but it is one of those shortest paths.

Figure 3 shows the geodesic distances between contiguous linguistic expressions in Example 1. Distances between non-contiguous linguistic expressions can be obtained as the sum of distances through shortest paths between them.

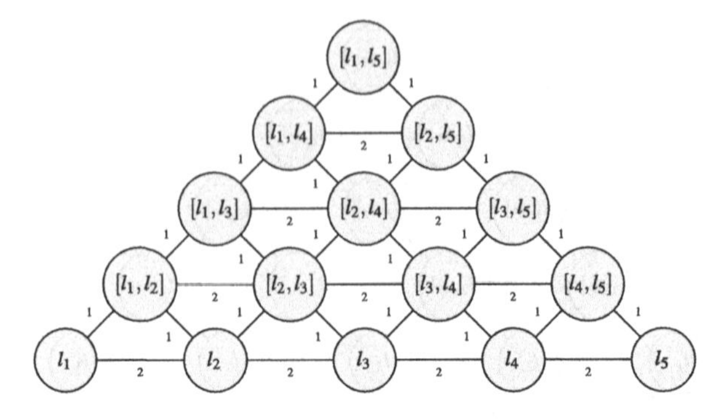

Fig. 3 Geodesic distances between contiguous linguistic expressions for $g = 5$

3 Penalizing The Imprecision

According to the geodesic metric d_G, the distance between two consecutive linguistic terms l_h and l_{h+1} is equal to 2. Imagine now an individual doubting about which one to choose (either l_h or l_{h+1}). If allowed, this individual may assess both of them, the linguistic expression $[l_h, l_{h+1}]$. This linguistic expression is in a geodesic distance of 1 from both l_h and l_{h+1}. In that sense, an individual confident about which linguistic term assesses is treated in the same way that an individual who assesses several linguistic terms.

In this chapter, we consider that precision in the assessments should be rewarded or, in a similar fashion, the imprecision should be penalized. That being said, we consider two kinds of penalization through two parameters α and β that must be chosen according to the penalization we want to impose.

Every time an agent assesses an additional linguistic term (i.e., the cardinality of the linguistic expression rises by 1), her level of imprecision increases. As we go up in the layers of Fig. 1, each linguistic expression is less precise than in the previous layer. So, the bottom layer has the highest precision (a single linguistic term), the second layer is less precise (two linguistic terms), the third one is even less precise (three linguistic terms), and so on.

Taking into account that the loss of precision should be penalized, we propose two different ways of penalization. First, for each linguistic term we add up, we increase the distance with a penalization of α: the distances from l_h to $[l_h, l_{h+1}]$ or $[l_{h-1}, l_h]$ are not 1, but $1+\alpha$. This penalization can be modeled by adding up $\alpha\, d_M(\#\mathscr{E}, \#\mathscr{F})$ to $d_G(\mathscr{E}, \mathscr{F})$.

Following this α-penalization, the distances between some linguistic expressions are as follows:

$$d(l_2, [l_2, l_3]) = d_G(l_2, [l_2, l_3]) + \alpha\, d_M(\#l_2, \#[l_2, l_3]) = 1 + \alpha$$
$$d([l_3, l_4], [l_3, l_5]) = d_G([l_3, l_4], [l_3, l_5]) + \alpha\, d_M(\#[l_3, l_4], \#[l_3, l_5]) = 1 + \alpha$$
$$d(l_1, [l_1, l_3]) = d_G(l_1, [l_1, l_3]) + \alpha\, d_M(\#l_1, \#[l_1, l_3]) = 2 + 2\alpha.$$

Until now, we have considered a penalization for the layer variation. Every additional linguistic term is penalized by α. Consequently, going from 2 linguistic terms up to 3 is the same than going from 3 up to 4. The second penalization takes into account not only how many linguistic terms the agent is using, but how far are from the maximum precision. What is the same, the higher the linguistic expression is in the layers, the more each added term should be penalized. For instance, the penalization from l_2 to $[l_2, l_3]$ should not be the same that from $[l_2, l_3]$ to $[l_2, l_4]$. In this regard, the β-penalization appears. This penalization increases as we climb the graph. That way, going up from 2 linguistic terms up to 3 is penalized by $1+\alpha+\beta$, and going up from 3 up to 4 by $1 + \alpha + 2\beta$. To model this β-penalization we introduce a new function $\rho : \mathbb{N} \times \mathbb{N} \longrightarrow \mathbb{N}$ defined as

$$\rho(a, b) = \frac{(a + b - 3)\,|a - b|}{2}.$$

Notice that $\rho(a, a + 1) = a - 1$ for every $a \in \mathbb{N}$.

If we apply the function ρ to the "linguistic expressions cardinality", we would obtain the number of times we should use the β-penalization. Taking into account that, as we climb up from the second layer to the top, we are increasing by β the penalization, the function ρ allows us to add the penalization of every layer. For instance, if we compare the linguistic expression $[l_2, l_3]$, which is on the second layer, and the linguistic expression $[l_1, l_5]$, which is on the fifth layer, we have to climb up a total of three layers. Climbing up form the second to the third layer it penalizes β, form the third to the fourth layer it penalizes 2β and from the fourth to the fifth layer it penalizes 3β. Adding all the β's we obtain $1 + 2 + 3 = 6$ or, similarly using the function ρ,

$$\rho(2, 5) = \frac{(2 + 5 - 3)\,|2 - 5|}{2} = 6.$$

We now introduce a family of parameterized metrics that agglutinates the geodesic metric and the mentioned penalizations.

Proposition 1 *For all* $\alpha, \beta \geq 0$, *the function* $d : \mathbb{L} \times \mathbb{L} \longrightarrow \mathbb{R}$ *defined as*

$$d(\mathscr{E}, \mathscr{F}) = \begin{cases} d_G(\mathscr{E}, \mathscr{F}) + \alpha\, d_M(\#\mathscr{E}, \#\mathscr{F}) + \beta\, \rho(\#\mathscr{E}, \#\mathscr{F}), & \text{if } \#\mathscr{E} + \#\mathscr{F} > 3 \\ d_G(\mathscr{E}, \mathscr{F}) + \alpha\, d_M(\#\mathscr{E}, \#\mathscr{F}), & \text{if } \#\mathscr{E} + \#\mathscr{F} \leq 3 \end{cases}$$

is a metric, and it is called the metric associated with (α, β).

Proof Since every linear combination of metrics is a metric, it is only necessary to check that ρ is a metric when it is restricted to $N = \{(a, b) \in \mathbb{N}^2 \mid a + b > 3\}$. Clearly, $\rho(a, b) \geq 0$, $\rho(a, b) = \rho(b, a)$, and $\rho(a, b) = 0$ if and only if $a = b$, for all $a, b \in N$. To prove the triangular inequality, $\rho(a, b) \leq \rho(a, c) + \rho(c, b)$ for all $a, b, c \in N$, six cases have to be considered: $a \leq b \leq c$, $a \leq c \leq b$, $b \leq a \leq c$, $b \leq c \leq a$, $c \leq a \leq b$ and $c \leq b \leq a$. Suppose $a \leq b \leq c$ (the other five cases are

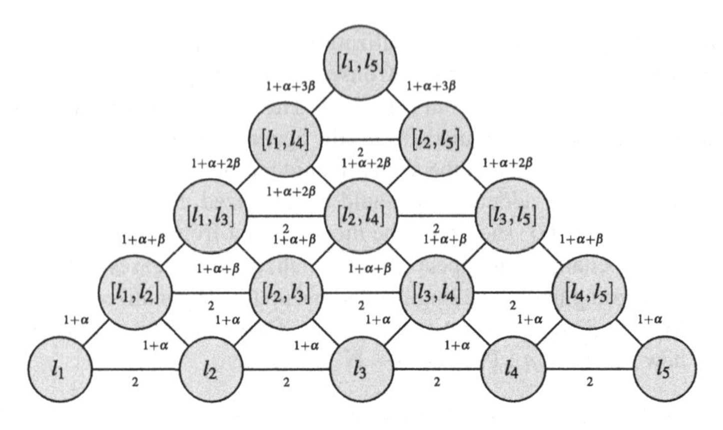

Fig. 4 Representation of the metric associated with (α, β) for $g = 5$

analogous). It is immediate to see that

$$\rho(a, b) = \frac{b^2 - a^2 + 3(a - b)}{2}$$
$$\rho(a, c) = \frac{c^2 - a^2 + 3(a - c)}{2}$$
$$\rho(c, b) = \frac{c^2 - b^2 + 3(b - c)}{2}.$$

Then, $\rho(a, b) \leq \rho(a, c) + \rho(c, b)$ is equivalent to $(c - b)(c + b - 3) \geq 0$, and it is obviously true for all $a, b, c \in N$. □

Figure 4 shows the distances between contiguous linguistic expressions for $g = 5$. Distances between non-contiguous linguistic expressions can be obtained as the sum of distances through shortest paths between them.

Remark 1 Some values of α and β can lead us into undesirable results. For instance, if $\alpha > 1$, we have $d(l_4, l_5) = 2 < 1 + \alpha = d([l_4, l_5], l_5)$. Analogously, if $\alpha + \beta > 1$, we have $d([l_3, l_4], l_5) = 3 + \alpha < 2 + 2\alpha + \beta = d([l_3, l_5], l_5)$. To avoid these paradoxes, we should impose some conditions over the values of α and β.

4 Ordering Linguistic Expressions

In the last section we have shown that is possible to obtain some strange orders among the linguistic expression. We now introduce an intuitive order, the canonical linear order. It ranks a linguistic expression over another if the sum of the subindices

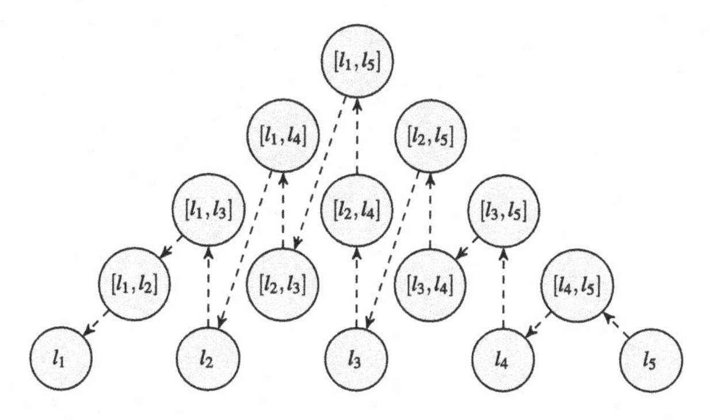

Fig. 5 Representation of the canonical linear order $\succcurlyeq_{\mathbb{L}}$ for $g = 5$

of the first one is higher than the sum of second one. If both linguistic expressions have the exact same addition, we should rank ahead the more precise one.

Definition 2 The *canonical order* on $\mathbb{L}$ is defined as

$$[l_h, l_k] \succcurlyeq_{\mathbb{L}} [l_{h'}, l_{k'}] \;\Leftrightarrow\; \begin{cases} h + k > h' + k' \\ \text{or} \\ h + k = h' + k' \;\text{ and }\; k - h \le k' - h'. \end{cases}$$

It is easy to see that $\succcurlyeq_{\mathbb{L}}$ is a linear order. Figure 5 shows this canonical linear order for $g = 5$.

Proposition 2 *For every metric* $d : \mathbb{L} \times \mathbb{L} \longrightarrow \mathbb{R}$*, the binary relation* $\succcurlyeq_d$ *on* $\mathbb{L}$ *defined as*

$$\mathscr{E} \succcurlyeq_d \mathscr{F} \quad\Leftrightarrow\quad d(\mathscr{E}, l_g) \le d(\mathscr{F}, l_g)$$

is a weak order.

Definition 3 Let T_g be the following triangles

- If g is odd

$$T_g = \left\{(\alpha, \beta) \in [0, \infty)^2 \mid \alpha + \frac{1}{2}\beta(g-1) < \frac{1}{g-2}\right\}.$$

- If g is even

$$T_g = \left\{(\alpha, \beta) \in [0, \infty)^2 \mid \alpha + \frac{1}{2}\beta(g-2) < \frac{1}{g-1}\right\}.$$

In Fig. 6 the triangle $T_5 = \{(\alpha, \beta) \in [0, \infty)^2 \mid \alpha + 2\beta < 1/3\}$ is showed.

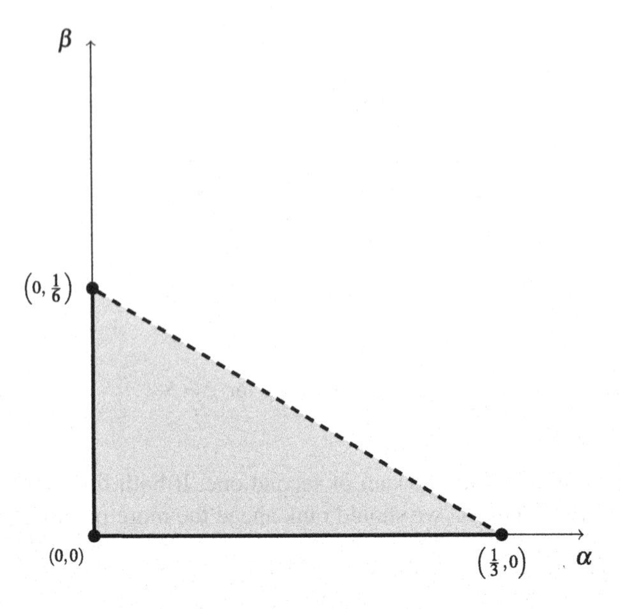

Fig. 6 Graphical representation of T_5

Proposition 3 *If* $d : \mathbb{L} \times \mathbb{L} \longrightarrow \mathbb{R}$ *is the metric associated with* (α, β)*, then* $\succcurlyeq_d \; = \; \succcurlyeq_{\mathbb{L}} \Leftrightarrow (\alpha, \beta) \in T_g$.

Proof Let us consider that g is odd.

$\Rightarrow$) Suppose that $\succcurlyeq_d \; = \; \succcurlyeq_{\mathbb{L}}$. By Definition 2, we have

$$[l_1, l_g] \succ_d \left[l_{\frac{g-1}{2}}, l_{\frac{g+1}{2}}\right]$$

i.e.,

$$d\left([l_1, l_g], l_g\right) < d\left(\left[l_{\frac{g-1}{2}}, l_{\frac{g+1}{2}}\right], l_g\right).$$

Taking into account

$$\begin{aligned} d\left([l_1, l_g], l_g\right) &= d_G\left([l_1, l_g], l_g\right) + \alpha\, d_M\left(\#[l_1, l_g], \#l_g\right) + \beta\, \rho\left(\#[l_1, l_g], \#l_g\right) \\ &= d_M((g-1, 0), (g-1, g-1)) + \alpha\, d_M(g, 1) + \beta\, \rho(g, 1) \\ &= g - 1 + \alpha\,(g-1) + \frac{1}{2}\beta(g-2)(g-1) \end{aligned}$$

and

$$d\left(\left[l_{\frac{g-1}{2}}, l_{\frac{g+1}{2}}\right], l_g\right) = d_G\left(\left[l_{\frac{g-1}{2}}, l_{\frac{g+1}{2}}\right], l_g\right) + \alpha\, d_M\left(\#\left[l_{\frac{g-1}{2}}, l_{\frac{g+1}{2}}\right], \#l_g\right)$$

$$= d_M\left(\left(\frac{g-1}{2}, \frac{g-3}{2}\right), (g-1, g-1)\right) + \alpha\, d_M(2, 1) = g + \alpha,$$

we have

$$[l_1, l_g] \succ_d \left[l_{\frac{g-1}{2}}, l_{\frac{g+1}{2}}\right] \Leftrightarrow g - 1 + \alpha\,(g-1) + \frac{1}{2}\beta(g-2)(g-1) < g + \alpha$$
$$\Leftrightarrow \alpha + \frac{1}{2}\beta(g-1) < \frac{1}{g-2}.$$

Consequently, $(\alpha, \beta) \in T_g$.

$\Leftarrow$) If $(\alpha, \beta) \in T_g$, it is a routine to check $\succcurlyeq_d = \succcurlyeq_{\mathbb{L}}$.

Let now us consider that g is even.

$\Rightarrow$) Suppose that $\succcurlyeq_d = \succcurlyeq_{\mathbb{L}}$. By Definition 2, we have

$$l_{\frac{g}{2}} \succ_d [l_1, l_g]$$

i.e.,

$$d\left(l_{\frac{g}{2}}, l_g\right) < d\left([l_1, l_g], l_g\right).$$

Taking into account

$$d\left(l_{\frac{g}{2}}, l_g\right) = d_G\left(l_{\frac{g}{2}}, l_g\right) + \alpha\, d_M\left(\#l_{\frac{g}{2}}, \#l_g\right)$$
$$= d_M\left(\left(\frac{g-2}{2}, \frac{g-2}{2}\right), (g-1, g-1)\right) + \alpha\, d_M\left(1, 1\right) = g + \alpha$$

and

$$d\left([l_1, l_g], l_g\right) = g - 1 + \alpha\,(g-1) + \frac{1}{2}\beta(g-2)(g-1),$$

we have

$$l_{\frac{g}{2}} \succ_d [l_1, l_g] \Leftrightarrow g - 1 + \alpha\,(g-1) + \frac{1}{2}\beta(g-2)(g-1) < g + \alpha$$
$$\Leftrightarrow \alpha + \frac{1}{2}\beta(g-2) < \frac{1}{g-1}.$$

Consequently, $(\alpha, \beta) \in T_g$.

$\Leftarrow$) If $(\alpha, \beta) \in T_g$, it is a routine to check $\succcurlyeq_d = \succcurlyeq_{\mathbb{L}}$. □

5 The Decision Process

Let $v_i^p \in \mathbb{L}$ be the linguistic expression assessed by voter $p \in V$ to alternative $x_i \in X$, and $\boldsymbol{v}_i = \left(v_i^1, \ldots, v_i^m\right) \in \mathbb{L}^m$ the *assessments vector* of alternative x_i.

A *profile* is a matrix $m \times n$ with coefficients in $\mathbb{L}$ whose columns contain the assessments vectors of the alternatives

$$\boldsymbol{v} = (\boldsymbol{v}_1 \mid \cdots \mid \boldsymbol{v}_i \mid \cdots \mid \boldsymbol{v}_n) = \begin{pmatrix} v_1^1 & \cdots & v_i^1 & \cdots & v_n^1 \\ \cdots & \cdots & \cdots & \cdots & \cdots \\ v_1^p & \cdots & v_i^p & \cdots & v_n^p \\ \cdots & \cdots & \cdots & \cdots & \cdots \\ v_1^m & \cdots & v_i^m & \cdots & v_n^m \end{pmatrix} = \left(v_i^p\right).$$

The set of profiles is denoted by $\mathbb{V}$.

For each $i \in \{1, \ldots, m\}$, the distance between the assessments vector of x_i and l_g is defined as

$$d(\boldsymbol{v}_i, l_g) = \sum_{p=1}^{m} d\left(v_i^p, l_g\right).$$

Proposition 4 *Given $\alpha, \beta \geq 0$, let d be the metric associated with (α, β). Then, the binary relation $\succcurlyeq_F$ on X defined as*

$$x_i \succcurlyeq_F x_j \Leftrightarrow d(\boldsymbol{v}_i, l_g) \leq d(\boldsymbol{v}_j, l_g)$$

is a weak order on X.

Proof It is straightforward. □

Definition 4 A *decision rule* is a mapping $F : \mathbb{V} \longrightarrow W(X)$ that satisfies the following properties

1. *Anonymity*: For every permutation π on $\{1, \ldots, m\}$ and every profile $\boldsymbol{v} = \left(v_i^p\right) \in \mathbb{V}$, it holds
$$F\left(\boldsymbol{v}^\pi\right) = F\left(\boldsymbol{v}\right),$$
where $\boldsymbol{v}^\pi = \left(v_i^{\pi(p)}\right)$.
2. *Neutrality*: For every permutation σ on $\{1, \ldots, n\}$ and every profile $\boldsymbol{v} = \left(v_i^p\right) \in \mathbb{V}$, it holds
$$F\left(\boldsymbol{v}_\sigma\right) = \left(F\left(\boldsymbol{v}\right)\right)_\sigma,$$
where $\boldsymbol{v}_\sigma = \left(v_{\sigma(i)}^p\right)$ and $x_{\sigma(i)}\ \left(F\left(\boldsymbol{v}\right)\right)_\sigma\ x_{\sigma(j)} \Leftrightarrow x_i\ F\left(\boldsymbol{v}\right)\ x_j$, i.e., $x_i\ \left(F\left(\boldsymbol{v}\right)\right)_\sigma\ x_j \Leftrightarrow x_{\sigma^{-1}(i)}\ F\left(\boldsymbol{v}\right)\ x_{\sigma^{-1}(j)}$.

3. *Independence*: For all pair of alternatives $x_i, x_j \in X$ and all pair of profiles $\boldsymbol{v} = \left(v_i^p\right), \boldsymbol{w} = \left(w_i^p\right) \in \mathbb{V}$, if $v_i^p = w_i^p$ and $v_j^p = w_j^p$ for every $p \in V$, it holds

$$x_i \, F(\boldsymbol{v}) \, x_j \;\Leftrightarrow\; x_i \, F(\boldsymbol{w}) \, x_j \quad \text{and} \quad x_j \, F(\boldsymbol{v}) \, x_i \;\Leftrightarrow\; x_j \, F(\boldsymbol{w}) \, x_i .$$

Proposition 5 *The mapping that assigns $\succcurlyeq_F$ to each profile is a decision rule.*

Proof The three conditions are trivially satisfied by $\succcurlyeq_F$ because of the commutativity of addition in $\mathbb{R}$ and the fact that the ranking between x_i and x_j provided by $\succcurlyeq_F$ only depends on $\boldsymbol{v}_i$ and $\boldsymbol{v}_j$. □

Definition 5 Given a weak order $\succcurlyeq$ on $\mathbb{L}$, a decision rule $F : \mathbb{V} \longrightarrow W(X)$ is *monotonic with respect to* $\succcurlyeq$ if for all pair of alternatives $x_i, x_j \in X$ and all pair of profiles $\boldsymbol{v} = \left(v_i^p\right), \boldsymbol{w} = \left(w_i^p\right) \in \mathbb{V}$, then if $w_i^p \succ v_i^p$ for some $p \in V$, $w_i^q = v_i^q$ for every $q \in V \setminus \{p\}$, and $w_j^q = v_j^q$ for every $q \in V$, it holds

$$x_i \, F(\boldsymbol{v}) \, x_j \;\Rightarrow\; x_i \, F(\boldsymbol{w}) \, x_j .$$

Proposition 6 *The mapping that assigns $\succcurlyeq_F$ to each profile is monotonic with respect to $\succcurlyeq_{\mathbb{L}}$.*

Proof

$$\begin{aligned}
x_i \, F(\boldsymbol{v}) \, x_j \;\Rightarrow\; d(\boldsymbol{v}_i, l_g) \le d(\boldsymbol{v}_j, l_g) \;&\Rightarrow\; \sum_{p=1}^{m} d\left(v_i^p, l_g\right) \le \sum_{p=1}^{m} d\left(v_j^p, l_g\right) \\
&\Rightarrow \sum_{q \in V \setminus \{p\}} d\left(v_i^q, l_g\right) + d\left(v_i^p, l_g\right) \le \sum_{p=1}^{m} d\left(v_j^p, l_g\right) \\
&\Rightarrow \sum_{q \in V \setminus \{p\}} d\left(w_i^q, l_g\right) + d\left(v_i^p, l_g\right) \le \sum_{p=1}^{m} d\left(w_j^p, l_g\right).
\end{aligned}$$

By means of the canonical order

$$w_i^p \succcurlyeq_{\mathbb{L}} v_i^p \;\Rightarrow\; d\left(w_i^p, l_g\right) \le d\left(v_i^p, l_g\right).$$

Then,

$$\begin{aligned}
&\sum_{q \in V \setminus \{p\}} d\left(w_i^q, l_g\right) + d\left(w_i^p, l_g\right) \le \sum_{p=1}^{m} d\left(w_j^p, l_g\right) \\
&\sum_{p=1}^{m} d\left(w_i^p, l_g\right) \le \sum_{p=1}^{m} d\left(w_j^p, l_g\right) \;\Rightarrow\; x_i \, F(\boldsymbol{w}) \, x_j .
\end{aligned}$$

□

6 Illustrative Examples

In this section we show different aspects of the proposed method of ranking through three toy examples. The first one shows how the method can provide the same ranking whenever imprecision is not penalized. The second example is about how different values of the parameters α and β can provide different rankings. And the third one shows that in some cases ties are obtained irrespectively of the values of the parameters α and β.

Example 3 Consider two alternatives x_1 and x_2 assessed by three voters through the set of linguistic terms $L = \{l_1, l_2, l_3, l_4, l_5\}$ whose meanings are given in Table 1. The assessments are shown in Table 2.

Using the metric d associated with (α, β), with $\alpha, \beta \geq 0$, we obtain

$$\begin{aligned} d(\boldsymbol{v}_1, l_5) &= d(v_1^1, l_5) + d(v_1^2, l_5) + d(v_1^3, l_5) \\ &= 4 + (2 + 2\alpha + \beta) + (7 + \alpha) = 13 + 3\alpha + \beta, \\ d(\boldsymbol{v}_2, l_5) &= d(v_2^1, l_5) + d(v_2^2, l_5) + d(v_2^3, l_5) \\ &= (3 + 3\alpha + 3\beta) + (5 + \alpha) + (5 + \alpha) = 13 + 5\alpha + 3\beta. \end{aligned}$$

Since $13 + 3\alpha + \beta < 13 + 5\alpha + 3\beta \Leftrightarrow \alpha + \beta > 0$, we have $x_1 \succ_F x_2 \Leftrightarrow \alpha + \beta > 0$ and, consequently, $x_1 \sim_F x_2 \Leftrightarrow \alpha = \beta = 0$. In other words, x_1 and x_2 are in a tie whenever imprecision is not penalized. But if it is, then x_1 always defeats x_2.

Example 4 Consider again two alternatives x_1 and x_2 now assessed by four voters through the set of linguistic terms $L = \{l_1, l_2, l_3, l_4, l_5\}$ with the meanings given in Table 1. Taking into account the assessments provided in Table 3, we can see how depending on the values of α and β, these alternatives are ranked in a different way.

Using the metric d associated with (α, β), with $\alpha, \beta \geq 0$, we obtain

Table 2 Assessments in Example 3

Alternative	Voter 1	Voter 2	Voter 3
x_1	l_3	$[l_3, l_5]$	$[l_1, l_2]$
x_2	$[l_2, l_5]$	$[l_2, l_3]$	$[l_2, l_3]$

Table 3 Assessments in Example 4

Alternative	Voter 1	Voter 2	Voter 3	Voter 4
x_1	$[l_1, l_5]$	$[l_1, l_5]$	l_3	l_3
x_2	$[l_1, l_4]$	$[l_2, l_5]$	$[l_1, l_3]$	$[l_3, l_5]$

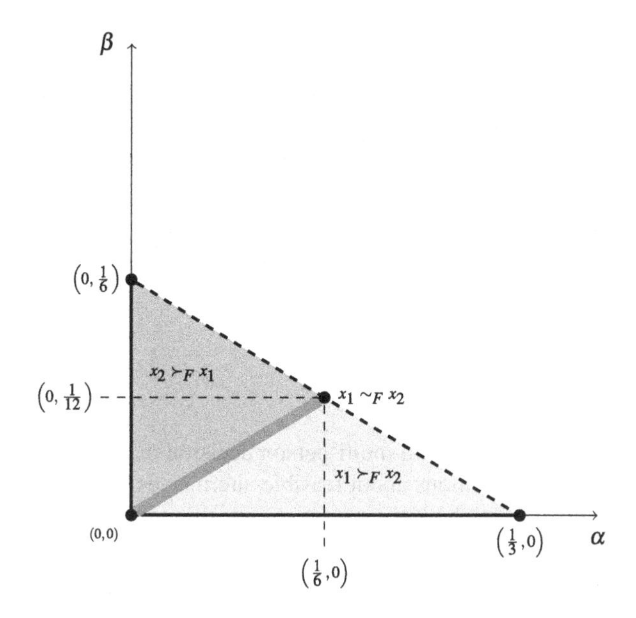

Fig. 7 Distribution of the rankings in Example 4

$$
\begin{aligned}
d(\boldsymbol{v}_1, l_5) &= d(v_1^1, l_5) + d(v_1^2, l_5) + d(v_1^3, l_5) + d(v_1^4, l_5) \\
&= (4 + 4\alpha + 6\beta) + (4 + 4\alpha + 6\beta) + 4 + 4 = 16 + 8\alpha + 12\beta, \\
d(\boldsymbol{v}_2, l_5) &= d(v_2^1, l_5) + d(v_2^2, l_5) + d(v_2^3, l_5) + d(v_2^4, l_5) \\
&= (5 + 3\alpha + 3\beta) + (3 + 3\alpha + 3\beta) + (6 + 2\alpha + \beta) + (2 + 2\alpha + \beta) \\
&= 16 + 10\alpha + 8\beta.
\end{aligned}
$$

Since $16 + 8\alpha + 12\beta < 16 + 10\alpha + 8\beta \Leftrightarrow \alpha > 2\beta$, we have $x_1 \succ_F x_2 \Leftrightarrow \alpha > 2\beta$, $x_2 \succ_F x_1 \Leftrightarrow \alpha < 2\beta$, and $x_1 \sim_F x_2 \Leftrightarrow \alpha = 2\beta$. Consequently, depending how imprecision is penalized, x_1 and x_2 are ordered in a different way. See Fig. 7.

Example 5 Consider again two alternatives x_1 and x_2 assessed by three voters through the set of linguistic terms $L = \{l_1, l_2, l_3, l_4, l_5\}$ with the meanings given in Table 1. The assessments are provided in Table 4.

Using the metric d associated with (α, β), with $\alpha, \beta \geq 0$, we obtain

Table 4 Assessments in Example 5

Alternative	Voter 1	Voter 2	Voter 3
x_1	l_3	$[l_2, l_3]$	$[l_1, l_4]$
x_2	l_4	$[l_1, l_2]$	$[l_1, l_4]$

$$\begin{aligned} d(v_1, l_5) &= d(v_1^1, l_5) + d(v_1^2, l_5) + d(v_1^3, l_5) \\ &= 4 + (5 + \alpha) + (5 + 3\alpha + 3\beta) = 14 + 4\alpha + 3\beta, \\ d(v_2, l_5) &= d(v_2^1, l_5) + d(v_2^2, l_5) + d(v_2^3, l_5) \\ &= 2 + (7 + \alpha) + (5 + 3\alpha + 3\beta) = 14 + 4\alpha + 3\beta. \end{aligned}$$

Despite of the values of α and β we choose, the result is the same for the two alternatives. Thus, $x_1 \sim_F x_2$ for all possible values of the parameters.

7 Concluding Remarks

In this paper, we have introduced a multi-person decision making procedure where agents may express their opinions about feasible alternatives by means of linguistic terms, if they are confident about their opinions, or through linguistic expressions composed by consecutive linguistic terms, in the case they are not confident about their opinions. The proposal allows to penalize the imprecision by means of two parameters.

As further research, it would be interesting to consider some breaking-tie processes, to analyze additional properties and advantages of the proposed method, and to apply the introduced multi-person decision making procedure to some real problems.

Acknowledgments The authors are grateful to Jorge Alcalde-Unzu and Ilan Fischer for their suggestions. The financial support of the Spanish Ministerio de Ciencia e Innovación (projects ECO2009-07332, ECO2009-12836, ECO2008-03204-E/ECON, TIN2010-20966-C02-01 and TIN2010-20966-C02-02, SENSORIAL Research Project TIN2010-20966-C02-01 and TIN2010-20966-C02-02), the Spanish Ministerio de Economía y Competitividad (project ECO2012-32178), and ERDF are also acknowledged.

References

1. Balinski, M., Laraki, R.: A theory of measuring, electing and ranking. Proc. Nat. Academy Sci. U.S.A. **104**, 8720–8725 (2007)
2. Balinski, M., Laraki, R.: Majority Judgment. Measuring, Ranking, and Electing. The MIT Press, Cambridge, (2011)
3. Balinski, M., Laraki, R.: Election by Majority Judgement: Experimental evidence. In: Dolez, B, Grofman, B, Laurent, A.(eds.), In Situ and Laboratory Experiments on Electoral Law Reform: French Presidential Elections. Studies in Public Choice vol. **25**, pp. 13–54. Springer, New York (2011)
4. de Borda, J.C, Mémorie sur les élections au scrutin, Historie de l'Academie Royale des Sciences, Paris, (1781)
5. Brams, S.J., Fishburn, P.C.: Approval Voting. Am. Political Sci. Rev. **72**, 831–847 (1978)
6. Brams, S.J., Fishburn, P.C.: Approval Voting. Birkhäuser, Boston (1983)
7. Falcó, E., García-Lapresta, J.L.: A distance-based extension of the majority judgement voting system. Acta Universitatis Matthiae Belii, series Mathematics **18**, 17–27 (2011)

8. Falcó, E., García-Lapresta, J.L.: Aggregating individual assessments in a finite scale. World Conference on Soft Computing, San Francisco (2011)
9. Felsenthal, D.S., Machover, M.: The Majority Judgement voting procedure: a critical evaluation. Homo Oeconomicus **25**, 319–334 (2008)
10. García-Lapresta, J.L.: A general class of simple majority decision rules based on linguistic opinions. Inf. Sci. **176**, 352–365 (2006)
11. García-Lapresta, J.L., Llamazares, B., Martínez-Panero, M.: A Social Choice analysis of the Borda rule in a general linguistic framework. Inte. J. Comput. Intell. Syst. **3**, 501–513 (2010)
12. García-Lapresta, J.L., Martínez-Panero, M.: Linguistic-based voting through centered OWA operators. Fuzzy Optim. Decis. Mak. **8**, 381–393 (2009)
13. García-Lapresta, J.L., Martínez-Panero, M., Meneses, L.C.: Defining the Borda count in a linguistic decision making context. Inf. Sci. **179**, 2309–2316 (2009)
14. Herrera, F., Martínez, L.: A 2-tuple fuzzy linguistic representation model for computing with words. IEEE Trans. Fuzzy Syst. **8**, 746–752 (2000)
15. Kacprzyk, J., Zadrożny, S.: Computing with words in decision making: through individual and collective linguistic choice rules. Int. J. Uncertain, Fuzziness Knowl. Based Syst. 9 (Suppl.), 89–102 (2001)
16. Ma, J., Ruan, D., Xu, Y., Zhang, G.: A fuzzy-set approach to treat determinacy and consistency of linguistic terms in multi-criteria decision making. Int. J. Approximate Reasoning **44**(2), 165–181 (2007)
17. Nurmi, H.: Voting Theory. In: Ríos-Insua, D, French, S, (eds.) e-Democracy. A Group Decision and Negotiation Perspective. Advances in Group Decision and Negotiation, vol. 5, pp. 101–124. Springer, Berlin, (2010)
18. Rodríguez, R., Martínez, L., Herrera, F.: Hesitant fuzzy linguistic terms sets for decision making. IEEE Trans. Fuzzy Syst. **20**(1), 109–119 (2012)
19. Roselló, L., Prats, F., Agell, N., Sánchez, M.: Measuring consensus in group decisions by means of qualitative reasoning. Int. J. Approx. Reason. **51**, 441–452 (2010)
20. Roselló, L., Prats, F., Agell, N., Sánchez, M.: A qualitative reasoning approach to measure consensus. In: Herrera-Viedma, E., García-Lapresta, J.L., Kacprzyk, J., Numi, H., Fedrizzi, M., Zadròzny, S. (eds.) Consensual Proceses, STUDFUZZ, vol. 267, pp. 235–261. Springer, Berlin (2011)
21. Roselló, L., Prats, F., Agell, N., Sánchez, M., Mazaira, F.A.: Using consensus and between generalized multiattribute linguistic assessments for group decision-making. Inform. Fusion (in press). doi:10.1016/j.inffus.2011.09.001
22. Smith, W.D., On Balinski and Laraki's Majority Judgement median-based range-like voting scheme. http://rangevoting.org/MedianVrange.html (2007)
23. Tang, Y., Zheng, J.: Linguistic modelling based on semantic similarity relation among linguistic labels. Fuzzy Sets Syst. **157**, 1662–1673 (2006)
24. Travé-Massuyès, L., Dague, P. (eds.): Modèles et Raisonnements Qualitatifs. Hermes Science, Paris (2003)
25. Travé-Massuyès, L., Piera, N.: The orders of magnitude models as qualitative algebras. Proceedings of the 11th International Joint Conference on Artificial Intelligence, Detroit, (1989)
26. Wallsten, T.S., Budescu, D.V., Rapoport, A., Zwick, R., Forsyth, B.: Measuring the vague meanings of probability terms. J. Exp. Psychol. **115**, 348–365 (1986)
27. Weber, A.: Über den Standort der Industrien. Erster Teil: Reine Theorie des Standorts. Verlag JCB Mohr Tübingen (1909)
28. Yager, R.R.: Centered OWA operators. Soft. Comput. **11**, 631–639 (2007)
29. Zadeh, L.A.: Fuzzy logic = computing with words. IEEE Trans. Fuzzy Syst. **4**, 103–111 (1996)
30. Zahid, M.A.: A new framework for elections. Ph.D. Dissertation, Tilburg (2012)
31. Zimmer, A.C.: Verbal vs. numerical processing of subjective probabilities. In: Scholz, R.W. (ed.) Decision Making Under Uncertainty. pp. 159–182. North-Holland, Amsterdam, (1983)

Risk Perception and Ambiguity in a Quantile Cumulative Prospect Theory

Marcello Basili

Abstract This chapter introduces a version of Cumulative Prospect Theory in a quantile utility model with multiple priors on possible events as proposed in [8]. The chapter analyzes the decision-maker's risk and ambiguity perception facing ordinary and exterme events. It is showed a new functional that models asymmetric attitude with respect to ambiguity on extreme events (optimism respects windfall gains and pessimism respects catastrophic events) and the decision-maker's attitude to consider maximization of entropy as a rule of inference. Finally, it is defined a simplified approach based on the epsilon contamination method of a probability distribution.

Keywords Ambiguity · Multiple priors · Quantiles · Entropy · Extreme events

1 Introduction

This chapter regards some recent development in Cumulative Prospect Theory, in particular how it is possible to consider the notion of distorted probabilities and extreme events without any assumption about the shape or properties of the decision-maker's utility function e.g., [32, 63]. This chapter is mainly focused on [8] who introduced

[1] In standard Quantile Utility Models [17, 53, 61] the decision-maker orders feasible alternatives with respect to the highest γth -*quantile* of the induced cumulative probability distribution over outcomes (if $\gamma = 0$ or $\gamma = 1$ the standard maxmin or maxmax decisional rule is obtained). This single statistic decisional rule produces very large classes of indifference because of a lottery is evaluated by the fixed γth -*quantile* and it is irrelevant what happens in the probability distribution outside it.

[2] Evidences are in [35].

M. Basili (✉)
Department of Economics and Statistics, University of Siena, Siena, Italy
e-mail: basili@unisi.it

P. Guo and W. Pedrycz (eds.), *Human-Centric Decision-Making Models for Social Sciences*, Studies in Computational Intelligence 502,
DOI: 10.1007/978-3-642-39307-5_6,

a new version of decision making rule under uncertainty based on quantiles of the probability distributions over consequences. Basili and Chateauneuf simply assumed an ambiguity averse decision-maker and used two quantiles that defines an interval of ordinary events (familiar), because they are considered more reliable and closer to her experienced life, and two tails where are included extreme events, or events with very small probabilities of occurring but very large consequences (windfall gains and catastrophic losses).[1] The new representation takes into account fat-tailed events, usually misvalued in standard approaches, due to cognitive insensitivity to small probability outcomes and also includes competence effect.[2]

Section 2 shows that axiomatization and characterization of an agent beliefs by a capacity made possible to represent her partial knowledge about future, indeed formal representation of decision making problems under ambiguity as opposed to risk. Capacities allowed to encode beliefs in subjective probability distributions when the decision-maker has *multiple priors*, *fuzzy prior* and *non-additive prior*.

Section 3 introduces a new decision rule under uncertainty based on entropy maximization. CPT is generalized by introducing a quantile representation with multiple prior on possible events. The set of priors reflect the decision-maker's assessment of the reliability of available information about the underlying uncertainty.

Section 4 exihibits the new functional form to evaluate prospects. Facing a risky situation the decision-maker adopts diversification, an intuitive and consistent strategy for reducing likely loss. Since entropy reflects the diversification degree of a portfolio of choices, the decision-maker applies the Maximum Entropy Principle to elicit that probability distribution. Because of tractability and theoretical soundness, the value functional of a prospect is defined by the Choquet integral of an appropriate quantile function.

Section 5 provides derivation of the new functional form when the decision-maker's ambiguity attitude is characterized through the ϵ-*contamination of confidence*. Cautiousness in her belief leads the decision-maker to elicit the closest to uniformity probability distribution in her information set, but lack of confidence in her opinion and awarness of a possible error forces her to consider an ϵ-*contamination* of the elicited probability distribution. Finally, it is considered a simpler approach where the decision-maker only considers her most credible probability distribution, even if not fully reliable, and the parameter ϵ is the value that captures error in her assessment.

2 Related Literature

In a celebrated paper Camerer observes "because economics is the science of how resources are allocated by individuals and by collective institutions like firms and markets, the psychology of individual behavior should underlie and inform economics, much as physics informs chemistry; archaeology informs anthropology; or neuroscience informs cognitive psychology. However, economists routinely—and proudly—use models that are grossly inconsistent with findings from psychology.

A recent approach, behavioral economics, seeks to use psychology to inform economics, while maintaining the emphases on mathematical structure and explanation of field data that distinguish economics from other social sciences. In fact, behavioral economics represents a reunification of psychology and economics, rather than a brand new synthesis, because early thinking about economics was shot through with psychological insight" [16].

The aspiration of behavioral economics is to increase the explanatory power of economic theories by deepening the psychological foundation of economic behavior through a greater realism that explains anomalies and counterexamples observed in financial and real markets. In this perspective a larger interaction between behavioral economics, experimental economics, and decision theory has been seen in the last decades. Empirical literature appears to confirm that gender, age, education and income may introduce new fascinating and intriguing explanations of the main anomalies of markets. Concurrently, psychology puts in evidence that specific individual characters such as the perceived control, affective and emotional factors may influence individual attitude towards of alternative choices [60, 70, 71].

As a matter of fact all previous considerations apply to individuals' risk perception since it involves judgment and choice. In fact, risk perception is how individuals understand and comprehend uncertain events. Even if risk has manifold different meanings that involve characterization of implied events: hazard, threat, dread, controllability, vulnerability, fatality, etc., there is a general consensus about the definition of risk as probability of occurrence and magnitude of consequences.

It is straightforward to note that risk perception that not only involves likelihood of probability and appraisement of consequences but also time consideration on uncertain prospects. Nonetheless, risk perception is a specific concept distinguished from expected return and attractiveness[3] [3, 26, 31]. Resting on subjective probability distributions about feasible future events, risk perception appears to be a subjective notion that encompasses interpretation (awareness, knowledge, information, familiarity, dread, etc.), and beliefs (chance) about possible consequences of a feasible option in an opportunity set. Crucially "perceptions of risk play a prominent role in the decisions people make, in the sense that differences in risk perception lie at the heart of disagreements about the best course of action between technical experts and members of the general public...Both individual and group differences in preference for risky decision alternatives and situational differences in risk preference have been shown to be associated with differences in perceptions of the relative risk of choice options, rather than with differences in attitude towards (perceived) risk, i.e., a tendency to approach or to avoid options perceived as riskier" [72].

It is possible to identify at least three different areas of research on people preferences in regard to perceived risk that emerged in facing the inadequacy of the dominant cognitive paradigm in experimental studies and behavioral analyses of the 1970s. The first approach, the Cultural Theory or Socio-cultural Paradigm originated in sociology and anthropology. The second approach is formulated in psychology,

---

[3] Expected return is concerned with evaluation rule; attractiveness involves evaluation and attitude toward risk.

it is known as the Psychometric Paradigm. The third approach is introduced in the decision theory, it concentrates on the relationship between lack of knowledge about future occurrence or systems, probability elicitation and optimal decision rules, it is known as the Axiomatic Measurement Paradigm.[4]

In cultural theory risk perception is informed by social and cultural determinants. Cultural theory is based on the group-grid analysis, a two dimensional scheme, that represents the degree of social activity and how the social context governs the individual behavior. Cultural theory claims a correspondence between individuals and particular ways of life and in the group-grid scheme and correspondence originates in five types of people: egalitarians, hierarchists, individualists, fatalists and hermits that can be used to"predict and explain what kinds of people will perceive which potential hazards to be how dangerous" [81, p. 42].[5]

On the contrary, the psychometric paradigm and the axiomatic measurement paradigm are based on the assumption that risk is a subjective notion, even if they context the main assumption of economic risky choice model, indeed that risk attitude, or curvature of the individual utility function, only determines individual risk-taking. In standard economic theory risk attitude is measured by the Arrow-Pratt absolute and relative risk aversion index [4, 57], but experimental psychology and behavioral economics put in evidence inconsistent behaviors under imperfect knowledge and complex decision making processes.

In the 1970s cognitive psychologists [27, 28, 70] define a research agenda to answer some crucial questions about: determinants of perceived risk, different risk perception, role of information and judgment in risk assessment, perception of benefit and acceptability of risk. Psychologists observed that the subjective nature of risk makes at least qualitative factors (ethical values, priorities, emotions, voluntariness of threat exposure, history, ideology, etc.) as important as quantitative characteristics in risk assessment, risk management and in evaluation of a risky decision [29]. Psychologists develop psychometric risk scales that assess individual risk attitude and behavior in dependence of how people process information, since risk perception is influenced by mental processes that govern information elaboration, indeed the associative system and the analytic process.[6] Combining different co-operating mental processes, psychometric paradigm gives a scheme to understand individuals' behavior.[7] Because of almost invariant structure and intuitive appeal of factors that affect risk perception, the psychometric paradigm obtained an enormous success, even if some critics point out that its explanatory power mainly derives from including dread among the relevant variables and misleading data analysis using means [68].

---

[4] A comprehensive review of the three paradigms is in [79].

[5] Since it is assumed that risk is a strictly subjective and psychologically determined notion, cultural theory is not considered in this review. A critical evaluation of cultural theory is in [56].

[6] Details are in [80].

[7] The psychometric paradigm is a taxonomic scheme that produces results that are usually summarized in a two(three)-dimensional space derived by factor analysis from the intercorrelation of given risk characteristics, where each factor is made up of a combination of characteristics, e.g., dread-controllability or known-observable [69, 72].

Decision theorists assume that ambiguity attitude emerges when individuals face vague and incomplete information. Ambiguity attitude refers to beliefs and preferences on feasible actions. Ambiguity influences perception of uncertain actions and induces human beings to elicit probabilities and apply decision rules that violate axioms of the rationality paradigm, based on the Bayesian approach, of the standard economic theory. Decision theory under uncertainty rests on [62] subjective expected utility theory, even if there exists a representation of uncertainty based on explicit probability approach, so that the objects of individuals' choices consisted of lottery prizes, with given objective probability distributions over outcomes [76]. Savage's approach centers around two fundamental assumptions. First, a complete list of possible future states of the world is available to the individual—a list that, in an interpersonal context, is common knowledge to all individuals. The individual is endowed with subjective beliefs over the state space (uncertain prospect). These beliefs are represented by a well-defined (additive) probability function. As a result, individuals in uncertain settings are supposed to be able to undertake expected cost/benefit analysis in information gathering, and hence to reach an informational optimum. In particular, in the tradition of the choice theoretic approach to subjective probability developed by Ramsey and De Finetti, Savage put forward an axiomatic framework in which the subjective additive probabilities of an individual are elicited from choices and satisfy the requirement of consistency (Dutch Book or Arbitrage in Gambling). The second fundamental assumption of probabilistic decision making has to do with the cognitive capabilities of the decision-maker. The processing of information consists of a Bayesian process of updating individual beliefs (prior probability distribution), when a signal is received on the realization of the state (Bayes's rule).

These assumptions follow from the implicit hypothesis that individuals are rational in a strong sense, namely that they have a complete knowledge of all possible states of the world and can manage to deduce all logical propositions contained in the axioms of the theory. Both assumptions have been questioned and abandoned in: bounded rationality models [66, 67] and non-expected utility models [33, 39, 59, 63, 84]. In the approach based on bounded rationality, decision-makers replace utility maximization by violating exponential discounting [50]. In non-expected utility models economic agents have distorted probabilities, contractions or expansions of prior linear probabilities, that are capable of accommodating individuals' perception of probabilities through weighting functions.

2.1 Behavioral Foundation of Risk Perception

The behavioral foundation of risk perception is in alternative representations of individual beliefs and rules for choice that were paramount in decision theory since the time of Savage's synthesis. From a theoretical point of view, Shackle [65] strongly objected to the use of probability functions on the grounds that the list of possible states of the world conditioning crucial entrepreneurial choices cannot be assumed as given. Decision theorists and statisticians like [36, 37, 77] set criteria for decisions under complete or partial ignorance that are alternatives to expected utility maximization [52]. Moreover, since the inception of experimental economics it was evident that expected utility theory was descriptively inadequate [58]. In particular, Edwards [21, p. 49] argued that "a great deal of experimental evidence that bears on the additivity of subjective probability is now available and it argues against the additive property so strongly that I do not see how it is possible any longer to defend that property. Fortunately, it may be possible to develop a utility-subjective probability model that is mathematically satisfactory and that does not require subjective probabilities to add to one or anything else". As well known, Edwards's experimental evidence constitutes the main starting point of Kahneman and Tversky's studies about the use of decision weights to represent the way decision makers feel about probabilities within the framework of a descriptive model of decision making under uncertainty. After Kahneman and Tversky [39] it has become usual to utilize weighted functions to represent how decision makers overweigh low probabilities and underweigh high probabilities, a pattern of behavior observed under uncertainty [82].

Real reports[8] and experimental evidence[9] confirmed that under uncertainty, a decision maker not only considers "the reliability, credibility or adequacy of information, experience, advice, intuition taken as a whole: not about the relative support it may give to one hypothesis as opposed to another, but about its ability to lend support to any hypothesis—any set of definite options—at all" [23, p. 192], but also "relative willingness to rely upon it in [her] decision-making; and various factors enter [her] decision criterion in linear combination" [23, p. 193]. Moreover, there is a strict reciprocal influence between beliefs and consequences, that is consequences induce a particular distortion of beliefs on the basis of amount (catastrophic, windfall, ordinary) and/or sign (gain and loss) and beliefs (low or high) modify perception of consequences.

2.2 Prospect Theory

Prospect Theory emerges as a result of thought experiments that verify the assumption that individual perception is reference-dependent, that is "the perceived attributes of a focal stimulus reflect the contrast between that stimulus and a context of prior and concurrent stimuli....Intuitive evaluations of outcomes are also reference-dependent"

[8] It is well known the situation in which the President of the USA and the White House Staff had to decide about the development of nuclear weapons (InterContinental Ballistic Missiles—ICBMs) in the face of the menace of USSR in the 60's. In a condition of ambiguity, characterized by a set of probability distributions, none of which were fully reliable, the President assumed the worst scenario (full pessimism), that is the USSR had hundreds of ICBM, and launched the Minuteman missiles Program. In the fall of 1961, a revised highly secret report set that "the missile gap favouring the Soviets had been a fantasy. There was a gap, but it was currently ten to one in our favour. Our 40 Atlas and Titan ICBMs were matched by 4 Soviet SS-6 ICBMs at one launching site at Plesetsk" [24, p. 32]

[9] Wakker [78].

[38, p. 1455], Specifically, an individual seems to dislike a decrease in wealth with respect to her initial endowment more than she likes a gain: this finding has been expressed by a value function that is kinked at the reference point (typically the current endowment) and loss averse, indeed a value function that shows diminishing sensitivity away from the refererence point. At the same time individuals are under the infuence of the psychological impact of the absolute and relative value of probabilities. In fact they show a general attitude to overweight low probability events and underweight large probability events, but at the same time they are more sensitive to small changes from zero probability (possibility effect) or to one probability (certainty effect), than when the same changes occur at intermediate probabilities. Kahneman and Tversky [39] proposed in the seminal paper the inverse S-shaped value function, i.e., convex on losses and concave on gains to render asimmetric risk attitudes, and the probability weighting function to represent diminishing sensitivity, possibility effect and certainty effect.[10]

Kahnemann and Tversky [40] later rationalized decision making under uncertainty through the cumulative version of prospect theory, by combining the empirical realism of their original prospect theory with the theoretical tractability of non-additive measures. Cumulative Prospect Theory introduced the notion of distort probability distributions, obtained through a probability weighting function attached cumulatively to outcomes, to maximize the overall value of a prospect. Cumulative Prospect Theory solved the violation of stochastic dominance that occurred in Prospect Theory by connect sign-dependence with Rank-Dependent Expected Utility [59].[11]

In 1980 Thaler introduces prospect theory in economics to explain violation of consumer theory, and in the last decade a lot of papers have tried to accommodate new evidence drawn from financial market observations into larger models of market or multi-agent experiments, following intuitions of behavioral economics. In particular, prospect theory has been introduced to resolve financial puzzles, such as the equity premium puzzle [54], and reconcile financial theory with observations [9–11, 20, 30, 34].[12]

[10] In cumulative prospect theory the general attitude toward uncertainty consists of risk attitude and ambiguity attitude. Risk attitude is described by the slope of the weighting function. Ambiguity attitude is summarized by the parameters of the weighting function: δ (elevation) that represents dominant attitude (pessimism or optimism) and γ (curvature) that shows sensitivity to partial information [14, 24].

[11] Cumulative Prospect Theory is a special class of Rank-Dependent Expected Utility where the reference point originates negative and positive ranks.

[12] Kahneman and Tversky [39] is one of the most quoted paper (more than 26000 quotations) and Prospect Theory has been applied almost everywhere in Economics, Insurance Economics, Law and Economics, Managment Science etc.

2.3 Familiarity, Competence and Extreme Events

In order to cope with some observed phenomena, such as preference reversal or counterfactual thinking, the literature on decision making under uncertainty has recently questioned the critical assumption of the single reference point assumed in prospect theory and cumulative prospect theory. The preference reversal phenomenon has been explained in models of context-free preferences, in which independence and/or reduction axioms [41, 64], transitivity of preferences [51] or the double matching axiom [83] are violated. Counterfactual thoughts have been considered a key determinant of the individual behavior in post-decisional regret frameworks, where regret is evoked whenever an obtained outcome compares unfavourably with an outcome that the investor could have obtained she had chosen differently [13]. Crucially, preference reversal and counterfactual thinking have been explained in variants of prospect theory by introducing lotteries as reference points [43, 44, 73] or simply by comparing obtained outcomes with unchosing ones (non-investment or inaction outcome, the best-performing outcome, the worst-performing outcome) to assess the investor regret [49, 75].

Basili et al. [6] provided a peculiar characterization of the decision maker's behavior under uncertainty introducing a sort of familiarity bias. In fact psychometric studies have put in evidence that familiarity with a risk alters the individual perception of its riskiness, with the result that "an accident that takes many lives may produce relatively little social disturbance (beyond that caused to the victims' families and friends) if it occurs as part of a familiar and well-understood system (e.g., a train wreck). However, a small incident in an unfamiliar system (or one perceived as poorly understood), such as a nuclear waste repository or a recombinant DNA laboratory, may have immense social consequences if it is perceived as a harbinger of future and possibly catastrophic mishaps" [72, p. 13].[13] Basili et al. [6] presented a representation theorem for a decision-maker that is ambiguity averse towards very large losses and ambiguity seeker towards unusual gains, but is assumed to be ambiguity neutral with respect to a set of ordinary outcomes. Crucially, they introduced a reference set, that is, an interval of outcomes that the decision-maker feels more familiar with than other possible outcomes, instead of assuming the existence of a unique reference point.[14]

Basili et al. [7] introduced a characterization of the precautionary principle[15] for a decision making process under uncertainty with catastrophic losses and/or windfall gains. Unlike Basili et al. [6], where it is assumed that people act as in stan-

[13] Assumption that decision-makers are affected by familiarity bias is confirmed in empirical researches, such as de Lara Resende and Wu (2010) who find that "the decision weighting function parameters estimated for losses show an elevation parameter for less familiar domains higher than the elevation parameter for more familiar domain".

[14] Baillon and Cabantous [5, p. 135] put in evidence that even "identical (convex hulls of) possible priors can be treated differently by the same individual depending on the source of uncertainty".

[15] Precautionary Principle is considered the rational guide to policy making in situations characterized by scientific uncertainty, irreversibility, and catastrophic events.

dard cumulative prospect theory, they suppose that the decision maker's behavior is characterized by the opposite ambiguity attitude with respect to extremely negative events, indeed pessimism in place of optimism (since the latter behavior could induce dissipative choices instead of conservative ones) and optimism instead of pessimism towards windfall gains. This is justified by real life observations and recent experimental papers [1, 25, 48], that provide some evidence that contrasts the assumption of standard cumulative prospect theory. Their value functional provides a consistent notion of the precautionary principle that could be used to contrast disregarded prevention of disasters and sloppy response operations in emergencies. Indeed whenever individuals show some difficulty in perception of ambiguous extreme risks, such as in the case of Hurricane Katrina, the human version of mad cow disease (CJDv), the possible pandemic of the human avian flu (A-H5N1 or A-H1N1 virus) or consequences induced by global warming.

3 Entropy and Quantile Functions for the Representation Theorem

In 2011, Basili and Chateauneuf introduced a multiple quantile utility model of CPT and a representation theorem that models not only asymmetric attitude with respect to ambiguity on extreme events (optimism respects windfall gains and pessimism respects catastrophic events), but also the decision-maker attitude to consider maximization of entropy as a rule of inference when information is ambiguous and scanty. Maximum entropy probability, which is a measure of conflict of evidence, is a measure of the diversification degree and a rational form of prudence. The maximum entropy principle is a general method to choose a probability distribution under uncertainty, indeed the method that elicits the most unbiased-uniform distribution among all the possible ones.

The maximum entropy principle was introduced by Jaynes [46, 47] in physics as a generalization of the classical Principle of Insufficient Reason of Laplace and it is a general method to choose a probability distribution under uncertainty, indeed the method that elicits the most unbiased-uniform distribution among all the possible ones. Basili and Chateauneuf obtained an inverse cumulative function, that is less peaked, through maximization of entropy, in intermediate quantiles, but has the fattest tails, simce it maximizes the minimum expected loss and the maximum expected gain. Crucially this result is coherent with evidence of Tsallis-Renyi entropy (*non-extensive entropy measure*) applications in economics, where the individual attitude with respect to ordinary and extreme events is considered within the context of non-extensive statistical mechanics.[16]

[16] Non-extensive entropy not only permits to represent a different behavior of the decision-maker with respect to ordinary and extreme events, but also to take into account the dependence between perception of ambiguity, competence and reliability of information [74].

Interesting enough, they introduced an alternative method to define the composite value function of a prospect that is suitable for using the class of ϵ-*contaminated* probabilities, where the parameter $\epsilon \in [0, 1]$ indicates the probability deviation from the prior, a basic methodology in Bayesian robustness analysis.

4 Determination of the Composed Inverse Cumulative Function

For simplicity suppose that S is a finite state space, $\mathcal{A}$ is a σ-*algebra* of events (e.g., the power set 2^S) and P is a set of probability distributions on $(S, \mathcal{A})$. A measure υ is a capacity on $(S, \mathcal{A})$ if $\upsilon : A \in \mathcal{A} \to \upsilon(A) \in \mathbb{R}$, where $\upsilon(\emptyset) = 0$, $\upsilon(S) = 1$ and $(A, B) \in \mathcal{A}^2$ such that $A \subseteq B \Longrightarrow \upsilon(A) \leq \upsilon(B)$. A capacity υ is convex if $\upsilon(A \cup B) + \upsilon(A \cap B) \geq \upsilon(A) + \upsilon(B)$, $\forall A, B \in \mathcal{A}$. The dual capacity[17] $\overline{\upsilon}$ of a capacity υ is defined by $\overline{\upsilon}(A) = 1 - \upsilon(A^C)$ $\forall A \in \mathcal{A}$, where $A^C = S \setminus A$ is the complement of the set A. It is considered a decision-maker facing uncertainty, where uncertainty is modeled through the *core* of a convex capacity υ, i.e., through the set $C(\upsilon)$ of probability distributions P on $(S, \mathcal{A})$ above υ, or $P(A) \geq \upsilon(A)$ $\forall A \in \mathcal{A}$.

For $X : S \to \mathbb{R}$, the cumulative distribution function F_X of X is defined by $x \in \mathbb{R} \to F_X(x) = P(X \leq x)$. A common pseudo-inverse of F_X denoted the quantile-function F_X^{-1} is defined by $p \in [0, 1] \to F_X^{-1}(p) = Inf\left\{x \in \overline{\mathbb{R}}, F_X(x) \geq p\right\}$, such that $F_X^{-1}(0) = -\infty$, $F_X^{-1}(1) = \underset{s \in S}{Max} X(s)$ and F_X^{-1} is non-decreasing and left-continuous.

For $X \in \mathbb{R}^S$ and υ a capacity on $\mathcal{A}$, the Choquet integral[18] of X w.r.t. υ denoted $\int X d\upsilon$ is defined by $\int X d\upsilon = \int_{-\infty}^{0} (\upsilon(X \geq t) - 1)dt + \int_0^{+\infty} \upsilon(X \geq t)dt$.

The cumulative distribution F_X^υ of X with respect to capacity υ is defined by $x \in \mathbb{R} \to F_X^\upsilon(x) = \overline{\upsilon}(X \leq x)$ and it is possible to define the quantile function $F_X^{\upsilon^{-1}}$ by $p \in [0, 1] \to F_X^{\upsilon^{-1}}(p) = Inf\left\{x \in \overline{\mathbb{R}}, F_X^\upsilon(x) \geq p\right\}$.

The the quantile function $F_X^{\upsilon^{-1}}(0) = -\infty$, $F_X^{\upsilon^{-1}}(1) = \underset{s \in S}{Max} X(s)$, $F_X^{\upsilon^{-1}}$ is non-decreasing, left-continuous and completely defined by its values on $(0, 1)$. It follows that:

Theorem 1 $\forall X \in \mathbb{R}^S : \displaystyle\int X d\upsilon = \int_0^1 F_X^{\upsilon^{-1}}(p)dp$.

Theorem 1 expresses the Choquet integral as an integral of a quantile function and resultant composed inverse cumulative function is capable of describing pessimism and optimism on extreme outcomes and ambiguity neutrality on ordinary outcomes.[19] In fact, given an act $X : S \to \mathbb{R}$ and $(\alpha, \beta) \in [0, 1]^2$, $\alpha \leq \beta$, such that $[\alpha, \beta]$ determines the interval of cumulative probability between which outcomes

[17] Note that if a capacity υ is covex its dual capacity $\overline{\upsilon}$ is concave.

[18] For $X \in \mathbb{R}^S$, $E_P(X) = \int_0^1 F_X^{-1}(p)dp$ is the mathematical expectation of X w.r.t. P.

[19] Proof is in [8, p. 1098].

can be considered as ordinary, the decision-maker values outcomes between these two quantiles in an ambiguity neutral way by probability distribution π. In this way she models pessimism in the lower tail $[0, \alpha]$, i.e., she models the attitude of the DM who minimizes the expectation of X on this quantile with respect to all $P = C(\upsilon)$, and symmetrically optimism in the upper tail $[\beta, 1]$, through maximization of the expectation of X in this quantile with respect to all $P = C(\upsilon)$.

The decision-maker defines the lower tail and the upper tail, by choosing $\alpha, \beta \in [0, 1]$, where $\alpha \leq \beta$, and then calculates the value of $X \in \mathbb{R}^S$ through $I(X) = I_1(X) + I_2(X) + I_3(X)$, where:

$I_1(X) = \int_0^\alpha F_X^{\upsilon^{-1}}(p)dp$; $I_2(X) = \int_\alpha^\beta F_X^{\pi^{-1}}(p)dp$ and $I_3(X) = \int_\beta^1 F_X^{\overline{\upsilon}^{-1}}(p)dp$.

Since[20] I satisfies monotonicity or $X \geq Y \Longrightarrow I(X) \geq I(Y)$, constant additivity or $I(X + a \cdot S^*) = I(X) + a \; \forall a \in \mathbb{R}$ and positive homogeneity or $I(aX) = aI(X)$ $\forall a \geq 0$ and $\forall X$, it is possible to characterize the pessimistic attitude of the DM with respect to outcomes in the lower tail and her optimistic attitude with respect to outcomes in the upper tail.[21]

Because of ambiguous information, the decision-maker solves the problem of assigning values to probabilities distributions in $C(\upsilon)$ by applying the *Maximum Entropy Principle* and focusing on the probability distribution π in $C(\upsilon)$. The probability distribution π is the '*less concentrated*' in $P = C(\upsilon)$.

Determination of the probability distribution $\pi \in core(\upsilon)$ that maximizes entropy is a crucial problem in the model. In fact the probability distribution π is the pivotal distribution (ambiguity neutrality) for the definition of the decision-maker's pessimism and optimism Basili and Chateauneuf [8] found that an efficient algorithm to obtain π, that is the less diffuse probability distribution in the core of a capacity, was proposed by [45]. Jaffray set that the natural candidate was the probability distribution closer to uniformity "in particular the smallest elementary probability should be as big as possible". To elicit $\pi \in C(\upsilon)$ consistent with the convex capacity υ, Jaffray considered the dual capacity $\overline{\upsilon}$. Assumed $A_0 = \varnothing$, Jaffray defined a family of disjoint non-empty subsets A_k of S, where $k \geq 1$, such that: $A_k \subseteq (A_0 \cup ... \cup A_{k-1})^c$ and $\frac{\overline{\upsilon}(A_0 \cup ... \cup A_{k-1} \cup A_k) - \overline{\upsilon}(A_0 \cup ... \cup A_{k-1})}{|A_k|} =$

$$= Min \left\{ \begin{array}{c} \frac{\overline{\upsilon}(A_0 \cup ... \cup A_{k-1} \cup E) - \overline{\upsilon}(A_0 \cup ... \cup A_{k-1})}{|E|}, \\ \varnothing \neq E \subset (A_0 \cup ... \cup A_{k-1})^c \end{array} \right\}.$$

Setting $\alpha_k = \frac{\overline{\upsilon}(A_0 \cup ... \cup A_k) - \overline{\upsilon}(A_0 \cup ... \cup A_{k-1})}{|A_k|}$, Jaffray showed that the maximum entropy probability π in $C(\upsilon)$ is clearly specified by $\pi(\{s\}) = \alpha_k$ when s belongs to A_k for $k \geq 1$.

Once the unique probability π has been selected, it is straightforward to define the DM pessimism and optimism with respect to outcomes in the tails.

---

[20] Proof and details in [8, p. 1099]

[21] Cardin [15] introduces a class of aggregation functional based on a multiple quantile model that generalizes the approach established in [8].

Definition 2 *The decision-maker is pessimistic with respect to the lower tail if* $I_1(X) \leq \int_0^{\alpha} F_X^{\pi^{-1}}(p)dp$ *(overestimated losses with respect to the reference probability* π*).*

Definition 3 *The decision-maker is optimistic with respect to the upper tail if* $I_3(X) \geq \int_{\beta}^{1} F_X^{\pi^{-1}}(p)dp$ *(overestimated gains with respect to the reference probability* π*).*

Proposition 4 *The decision-maker is pessimistic in the lower tail and optimistic in the upper tail.*[22]

5 Entropy and ϵ-Contamination

Ellsberg [22] suggested for the solution of his famous Paradox, about two urns containing red and black balls, the functional form $\left[\rho q^0 + (1-\rho)q_X^{\min}\right](X)$, where q^0 is the estimated probability vector and $q_X^{\min}$ the probability vector in Δ, such that "in the case of the red, yellow and black balls, supposing no samples and no explicit information except that $1/3$ of the balls are red" [22, p. 665], corresponding to min, for the act X. The set of probability distributions Δ on the act is exogenous and the parameter $\rho \in [0, 1]$ reflects the subjective "degree of confidence, in a given state of information or ambiguity, in the estimated distribution q^0, which in turn reflects all of his judgments on the relative likelihood of distributions, including judgments of equal likelihood" [22, p. 664].

In robust Bayesian analyses the class Γ of prior distributions is an *ϵ-contaminated class* where, given an initial believable prior q_0, the parameter $\epsilon \in [0, 1]$ is the amount of error that is deemed possible for that prior. For $A \subset S$, if the decision-maker is not certain about sureness (reliability) of q_0, she considers a perturbation (contamination) of q_0 through some probability distribution q in the set of all priors with weight ϵ, that is $\Gamma = \{q : q = (1-\epsilon)q_0 + \epsilon q, q \in Q\}$. "Stated another way, further reflection might to alterations of probability judgements by an amount ϵ. Hence, possible priors involving such alterations should be included in Γ" [12, p. 462] and the class Γ of beliefs is only a special class of capacity (a particular case of Dempster-Shafer belief functions).

Previous considerations induce quite straightforward to represent ambiguity through *ϵ-contamination*, when the decision-maker faces multiple priors [19, 42, 55]. In this perspective Basili and Chateauneuf [8] considered the capacity υ obtained by the *ϵ-contamination* of the probability P_0, such that $\upsilon(A) = (1-\epsilon)P_0(A)$, $\forall A \neq S$.[23] They showed that for the particular convex case of *ϵ-contamination* of a given probability P_0, Jaffray's algorithm is particularly efficient [8, p. 1100]:

[22] Proof is in [8, p. 1099]
[23] If $A = S$ then $\upsilon(A) = 1$.

Theorem 5 *Let υ be the ϵ-contamination of P_0, and let π denote the maximum entropy probability in $C(\upsilon)$. If $\epsilon = 0$, $\pi = P_0$; if $\epsilon = 1$, π is the uniform distribution. Finally if $\epsilon \in (0, 1)$, π is defined by: $\pi(\{s\}) = \frac{\overline{\upsilon}(A_1)}{|A_1|}$, for any $s \in A_1$, where $\frac{\overline{\upsilon}(A_1)}{|A_1|} = Min\left\{\frac{\overline{\upsilon}(E)}{|E|}, \varnothing \neq E \subseteq S\right\}$ (1), if $A_1 = S$, π is therefore the uniform distribution; if $A_1^c \neq \varnothing$ then π is furthermore defined by $\pi(\{s\}) = (1-\epsilon)P_0(\{s\})$, $\forall s \in A_1^c$ (2). Note that $\frac{\overline{\upsilon}(E)}{|E|} = \frac{\epsilon + (1-\epsilon)P_0(E)}{|E|}$ for any $\varnothing \neq E \subseteq S$.*[24]

5.1 A Simplified Model

Jaffray's algorithm is a general procedure to identified the maximum entropy probability distribution in $C(\upsilon)$, but it is not immediate. If the decision-maker has some competence or confidence about consequence induced by a given act, she could attach a different reliability to estimation of events in lower tail and upper tail when information is vague or incomplete.[25] This scenario can be represented through an asymmetric ambiguity attitude respect extreme events and synthethized in more than one ϵ-*contamination* of the given probability, inducing in such a way a sort of skewness in the composite inverse cumulative function. Basili and Chateauneuf [8] set the following simple procedure. Given an enough reliable and credible probability measure $P_0 \in C(\upsilon)$ and *asymmetric* pessimism and optimism on extreme events, summarized by ϵ_1 (degree of pessimism) on the lower tail and ϵ_2 (degree of optimism) on the upper tail, such that $\epsilon_1 \neq \epsilon_2$ and $\epsilon_1 + \epsilon_2 \leq 1$. Denoted υ_1 the ϵ_1-*contamination* of P_0 and υ_2 the ϵ_2-*contamination* of P_0, the value of the act X is defined as:

$I(X) = I_1(X) + I_2(X) + I_3(X)$, where

$I_1(X) = \int_0^\alpha F_X^{\upsilon_1^{-1}}(p)dp$; $I_2(X) = \int_\alpha^\beta F_X^{P_0^{-1}}(p)dp$ and $I_3(X) = \int_\beta^1 F_X^{\overline{\upsilon_2}^{-1}}(p)dp$.

This new functional satisfies monotonicity, constant additivity and positive homogeneity. The new procedure allows the expression of simultaneous different degrees of ambiguity attitude expressed by more than one contamination. In the new approach $(1 - \epsilon_1 - \epsilon_2)$ captures the notion of reliability of the chosen probability distribution P_0 whereas ϵ_1 and ϵ_2 measure the error in the elicitation of P_0 in $C(\upsilon)$, weighted by the asymmetric confidence with respect to reliability of extreme outcomes.[26] It is worth noticing that assuming $\epsilon_1 \geq \epsilon_2$ it is possible to obatin a more conservative version of the precautionary principle that always combines conservative and dissipative behavior, given the reliability of the probability distribution P_0, and overcomes the failure of the full conservative measure, e.g., maxmin decisional rule.

---

[24] Proof is in [8, p. 1100].

[25] As an example consider the collection of probabilistic opinions given by experts and the DM is relatively less optimistic on upper tail events.

[26] This result, that arranges a mixture of pessimism and optimism, is coherent with neo-additive capacities modelled by Chambers [17].

References

1. Abdellaoui, M., Bleichrodt, H., Paraschiv, C.: Loss aversion under prospect theory: a parameter-free measurement. Manage. Sci. **53**, 1659–1674 (2007)
2. Abdellaoui, M., L'Haridon, O., Zank, H.: Separating curvature and elevation: a parametric probability weighting function. J. Risk Uncertain. **41**, 39–65 (2010)
3. Alhakami, A., Slovic, P.: A psychological study of the inverse relationship between perceived risk and perceived benefit. Risk Anal. **14**, 1085–1096 (1994)
4. Arrow, K.: Aspects of the Theory of Risk-Bearing. Yrjo Jahnsson Lecture, Helsinki (1965)
5. Baillon, A., Cabantous, L.: Aggregating imprecise or conflicting beliefs: an experimental investigation using modern ambiguity theories. J. Risk Uncertain. **44**, 115–147 (2012)
6. Basili, M., Chateauneuf, A., Fontini, F.: Choices under ambiguity with familiar and unfamiliar outcomes. Theor. Decis. **58**, 195–207 (2005)
7. Basili, M., Chateauneuf, A., Fontini, F.: Precautionary principle as a rule of choice with optimism on windfall gains and pessimism on catastrophic losses. Ecol. Econ. **67**, 485–491 (2008)
8. Basili, M., Chateauneuf, A.: Extreme events and entropy: a multiple quantile utility model. Int. J. Approx. Reason. **52**, 1095–1102 (2011)
9. Barberis, N., Huang, M., Santos, T.: Prospect theory and asset prices. Q. J. Econ. **116**, 1–53 (2001)
10. Barberis, N., Huang, M., Thaler, R.: Individual preferences. Monetary gambles and the equity premium. Am. Econ. Rev. **96**, 1069–1090 (2006)
11. Barberis, N., Huang, M.: The loss aversion / narrow framing approach to the equity premium puzzle. In: Mehra, R. (ed.) Handbook of Investments: Equity Premium. North Holland, Amsterdam (2007)
12. Berger, J., Berliner, M.: Robust bayes and empirical Bayes analysis with epsilon-contaminated priors. Ann. Stat. **14**, 461–486 (1986)
13. Boles, T., Messick, D.: A reverse outcome bias: the influence of multiple reference points on the evaluation of outcomes and decisions. Organ. Behav. Hum. Decis. Process. **61**, 262–275 (1995)
14. Cañadas E., Rodríguez-Bailón, R., Milliken, B., Lupiáñez, J.: Social categories as a context for the allocation of attentional control. J. Exp. Psychol. (2012). doi:*10.1037/a0029794*
15. Cardin, M.: A quantile approach to integration with respect to non-additive measures. In: Modeling Decisions for Artificial Intelligence. Lecture Notes in Computer Science, vol. 7647, pp. 139–148. Springer, Berlin (2012)
16. Camerer, C.: Behavioral economics: reunifying psychology and economics. Proc. Nat. Acad. Sci. **96**, 10575–10577 (1999)
17. Chambers, C.: An axiomatization of quantiles on the domain of distribution functions. Math. Financ. **19**, 335–342 (2009)
18. Chateauneuf, A., Eichberger, J.: Choice under uncertainty with the best and worst in mind: neo-additive capacities. J. Econ. Theor. **137**, 538–567 (2007)
19. Chateauneuf, A., Gajdos, T., Jaffray, J.Y.: Regular updating. Theor. Decis. **71**, 111–128 (2011)
20. Giorgi E. D., Hens T., M., Rieger (2007). Financial Market Equilibriawith Cumulative Prospect Theory, Swiss Finance Institute Research Paper 07-21, Geneva, Switzerland
21. Edwards, W.: The theory of decision making. Psychol. Bull. **5**, 380–417 (1954)
22. Ellsberg, D.: Risk, ambiguity, and the savage axioms. Q. J. Econ. **75**, 643–669 (1961)
23. Ellsberg, D.: Risk Ambiguity and Decis. Garland Publishing Inc., New York (2001)
24. Ellsberg, D.: Secrets: A Memory of Vietnam and The Pentagon Papers. Penguin Books, New York (2002)
25. Etchart-Vincent, N.: Is probability weighting sensitive to the magnitude of consequences? An Exp. invest. on losses J. Risk Uncertain. **28**, 217–235 (2004)
26. Finucane, M., Alhakami, A., Slovic, P.: The affect heuristic in judgment of risks and benefits. J. Beh. Decis. Mak. **13**, 1–17 (2000)
27. Fischhoff, B., Slovic, P.: Knowing with certainty: the appropriateness of extreme confidence. J. Exp. Psychol. Hum. Percept. Perform. **3**, 552–564 (1977)

28. Fischhoff, B., Slovic, P., Lichtenstein, S., Read, S.: How safe is safe enough? A psychometric study of attitudes towards technol. risks and benefits, Policy Sci. **9**, 127–152 (1978)
29. Fischoff, B.: Risk perception and communication unplugged: twenty tears of process. Risk Anal. **15**, 137–145 (1995)
30. Frydman, C., Barberis, N., Camerer, C., Bossaerts, P., Rangel, A.: Testing theories of investor behavior using neural data. Working paper, California Institute of Technology (2011)
31. Ganzach, Y., Ellis, S., Pazy, A.: On the perception and operationalization of risk perception. Judgm. Decis. Mak. **3**, 317–324 (2008)
32. Gilboa, I., Schmeidler, D.: Maxmin expected utility with a non-unique prior. J. Math. Econ. **18**, 141–153 (1989)
33. Gilboa, I., Schmeidler, D.: Additive representations of non-additive measures and the Choquet integral. Ann. Oper. Res. **52**, 43–65 (1994)
34. Green, T., Hwang, B.: Initial public offerings as lotteries: skewness preference and first-day returns. Manage. Sci. **58**, 432–444 (2012)
35. Hayashi, T., Wada, R.: Choice with imprecise information: an experimental approach. Theor. Decis. **69**, 355–373 (2010)
36. Hodges, J., Lehmann, E.: The uses of previous experience in reaching statistical decision. Ann. Math. Stat. **23**, 396–407 (1952)
37. Hurwicz, L.: Some specification problems and application to econometric models (abstract). Econometrica **19**, 343–344 (1951)
38. Kahneman, D.: Maps of bounded rationality: psychology for behavioral economics. Am. Econ. Rev. **93**, 1449–1475 (2003)
39. Kahneman, D., Tversky, A.: Prospect theory: an analysis of decision under risk. Econometrica **47**, 263–292 (1979)
40. Kahnemann, D., Tversky, A.: Advances in prospect theory: cumulative representation of uncertainty. J. Risk Uncertain. **5**, 297–323 (1992)
41. Karni, E., Safra, Z.: Reference reversals and the observability of preferences by experimental methods. Econometrica **55**, 675–685 (1987)
42. Kopylov, I.: Choice deferral and ambiguity aversion. Theor. Econ. **4**, 199–225 (2009)
43. Koszegi, B., Rabin, M.: A model of reference-dependent preferences. Q. J. Econ. **121**, 1133–1165 (2006)
44. Koszegi, B., Rabin, M.: Reference-dependent risk attitudes. Am. Econ. Rev. **97**, 1047–1073 (2007)
45. Jaffray, J.Y.: On the maximum entropy probability which is consistent with a convex capacity. J. Uncertain. Fuzziness Knowl. Based Syst. **3**, 27–33 (1995)
46. Jaynes, E.: Information theory and statistical mechanics I. Phy. Rev. **106**, 620–630 (1957a)
47. Jaynes, E.: Information theory and statistical mechanics II. Phys. Rev. **108**, 171–190 (1957b)
48. Levy, M., Levy, H.: Prospect theory: much ado about nothing? Manage. Sci. **48**, 1334–1349 (2002)
49. Lin, C., Huang, W., Zeelenberg, M.: Multiple reference points in investor regret. J. Econ. Psychol. **27**, 781–792 (2006)
50. Loewenstein, G., Prelec, D.: Anomalies in intertemporal choice: evidence and an interpretation. Q. J. Econ. **107**, 573–597 (1992)
51. Loomes, G., Starmer, C.: Preference reversal: information processing or rational non-transitive choice? Econ. J. **99**, 140–151 (1989)
52. Luce, D., Raiffa, H.: Games and Decisions. Wiley, New York (1957)
53. Manski, C.: Ordinal utility models of decision making under uncertainty. Theor. Decis. **25**, 79–104 (1988)
54. Mehra, R., Prescott, E.: The equity premium: a puzzle. J. Monetary Econ. **15**, 145–161 (1985)
55. Nishimura, K., Ozaki, H.: An axiomatic approach to epsilon-contamination. Econ. Theor. **27**, 333–340 (2006)
56. Oltedal, S., Moen, B., Klempe, H., Rundmo, T.: Explaining risk perception: an evaluation of cultural theory. Rotunde 85. Norwegian University of Science and Technology, Department of Psychology, Norway (2004)

57. Pratt, J.: Risk aversion in the small and in the large. Econometrica **32**, 122–136 (1964)
58. Preston, G., Baratta, P.: An experimental study of the auction-value of an uncertain outcome. Am. J. Psychol. **61**, 183–193 (1948)
59. Quiggin, J.: A theory of anticipated utility. J. Econ. Behav. Organ. **3**, 323–343 (1982)
60. Rohrmann, B., Renn, O.: Risk perception research: an introduction. In: Rohrmann, B., Renn, O. (eds.) Cross-Cultural Risk Perception: A Survey of Empirical Studies. Kluwer Academic Publishers, Boston (2000)
61. Rostek, M.: Quantile maximization in decision theory. Rev. Econ. Stud. **77**, 339–371 (2010)
62. Savage L.: The Foundation of Statistics. New York, Wiley (1954). 1972, 2nd ed.
63. Schmeidler, D.: Subjective probability and expected utility without additivity. Econometrica **57**, 571–587 (1989)
64. Segal, U.: Does the preference reversal phenomenon necessarily contradict the independence axiom? Am. Econ. Rev. **78**, 233–236 (1988)
65. Shackle G.: Expectations in Economics. Cambridge University Press, Cambridge (1949). 1952, 2nd edn.
66. Simon, H.: Rational choice and the structure of the environment. Psychol. Rev. **63**, 129–138 (1956)
67. Simon, H.: Models of Bounded Rationality. MIT Press, Cambridge (1982)
68. Sjöberg, L., Moen, B., Torbjørn, R.: Explaining Risk Perception. An valuation of the psychometric paradigm in risk perception research. Rotunde Publikasjoner 84. Editor: Torbjørn Rundmo Norwegian University of Science and Technology, Department of Psychology, (2004)
69. Slovic, P.: Perception of risk. Science **236**, 280–285 (1987)
70. Slovic, P., Fischhoff, B., Lichtenstein, S.: Rating the risks. Environment **21**(14–20), 36–39 (1979)
71. Slovic, P., Fischhoff, B., Lichtenstein, S.: Facts and fears: understanding perceived risk. In: Schwing, R., Albers, W. (eds.) Societal Risk Assessment: How Safe is Safe Enough? Plenum Press, New York (1980)
72. Slovic, P., Weber, E.: Perception of risk posed by extreme events. Paper Prepared for Discussion at the Conference 'Risk Management Strategies in an Uncertain World', Palisades, New York (2002)
73. Starmer, C., Schmidt, U.: Third-generation prospect theory. J. Risk Uncertain. **36**, 203–223 (2008)
74. Tsallis, C., Anteneodo, C., Borland, L.: Nonextensive statistical mechanics and economics. Phys. A **324**, 89–100 (2003)
75. Tsiros, M.: Effect of regret on post-choice valuation: the case of more than two alternatives. Organ. Behav. Hum. Decis. Process. **76**, 48–69 (1998)
76. von Neumann, J., Morgenstern, O.: Theory of Games and Economic Behaviour. Princeton University Press, Princeton (1944)
77. Wald, A.: Statistical decision functions which minimize the maximum risk. Ann. Math. **46**, 265–280 (1945)
78. Wakker, P.: Prospect Theory for Risk and Ambiguity. Cambridge University Press, Cambridge (2010)
79. Weber, E.: Decision and choice: risk, empirical studies. In: Smelser, N., Baltes, P., (eds.) International Encyclopedia of the Social and Behavioral Sciences Elsevier. Science Limited, Oxford (2001)
80. Weber, E.: Experience-based and description-based perceptions of long-term risk: why global warming does not scare us (yet). Clim. Change **77**, 103–120 (2006)
81. Wildavsky, A., Dake, K.: Theories of risk perception: who fears what and why? Daedalus **119**, 41–50 (1990)
82. Wu, G., Gonzales, R.: Non linear decision weights in choice under uncertainty. Manage. Sci. **45**, 74–85 (1999)
83. Wu, G., Markle, A.: An empirical test of gain-loss separability in prospect theory. Manage. Sci. **54**, 1322–1335 (2008)
84. Yaari, M.: The dual theory of choice under risk. Econometrica **55**, 95–115 (1987)

Effective Decision Making in Changeable Spaces, Covering and Discovering Processes: A Habitual Domain Approach

Moussa Larbani and Po Lung Yu

Abstract This chapter proposes a model of covering and discovering processes for solving non-trivial decision making problems in changeable spaces, which encompass most of the decision making problems that a person or a group of people encounter at individual, family, organization and society levels. The proposed framework fully incorporates two important aspects of the real-decision making process that are not fully considered in most of the traditional decision theories: the cognitive aspect and the psychological states of the decision makers and their dynamics. Moreover, the proposed model does not assume that the set of alternatives, criteria, outcomes, preferences, etc. are fixed or depend on some probabilistic and/or fuzzy parameter with known probability distribution and/or membership function. The model allows the creation of new ideas and restructuring of the decision parameters to solve problems. Therefore, it is called *decision making/optimization in changeable spaces* (DM/OCS). DM/OCS is based on Habitual Domain theory, the decision parameters, the concept of competence set and the mental operators 7-8-9 *principles of deep knowledge*. Some illustrative examples of challenging problems that cannot be solved by traditional decision making/optimization techniques are formulated as DM/OCS problems and solved. Finally, some directions of research are provided in conclusion.

AMS classification: 90C99, 91B06, 90B50 and 91E99.

M. Larbani (✉)
Department of Business Administration, IIUM University, Jalan Gombak, 53100 Kuala Lumpur, Malaysia
e-mail: larbani61@hotmail.com

P. L. Yu
Institute of Information Management, National Chiao Tung University, Hsinchu 30010, Taiwan

P. L. Yu
School of Business, University of Kansas, Lawrence, KS66045, USA
e-mail: yupl@mail.nctu.edu.tw

P. Guo and W. Pedrycz (eds.), *Human-Centric Decision-Making Models for Social Sciences*, Studies in Computational Intelligence 502,
DOI: 10.1007/978-3-642-39307-5_7,

Keywords Habitual domains · Decision making · Changeable spaces · Parameters · Covering · Discovering · Competence set · Decision blinds · Decision traps

1 Introduction

The study of decision making process started few centuries ago. It has evolved through four stages. (1) Preoccupation with the rational, (2) critiques and extensions of the rational tradition, (3) creation of fully articulated alternatives to the rational and, finally, (4) a multi-perspective view of decision making [2]. The rational theory of decision making is based on utility function. In this model it is assumed that a set of alternatives X is available and fixed and a utility function $u(.) : X \rightarrow R$ (R is the real line) that represents the preferences of the Decision Maker (DM) is constructed. To each alternative x the function associates a value or utility $u(x)$. An alternative y is preferred to an alternative x if $u(y) > u(x)$. The decision having the largest utility is chosen as the optimal decision. In other words, the optimal decision is selected by solving the optimization problem $\max_{x \in X} u(x)$. Such a model assumes that (i) the DM is rational, that is, his preferences, tastes, etc. are consistent with each other and (ii) he has the capability of looking at all possible choices and outcomes, weighing each, and then making an optimal decision based upon these deliberations.

When a decision making problem involves uncertainty, the expected utility function is used instead of the utility function [30]. The DM weighs the different scenarios of each alternative by probabilities and computes the expected value of each alternative by: first multiplying the utility of each of its scenarios by the corresponding probability, then summing up over all the alternative's scenarios. The procedure ends by selecting the best alternative as the one presenting the highest expected value. The existence of utility function requires the preferences of DM to be transitive, complete, and continuous (convex) [30].

Most of the early quantitative models and theories of decision making were based on the optimization model $\max_{x \in X} u(x)$. These models were later generalized to decision making problems involving multiple criteria [26, 36]. To better represent real-decision making problems, both single criterion and multiple criteria decision making models were further extended to models incorporating stochastic or fuzzy [4] or fuzzy-stochastic parameters with known probability distribution and/or membership functions [21, 22, 24, 43]. Numerous methods were developed to solve optimization problems as linear programming [9], nonlinear programming methods [3], stochastic programming methods [22], multiple criteria optimization methods [11, 24, 26], and so on. It is important to note also that the advances made in computer sciences considerably contributed to the development of decision making software that facilitate decision making.

Critiques and extensions of the rational tradition began to appear in the next stage of decision making literature as scholars realized that neither man nor his organizations were capable of making decisions which took into account all possible alternatives, assessed all possible outcomes, and selected the optimal among such alternatives. Simon [23] stated that man does not make decisions by "optimizing" principle rather he uses the "satisfycing" principle. The satisfycing or bounded rationality based models of decision making assume that the DM chooses certain levels of satisfaction, and then a decision is selected as soon as it achieves the fixed levels of satisfaction or goes beyond.

Later, Tversky and Kahneman [14, 27–29] provided experimental evidence of the limitations of the rational model of decision making. Tversky [27] showed that the preferences of a DM may not be transitive. Tversky and Kahneman [28, 29] provided experimental evidence of violation of the expected utility model when a DM is faced with a decision making problem involving states of nature. These works led the latter two scholars to the introduction of an extension of expected utility theory, the Prospect theory [14]. Prospect theory is a descriptive model of decision making that attempts to describe how we make decisions and why our decisions violate the expected utility model. This theory predicts that people will be especially averse to loss and will show differences in preferences depending on how alternatives are presented, or framed.

Busemeyer and Towsend [6] extended the static expected utility theory to the Decision Field theory via a dynamic approach where a probability function that maps each pair of actions into the interval [0,1] describes the variation of the preferences with respect to deliberation time.

In the third stage, full-fledged alternative views to the classical rational tradition were developed. Cyert and March [8], for example, introduced an organizational theory of decision making in the book "A Behavioral Theory of the Firm" that was to replace the neo-classical economic theory of the firm. Several decidedly non-rational views of the decision making process grew out of the literature of cognitive and perceptual psychology as the Attribution theory [15]. The Attribution theory is a decision making theory that is based on schemata and heuristics. A schemata is a working hypothesis about some aspect of the environment and may be a concept of the self (self schema), other individuals (person schemata), groups (role schemata), or sequence of events in the environment (scripts). In addition to using them to organize their interpretation of their environment, people use schemata to develop scripts for action. Heuristics consist of rules people use to test their schemata and facilitate the processing of information.

From stages one to three, it appears that the decision making process involves many aspects as psychology, sociology, rationality, etc. Moreover, different decision making models may lead to different outcomes.

In the fourth, stage attempts were made to develop a multiple-perspective approach to the study of decision making. For instance, in 1971, Allison made an explicit elaboration of the multiple-perspectives idea in the book "The Essence of Decision; Explaining the Cuban Missile Crisis" [1]. Steinbruner [25] elaborated the Cybernetic Theory of Decision based on the multiple-perspective approach. However, there is no theory that encompasses all the multiple aspects of decision making process as cognitive processes, rationality, emotions, group behavior, etc.

Dynamic Decision Making is a model that encompasses decision making problems under conditions which require a series of decisions, where the decisions are not independent, where the state of the world changes, both autonomously and as a consequence of the DM's actions, and where the decisions have to be made in real time [5, 10]. According to Brehmer [5], it is difficult to find useful normative theories for these kinds of decision making problems, and research thus has to focus on descriptive issues. A general approach, based on control theory, is proposed as a means to organize research in the area. An experimental paradigm for the study of dynamic decision making, the computer stimulated microworlds (i.e., Decision Making Games, DMgames), has been introduced to study decision performance. In fact, research in dynamic decision-making is mostly laboratory-based [12].

It is important to note that the existing theories of decision making have achieved tremendous results in both theory and applications [11–13, 21, 22, 24, 26, 36, 43]. However, most of them cannot handle problems involving parameters with unknown shapes and behavior i.e. *unstructured uncertainty*. Indeed, in real decision making, some parameters may even be intangible. Without special efforts, we may not be aware of their existence. Even when they are noticed, their dimensions, ranges and shapes may not be easily predetermined or assumed as in probabilistic and/or fuzzy models. Often, real-life decision making problems also involve parameters that are changeable, including the set of alternatives, the criteria and the DMs, as situations and psychological states of the DMs change. Discovering and controlling the change of these parameters is a vital part of the process of solving challenging decision making problems. A decision making problem involving changeable parameters is called a *decision making in changeable spaces* (DMCS) problem. DMCS problems have been introduced and discussed in [18, 38]. A broad class of decision making problems involving DMCS problems are *covering* and/or *discovering* problems. Let us explain these two problems by examples.

Why a professional tiger hunter and a newborn are not afraid of a tiger when they see it, while common people are afraid of it? It is because the hunter has the competence to deal with the tiger, while the newborn has no idea about the danger when dealing with it; the common people are afraid of the tiger because they know its danger but they don't have the necessary competence to deal with it. For effective decision making problem solving, one need to have the necessary competence it requires. Given a decision making problem, the *competence set* associated to this problem is defined as the set of knowledge, ideas, skills, know-how, resources, efforts, etc. necessary to effectively solve it [37]. Covering and discovering are two non-trivial decision making processes that people encounter in their socio-economic activities at individual, family, organization and society levels. The covering problem can be defined as "how to transform a given competence set to cover a targeted competence set". A covering problem may be difficult to solve when the Decision Makers (DMs) do not know exactly their actual competence sets and/or the target competence set and/or the way they should transform their competence sets to acquire the target competence set.

Given a competence set, what is the best way to make use of it to solve unsolved problems or to create value? The process underlying this problem solving or value creation involves discovering. Thus, a discovering process can be defined as identifying how to use available tangible and intangible skills, and resources to solve an unsolved problem or to produce new ideas, concepts, products or services that satisfy some newly-emerging needs of people. From the definitions of covering and discovering, it appears clearly that these two classes of non-trivial decision making problems encompass most of the non-trivial problems people face in their economic, social and academic activities. Let us give a pair of examples of covering and discovering problems for illustration.

Example 1.1 (Horse Race [31]). A retiring corporate chairman invited to his ranch two finalists (A and B) from whom he would select his replacement using a horse race. A and B, equally skillful in horseback riding, were given a black and white horse, respectively. The chairman laid out the course for the horse race and said, "Starting at the same time now, whoever's horse is slower in completing the course will be selected as the next chairman!" The two candidates were puzzled at the beginning because the rule of the race is not the commonly known rule that the first who crosses the finishing line is the winner. They need to discover new strategies to win the race themselves. Thus, the candidates face a discovering and covering problem. Discovering strategies and executing them to win the race (cover their objective).

In the sequel, when a decision making problem involves discovering and covering, we say that it involves a *dis/covering* problem.

Example 1.2 (Emerson's Calf [31]) One very cold weather day, Ralph W. Emerson and his son had an extremely difficult time pushing a young calf into their barn for shelter. As Ralph pulled from the head of the calf and his son pushed with determination from behind, the calf stiffed its legs and met his action with strong resistance. The calf's survival instinct was aroused by the action of Emerson and his son. Here, Emerson and his son tried their best using their knowledge and physical force (resource) to bring the calf into the barn (target), they could not. This means that they need to transform their set of knowledge and resources to reach or cover their objective, that is, to bring the calf into the barn. In other words, they face a dis/covering problem.

Later we will see how these two DMCS problems were solved within the introduced model. It is important to note that most of decision making problems involve a dis/covering process. Indeed, the discovering process requires an objective to reach, that is, covering, while in the covering process, the DMs may need to generate new ideas and concepts, that is, discovering. For instance, in the Example 1.1, the candidates have a target to cover, win the race; however, they have to discover winning strategies. Thus, to solve a non-trivial decision problem, we are generally involved with a dis/covering problem or process.

From the foregoing discussion, it appears that the existing decision making theories can be categorized into rational theories and cognitive theories. The rational theories are normative, however, they do not reflect the process of decision making with respect to preferences of the DM as pointed out by Tversky and Kahenman [14, 27–29], while cognitive theories are basically descriptive and lack a well-developed theory. Moreover, both rational and cognitive theories cannot handle DMCS problems and do not allow restructuring of the decision parameters to solve problems.

Following the multiple-perspective approach to decision making problems, in this chapter, we propose a model of decision making based on the Habitual Domain (HD) theory [31–35]. It is a comprehensive mathematical framework in the sense that it incorporates the psychological and the cognitive aspects of decision making. Compared to existing decision making models, the introduced model has the following unique features: (i) it can handle dynamic decision making problems involving unstructured uncertainty i.e. DMCS problems at individual, group and organizational levels, and in conflict resolution, (ii) it is at the same time descriptive and normative in the sense that it suggests courses of action to solve decision making problems, including restructuring their parameters (iii) it introduces the competence set as a tool in decision process and (iv) it formulates decision making problems as dis/covering problems.

Specifically, we introduce a mathematical model of DMCS problem focusing on dis/covering, called *decision making /optimization in changeable spaces* (DM/OCS). DM/OCS is based on HD theory, the decision parameters, the concept of competence set and the mental operators 7-8-9 *principles of deep knowledge*.

The rest of the chapter is organized as follows. Section 2 briefly presents the decision making process from HD theory perspective and the decision parameters. A detailed discussion of these parameters can be found in [38]. Section 3 presents a brief competence set analysis and the mental operators 7-8-9 principles of deep knowledge. A more detailed competence set analysis and explanation of the 7-8-9 principles are provided in [37] and [31, 32, 34], respectively. Section 4 formally defines covering and discovering, introduces OCS and provides necessary and sufficient conditions for dis/covering completion. Section 5 provides some applications. Section 6 concludes the chapter and provides further research directions.

2 Habitual Domains and Decision Making in Changeable Spaces

As mentioned in the introduction, the traditional framework for decision making is not appropriate for the formulation and resolution of DMCS problems because it does not take into account the psychological states of the DMs, does not handle parameters with unknown shapes and ranges, and does not allow the restructuring of the decision parameters during the decision process. In this chapter, we demonstrate that HD theory [31–35] can be used as a basis to develop a comprehensive framework for DMCS problems. Thus, we essentially use HD theory as a general framework to develop our model of dis/covering. In this section, because of space constraint, we briefly introduce the HD theory.

The collection of ideas and actions (including ways of perceiving, thinking, responding, acting, and memory) in our brain, together with their formation, dynamics, and basis in experience and knowledge, is called our *Habitual Domain* (HD) [31–35]. Over time, unless extraordinary or purposeful effort is exerted, our HD will become stabilized within a certain domain. This can be proven mathematically [7]. As a consequence, we observe that each of us has habitual ways of eating, dressing, speaking, reacting to specific situations, etc.

The concept of an individual's HD can be extended to other living entities, such as companies, social organizations, and groups in general. The following are the basic elements of HD.

(i) The *potential domain* (PD): the collection of ideas and actions that can potentially be activated to occupy our attention.
(ii) The *actual domain* (AD): the set of ideas and actions that are actually activated or which occupy our attention.
(iii) The *activation probabilities* (AP): the probabilities that ideas or actions in PD also belong to AD.
(iv) The *Reachable Domain* (RD): the set of ideas and actions that can be attained from a given set in AD.

Thus, the habitual domain can be formally formulated as

$$\mathrm{HD}_t = \{\mathrm{PD}_t,\ \mathrm{AD}_t,\ \mathrm{AP}_t,\ \mathrm{RD}_t\}, \tag{1}$$

where t represents time. The theory of HD is based on eight hypotheses H1-H8 [31, 32, 34]: (H1) Circuit Pattern Hypothesis, (H2) Unlimited Capacity Hypothesis, (H3) Efficient Restructuring Hypothesis,(H4) Analogy and Association Hypothesis, (H5) Goal Setting and State Evaluation Hypothesis, (H6) Charge Structure and Attention Allocation Hypothesis, (H7) Discharge Hypothesis, and (H8) Information Input Hypothesis.

The hypotheses H1-H8 are explained in Appendix. Note that hypotheses H1-H4 describe how the brain operates, while hypotheses H5-H8 describe how the mind operates. Moreover, a high level of charge can be a drive for active problem solving or can create a mental stress if no action is taken. In fact, it is humans that make decisions; therefore, understanding the human behavioral system plays a vital role in making good decisions. The complex processes of human behaviors have a common denominator resulting from a common behavior mechanism. The mechanism depicts the dynamics of human behavior. Based on the literature of psychology, neural physiology, dynamic optimization theory, and system science, Yu [31, 34, 36] described the dynamic human behavior mechanism as presented in Fig. 1, which is briefly explained below:

(i) **Box (1)** is our brain and its extended nervous system. Its functions may be described by the four hypotheses (**H1–H4**) in Appendix.
(ii) **Boxes (2)–(3)** represent two basic functions of our mind, Goal Setting and State Evaluation, explained by **H5** in Appendix.
(iii) **Boxes (4)–(6)** represent how we allocate our attention to various events, described by **H6** in Appendix.

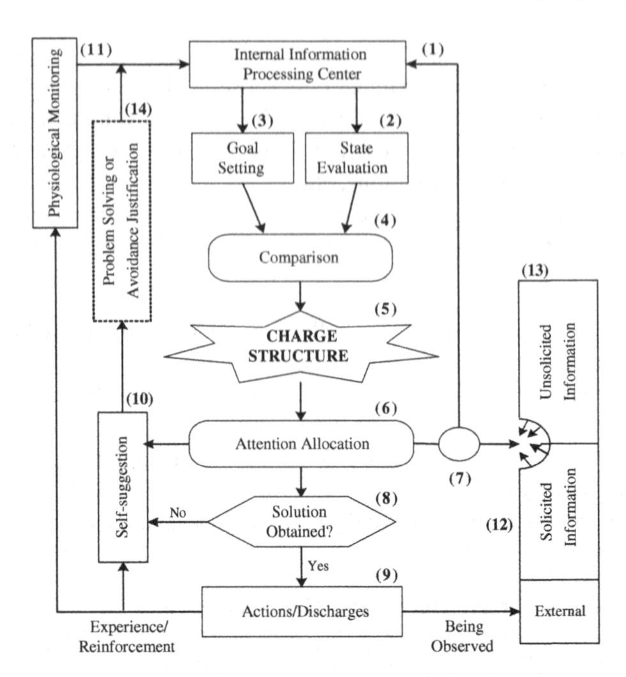

Fig. 1 The human behavior mechanism

(iv) **Boxes (8)–(9), (10)** and **(14)** represent the *least resistance principle* which humans use to release their charges (precursors of mental stress), described by **H7** in Appendix.

(v) **Boxes (7), (12)–(13)** and **(11)** represent the information input into our information processing center (**Box (1)**). **Boxes (10) and (11)** are two important functions of human thinking and information processing. **Boxes (7), (12)–(13)** represent external information inputs, an important parameter in decision making, which are explained in **H8** in Appendix.

2.1 Decision Making Parameters

Dis/covering problems are fundamentally DMCS problems. Therefore, a complete description of the real decision making process is a prerequisite to formulation of a model of dis/covering problems. Decision making in non-trivial situations is a complex process that involves two interacting groups of *parameters*, namely, the *decision elements* and *environmental facets* [38]. Moreover, many non-trivial decision problems involve uncertainty and the unknown, which can lead to judgmental fuzziness and decision failure. The unknowns and uncertainty may be due to the changing nature and/or unawareness of the relevant parameters. In the next section, we briefly describe the decision parameters.

2.1.1 Decision Elements

In general, there are five basic elements involved in a decision making process. These are (i) decision alternatives, (ii) decision criteria, (iii) decision outcomes, (iv) decision preferences, and (v) decision information inputs. In existing decision theories, these elements are implicitly assumed to be fixed or vary according to some known patterns (e.g. probability distribution). In real decision making problems, these elements are not fixed, they may change unpredictably over time depending on events, information input and the psychological states of DMs, especially, when the decision process is in a transition state. In other words, each of these elements is a changeable space. They can also be considered as HDs, they tend to stabilize if no relevant event and/or information arrives.

2.1.2 Decision Environmental Facets

Decision environments may be described by four facets: (i) decisions as a part of the human behavior mechanism, (ii) stages of the decision making processes, (iii) players in the decision making processes and (iv) unknowns in the decision making processes. Here also, the existing decision theories implicitly assume that the facets (ii)–(iv) are fixed or change according to some structured way as in dynamic programming, dynamic stochastic decision making models and games in extensive form. However, in real decision making problems, they may change in an unpredictable way over time depending on the psychological states of DMs and the arriving events and information. As for (i), in most of the decision theories, it is not incorporated. Let us elaborate more on the unknowns in the decision making process [38].

Knowing the unknowns and how to manage them may add satisfaction to our decision processes; otherwise, they may create fear, frustration and bitterness. Unknowns may exist in any decision element. Because of HDs and being unaware of the decision parameters and their changing nature, people would easily have *decision blinds* or even get into *decision traps*. When blinds or traps occur, it is hard to see the problem clearly, let alone to solve it effectively and efficiently. Note that the blinds and traps can cause fuzziness or unknown in decision making. Conversely, fuzziness and unknowing about the nature of problems can lead to decision blinds, traps and wrong decisions.

Formally, a DMCS problem can be represented by the following collection (from a single DM perspective) $\{X_t, C_t, F_t, D_t, J_t, I_t, \mathrm{HD}_t, , U_t, Q_t\}$, where the time t varies in a certain interval $[0, L]$, representing the allowable time for solving the problem, X_t*is* the available alternative set at time t; C_t is the criteria set at time t; F_t is the outcome measured in terms of the criteria at time t; D_t is the preference of DM at time t, J_t is the set of possible outcomes at time t, I_t is the information inputs at time t, HD_t is the DM's HD at time t, U_t is the set of unknowns at time t and Q_t is the set of the involved DMs at time t. It is important to note that the described decision parameters not only vary with time, but also mutually interact with each other through time. *Time optimality* and *time satisficing* solutions (optimal or satisficing as perceived by DMs during certain period of time, see [31, 34]) become important solution concepts.

3 Competence Set Analysis

Competence Set Analysis began with Yu in 1989 [35], as a derivative of HD theory. Its mathematical foundation was built by Yu and Zhang [40–42]. The competence set (*CS*) for a given decision problem is defined as the collection of ideas, knowledge, skills, know-how, efforts and resources required for its effective resolution [37]. Therefore, knowing the characteristics and dynamics of this set is essential for successfully solving challenging problems. When the decision maker thinks he/she has already acquired and mastered the *CS* as perceived, he/she would feel comfortable making the decision and/or undertaking the challenge.

3.1 Decision Blinds and Decision Traps from a Competence Set Perspective

The competence set *CS*(*E*) of a problem *E* is, in fact, a projection of the DMs' HDs onto the problem. Implicitly, it contains Actual Domain, Potential Domain, Reachable Domain, and Activation Probabilities (1). For simplicity, assume that *CS* (*E*) is constant and denote by $CS^t(E)$, the competence set of the DMs at any time t. When$CS(E) \subset CS^t(E)$, for some time t, the DMs are able to solve the problem *E*, while when $CS(E) \not\subset CS^t(E)$, the DMs are not able to completely solve the problem *E* at time t, then $CS(E)\backslash CS^t(E)$ would be the *decision blinds*, it is the set of all the competencies required but not seen by the decision makers at time t. See the illustration in Fig. 2.

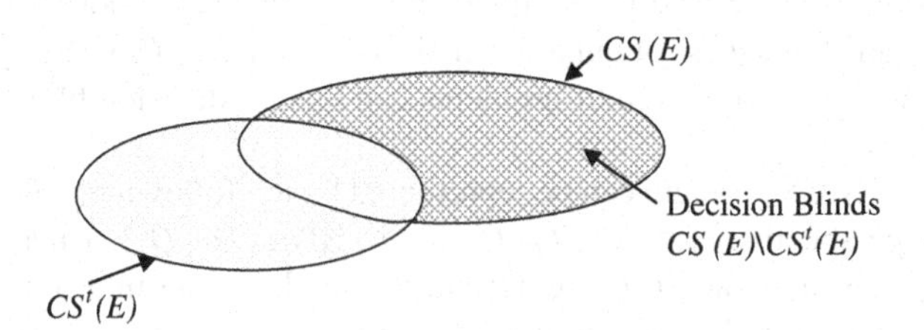

Fig. 2 Decision blinds

Note that the larger the decision blind is, the more likely it is that the DMs might make important mistakes [37].

Suppose that $CS^t(E)$ is fixed or contained in a certain domain for some period of time and $CS(E)\backslash CS^t(E)$ is large and significant, then we tend to make mistakes in decisions and we are in a *decision trap* during the considered period. Note that $CS^t(E)$ being contained or fixed in a certain domain is equivalent to the corresponding Actual Domain and Reachable Domain being fixed or trapped in a certain domain. This can occur when we are in a very highly charged state of mind or when we are over confident, which makes us respond quickly and unthinkingly and to habitually commit the behavior of decision traps. In Fig. 3, one can see that decision blinds reduce as we move our Actual Domain from A to B then to C. By changing our Actual Domain, we can change and expand our Reachable Domain. We can reduce decision blinds and/or avoid decision traps by systematically changing the Actual Domain. For illustration, assume that *CS(E)* and the Reachable Domain are given, as depicted in Fig. 3.

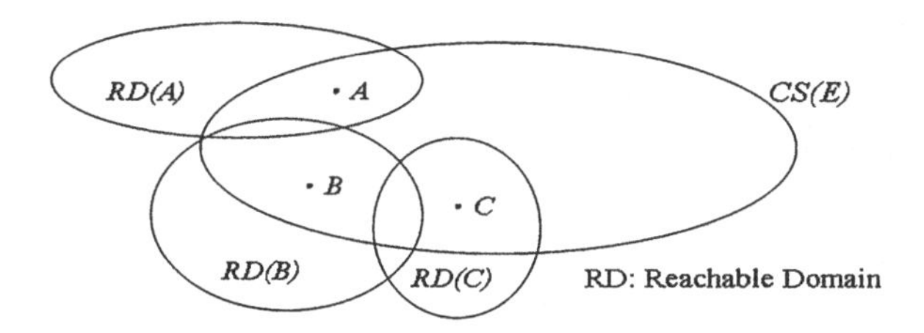

Fig. 3 Reducing decision blinds and/or avoiding decision traps

Then, as we move the Actual Domain from A to B, then to C, our decision blinds reduce progressively from $CS(E)\backslash RD(A)$ to $CS(E)\backslash(RD(A) \cup RD(B))$ then to $CS(E)\backslash(RD(A) \cup RD(B) \cup RD(C))$.

For challenging decision problems, we can treat the 9 decision parameters of Sect. 2.1 as 9 points of the Actual Domain. Systematically moving over these 9 parameters and pondering their possible Reachable Domains can expand our Reachable Domain for dealing with the challenging problems. As a consequence, $CS^t(E)$ is expanded and our decision blinds,$CS(E)\backslash CS^t(E)$, reduced. In the next section, we will introduce HD tools to enrich and expand our HD and $CS^t(E)$ to reduce the decision blinds and avoid decision traps.

Let us illustrate the concepts of decision trap and decision blinds.

Example 3.1 Let us consider Example 1.2 (Emerson's Calf). At the beginning, there were three DMs involved in the situation, Emerson, his son and the Calf. Emerson and his son wanted to bring the Calf to the barn because the weather was very cold. Thus, the targeted competence set *CS* is a set that encompasses the skill of being friendly with animals and make them obey to do what we want them to do, in the present case, going into the barn. The initial competence set of Emerson and his son, CS^*, is limited to their skills and competencies derived from their experiences and education. Emerson and his son adopted the strategy of pushing the Calf into the barn. Although the intention of Emerson and his son was good, the strategy did not work. Let us explain this situation by HD theory. In fact, the Calf perceived the action of Emerson and his son as a threat to its life (see Hypothesis H5, goal setting and state evaluation, in Sect. 2). This threat created a large deviation from the ideal position of the goal representing its life, which triggered immediate action (see Hypotheses H6 and H7) from the Calf in the form of strong resistance to the action of Emerson and his son. Since Emerson and his son maintained the same strategy, the reaction of Calf became even stronger. We can conclude that Emerson and his son were definitely in a decision trap, that is, CS^t was constant and did not evolve to cover *CS*. Although Emerson was a knowledgeable and famous writer, his habitual domain and competence set did not include the skills and knowledge to make the Calf accept to come to the barn peacefully. Obviously, his son too did not have the needed skills to solve the problem with his limited experience and exposure to animals. Thus, the decision blind, $CS \backslash CS^t$ was nonempty and significant and included the skills of dealing with animals and making them accept to do what one wants them to do.

3.2 Clarifying Fuzziness and Unknown in Decision Making Using HD Tools

Uncertainty pervades decision making process. It may be structured as stochastic, fuzzy, fuzzy-stochastic, etc. or unstructured, that is, represented by parameters with unknown shapes, dimensions and behavior. Uncertainty results mainly from the cognitive limitations of human beings and the complexity and unpredictability of the environment behavior. It is the main source of blinds and traps in the decision making process. To reduce the blinds and avoid traps, in addition to being aware of the decision parameters and their changing nature, we need HD tools to expand and enrich our Actual Domain and Reachable Domain and look into the depth of the Potential Domains. The HD tools can also expand and enrich our perception of the decision problem and its related parameters. Here, we present three toolboxes: the *seven empowering operators*, the *eight basic methods for expanding HD* and the *nine principles of deep knowledge*. Note that all the HD tools in all three boxes may be used or ignored by DMs. In fact, the more we use them, the more powerful they will be in our brain and the more they will be ready to help us expand and enrich our HD. We call them the *7-8-9 principles of deep knowledge* (referred to as the *7-8-9 principles* in the sequel) [31, 32, 34]. These basic general principles can be used individually or combined to create new ideas to solve dis/covering problems.

3.2.1 Seven Empowerment Operators

The seven empowering operators listed in Table 1 can make our minds think positively, with great hope and confidence to explore our world as to achieve our goals.

As a result, whenever these operators occupy our minds, we could expand and enrich our HD.

Table 1 The seven empowering operators

M_1	Everyone is a priceless living entity. We are all unique creations who carry the spark of the divine
M_2	Clear, specific and challenging goals produce energy for our lives. I am totally committed to doing and learning with confidence. This is the only way I can reach the goals
M_3	There are reasons for everything that occurs. One major reason is to help us grow and develop
M_4	Every task is part of my life mission. I have the enthusiasm and confidence to accomplish this mission
M_5	I am the master of my living domain. I take responsibility for everything that happens in it
M_6	Be appreciative and grateful and don't forget to give back to society
M_7	Our remaining lifetime is our most valuable asset. I will enjoy it fully and make a 100 % contribution to society in each moment of my remaining life

3.2.2 Eight Basic Methods for Expanding HD

These eight basic principles, as listed in Table 2, are almost self-explanatory. They can expand our HDs. These principles, usually, through self-suggestion, will enable us to generate new ideas, new concepts and, consequently, to expand our HDs [31, 32, 34].

Table 2 Eight methods for expanding and enriching HDs

M_8	Learning actively
M_9	Projecting from a higher position
M_{10}	Active association
M_{11}	Changing the relevant parameters
M_{12}	Changing the environment
M_{13}	Brainstorming
M_{14}	Retreating in order to advance
M_{15}	Praying or meditating

Table 3 Nine principles of deep knowledge

$\mathbf{M_{16}}$	The deep and down principle
$\mathbf{M_{17}}$	The alternating principle
$\mathbf{M_{18}}$	The contrasting and complementing principle
$\mathbf{M_{19}}$	The revolving and cycling principle
$\mathbf{M_{20}}$	The inner connection principle
$\mathbf{M_{21}}$	The changing and transforming principle
$\mathbf{M_{22}}$	The contradiction principle
$\mathbf{M_{23}}$	The cracking and ripping principle
$\mathbf{M_{24}}$	The void principle

3.2.3 Nine Principles of Deep Knowledge

The nine principles for deep knowledge, as listed in Table 3, not only allow us to understand and expand our HDs, but help us on how to use our own HDs and other people's HDs to solve our problems as well.

It is important to note that there is no guarantee that the 7-8-9 principles are enough to handle all transformations of a given competence set CS. We do not pretend that this list is exhaustive. These principles are derived from HD theory and a thorough analysis of literature on psychology, problem solving, creativity, innovation, scientific discovery and critical thinking. Moreover, more principles can be derived from them for specific problems in specific areas. For instance, one may consider the mathematical transformations under the principle M_{21}, the "changing and transforming principle". Moreover, their combination generates more principles.

Remark 3.1 One can, by analogy, see the cognitive process of a DM as an internal light of the mind that illuminates part of his/her Potential Domain. Then the efficiency and effectiveness of solving a DMCS problem depends on the *brightness, intensity* and*orientation* (*flexibility*) of the internal light. It is important to note that the 7-8-9 principles of deep knowledge are tools that the DM can use to go deeper into his/her Potential Domain, to change and enlarge the scope of his/her Reachable and Actual domains. Therefore, these principles have a direct effect on the guidance and control of DM's internal light. The more they are used skillfully and frequently, the brighter, intense and flexible (in terms of direction or orientation) is the DM's internal light. The brightness of the internal light allows illuminating or shining the area it points to, its intensity helps to penetrate into the depth of Potential Domains in the direction it focuses and its flexibility makes it possible to change direction it points to. For instance, consider the problem of covering the points A, B and C of the targeted competence set *CS*(E) in Fig. 3. By using one or more of the 7-8-9 principles, the DM may increase the brightness of the light to enlarge the scope of his/her Reachable Domain to reach the point A. Next, using some of the 7-8-9 principles, he/she may increase the intensity of the light so as to reach deeper parts of the Potential Domain, then reach the point B. The DM may use the 7-8-9 principles to increase the brightness, intensity and change the direction (flexibility) of the internal light, at the same time, to reach the point C. Indeed, referring to Fig. 2, a person who diligently and

repeatedly uses the 7-8-9 principles increases the brightness, intensity and flexibility of his/her internal light thereby enlarging his/her Reachable Domain in all directions. Thus, enlarging the HDs by 7-8-9 principles increases the potential of DMs to cover a larger amount of ideas, skills and concepts that are part of the targeted competence set $CS(E)$, that is, to reduce the decision blinds and avoid decision traps.
The cognitive process in dis/covering may be described by the situation of a person that enters a dark and open area for the first time, equipped with a torch with variable brightness and intensity, to look for something he/she believes it is there. The dark area would be some knowledge space, while the area covered by the torch's light would be the Reachable Domain, and the torch's light is his/her internal light. Falling in a decision trap can be similarly represented by a person that enters a cave and doesn't know how to get out. He/she has a torch with variable brightness and intensity to look for the way out. Thus, completing dis/covering depends essentially on the brightness, intensity and flexibility of the torch's light.

4 Optimization in Changeable Spaces

In this section, we present a new optimization paradigm to formulate and solve DMCS problems, *the optimization in changeable spaces* (OCS) or *Second Order Optimization*. The OCS model introduces psychological and cognitive aspects and the possibility to restructure the decision parameters into optimization, which have never been considered in existing optimization models [3, 6, 12–15, 19, 21, 22, 24, 26, 43]. OCS is formulated in the framework of HD theory (Sect. 2). The fundamental elements of OCS are the decision parameters including the five decision making elements and the four decision making environmental facets (see Sect. 2), the 7-8-9 principles of deep knowledge (see Sect. 3.2) and the concept of competence set presented in Sect. 3. In our formulation of OCS problems, we focus on one general problem, namely, the dis/covering problem. Before that, we need to introduce some mathematical notations and tools. The 7-8-9 principles, $M_1, M_2, \ldots, M_{24}$, can help generate new ideas to solve problems, therefore, mathematically, they can be thought of operators that transform a given set of ideas into another set of ideas. Thus, the domain of these operators is the Ω_1-space of all the knowledge, know-how and skills that the whole humanity has reached so far. The operators $M_1, M_2, \ldots, M_{24}$ are set-to-set functions with domain Ω_1-space and range Ω-space, such that for any subset A of Ω_1, $M_i(A) \subset \Omega$, $i = 1, \ldots, 24$. The space Ω is the space of all the knowledge, know-how and skills that the whole humanity has reached so far and the knowledge, know-how and skills it will reach in the future. The Ω -space is not a set in the traditional or fuzzy sense because its boundaries are not known and change over time in an uncertain way. For presentation convenience, denote by $M = \{M_1, M_2, \ldots, M_{24}\}$ the set of 7-8-9 principles. In the decision making process, the DMs may apply these principles individually at some times or use a sequence of them at some other times; an individual principle may also be repeatedly applied in some period of time. A finite compound of principles $M_{i(1)} o M_{i(2)} o \ldots o M_{i(s)}$ from

M is called *ideas generation operator* or *IG*-operator. For any part A of the Ω_1-space, the DMs could generate new ideas by the operation $M_{i(1)} o M_{i(2)} o \ldots o M_{i(s)}(A)$ that we call *ideas generation operation* or *IG-operation.* Let us denote the set of all *IG*-operators based on M by

$$CM = \{M_{i(1)} o M_{i(2)} o \ldots o M_{i(s)} / M_{i(j)} \in M,\ j = 1, 2, \ldots, s,\ s \in \{1, 2, \ldots\}\}. \tag{2}$$

4.1 Covering

In Sect. 1, we have literally defined covering problem as "how to transform a given competence set CS^* into a set that contains a targeted competence set *CS*". In fact, at any time, the competence set is just a projection of DMs' Habitual Domains (1); hence, it has an Actual Domain, a Reachable Domain, a Potential Domain and Activation Probabilities. In this chapter, we focus only on the Reachable Domain part of competence set, that is, the skills, know-how and resources that can be reached from the Actual Domain. Generally speaking, at any time, the competence set of DMs may include part or all of the decision parameters, skills and resources related to the decision making problem. The process of transformation from one competence set to another can occur when there is a change in Actual Domain or the Reachable Domain is expanded to deeper parts of the Potential Domain or some new ideas are acquired from outside of the DMs' HDs. To realize such transformation, the use of the 7-8-9 principles of deep knowledge is very useful. We will use *IG*-operators as tools of competence set transformation.

4.1.1 Feasibility and Minimum Time and/or Cost Covering

Assume that a time frame, [0, *L*], for completing the covering is given. The first question that arises is: Is there an *IG*-operator *H* that can complete covering within the allowed time? In other words, is the covering problem feasible? This problem can be formulated as follows

$$\text{Find } H,\ H \in CM\ , t(H) \leq L \text{ and } CS \subset H(CS^*), \tag{3}$$

here and in the rest of the chapter, $t(H)$ is the duration of the transformation, by IG-operator *H*, of the given competence set including the time spent for finding or selecting the *IG-operator* H; $CS \subset H(CS^*)$ means that the target competence set *CS* is covered by the new competence set $H(CS^*)$ resulting from the transformation of CS^* through the operation *H*. We will deal with covering feasibility problem in Sect. 4.3. Assume that the covering problem is feasible, i.e. there exists at least one *IG*-operator that can lead to *CS* covering within the allowed time. Then the minimum time covering problem can be formulated as follows.

$$\min t(H), \quad Subject\ to \quad H \in CM,\ t(H) \leq L,\ CS \subset H(CS^*). \tag{4}$$

In the constraints of (4), the set CS^* can be replaced by Ω_1-space. As far as the authors know such a problem has not been discussed in literature. The unique feature of this problem is that it involves the mental operator *H* in *CM* that is defined in a domain that is not endowed with some known mathematical structure to be tractable with traditional optimization methods. Some new mathematical structures are suitable to solve this problem in its general form. This could be a worthy direction of research. For the time being, in order to make the problem (4) tractable by traditional methods, some further restrictive assumptions should be made. The optimization problem (4) can be formulated in a way that takes into account the stages or steps of the decision making process as follows. Assuming that there are $p+1$ stages, we have

$$\begin{aligned} \min T &= \sum_{i=0}^{p} t(H(i)) \\ Subject\ to \quad H(i)(CS^i) &= CS^{i+1}, \quad H(i) \in CM, \quad i = 0, \ldots, p, \\ CS^0 &= CS^*,\ CS \subset CS^{p+1}, \quad T \leq L, \end{aligned} \tag{5}$$

where *H*(*i*) is the *IG*-operator implemented at stage *i*. This problem looks like an optimal control problem, however, it has a fundamental difference, $H(i)$ is not a traditional control function, it is an ideas generating operator and the dynamics of CS^t is not governed by a differential equation or difference-equation.

Assume that the DMs can provide an estimate *c*(*H*) of the cost of any *IG*-operator $H \in CM$, then the minimum cost covering problem can be formulated as follows

$$\min c(H),\ Subject\ to \quad H \in CM,\ CS \subset H(CS^*). \tag{6}$$

When the DMs are interested in time and cost efficiency at the same time, a multiple criteria formulation is more suitable

$$\min c(H),\ \min t(H),\ Subject\ to \quad H \in CM,\ t(H) \leq L,\ CS \subset H(CS^*). \tag{7}$$

The reader may derive more OCS problems from the previous models. In Sect. 5, we present some applications of the models (3)–(7).

Remark 4.1 In traditional optimization, the term "minimization" is about finding the absolute minimum of some objective function subject to some constraints. Absolute minimum may not be reached when a problem involves human psychology and changeable spaces. Therefore, in HD theory "minimization" is about reducing the charge level of the DM to a satisfactory or acceptable level. Thus, minimization in problems (4)–(7) and in the problems that appear in the rest of the chapter should be understood in the HD theory sense not in traditional sense. Moreover, existing dynamic models of decision making as dynamic programming and dynamic stochastic models assume a structured uncertainty, while the problems (4), (5) and

(7) are formulated in changeable spaces, that is, they can incorporate unstructured uncertainty.

4.2 Discovering

In this section, we present the discovering problem as an OCS problem. In Sect. 1, we have seen that discovering is the transformation of a given competence set CS^* into a new competence set so as to solve an unsolved problem or create value. In terms of HD theory, discovering contributes to reducing the charge level (see Appendix and Sect. 2) or relieving the pain of some targeted people. Thus, the general formulation of discovering as an OCS problem is

$$\min ch(H(CS^*)), \; Subject\ to \quad H \in CM \tag{8}$$

where $ch(H(CS^*))$ is the resulting charge level after implementation of the *IG*-operator *H*. When discovery time is limited, a multiple criteria formulation is more suitable

$$\min ch(H(CS^*)), \; \min t(H), \; Subject\ to \quad H \in CM, \; t(H) \leq L \tag{9}$$

where *t*(*H*) is the duration of the operation $H(CS^*)$. In fact, the traditional mathematical programming problem

$$\min \; f(x), \; Subject\ to \quad g_i(x) = b_i, i = 1, 2, \ldots, m, x \in X \tag{10}$$

is a special case of the problem (8). On the one hand the objective function *f*(*x*) of (10) represents profit or cost in general, which is a measure of the most two important goals in economic activities: profit maximizing or cost minimizing. In most of the time, they are the only goals in economic activities. Therefore, deviations from the maximum profit or minimum cost significantly contribute to the charge level of DMs. In most of the time they are its determinants. Thus, the objective function of (10) can be seen as a special case of charge level. On the other hand, generally, the constraints of (10) express the available resources and how they are used for a given decision x. In other words, these constraints express the existing resources (competence set) and how they are used or transformed to create value for a given decision x.

Remark 4.2 The problem (10) could be an optimal control problem i.e. $f(x)$ could be an integral functional, the constraints could be a system of differential equations or difference equations and *x* a time dependent control function *u*(*t*).

The traditional optimization problem reduces to a dis/covering problem. Formulating (10) as an OCS problem, we obtain the problem

$$\min \; c(H), \; Subject\ to \quad H(CS^*) \cap CS \neq \emptyset, H \in CM, \tag{11}$$

where $CS^* = \{x/x \in X, g_i(x) = b_i, i = 1, 2, .., m\}$ is the fixed set of available alternatives derived from the constraints of (10); $CS = \{x^*/x* \in CS^*, f(x^*) = \min_{x \in CS^*} f(x)\}$, is the set of optimal solutions of (10), H is an IG-operator and c(H) is the cost of finding and implementing H, e.g. time cost. In terms of the problem (10), the operator H in (11) may be interpreted as an optimization method or algorithm, e.g. gradient method. Here, covering *CS* means covering at least one element from *CS*. Generally, a solution H of (11), involves mathematical transformations that can be represented mainly by the principle M_{21}, the "Transforming and Changing Principle" and other principles from the 7-8-9 principles. Thus, the formulation of a traditional optimization problem alone is not an OCS problem because its decision parameters are fixed. However, considering the search for its optimal solution, it becomes an OCS problem. In general, (10) may appear at the late stages of the resolution process of an OCS problem once the decision blinds and decision traps are eliminated.

4.3 Necessary and Sufficient Conditions for Covering

In this subsection, we present the dis/covering process in a way that allows us to derive some necessary and sufficient conditions for its completion within the allowable time $[0, L]$. We consider an approach based on the cardinality of the competence set CS^t at any time t.

The resolution of the covering problem depends on the awareness of the DMs of the decision parameters related to the problem, the 7-8-9 principles and the dynamics of their HDs. Naturally, in covering problems, the HDs and competence sets of the DMs should not stay trapped for long time in some area before covering is completed, especially, when the available time is limited and short. Thus, by avoiding long-lasting decision traps within the allowed time $[0, L]$, the covering problem could be solved. Generally speaking, if the allowed time is large enough, the covering problem could be solved when there is a continuous acquiring (up to some period Δ) of new elements from the targeted competence set *CS*. In order to get some practical results, we make two general assumptions. Assumption 4.1 is valid for this section only, while Assumption 4.2 is valid for the rest of the chapter.

Assumption 4.1 At any time t of the dis/covering process, the DMs have a correct perception about their actual competence set and the actual targeted competence set. Moreover, the targeted competence set *CS* is constant.

Assumption 4.2 One may generally assume that when ideas or skills or resources are acquired they are not lost in the future, that is, the sequence of competence sets CS^t is non decreasing, that is, $CS^t \subset CS^{t'}$ for all t, t' such that $t < t'$.

It is important to note the difference between the target competence set CS and the problem that DMs face. The DMs know the problem to be solved, but they may completely or partially ignore the target competence set, CS, required to solve it. For instance, in the Example 2.1, the two candidates know the problem to be solved:

make sure one's horse passes the finishing line last, while they do not know how the achieve this, that is, they do not know the target competence set, CS. Later we will see how candidate A discovered CS.

4.3.1 Cardinality Approach to Covering

Let us assume that the initial competence set CS^* and the targeted competence set *CS* are finite (it is generally the case), i.e. $CS^* = \{a_1, a_2, \ldots, a_n\}$ and $CS = \{b_1, b_2, \ldots, b_m\}$.

Definition 4.1 Let $CS^0 = CS^*$, then the covering problem can be formulated as follows.

$$\text{Find the first time } t^{**} \leq L \text{ such that } CS \subset CS^{t^{**}}, \tag{12}$$

where *L* is the maximum time allowed to solve the covering problem.

Denote by $AQ^t = CS^t \cap CS$ the acquired set of ideas, skills and resources from the targeted competence set *CS* at time *t* and let $q_t = Card\{AQ^t\}$ be the cardinality of AQ^t . Then, the covering problem (12) can be simply formulated as follows

$$\text{Find the smallest time } t^{**} \leq L \text{ such that } q_{t^{**}} = m = Card\{CS\}. \tag{13}$$

Definition 4.2 (*Decision trap*) We say that the decision makers are in a decision trap iff there exists some time t^0 such that q_t is constant for all *t* in the interval $[t^0, t^1]$, where $t^1 > t^0 + \Delta$. The time Δ, $0 < \Delta < L$ depends on the allowable covering time [0, *L*], it can be subjectively set by the decision makers. Δ is called *decision trap threshold*.

Definition 4.2 means that, the DMs cannot acquire any new elements from $CS \backslash CS^{t^0}$ to be added to CS^t during the period $[t^0, t^1]$. In terms of HD theory, Δ depends on the DMs' charge level. If it is high, the DMs tend to take a small value for Δ, whereas when it is low, the DMs tend to take a large value for Δ. It is important to note that it often happens that the DMs fall in a decision trap without being aware of it. In this case, the period Δ is not relevant and the covering process may not be completed within the allowed time. In this chapter, we assume that when the DMs are in a decision trap, they are aware of it. A covering process that does not involve decision traps can be formally defined as follows.

Definition 4.3 We say that the covering process is *operational* iff for all *t* such that $CS \backslash CS^t \neq \emptyset$, there exists some period $r(t) \leq \Delta$ such that $t + r(t) \leq L$ and $q_{t+r(t)} - q_t \geq 1$.

In other words, starting from any time *t*, the competence set CS^t, does not stop acquiring additional elements from *CS* for a period longer than Δ, which means that an operational covering process is a process that does never falls in a decision trap. The condition $CS \backslash CS^t \neq \emptyset$ means that the covering process is not completed at time *t*, while $t + r(t) \leq L$ and $q_{t+r(t)} - q_t \geq 1$ mean that at least one element from

$CS \backslash CS^t$ is acquired by DMs within the remaining time [t, L], without falling in a decision trap. Consequently, we have the following necessary condition for covering completion within the allowable time.

Necessary conditions: There are two necessary conditions for completing the covering.

(i) In case the covering process is expected to be operational, a necessary and sufficient condition is that $t^* = \max\{t(b_i), i = 1, 2, \ldots, m\} \leq L$, where $m = Card\{CS\}$ and $t(b_i)$ is the smallest time t such that $b_i \in CS^t$, i.e. the first time b_i is acquired by the DMs. Let us assume that the acquiring of elements from *CS* takes place sequentially, that is, at any time t, the covering process is dedicated fully to the acquisition of only one element from *CS*. Consequently, since $m = Card\{CS\}$, in the extreme case when each element of *CS* requires the maximum time period Δ to be covered in an operational covering process, the necessary condition is $m\Delta \leq L$ (in Proposition 4.1, we prove that, in general, it is also a sufficient condition). Thus, the period of time $m\Delta$ can be taken as an upper bound for completing the covering process on time in an operational covering, provided that $m\Delta \leq L$. In case the DMs expects only a certain number s of problems that will take the maximum resolution period of time Δ in an operational covering process, then the necessary condition to complete the covering process is $s\Delta \leq L$. It is important to note that this necessary condition is not valid if the acquisition of elements from *CS* is not sequential (i.e. it can be parallel).

(ii) In case the DMs expect to fall in a certain number of decision traps (challenging problems), then the necessary condition to complete the covering process is $t_{\max} \leq L$, where

$$t_{\max} = \max\{t / \text{ the DMs are in a decision trap at time } t\}.$$

Sufficient conditions. The following two propositions establish the feasibility or sufficient condition for a covering within the allowable covering time, when the process is operational.

Proposition 4.1 Assume that the acquisition of elements from *CS* is sequential, $m\Delta \leq L$ and the covering process is operational. Then the covering of *CS* can be achieved within the allowable covering time [0, L].

Proof Let us recall that based on (13) the covering problem is solved when $q_t = m$ for some time t. Assume the worst case, that for all t such that $t \leq L - \Delta$, we have $r(t) = \Delta$, since the process is operational, we have $q_{t+\Delta} - q_t \geq 1$, for all $t \leq L - \Delta$. Therefore, $q_{m\Delta} = (q_{m\Delta} - q_{(m-1)\Delta}) + (q_{(m-1)\Delta} - q_{(m-2)\Delta}) + \ldots. + (q_\Delta - q_0) + q_0 \geq m$. Then either $q_{s\Delta} = m$ for some $s < m$, then $q_{m\Delta} = m$ because q_t is non decreasing by Assumption 4.2 or the process continues until the time $m\Delta$, then $q_{m\Delta} = m$ because the process is operational.

Let us now turn to the difficult case when the process encounters decision traps. Assume that the DMs enter a decision trap at some time t^0 and consider the problem at some time $t^1 > t^0 + \Delta$. As stated above, this means that the DMs are stuck and recognize that they cannot reach or acquire any additional point from *CS* after trying during the period Δ. In terms of HD theory, this means that the Actual Domains and/or the Reachable Domains of the DMs are trapped in some area. In order to reach an additional element from *CS*, the DMs need either to change their Actual Domains and/or expand their Reachable Domains towards *CS*. As for the change of Actual Domains, the DMs have to consider decision parameters (see Sects. 2.1.1–2.1.2) and go deeper into their Potential Domains to expand their Reachable Domains, to come out with new ideas and concepts. For this purpose, the 7-8-9- principles presented in Sect. 3.2 can be very useful.

Definition 4.5 At any time t^1, an *IG*-operation using an *IG*-operator *H* is said to be *successful* in advancing the covering process, if there exists a subset *A* of Ω_1-space such that $H(A) \cap (CS \backslash CS^{t^1}) \neq \emptyset$. Otherwise, it is said unsuccessful.

Most likely, the subset *A* of the Ω_1-space could beCS^{t^1} itself or part of CS^{t^1} or contain only part of it. If *H* is successful, the new competence set is $CS^{t^2} = CS^{t^1} \cup (H(A) \cap (CS \backslash CS^{t^1}))$, which is larger than CS^{t^1}. In terms of the sequence q_t introduced above, an *IG*-operation that starts at time t^1 and ends at time t^2 is said to be successful if $q_{t^1} < q_{t^2}$. Thus, we have the following sufficient condition for solving the covering problem.

Proposition 4.2 Assume that the following conditions are satisfied:

(i) $t_{max} \leq L$
(ii) $\Delta(m - q_{t_{\max}}) \leq L - t_{\max}$
Then the covering problem can be solved within the given time frame [0, *L*].

Proof The condition (i) implies that, each time the DMs get into a decision trap, they are able to get out of it by acquiring at least one additional element from *CS* within the time frame. After the time t_{max}, the covering process becomes operational, i.e. no decision trap is expected. Then DMs will acquire at least one new element from *CS* within each period of time not exceedingΔ. Since at time t_{max} the number of acquired elements from *CS* is $q_{t_{\max}}$, then the number of uncovered elements is $m - q_{t_{\max}}$. If $m - q_{t_{\max}} = 0$, the covering is completed. Assume that $m - q_{t_{\max}} > 0$. Taking into account Assumption 4.2, the number $m - q_{t_{\max}}$ of elements would require a maximum covering time of $\Delta(m - q_{t_{\max}})$. Therefore, to complete the covering process, the time that remains after t_{max}, i.e.$(L - t_{\max})$, should be larger than $\Delta(m - q_{t_{\max}})$, which is guaranteed by condition (ii).

5 Applications

In Sects. 1–3, we have seen that a DMCS problem is a challenging problem that involves decision parameters that the DMs may ignore or have to discover. Moreover, we have pointed out that such problems cannot be solved by existing decision making models or optimization techniques. In Sect. 3, the basic mental operators,

7-8-9 principles, were presented as tools that can help expand the HDs of DMs so as to reduce the blinds and get out of decision traps to solve dis/covering problems. In Sect. 4 the covering and discovering problem were formulated as OCS problems. In this section, to illustrate this new class of optimization problems, we formulate the DMCS problems presented in Examples 1.1–1.2 as OCS problems using the models (4)–(9). Moreover, we provide a mathematical expression of some IG-operators from (2) to show the possibility of solving OCS problems mathematically.

Example 5.1 In Example 1.2 (Horse Race), at the beginning, the decision making parameters of the problem including the decision elements and decision environment facets as described in Sect. 2 are as follows. As for the decision elements, the set of alternatives is empty for both players because they don't know how to solve the problem. There is only one criterion in this game situation, the ranking of the candidate's horse when it crosses the finishing line. The outcomes of the horse race for each candidate are either his horse crosses the finishing line first or second. Each candidate prefers to make his horse cross the finishing line last. Finally, information input consists of the rules of the race given by the president of the company and any information that each candidate could get about the other. As for the environmental facets, the situation involves two players, candidates A and B that are involved in a horse race with very specific rules, therefore, each of them needs to understand and monitor the behavior of the other and devise strategies accordingly.

The race finished with A as a winner as follows. Candidate A jumped on B's horse and rode as fast as he could to the finishing line, while leaving his own horse behind. By the time B realized what was going on, it was already too late! Naturally, A became the new chairman.

This game involves two stages. The first stage is the decision trap period, when the two candidates did not know what to do. The second stage covers the resolution process to get out of the decision trap implemented by the candidate A. The unknown in such situations is the behavior and the competence set of the other candidate. Here, the use of the eight hypotheses H1-H8 of HD theory and the derived behavioral mechanism [31, 34] (see Fig. 1) is essential to understand the behavior of the involved DMs. In order to win the race, a candidate has to understand the decision elements and the environmental facets of the race situation, then based on his understanding of this situation, he needs to evaluate his competence set CS^* and the required competence set CS, if possible. Finally, by using the 7-8-9 principles, he may expand his competence set CS^* to cover CS or discover a CS (in case CS is unknown) that will make him feel confident to win the race.

Let us now formulate the problem as an OCS problem from the perspective of candidate A, a similar OCS problem can be formulated for candidate B. Once the two candidates have been told the details and the rules of the game, they were puzzled, i.e. they were in a decision trap. The reason is that the rules of the game are not the commonly known rules of horse race: generally, the rule is that the first candidate crossing the finishing line is the winner. The candidates faced a discovery problem. Thus, the initial competence sets $CS^*(\mathrm{A})$ and $CS^*(\mathrm{B})$ of A and B, respectively, consisted of the traditional knowledge about the rules of horse races and the individual skills of riding a horse. Here, since the candidates are in a game situation or competition, staying puzzled for few seconds may be considered as being in a decision trap. In game situations, falling in a decision trap, generally, leads to losing the game. The targeted competence set CS of each candidate is the set of competencies needed to win over the other candidate by making sure that his own horse crosses the finishing line last. Thus, we obtain the following OCS problem for candidate A

$$\min t_A(H), \ Subject\ to \quad H \in CM\,, CS \subset H(CS^*(A)), \tag{14}$$

where $t_A(H)$ is the duration of the *IG*-operation $H(CS^*(\text{A}))$ from A's perspective. In this problem, there is no specific time limitation. The game ends when one of the two horses crosses the finishing line. Let us now see how the candidate A developed a solution of (14) in steps.

Step 1. The candidate A analyzed the rules of the game using the principles "Deep and Down Principle", M_{16} , and "Projecting from a Higher Position", M_9, to activate new ideas from his Potential Domain to his Actual Domain or to his Reachable Domain, he then determined the most important objects that are involved in the horse race as the two pairs (A, H1) and (B, H2), that is, candidate A and his horse and candidate B and his horse. This operation resulted in a new competence set $CS^{t_1}(A)$ consisting of $CS^*(\text{A})$ and the pairs (A, H1) and (B, H2) as the main focus. The transformation from CS^* (A) to $CS^{t_1}(A)$ by using the operators M_{16} and M_9 can be expressed as follows

$$\text{M}_9 \,\text{o}\, \text{M}_{16}\,(CS^*(\text{A})) = CS^{t_1}(A). \tag{15}$$

Let us elaborate more on (15). At the beginning both candidates were in a stressful state. In terms of HD theory, we say that their charge level was high. In such psychological state, the Actual Domains and Reachable Domains of the candidates become very narrow; only ideas with strong circuit pattern can be activated from the Potential Domain. Ideas with weak circuit patterns cannot capture their attention. Such ideas can be activated only when the charge level is low. It could be that some very valuable weak circuit pattern ideas in solving the problem cannot be activated from the Potential Domain to the Actual Domain because of the high charge level of the candidates. The "Deep and Down Principle", M_{16} suggests reducing the charge level by relaxing, then thoughts or ideas that require lower charge for their activation can be activated to the Actual Domain. As a result, more relevant and good ideas for solving the problem could be activated to the Actual Domain. By applying this principle, candidate A could activate the idea of applying the principle of "Projecting from a Higher Position", M_9. When faced with a problem in a given system, we tend to look for the best solution within that system and pay minimal attention to other systems. Projecting from higher position to solve the problem means considering it from a different system or from higher perspective. In the problem at hand, candidate A used the principle M_9 to consider the situation from CEO's Position. He then came to the conclusion that the problem could be solved by concentrating his attention on the two pairs (A, H1) and (B, H2) before doing any physical effort: in general, a CEO is interested in managerial and problem solving skills than in horse riding skills. As a result of application of the two principles M_{16} and M_9 sequentially, candidate A transformed his initial competence CS^*(A) into a new competence set $CS^{t_1}(A)$ including the initial competence set and the focus on the pairs (A, H1) and (B, H2).

Step 2. Next, applying the "Alternating Principle", M_{17}, to the pairs (A, H1) and (B, H2), candidate A alternated the horses to obtain the pairs (A, H2) and (B, H1) by jumping on B's horse. This operation changed the rule of the game to "whoever crosses the finishing line first will be the winner". Then, riding B's horse as fast as he could to the finishing line, his horse will be definitely the last to cross the finishing line because candidate B was still trapped in CS^*(B) and did not understand on time what is going on. When B realized what was going on, it was too late! Candidate A won the race. The alternating operation resulted in a new competence set $CS^{t_2}(A)$ for candidate A that obviously includes the needed competence set *CS* for solving the game, which completes the dis/covering problem. The alternating operation can be represented as follows

$$CS \subset CS^{t_2}(A) = M_{17}(CS^{t_1}(A)). \tag{16}$$

One may summarize the whole dis/covering process (15)–(16), of the solution, as follows

$$CS \subset CS^{t_2}(A) = M_{17}oM_9oM_{16}(CS^*(A)).$$

Now, let us formulate mathematically the two steps of the resolution process, then derive the specific OCS problem of the form (14). In terms of HD theory, the operation of Step 1 means that candidate A has brought the two pairs (A, H1) and (B, H2) into his Actual Domain. In terms of the Activation Probability from the Potential Domain to the Actual Domain, this means that the Activation Probability of the pairs (A, H1) and (B, H2) became 1. Mathematically, one can formulate the operation (15) as follows. Assume that the initial competence set of candidate A is $CS^*(A) = \{a_1, a_2, \ldots, a_n\}$. Let $P^0(a_j)$, $j = 1, 2, \ldots, n$, be the initial activation probabilities of $a_1, a_2, \ldots, a_n$, respectively, from his Reachable Domain to his Actual Domain. Generally speaking, the Activation Probability can be seen as a time dependent function with domain as the Potential Domain and range [0,1], that is, $P^t(.)$: PD $\rightarrow$ [0, 1]. Then the operation (15) reduces to the transformation of the initial activation probabilities $P^0(a_j)$, $j = 1, 2, \ldots, n$ to a new set of probabilities $P^{t_1}(a_j)$, $j = 1, 2, \ldots, n$ such that $P^{t_1}((A, H1)) = P^{t_1}((B, H2)) = 1$, where t_1 is the duration of operation (15). This transformation can be expressed as a function S : $\{P^0(a_j), j = 1, 2, \ldots, n\} \rightarrow$ [0, 1], such that $S(P^0(a_j)) = P^{t_1}(a_j)$, $j = 1, 2, \ldots, n$.

Next, one may mathematically formulate the operation (16) as follows. Let us represent the elements A, H1, B and H2 by the numbers 1, 2, 3, and 4, respectively and let the vector d = (1, 2, 3, 4) represent the pairs (A, H1) and (B, H2). Then alternating horses through the "Alternating Principle " , M_{17} , by candidate A, can be identified with the following linear transformation $Fd = d'$, where d = (1, 2, 3, 4), $d' = (1, 4, 3, 2)$ and F is the matrix

$$F = \begin{pmatrix} 1\,0\,0\,0 \\ 0\,0\,0\,1 \\ 0\,0\,1\,0 \\ 0\,1\,0\,0 \end{pmatrix}. \tag{17}$$

Clearly, $d' = (1, 4, 3, 2)$ expresses the desired change of horses represented by the pairs (A, H2) and (B, H1). Thus, by alternating the horses, candidate A is now confident that he can win the race, that is, he has covered the required competence set *CS*, i.e. $CS \subset M_{17}(CS^{t_1}(A)) = CS^{t_2}(A)$, where $t_2 - t_1$ is the duration of operation $M_{17}(CS^{t_1}(A))$. Therefore, min $t_A(H) = t_A(M_{17}oM_9oM_{16}) = t_2$; the discovery problem (15)–(16) can be formulated mathematically as follows.

$$\min t(Fof),\ Subject\ to\ CS \subset F\,o\,f\,(CS^*(A)),\ f \in V,\ F \in W,$$

where *t(Fof)* is the duration of the operation Fof, $V = \{f/f : PD \rightarrow [0, 1]\}$ is the set of functions that assign activation probability from the Potential Domain to the Actual Domain and

$$W = \{F/\,F \text{ is a matrix alternating the elements of a finite ordered set}\},$$

the elements of W are similar to the matrix F in (17). The functions in *V* help to identify or select the elements of CS^* that the DMs should focus or concentrate on their attention.

Example 5.2 Let us go back to Example 1.2 (Emerson's Calf). In Example 3.1, we have showed that the decision blind $CS \backslash CS^t$ was nonempty and significant and Emerson and his son were in a decision trap. Since the weather was very cold, they had to solve the problem within a reasonable time as soon as possible. Emerson started to look around for help (See Hypothesis H8, information input, in Sect. 2). He saw his female home-maid, he looked at her perplexed, as if he was giving up and asking for possible help. She smiled and told him: you men come in to relax; I will solve the problem within few minutes. Indeed, she just put her forefinger into the mouth of the Calf, which immediately started to suck it. The maid, started to move into the barn, the Calf followed her peacefully. Note that it is by analogy and association (Hypothesis H4) between sucking the maid's finger and sucking its mother's teat that the Calf followed the maid peacefully.
In terms of HD theory, the maid could solve the problem because as women, by analogy and association (see hypothesis H4 in Sect. 2) to human behavior, she knows that babies like to suck their or other person finger when they are hungry. Moreover, as a mother, this knowledge is in the core of her HD, that is, the corresponding circuit pattern is so strong that it can be almost surely activated, that is, with probability 1 or close to 1 (see the concept of Activation Probability of the HD theory, in Sect. 2) when the situation requires it. Thus, the maid solved the problem because her competence set covers completely the required competence set *CS* to solve it.

Let us now formulate this DMCS problem as OCS problem and construct its solution. Clearly, the problem consists of two phases. Since Emerson and his son wanted to bring the Calf into the barn in a very cold day, naturally, achieving this as soon as possible would be better. Thus, the time minimization model

$$\min t_A(H),\ Subject\ to \quad H \in CM,\ CS \subset H(CS^*(A)) \tag{18}$$

is the most suitable to formulate the corresponding OCS problem. Let us now construct the *IG*-operator $H^* \in CM$ that solves the OCS problem (18). The resolution process consists of two phases. The first phase covers the time before the maid came into picture; the second covers the time after the maid intervened. In the first phase, Emerson and his son were in a decision trap. During this phase, we have $CS^t = CS^*$, that is, CS^t is constant. In the second phase, using the "Contradiction Principle", M_{22}, Emerson came to the conclusion that the strategy using force has to be stopped, however, he didn't know what to do to solve the problem. Then using the "Changing and Transforming Principle", M_{21}, and the Hypothesis H8 (Information Input Hypothesis, see Sect. 2 and Appendix) of HD theory, Emerson could change one of the parameters of the game situation (see Sect. 2.1.1), the set of players, by implicitly appealing to the maid as a new player that may have the potential to solve the problem. In terms of HD theory, a union or integration of two HDs has taken place to solve the same problem. Naturally, the resulting competence set would be larger than the individual competence sets. This phase of the game can be characterized by the following transformation

$$\mathrm{M}_{21} \circ \mathrm{M}_{22}\,(CS^*) = CS^1(\mathrm{ES} \cup \mathrm{MA}), \tag{19}$$

where ES means Emerson and son, MA means Maid, and $CS^1(\mathrm{ES} \cup \mathrm{MA})$ is the competence set resulting from aggregation of the competence set of Emerson and his son and the competence set of the maid. Next, by the "Principle of Active Association", M_{10}, and the hypothesis H4 of HD theory, the maid related the behavior of the Calf to the behavior of a human baby: they both like to suck milk from the teat / breast of their mother. Since sucking the forefinger by babies is in the core of HD of the maid, as mentioned above, she immediately activated it to her Actual Domain, for decision making. Thus, she was confident that giving her forefinger to the Calf to suck, it would follow her immediately to the barn, that is, she was confident that her competence set covers completely the required competence set to solve the problem i.e.

$$CS \subset CS^2 = M_{10}(CS^1(ES \cup MA)). \tag{20}$$

Thus, from (19) to (20), we conclude that the *IG*-operator $H^* = \mathrm{M}_{10} \circ \mathrm{M}_{21} \circ \mathrm{M}_{22}$ is a solution to the OCS problem (18). It is important to note that this example shows the importance of aggregating competence sets for solving DMCS problems. One can, similar to Example 5.2, construct the mathematical operations corresponding to the IG-operator H^*.

6 Conclusion and Further Research Directions

In this book chapter, we have presented a new general model for solving DMCS problems, focusing on dis/covering process, the OCS. The proposed model is a considerable departure from traditional decision theory framework for it incorporates human psychology and its dynamics, the cognitive aspect and the possibility of restructuring the parameters of the decision problem. The theoretical framework of this model is HD theory. The basic components of this model are the decision parameters (decision elements and environmental facets), the 7-8-9 principles and the concept of competence set. HD theory makes it possible to model the dynamics of the psychological states of the DMs and the decision making process, while the 7-8-9 principles offer the possibility to restructure the decision problem, and the generation of new ideas and strategies to get out of decision traps and/or to cover some targeted competence set. This aspect has never been taken into account in such a comprehensive way, in traditional decision and optimization models. Thus, the introduced model offers new possibilities to decision makers, managers and executives in solving real-world challenging decision problems effectively and efficiently.

Mathematical formulation of covering and discovering processes as well as the necessary and sufficient conditions for their completion are presented. Minimum time and/or cost covering and discovering problems are formulated as new types of optimization problems that can be called *optimization in changeable spaces problems* because of the presence of the mental operators 7-8-9 principles that operate on sets with unknown shapes or boundaries and dynamics. The new optimization models (3)–(9) we presented in this work open new directions of research such as (i) formulation and analysis of the innovation process [37] using DMCS and OCS (ii) the use of DMCS and OCS models in management, conflict resolution and game theory [16, 17, 39], planning and decision making, (iii) the use of DMCS and OCS in artificial intelligence: introduction of the new optimization models (4)–(9) in the emerging discipline of artificial economics [20] and e-economy, (iv) the use of DMCS and OCS in scientific discovery and (v) the use of DMCS and OCS in knowledge extraction (data mining).

The validation of the developed model has not been addressed in this chapter. We suggest the following approach. Consider an unsolved challenging problem and present it to two groups of people having similar expertise. One of the groups will be trained on HD theory, competence set analysis and on the use of the 7-8-9 principles, while the other not. The two groups are given the same period of time to solve the problem. After this period, an evaluation of the performance of the two groups on solving the given problem is made. This experiment has to be repeated enough times with different pairs of groups to allow for a statistical test to be conducted to reach a conclusion about the validity of the model. Alternatively, the same method can be used with the following change: instead of taking two different groups in each experiment, one can take the same group to which two different challenging problems are presented; one before training on HD theory, competence set analysis and the use of the 7-8-9 principles and the other after.

Appendix

Table A.1 Four hypotheses of brain operation

	Hypotheses	Descriptions
H1	Circuit pattern hypothesis	*Thoughts, concepts or ideas are represented by circuit patterns of the brain. The circuit patterns will be reinforced when the corresponding thoughts or ideas are repeated. Furthermore, the stronger the circuit patterns, the more easily the corresponding thoughts or ideas are retrieved in our thinking and decision making processes*
H2	Unlimited capacity hypothesis	*Practically every normal brain has the capacity to encode and store all thoughts, concepts and messages that one intends to*
H3	Efficient restructuring hypothesis	*The encoded thoughts, concepts and messages (H1) are organized and stored systematically as data bases for efficient retrieving. Furthermore, according to the dictation of attention they are continuously restructured so that relevant ones can be efficiently retrieved to release charges (Precursors of mental stress, see H6)*
H4	Analogy/association hypothesis	*The perception of new events, subjects, or ideas can be learned primarily by analogy and/or association with what is already known. When faced with a new event, subject, or idea, the brain first investigates its features and attributes in order to establish a relationship with what is already known by analogy and/or association. Once the right relationship has been established, the whole of the past knowledge (preexisting memory structure) is automatically brought to bear the interpretation and understanding of the new event, subject or idea*
H5	Goal setting and state evaluation hypothesis	*Each one of us has a set of goal functions and for each goal function we have an ideal state or equilibrium point to reach and maintain (goal setting). We continuously monitor, consciously or subconsciously, where we are relative to the ideal state or equilibrium point (state evaluation)*
H6	Charge structure and attention allocation hypothesis	*Each event is related to a set of goal functions. When there is an unfavorable deviation of the perceived value from the ideal, each goal function will produce various levels of charge (a precursor of mental stress). The totality of the charges by all goal functions is called the* ***charge structure*** *and it can change dynamically. At any point in time, our attention will be paid to the event which has the most influence on our charge structure*

Table A.2 Four hypotheses of mind operation

	Hypotheses	Descriptions
H7	Discharge hypothesis	*To release charges, we tend to select the action which yields the lowest remaining charge (the remaining charge is the resistance to the total discharge) and this is called the least resistance principle*
H8	Information inputs hypothesis	*Humans have innate needs to gather external information. Unless attention is paid, external information inputs may not be processed*

References

1. Allison, G.T.: The Essence of Decision: Explaining the Cuban Missile Crisis. Little, Brown and Company, Boston (1971)
2. Andersen D.L., Andersen D.F.: Theories of decision making: an annotated bibliography. Working Paper WP 943–77, Alfred P. Sloan School of Management, Massachusetts Institute of Technology (1977)
3. Bazaraa, M.S., Sherali, H.D., Shetty, C.M.: Nonlinear programming: theory and algorithms. Wiley-Interscience, New York (2006)
4. Bellman, R., Zadeh, L.A.: Decision-making in a fuzzy environment. Manag. Sci. **17**(4), 141–164 (1970)
5. Brehmer, B.: Dynamic decision making: human control of complex systems. Acta Psychol. **81**(3), 211–241 (1992)
6. Busemeyer, J.R., Towsend, J.T.: Decision field theory: a dynamic-cognitive approach to decision making in an uncertain environment. Psychol. Rev. **100**(3), 432–459 (1993)
7. Chan, S.J., Yu, P.L.: Stable habitual domains: existence and implications. J. Math. Anal. Appl. **110**(2), 469–482 (1985)
8. Cyert, R.M., March, J.G.: A behavioral theory of the firm. Prentice-Hall, Englewood Cliffs (1955)
9. Dantzig, G.B.: Linear programming and extensions. Princeton University Press, Princeton (1998)
10. Edwards, W.: Dynamic decision theory and probabilistic information processing. Hum. Factors **4**, 59–73 (1962)
11. Ehrgott, M., Gandibleux X.: Multiple criteria optimization: state of the art annotated bibliographic surveys. Kluwer Academic, Dordrecht (2003)
12. Gonzalez, G., Polina Vanyukov, P., Martin, M.K.: The use of microworlds to study dynamic decision making. Comput. Hum. Behav. **21**, 273–286 (2005)
13. Hsiao, N., Richardson, G.P.: In search of theories of dynamic decision making: a literature review, Proceedings of the 1999 International System Dynamics Conference, Wellington, New Zealand, August 1999. System Dynamics Society, Albany (1999)
14. Kahneman, D., Tversky, A.: Choices, prospect theory; an analysis of decision under risk. Econometrica **47**, 313–327 (1979)
15. Kelley, H.H.: Attribution theory in social psychology. Nebr. Symp. Motiv. **15**, 192–238 (1967)
16. Larbani, M., Yu, P.L.: n-person second-order games: a paradigm shift in game theory. J. Optim. Theory Appl. **149**(3), 447–473 (2011)
17. Larbani, M., Yu, P.L.: Two-person second-order games. Part II: restructuring operations to reach a win-win profile. J. Optim. Theory Appl. **141**, 641–659 (2009)
18. Larbani, M., Yu, P.L.: Decision making and optimization in changeable spaces, a new paradigm. J. Optim. Theory Appl. (JOTA) **155**(3), 727–761 (2012)

19. Luce, R.D., Raiffa, H.: Games and decisions: introduction and critical survey. Wiley, New York (1957)
20. Osinga, S., Hofstede, G.J., Verwaart, T.: Emergent results of artificial economics. Lecture Notes in Economics and Mathematical Systems. Springer, Berlin (2011)
21. Sakawa M., Nishizaki I., Katagiri H.: Fuzzy stochastic multiobjective programming. International series in Operations Research and Management Sciences, vol. 159. Springer, Heidelberg (2011)
22. Shapiro, A., Dentcheva, D., Ruszczyński A.: Lectures on stochastic programming: modeling and theory. MPS/SIAM Series on Optimization. 9, Society for Industrial and Applied Mathematics, Philadelphia (2009)
23. Simon, H.A.: Models of man: social and rational, mathematical essays on rational human behavior in a social Setting. Wiley, New York (1957)
24. Slowinski, R., Teghem, J.: Stochastic versus fuzzy approaches to multiobjective mathematical programming under uncertainty. Kluwer Academic, Dordrecht (1990)
25. Steinbruner, J.D.: The cybernetic theory of decision: new dimensions of political analysis. Princeton University Press, Princeton (1974)
26. Steuer, R.E.: Multiple criteria optimization: theory, computation and application. Wiley, New York (1986)
27. Tversky, A.: Intransitivity of preferences. Psychol. Rev. **76**(1), 31–48 (1969)
28. Tversky A., Kahneman D.: The Framing of decisions and the psychology of choice. Science **211**(4481), 453–458 (1981)
29. Tversky A., Kahneman D., Rational choice and the framing of decisions. J. Bus. **59**(4) Part 2: The Behavioral Foundations of, Economic Theory, S251–S278 (1986)
30. Von Neumann, J., Morgenstern, O.: Theory of games and economic behavior. Princeton University Press, Princeton (1947)
31. Yu, P.L.: Habitual Domains and forming winning strategies. NCTU Press, Taiwan (2002)
32. Yu, P.L.: Habitual domains: freeing yourself from the limits on your life. Highwater Editions, Kansas (1995)
33. Yu, P.L.: Habitual domains. Oper. Res. **39**(6), 869–876 (1991)
34. Yu, P.L.: Forming winning strategies, an integrated theory of habitual domains. Springer, New York (1990)
35. Yu, P.L.: Understanding behaviors and forming winning strategies. Monograph. School of Business, University of Kansas, Kansas (1989)
36. Yu, P.L.: Multiple criteria decision making: concepts, techniques and extensions. Plenum Press, New York (1985)
37. Yu, P.L., Chen, Y.C.: Dynamic multiple criteria decision making in changeable spaces: from habitual domains to innovation dynamics. Ann. Oper. Res. **197**, 201–220 (2012)
38. Yu, P.L., Chen, Y.C.: Blinds, fuzziness and habitual domain tools in decision making with changeable spaces. Hum. Syst. Manag. **29**(4), 231–242 (2010)
39. Yu, P.L., Larbani, M.: Two-person second-order games, Part 1: formulation and transition anatomy. J. Optim. Theory Appl. **141**(3), 619–639 (2009)
40. Yu, P.L., Zhang, D.: A marginal analysis for competence set expansion. J. Optim. Theory Appl. **76**(1), 87–109 (1993)
41. Yu, P.L., Zhang, D.: A foundation for competence set analysis. Math. Soc. Sci. **20**(3), 251–299 (1990)
42. Yu, P.L., Zhang, D.: Competence set analysis for effective decision making. Control Theory Adv. Technol. **5**(4), 523–547 (1989)
43. Zimmermann, H.J.: Fuzzy set theory and its application. Kluwer Academic, New York (2001)

Decision Making Under Interval Uncertainty (and Beyond)

Vladik Kreinovich

Abstract To make a decision, we must find out the user's preference, and help the user select an alternative which is the best—according to these preferences. Traditional utility-based decision theory is based on a simplifying assumption that for each two alternatives, a user can always meaningfully decide which of them is preferable. In reality, often, when the alternatives are close, the user is often unable to select one of these alternatives. In this chapter, we show how we can extend the utility-based decision theory to such realistic (interval) cases.

Keywords Decision Making · Interval Uncertainty · Utility

1 Introduction

To make a decision, we must:

- find out the user's preference, and
- help the user select an alternative which is the best—according to these preferences.

Traditional utility-based decision theory is based on a simplifying assumption that for each two alternatives A' and A'', a user can always meaningfully decide which of them is preferable. In reality, often, when the alternatives are close, the user is often unable to select one of these alternatives. How can we extend the utility-based decision theory to such realistic cases?

In this chapter, we provide an overview of such an extension. This paper is structured as follows: first, we recall the main ideas and results of the traditional utility-based decision theory. We then consider the case when in addition to deciding which of the two alternatives is better, the user can also reply that he/she is unable to decide between the two close alternatives; this leads to interval uncertainty.

V. Kreinovich (✉)
Department of Computer Science, University of Texas at El Paso, 500 W. University, El Paso, TX 79968, USA
e-mail: vladik@utep.edu

P. Guo and W. Pedrycz (eds.), *Human-Centric Decision-Making Models for Social Sciences*, Studies in Computational Intelligence 502,
DOI: 10.1007/978-3-642-39307-5_8,

Comment. Some of the results presented in this paper were previously reported at conferences [1, 23].

2 Traditional Utility-Based Decision Theory: Brief Reminder

Following [8, 27, 35], let us describe the main ideas and results of the traditional decision theory.

Main assumption behind the traditional utility-based decision theory. Let us assume that for every two alternatives A' and A'', a user can tell:

- whether the first alternative is better for him/her; we will denote this by $A'' < A'$;
- or the second alternative is better; we will denote this by $A' < A''$;
- or the two given alternatives are of equal value to the user; we will denote this by $A' = A''$.

Comment. In mathematical terms, we assume that the preference relation $<$ is a linear (total) order; in economics, this property of the preference relation is also known as *completeness.*

The notion of utility. Under the above assumption, we can form a natural numerical scale for describing attractiveness of different alternatives. Namely, let us select a very bad alternative A_0 and a very good alternative A_1, so that most other alternatives are better than A_0 but worse than A_1.

Since we assumed that the alternatives between which we need to choose are linearly ordered, there exists the best one—which can be selected as A_1, and the worst one—which can be selected as A_0. However, since one of the main objectives of this paper is to go beyond this simplifying linearity assumption, it is better to select A_1 and A_0 beyond the available alternatives. For example, we can choose, as A_1, an alternative "I win a billion dollars"—we do not have this alternative in our decision, but this alternative is easy to imagine. Similarly, as A_0, we can select a really bad alternative—and it is OK if this alternative is not a possible outcome of our current decision-making process.

Then, for every probability $p \in [0, 1]$, we can form a lottery $L(p)$ in which we get A_1 with probability p and A_0 with the remaining probability $1 - p$.

When $p = 0$, this lottery simply coincides with the alternative A_0: $L(0) = A_0$. The larger the probability p of the positive outcome increases, the better the result, i.e., $p' < p''$ implies $L(p') < L(p'')$. Finally, for $p = 1$, the lottery coincides with the alternative A_1: $L(1) = A_1$. Thus, we have a continuous scale of alternatives $L(p)$ that monotonically goes from A_0 to A_1.

We have assumed that most alternatives A are better than A_0 but worse than A_1: $A_0 < A < A_1$. Since $A_0 = L(0)$ and $A_1 = L(1)$, for such alternatives, we thus get $L(0) < A < L(1)$. We assumed that every two alternatives can be compared. Thus, for each such alternative A, there can be at most one value p for which $L(p) = A$;

for others, we have $L(p) < A$ or $L(p) > A$. Due to monotonicity of $L(p)$ and transitivity of preference, if $L(p) < A$, then $L(p') < A$ for all $p' \le p$; similarly, if $A < L(p)$, then $A < L(p')$ for all $p' > p$. Thus, the supremum (= least upper bound) $u(A)$ of the set of all p for which $L(p) < A$ coincides with the infimum (= greatest lower bound) of the set of all p for which $A < L(p)$. For $p < u(A)$, we have $L(p) < A$, and for for $p > u(A)$, we have $A < L(p)$. This value $u(A)$ is called the *utility* of the alternative A.

It may be possible that A is equivalent to $L(u(A))$; however, it is also possible that $A \not\equiv L(u(A))$. However, the difference between A and $L(u(A))$ is extremely small: indeed, no matter how small the value $\varepsilon > 0$, we have $L(u(A) - \varepsilon) < A < L(u(A) + \varepsilon)$. We will describe such (almost) equivalence by $\equiv$, i.e., we write that $A \equiv L(u(A))$.

How can we actually find utility values. The above definition of utility is somewhat theoretical, but in reality, utility can be found reasonably fast by the following iterative bisection procedure.

We want to find the probability $u(A)$ for which $L(u(A)) \equiv A$. On each stage of this procedure, we have the values $\underline{u} < \overline{u}$ for which $L(\underline{u}) < A < L(\overline{u})$. In the beginning, we have $\underline{u} = 0$ and $\overline{u} = 1$, with $|\overline{u} - \underline{u}| = 1$.

To find the desired probability $u(A)$, we compute the midpoint $\widetilde{u} = \dfrac{\underline{u} + \overline{u}}{2}$ and compare the alternative A with the corresponding lottery $L(\widetilde{u})$. Based on our assumption, there are three possible results of this comparison:

- if the user concludes that $L(\widetilde{u}) < A$, then we can replace the previous lower bound $\underline{u}$ with the new one $\widetilde{p}$;
- if the user concludes that $A < L(\widetilde{u})$, then we can replace the original upper bound $\overline{u}$ with the new one $\widetilde{u}$;
- finally, if $A = L(\widetilde{u})$, this means that we have found the desired probability $u(A)$.

In this third case, we have found $u(A)$, so the procedure stops. In the first two cases, the new distance between the bounds $\underline{u}$ and $\overline{u}$ is the half of the original distance. By applying this procedure k times, we get values $\underline{u}$ and $\overline{u}$ for which $L(\underline{u}) < A < L(\overline{u})$ and $|\overline{u} - \underline{u}| \le 2^{-k}$. One can easily check that the desired value $u(A)$ is within the interval $[\underline{u}, \overline{u}]$, so the midpoint $\widetilde{u}$ of this interval is an $2^{-(k+1)}$-approximation to the desired utility value $u(A)$.

In other words, for any given accuracy, we can efficiently find the corresponding approximation to the utility $u(A)$ of the alternative A.

How to make a decision based on utility values. If we know the utilities $u(A')$ and $u(A'')$ of the alternatives A' and A'', then which of these alternatives should we choose?

By definition of utility, we have $A' \equiv L(u(A'))$ and $A'' \equiv L(u(A''))$. Since $L(p') < L(p'')$ if and only if $p' < p''$, we can thus conclude that A' is preferable to A'' if and only if $u(A') > u(A'')$.

In other words, we should always select an alternative with the largest possible value of utility.

Comment. Interval techniques can help in finding the optimizing decision; see, e.g., [28].

How to estimate utility of an action: why expected utility. To apply the above idea to decision making, we need to be able to compute utility of different actions. For each action, we usually know possible outcomes $S_1, \ldots, S_n$, and we can often estimate the probabilities $p_1, \ldots, p_n$, $\sum\limits_{i=1}^{n} p_i = 1$, of these outcomes. Let $u(S_1), \ldots, u(S_n)$ be utilities of the situations $S_1, \ldots, S_n$. What is then the utility of the action?

By definition of utility, each situation S_i is equivalent (in the sense of the relation $\equiv$) to a lottery $L(u(S_i))$ in which we get A_1 with probability $u(S_i)$ and A_0 with the remaining probability $1 - u(S_i)$. Thus, the action in which we get S_i with probability p_i is equivalent to complex lottery in which:

- first, we select one of the situations S_i with probability p_i: $P(S_i) = p_i$;
- then, depending on the selected situation S_i, we get A_1 with probability $u(S_i)$ and A_0 with probability $1 - u(S_i)$: $P(A_1 \mid S_i) = u(S_i)$ and $P(A_0 \mid S_i) = 1 - u(S_i)$.

In this complex lottery, we end up either with the alternative A_1 or with the alternative A_0. The probability of getting A_1 can be computed by using the complete probability formula:

$$P(A_1) = \sum_{i=1}^{n} P(A_1 \mid S_i) \cdot P(S_i) = \sum_{i=1}^{n} u(S_i) \cdot p_i.$$

Thus, the original action is equivalent to a lottery in which we get A_1 with probability $\sum\limits_{i=1}^{n} p_i \cdot u(S_i)$ and A_0 with the remaining probability. By definition of utility, this means that the utility of our action is equal to $\sum\limits_{i=1}^{n} p_i \cdot u(S_i)$.

In probability theory, this sum is known as the expected value of utility $u(S_i)$. Thus, we can conclude that the utility of each action is equal to its expected utility; in other words, among several possible actions, we should select the one with the largest value of expected utility.

Non-uniqueness of utility. The above definition of utility depends on a selection of two alternatives A_0 and A_1. What if we select different alternatives A'_0 and A'_1? How will utility change? In other words, if A is an alternative with utility $u(A)$ in the scale determined by A_0 and A_1, what is its utility $u'(A)$ in the scale determined by A'_0 and A'_1?

Let us first consider the case when $A'_0 < A_0 < A_1 < A'_1$. In this case, since A_0 is in between A'_0 and A'_1, there exists a probability $u'(A_0)$ for which A_0 is equivalent to a lottery $L'(u'(A_0))$ in which we get A'_1 with probability $u'(A_0)$ and A'_0 with the remaining probability $1 - u'(A_0)$. Similarly, there exists a probability $u'(A_1)$ for which A_1 is equivalent to a lottery $L'(u'(A_1))$ in which we get A'_1 with probability $u'(A_1)$ and A'_0 with the remaining probability $1 - u'(A_1)$.

By definition of the utility $u(A)$, the original alternative A is equivalent to a lottery in which we get A_1 with probability $u(A)$ and A_0 with the remaining probability

$1 - u(A)$. Here, A_1 is equivalent to the lottery $L'(u'(A_1))$, and A_0 is equivalent to the lottery $L'(u'(A_0))$. Thus, the alternative A is equivalent to a complex lottery, in which:

- first, we select A_1 with probability $u(A)$ and A_0 with probability $1 - u(A)$;
- then, depending on the selection A_i, we get A'_1 with probability $u'(A_i)$ and A'_0 with the remaining probability $1 - u'(A_i)$.

In this complex lottery, we end up either with the alternative A'_1 or with the alternative A'_0. The probability $u'(A) = P(A'_1)$ of getting A'_1 can be computed by using the complete probability formula:

$$\begin{aligned} u'(A) = P(A'_1) = & P(A'_1 \mid A_1) \cdot P(A_1) + P(A'_1 \mid A_0) \cdot P(A_0) = \\ & u'(A_1) \cdot u(A) + u'(A_0) \cdot (1 - u(A)) = \\ & u(A) \cdot (u'(A_1) - u'(A_0)) + u'(A_0). \end{aligned}$$

Thus, the original alternative A is equivalent to a lottery in which we get A'_1 with probability $u'(A) = u(A) \cdot (u'(A_1) - u'(A_0)) + u'(A_0)$. By definition of utility, this means that the utility $u'(A)$ of the alternative A in the scale determined by the alternatives A'_0 and A'_1 is equal to $u'(A) = u(A) \cdot (u'(A_1) - u'(A_0)) + u'(A_0)$.

Thus, in the case when $A'_0 < A_0 < A_1 < A'_1$, when we change the alternatives A_0 and A_1, the new utility values are obtained from the old ones by a linear transformation. In other cases, we can use auxiliary events A''_0 and A''_1 for which $A''_0 < A_0, A'_0$ and $A_1, A'_1 < A''_1$. In this case, as we have proven, transformation from $u(A)$ to $u''(A)$ is linear and transformation from $u'(A)$ to $u''(A)$ is also linear. Thus, by combining linear transformations $u(A) \to u''(A)$ and $u''(A) \to u'(A)$, we can conclude that the transformation $u(A) \to u'(A)$ is also linear.

So, in general, utility is defined modulo an (increasing) linear transformation $u' = a \cdot u + b$, with $a > 0$.

Comment. So far, once we have selected alternatives A_0 and A_1, we have defined the corresponding utility values $u(A)$ only for alternatives A for which $A_0 < A < A_1$. For such alternatives, the utility value is always a number from the interval $[0, 1]$.

For other alternatives, we can define their utility $u'(A)$ with respect to different pairs A'_0 and A'_1, and then apply the corresponding linear transformation to re-scale to the original units. The resulting utility value $u(A)$ can now be an arbitrary real number.

Subjective probabilities. In our derivation of expected utility, we assumed that we know the probabilities p_i of different outcomes. In practice, we often do not know these probabilities, we have to rely on a subjective evaluation of these probabilities. For each event E, a natural way to estimate its subjective probability is to compare the lottery $\ell(E)$ in which we get a fixed prize (e.g., \$1) if the event E occurs and 0 is it does not occur, with a lottery $\ell(p)$ in which we get the same amount with probability p. Here, similarly to the utility case, we get a value $ps(E)$ for which $\ell(E)$ is (almost) equivalent to $\ell(ps(E))$ in the sense that $\ell(ps(E) - \varepsilon) < \ell(E) < \ell(ps(E) + \varepsilon)$ for every $\varepsilon > 0$. This value $ps(E)$ is called the *subjective probability* of the event E; see, e.g., [5, 25, 27, 37].

For each event E, we can efficiently find its subjective probability by using a bisection procedure which is similar to how we can find utilities.

From the viewpoint of decision making, each event E is equivalent to an event occurring with the probability $ps(E)$. Thus, if an action has n possible outcomes $S_1, \ldots, S_n$, in which S_i happens if the event E_i occurs, then the utility of this action is equal to $\sum\limits_{i=1}^{n} ps(E_i) \cdot u(S_i)$.

3 Towards a More Realistic Way to Describe User Preference: Interval Uncertainty

Beyond traditional utility-based decision making: towards a more realistic description. Previously, we assumed that a user can always decide which of the two alternatives A' and A'' is better:

- either $A' < A''$,
- or $A'' < A'$,
- or $A' \equiv A''$.

In practice, a user is sometimes unable to meaningfully decide between the two alternatives A' and A''; see, e.g., [9, 27]. We will denote this option by $A' \parallel A''$.

In mathematical terms, this means that the preference relation is no longer a *total* (linear) order, it can be a *partial* order.

From utility to interval-valued utility. Similarly to the traditional utility-based decision making approach, we can select two alternatives $A_0 < A_1$ and compare each alternative A which is better than A_0 and worse than A_1 with lotteries $L(p)$. The main difference is that here, the supremum $\underline{u}(A)$ of all the values p for which $L(p) < A$ is, in general, smaller than the infimum $\overline{u}(A)$ of all the values p for which $A < L(p)$. Thus, for each alternative A, instead of a single value $u(A)$ of the utility, we now have an *interval* $[\underline{u}(A), \overline{u}(A)]$ such that:

- if $p < \underline{u}(A)$, then $L(p) < A$;
- if $p > \overline{u}(A)$, then $A < L(p)$; and
- if $\underline{u}(A) < p < \overline{u}(A)$, then $A \parallel L(p)$.

We will call this interval the *utility* of the alternative A.

How to efficiently elicit the interval-valued utility from the user. To elicit the corresponding utility interval from the user, we can use a slightly modified version of the above bisection procedure. At first, the procedure is the same as before: namely, we produce a narrowing interval $[\underline{u}, \overline{u}]$ for which $L(\underline{u}) < A < L(\overline{u})$.

We start with the interval $[\underline{u}, \overline{u}] = [0, 1]$, and we repeatedly compute the midpoint $\widetilde{u} = \dfrac{\underline{u} + \overline{u}}{2}$ and compare A with $L(\widetilde{u})$. If $L(\widetilde{u}) < A$, we replace $\underline{u}$ with $\widetilde{u}$; if $A < L(\widetilde{u})$, we replace $\overline{u}$ with $\widetilde{u}$. If we get $A \parallel L(\widetilde{p})$, then we switch to the new second stage of the iterative algorithm. Namely, now, we have *two* intervals:

- an interval $[\underline{u}_1, \overline{u}_1]$ (which is currently equal to $[\underline{u}, \widetilde{u}]$) for which $L(\underline{u}_1) < A$ and $L(\overline{u}_1) \parallel A$, and
- an interval $[\underline{u}_2, \overline{u}_2]$ (which is currently equal to $[\widetilde{u}, \overline{u}]$) for which $L(\underline{u}_2) \parallel A$ and $A < L(\overline{u}_2)$.

Then, we perform bisection of each of these two intervals. For the first interval, we compute the midpoint $\widetilde{u}_1 = \dfrac{\underline{u}_1 + \overline{u}_1}{2}$, and compare the alternative A with the lottery $L(\widetilde{u}_1)$:

- if $L(\widetilde{u}_1) < A$, then we replace $\underline{u}_1$ with $\widetilde{u}_1$;
- if $L(\widetilde{u}_1) \parallel A$, then we replace $\overline{u}_1$ with $\widetilde{u}_1$.

As a result, after k iterations, we get the value $\underline{u}(A)$ with accuracy 2^{-k}.

Similarly, for the second interval, we compute the midpoint $\widetilde{u}_2 = \dfrac{\underline{u}_2 + \overline{u}_2}{2}$, and compare the alternative A with the lottery $L(\widetilde{u}_2)$:

- if $L(\widetilde{u}_2) \parallel A$, then we replace $\underline{u}_2$ with $\widetilde{u}_2$;
- if $A < L(\widetilde{u}_2)$, then we replace $\overline{u}_2$ with $\widetilde{u}_2$.

As a result, after k iterations, we get the value $\overline{u}(A)$ with accuracy 2^{-k}.

Comment. Similar to the case of exactly known utilities, when we replace alternatives A_0 and A_1 with alternatives A'_0 and A'_1, the new values $\underline{u}'$ and $\overline{u'}$ are related to the original values $\underline{u}$ and $\overline{u}$ by the same linear transformation $u' = a \cdot u + b$: $\underline{u}' = a \cdot \underline{u} + b$ and $\overline{u'} = a \cdot \overline{u} + b$.

Interval-valued subjective probability. Similarly, when we are trying to estimate the probability of an event E, we no longer get a single value $ps(E)$, we get an interval $[\underline{ps}(E), \overline{ps}(E)]$ of possible values of probability.

By using bisection, we can feasibly elicit the values $\underline{ps}(E)$ and $\overline{ps}(E)$; alternative ways of eliciting interval-valued probabilities are described in [13, 14].

4 Decision Making Under Interval Uncertainty

Need for decision making under interval uncertainty. In the traditional utility-based approach, for each alternative A, we produce a number $u(A)$—the utility of this alternative. Then, an alternative A' is preferable to the alternative A'' if and only if $u(A') > u(A'')$.

How can we make a similar decision in situations when we only know interval-valued utilities?

Comment. Several approaches have been proposed for such decision-making; for example, several approaches for decision making under interval-valued probabilities are described and compared in [42]. In this chapter, we concentrate on approaches which naturally extend the above utility approach.

How to make a decision under interval uncertainty: a natural idea. For each possible decision d, we know the interval $[\underline{u}(d), \overline{u}(d)]$ of possible values of utility. Which decision shall we select? A seemingly natural idea is to select all decisions d_0 that *may* be optimal, i.e., which are optimal for some function $u(d) \in [\underline{u}(d), \overline{u}(d)]$. There is a minor problem with this definition: that checking all possible functions is not feasible. However, this problem is easy to solve, since this condition can be reformulated in simpler equivalent terms.

Let us describe this reformulation.

Definition 1. *Let D be a set; its elements will be called* possible decisions. *Let $\mathbf{u}$ be a function that assigns, to each possible decision $d \in D$, an interval $\mathbf{u}(d) = [\underline{u}(d), \overline{u}(d)]$. A function u which maps D into real numbers is called a* possible utility function *if $\underline{u}(d) \le u(d) \le \overline{u}(d)$ for all d. We say that a decision d_0 is* possibly optimal *if $u(d_0) = \max\limits_{d \in D} u(d)$ for some possible utility function u.*

Proposition. *A decision d_0 is possibly optimal if and only if*

$$\overline{u}(d_0) \ge \max_d \underline{u}(d).$$

Comment. This equivalent inequality is indeed easy to check.

Proof. If d_0 is possibly optimal, then $u(d_0) \ge u(d)$ for all d. Thus, from $\overline{u}(d_0) \ge u(d_0) \ge u(d) \ge \underline{u}(d)$, we conclude that $\overline{u}(d_0) \ge \underline{u}(d)$ for all d. Hence, we get $\overline{u}(d_0) \ge \max\limits_d \underline{u}(d)$.

Vice versa, suppose that $\overline{u}(d_0) \ge \max\limits_d \underline{u}(d)$, i.e., that $\overline{u}(d_0) \ge \underline{u}(d)$ for all d. Then, we can take the following possible utility function u: $u(d_0) = \overline{u}(d_0)$ and $u(d) = \underline{u}(d)$ for all $d \ne d_0$. For this possible utility function, $u(d_0) \ge u(d)$ for all d, so d_0 is indeed a possibly optimal decision. The equivalence is proven.

Comment. Interval computations can help in describing the range of all such d_0; see, e.g., [28].

Need for definite decision making. In practice, we would like to select *one* decision; which one should be select?

At first glance, the situation may sound straightforward: if $A' \parallel A''$, it does not matter whether we select A' or A''. However, this is *not* a good way to make a decision. For example, let us assume that there is an alternative A about which we know nothing. In this case, we have no reason to prefer A or $L(p)$, so we have $A \parallel L(p)$ for all p. By definition of $\underline{u}(A)$ and $\overline{u}(A)$, this means that we have $\underline{u}(A) = 0$ and $\overline{u}(A) = 1$, i.e., the alternative A is characterized by the utility interval [0, 1].

In this case, the alternative A is indistinguishable both from a good lottery $L(0.999)$ (in which the good alternative A_1 appears with probability 99.9 %) and from a bad lottery $L(0.001)$ (in which the bad alternative A_0 appears with probability 99.9 %). If we recommend, to the user, that A is equivalent both to to $L(0.999)$ and $L(0.001)$, then this user will feel comfortable exchanging his chance to play in the good lottery with A, and then—following the same logic—exchanging A with a chance to play in a bad lottery. As a result, following our recommendations, the user switches from a very good alternative to a very bad one.

This argument does not depend on the fact that we assumed complete ignorance about A. Every time we recommend that the alternative A is equivalent to $L(p)$ and $L(p')$ with two different values $p < p'$, we make the user vulnerable to a similar switch from a better alternative $L(p')$ to a worse one $L(p)$. Thus, there should be only a single value p for which A can be reasonably exchanged with $L(p)$.

In precise terms: we start with the utility interval $[\underline{u}(A), \overline{u}(A)]$, and we need to select a single utility value u for which it is reasonable to exchange the alternative A with a lottery $L(u)$. How can we find this value u?

How to make decisions under interval uncertainty: Hurwicz optimism-pessimism criterion. The problem of decision making under such interval uncertainty was first handled by the future Nobelist L. Hurwicz in [16].

We need to assign, to each interval $[\underline{u}, \overline{u}]$, a utility value $u(\underline{u}, \overline{u})$.

No matter what value u we get from this interval, this value will be larger than or equal to $\underline{u}$ and smaller than or equal to $\overline{u}$. Thus, the equivalent utility value $u(\underline{u}, \overline{u})$ must satisfy the same inequalities: $\underline{u} \leq u(\underline{u}, \overline{u}) \leq \overline{u}$. In particular, for $\underline{u} = 0$ and $\overline{u} = 1$, we get $0 \leq \alpha_H \leq 1$, where we denoted $\alpha_H \stackrel{\text{def}}{=} u(0, 1)$.

We have mentioned that the utility is determined modulo a linear transformation $u' = a \cdot u + b$. It is therefore reasonable to require that the equivalent utility does not depend on what scale we use, i.e., that for every $a > 0$ and b, we have

$$u(a \cdot \underline{a} + b, a \cdot \overline{u} + b) = a \cdot u(\underline{u}, \overline{u}) + b.$$

In particular, for $\underline{u} = 0$ and $\overline{u} = 1$, we get

$$u(b, a + b) = a \cdot u(0, 1) + b = a \cdot \alpha_H + b.$$

So, for every $\underline{u}$ and $\overline{u}$, we can take $b = \underline{u}$, $a = \overline{u} - \underline{u}$, and get

$$u(\underline{u}, \overline{u}) = \underline{u} + \alpha_H \cdot (\overline{u} - \underline{u}) = \alpha_H \cdot \overline{u} + (1 - \alpha_H) \cdot \underline{u}.$$

This expression is called *Hurwicz optimism-pessimism criterion*, because:

- when $\alpha_H = 1$, we make a decision based on the most optimistic possible values $u = \overline{u}$;
- when $\alpha_H = 0$, we make a decision based on the most pessimistic possible values $u = \underline{u}$;
- for intermediate values $\alpha_H \in (0, 1)$, we take a weighted average of the optimistic and pessimistic values.

So, if we have two alternatives A' and A'' with interval-valued utilities $[\underline{u}(A'), \overline{u}(A')]$ and $[\underline{u}(A''), \overline{u}(A'')]$, we recommend an alternative for which the equivalent utility value is the largest. In other words, we recommend to select A' if $\alpha_H \cdot \overline{u}(A') + (1-\alpha_H) \cdot \underline{u}(A') > \alpha_H \cdot \overline{u}(A'') + (1-\alpha_H) \cdot \underline{u}(A'')$ and A'' otherwise.

Which value α_H should we choose? An argument in favor of $\alpha_H = 0.5$. Which value α_H should we choose?

To answer this question, let us take an event E about which we know nothing. For a lottery L^+ in which we get A_1 if E and A_0 otherwise, the utility interval is $[0, 1]$, thus, from a decision making viewpoint, this lottery should be equivalent to an event with utility $\alpha_H \cdot 1 + (1 - \alpha_H) \cdot 0 = \alpha_H$.

Similarly, for a lottery L^- in which we get A_0 if E and A_1 otherwise, the utility interval is $[0, 1]$, thus, this lottery should also be equivalent to an event with utility $\alpha_H \cdot 1 + (1 - \alpha_H) \cdot 0 = \alpha_H$.

We can now combine these two lotteries into a single complex lottery, in which we select either L^+ or L^- with equal probability 0.5. Since L^+ is equivalent to a lottery $L(\alpha_H)$ with utility α_H and L^- is also equivalent to a lottery $L(\alpha_H)$ with utility α_H, the complex lottery is equivalent to a lottery in which we select either $L(\alpha_H)$ or $L(\alpha_H)$ with equal probability 0.5, i.e., to $L(\alpha_H)$. Thus, the complex lottery has an equivalent utility α_H.

On the other hand, no matter what is the event E, in the above complex lottery, we get A_1 with probability 0.5 and A_0 with probability 0.5. Thus, this complex lottery coincides with the lottery $L(0.5)$ and thus, has utility 0.5. So, we conclude that $\alpha_H = 0.5$.

Comment. The fact that people with too optimistic attitude often make suboptimal decisions is experimentally confirmed, e.g., in [15].

Which action should we choose? Suppose that an action has n possible outcomes $S_1, \ldots, S_n$, with utilities

$$[\underline{u}(S_i), \overline{u}(S_i)],$$

and probabilities $[\underline{p}_i, \overline{p}_i]$. How do we then estimate the equivalent utility of this action?

We know that each alternative is equivalent to a simple lottery with utility $u_i = \alpha_H \cdot \overline{u}(S_i) + (1-\alpha_H) \cdot \underline{u}(S_i)$, and that for each i, the i-th event is—from the viewpoint of decision making—equivalent to $p_i = \alpha_H \cdot \overline{p}_i + (1 - \alpha_H) \cdot \underline{p}_i$. Thus, from the viewpoint of decision making, this action is equivalent to a situation in which we get utility u_i with probability p_i. We know that the utility of such a situation is equal to $\sum\limits_{i=1}^{n} p_i \cdot u_i$. Thus, the equivalent utility of the original action is equivalent to

$$\sum_{i=1}^{n} p_i \cdot u_i = \sum_{i=1}^{n} (\alpha_H \cdot \overline{p}_i + (1 - \alpha_H) \cdot \underline{p}_i) \cdot (\alpha_H \cdot \overline{u}(S_i) + (1 - \alpha_H) \cdot \underline{u}(S_i)).$$

Comment. One can easily see that if we replace the selected values A_0 and A_1 with A'_0 and A'_1, so that the utilities change linearly $u \to u' = a \cdot u + b$, then the above equivalent utility u_{equiv} also changes according to the same linear transformation $u'_{\text{equiv}} = a \cdot u_{\text{equiv}} + b$.

Discussion. We started with the situation in which a decision maker cannot decide between A' and A''. In this case, it is possible that A' is better, and it is also possible that A'' is better. In terms of interval-valued utilities $[\underline{u}(A'), \overline{u}(A')]$ and $[\underline{u}(A''), \overline{u}(A'')]$, this means that:

- there exists values $u(A') \in [\underline{u}(A'), \overline{u}(A')]$ and $u(A'') \in [\underline{u}(A''), \overline{u}(A'')]$ for which $u(A') > u(A'')$, and
- there exists values $u(A') \in [\underline{u}(A'), \overline{u}(A')]$ and $u(A'') \in [\underline{u}(A''), \overline{u}(A'')]$ for which $u(A') < u(A'')$.

In this case, the above approach recommends selecting one of the alternatives A' and A'':

- we recommend to select A' if

$$\alpha_H \cdot \overline{u}(A') + (1 - \alpha_H) \cdot \underline{u}(A') \geq \alpha_H \cdot \overline{u}(A'') + (1 - \alpha_H) \cdot \underline{u}(A'');$$

- we recommend to select A'' if

$$\alpha_H \cdot \overline{u}(A') + (1 - \alpha_H) \cdot \underline{u}(A') < \alpha_H \cdot \overline{u}(A'') + (1 - \alpha_H) \cdot \underline{u}(A'').$$

In this case, from the viewpoint of *descriptive* preference, we have uncertainty—we cannot decide between A' and A''. In this case, we make a recommendation. The recommended *prescriptive* (*normative*) preference will enable the user to make a good decision in a situation when this user is unsure which decision is better—this is exactly the type of situation in which user seek advise of specialists in decision making.

Observation: the resulting decision depends on the level of detail. We make a decision in a situation when we do not know the exact values of the utilities and when we do not know the exact values of the corresponding probabilities. Clearly, if gain new information, the equivalent utility may change. For example, if we know nothing about an alternative A, then its utility is $[0, 1]$ and thus, its equivalent utility is α_H. Once we narrow down the utility of A, e.g., to the interval $[0.5, 0.9]$, we get a different equivalent utility $\alpha_H \cdot 0.9 + (1 - \alpha_H) \cdot 0.5 = 0.5 + 0.4 \cdot \alpha_H$. On this example, the fact that we have different utilities makes perfect sense.

However, there are other examples where the corresponding difference is not as intuitively clear. Let us consider a situation in which, with some probability p, we gain a utility u, and with the remaining probability $1 - p$, we gain utility 0. If we know the exact values of u and p, we can then compute the equivalent utility of this situation as the expected utility value $p \cdot u + (1 - p) \cdot 0 = p \cdot u$.

Suppose now that we only know the interval $[\underline{u}, \overline{u}]$ of possible values of utility and the interval $[\underline{p}, \overline{p}]$ of possible values of probability. Since the expression $p \cdot u$ for the expected utility of this situation is an increasing function of both variables:

- the largest possible utility of this situation is attained when both p and u are the largest possible: $u = \overline{u}$ and $p = \overline{p}$, and
- the smallest possible utility is attained when both p and u are the smallest possible: $u = \underline{u}$ and $p = \underline{p}$.

In other words, the resulting amount of utility ranges from $\underline{p} \cdot \underline{u}$ to $\overline{p} \cdot \overline{u}$.

If we know the structure of the situation, then, according to our derivation, this situation has an equivalent utility

$$u_k = (\alpha_H \cdot \overline{p} + (1 - \alpha_H) \cdot \underline{p}) \cdot (\alpha_H \cdot \overline{u} + (1 - \alpha_H) \cdot \underline{u})$$

(*k* for *k*now). On the other hand, if we do not know the structure, if we only know that the resulting utility is from the interval $[\underline{p} \cdot \underline{u}, \overline{p} \cdot \overline{u}]$, then, according to the Hurwicz criterion, the equivalent utility is equal to

$$u_d = \alpha_H \cdot \overline{p} \cdot \overline{u} + (1 - \alpha_H) \cdot \underline{p} \cdot \underline{u}$$

(*d* for *d*on't know). One can check that

$$u_d - u_k =$$

$$\alpha_H \cdot \overline{p} \cdot \overline{u} + (1 - \alpha_H) \cdot \underline{p} \cdot \underline{u} - \alpha_H^2 \cdot \overline{p} \cdot \overline{u} - \alpha_H \cdot (1 - \alpha_H) \cdot (\underline{p} \cdot \overline{u} + \overline{p} \cdot \underline{u}) - (1 - \alpha_H)^2 \cdot \underline{p} \cdot \underline{u} =$$

$$\alpha_H \cdot (1 - \alpha_H) \cdot \overline{p} \cdot \overline{u} + \alpha_H \cdot (1 - \alpha_H) \cdot \underline{p} \cdot \underline{u} - \alpha_H \cdot (1 - \alpha_H) \cdot (\underline{p} \cdot \overline{u} + \overline{p} \cdot \underline{u}) =$$

$$\alpha_H \cdot (1 - \alpha_H) \cdot (\overline{p} - \underline{p}) \cdot (\overline{u} - \underline{u}).$$

This difference is always positive, meaning that additional knowledge decreases the utility of the situation. (This is maybe what the Book of Ecclesiastes means by "For with much wisdom comes much sorrow"?)

Comment. A similar example has been recently described in [12].

5 From Intervals to Arbitrary Sets

In the ideal case, we know the exact situation s in all the detail, and we can thus determine its utility $u(s)$. Realistically, we have an imprecise knowledge, so instead of a single situation s, we only know a *set* S of possible situations s. Thus, instead of a single value of the utility, we only know that the actual utility belongs to the set $U = \{u(s) : s \in S\}$. If this set U is an interval $[\underline{u}, \overline{u}]$, then we can use the above arguments to come up with its equivalent utility value $\alpha_H \cdot \overline{u} + (1 - \alpha_H) \cdot \underline{u}$.

What is U is a not an interval? For example, we can have a 2-point set $U = \{\underline{u}, \overline{u}\}$. What is then the equivalent utility?

Let us first consider the case when the set U contains both its infimum $\underline{u}$ and its supremum $\overline{u}$. The fact that we only know the set of possible values and have no other information means that *any* probability distribution on this set is possible (to be more precise, it is possible to have any probability distribution on the set of possible situations S, and this leads to the probability distribution on utilities). In particular, for each probability p, it is possible to have a distribution in which we have $\overline{u}$ with probability p and $\underline{u}$ with probability $1 - p$. For this distribution, the expected utility is equal to $p \cdot \overline{u} + (1 - p) \cdot \underline{u}$. When p goes from 0 to 1, these values fill the whole interval $[\underline{u}, \overline{u}]$. Thus, every value from this interval is the possible value of the expected utility. On the other hand, when $u \in [\underline{u}, \overline{u}]$, the expected value of the utility also belongs to this interval—no matter what the probability distribution. Thus, the set of all possible utility values is the whole interval $[\underline{u}, \overline{u}]$ and so, the equivalent utility is equal to $\alpha_H \cdot \overline{u} + (1 - \alpha_H) \cdot \underline{u}$.

When the infimum and/or supremum are not in the set U, then the set U contains points as close to them as possible. Thus, the resulting set of possible values of utility is as close as possible to the interval $[\underline{u}, \overline{u}]$—and so, it is reasonable to assume that the equivalent utility is as close to $u_0 = \alpha_H \cdot \overline{u} + (1 - \alpha_H) \cdot \underline{u}$ as possible—i.e., coincides with this value u_0.

6 Beyond Interval and Set Uncertainty: Partial Information About Probabilities

Formulation of the problem. In addition to the interval $\mathbf{x}$, we may also have *partial* information about the probabilities of different values $x \in \mathbf{x}$. How can we describe this partial information?

An *exact* probability distribution can be described, e.g., by its cumulative distribution function (cdf) $F(z) = \text{Prob}(x \le z)$. A *partial* information means that for each z, instead of knowing the exact value $F(z)$, we only know the bounds on $F(z)$, i.e., we only know the interval $\mathbf{F}(z) = [\underline{F}(z), \overline{F}(z)]$. Such an interval-valued cdf is known as a *p-box*; see, e.g., [7, 32]. Once we know the p-box, we consider all possible distributions for which, for all z, we have $F(z) \in \mathbf{F}(z)$.

The problem is that there are many ways to represent a probability distribution, and each leads to a different way to represent partial information. Which of these ways should we choose?

Which is the best way to describe the corresponding probabilistic uncertainty? One of the main objectives of data processing is to make decisions. A standard way of making a decision is to select the action a for which the expected utility (gain) is the largest possible. This is where probabilities are used: in computing, for every possible action a, the corresponding expected utility. To be more precise, we usually know, for each action a and for each actual value of the (unknown) quantity x, the corresponding value of the utility $u_a(x)$. We must use the probability distribution for x to compute the expected value $E[u_a(x)]$ of this utility.

In view of this application, the most useful characteristics of a probability distribution would be the ones which would enable us to compute the expected value $E[u_a(x)]$ of different functions $u_a(x)$.

Which representations are the most useful for this intended usage? General idea. Which characteristics of a probability distribution are the most useful for computing mathematical expectations of different functions $u_a(x)$? The answer to this question depends on the type of the function, i.e., on how the utility value u depends on the value x of the analyzed parameter.

Smooth utility functions naturally lead to moments. One natural case is when the utility function $u_a(x)$ is smooth. We have already mentioned, in the previous text, that we usually know a (reasonably narrow) interval of possible values of x. So, to compute the expected value of $u_a(x)$, all we need to know is how the function $u_a(x)$ behaves on this narrow interval. Because the function is smooth, we can expand it into Taylor series. Because the interval is narrow, we can consider only linear and quadratic terms in this expansion and safely ignore higher-order terms: $u_a(x) \approx c_0 + c_1 \cdot (x - x_0) + c_2 \cdot (x - x_0)^2$, where x_0 is a point inside the interval. Thus, we can approximate the expected value of this function by the expected value of the corresponding quadratic expression: $E[u_a(x)] \approx E[c_0+c_1 \cdot (x-x_0)+c_2 \cdot (x-x_0)^2]$, i.e., by the following expression: $E[u_a(x)] \approx c_0+c_1 \cdot E[x-x_0]+c_2 \cdot E[(x-x_0)^2]$. So, to compute the expectations of such utility functions, it is sufficient to know the first and second moments of the probability distribution.

In particular, if we use, as the point x_0, the average $E[x]$, the second moment turns into the variance of the original probability distribution. So, instead of the first and the second moments, we can use the mean E and the variance V.

In decision making, non-smooth utility functions are common. In decision making, not all dependencies are smooth. There is often a threshold x_0 after which, say, a concentration of a certain chemical becomes dangerous.

This threshold sometimes comes from the detailed chemical and/or physical analysis. In this case, when we increase the value of this parameter, we see the drastic increase in effect and hence, the drastic change in utility value. Sometimes, this threshold simply comes from regulations. In this case, when we increase the value of this parameter past the threshold, there is no drastic increase in effects, but there is a drastic decrease of utility due to the necessity to pay fines, change technology, etc. In both cases, we have a utility function which experiences an abrupt decrease at a certain threshold value x_0.

Non-smooth utility functions naturally lead to cumulative distribution functions (cdfs). We want to be able to compute the expected value $E[u_a(x)]$ of a function $u_a(x)$ which

- changes smoothly until a certain value x_0,
- then drops it value and continues smoothly for $x > x_0$.

We usually know the (reasonably narrow) interval which contains all possible values of x. Because the interval is narrow and the dependence before and after the threshold is smooth, the resulting change in $u_a(x)$ before x_0 and after x_0 is much smaller than the change at x_0. Thus, with a reasonable accuracy, we can ignore the small changes before and after x_0, and assume that the function $u_a(x)$ is equal to a constant u^+ for $x < x_0$, and to some other constant $u^- < u^+$ for $x > x_0$.

The simplest case is when $u^+ = 1$ and $u^- = 0$. In this case, the desired expected value $E[u_a^{(0)}(x)]$ coincides with the probability that $x < x_0$, i.e., with the corresponding value $F(x_0)$ of the cumulative distribution function (cdf). A generic function $u_a(x)$ of this type, with arbitrary values u^- and u^+, can be easily reduced to this simplest case, because, as one can easily check, $u_a(x) = u^- + (u^+ - u^-) \cdot u^{(0)}(x)$ and hence, $E[u_a(x)] = u^- + (u^+ - u^-) \cdot F(x_0)$.

Thus, to be able to easily compute the expected values of all possible non-smooth utility functions, it is sufficient to know the values of the cdf $F(x_0)$ for all possible x_0.

Describing the cdf is equivalent to describing the inverse *quantile* function—a function that assigns, to every possible probability $p \in [0, 1]$, the value $x = x(p)$ for which $F(x) = p$. For example, the quantile corresponding to $p = 0.5$ is the *median* of the probability distribution.

Summarizing: which statistical characteristics we select. Our analysis shows that the most appropriate characteristics are the moments and the values of the cdf (or, equivalently, the values of the quantiles).

Comment. How to estimate the values of the selected statistical characteristics? How to propagate these values via data processing? For answers to these questions, see [7, 32] and references therein.

7 What if We Cannot Even Elicit Interval-Valued Uncertainty: Symmetry Approach

Case study. In some situations, it is difficult to elicit even interval-valued utilities. As a case study, we consider the problem of selecting the best location for a meteorological tower.

In many applications involving meteorology and environmental sciences, it is important to measure fluxes of heat, water, carbon dioxide, methane and other trace gases that are exchanged within the atmospheric boundary layer. Air flow in this boundary layer consists of numerous rotating eddies, i.e., turbulent vortices of various sizes, with each eddy having horizontal and vertical components. To estimate the flow amount at a given location, we thus need to accurately measure wind speed (and direction), temperature, atmospheric pressure, gas concentration, etc., at different heights, and then process the resulting data. To perform these measurements, researchers build up vertical towers equipped with sensors at different heights; these tower are called *Eddy flux towers*.

When selecting a location for the Eddy flux tower, we have several criteria to satisfy; see, e.g., [2, 19].

- For example, the station should not be located too close to a road, so that the gas flux generated by the cars does not influence our measurements of atmospheric fluxes; in other words, the distance x_1 to the road should be larger than a certain threshold t_1: $x_1 > t_1$, or $y_1 \stackrel{\text{def}}{=} x_1 - t_1 > 0$.
- Also, the inclination x_2 at the station location should be smaller than a corresponding threshold t_2, because otherwise, the flux will be mostly determined by this inclination and will not be reflective of the atmospheric processes: $x_2 < t_2$, or $y_2 \stackrel{\text{def}}{=} t_2 - x_2 > 0$.

General case. In general, we have several such differences $y_1, \ldots, y_n$ all of which have to be non-negative. For each of the differences y_i, the larger its value, the better. Based on the above, our problem is a typical setting for *multi-criteria optimization*; see, e.g., [6, 38, 40].

Practical problem: reminder. We want to select the best location based on the values of the differences $y_1, \ldots, y_n$. For each of the differences y_i, the larger its value, the better.

Weighted average: a natural approach for solving multi-criterion optimization problems, and limitations of this approach. The most widely used approach to multi-criteria optimization is *weighted average*, where we assign weights $w_1, \ldots, w_n > 0$ to different criteria y_i and select an alternative for which the weighted average $w_1 \cdot y_1 + \ldots + w_n \cdot y_n$ attains the largest possible value.

This approach has been used in many practical problems ranging from selecting the lunar landing sites for the Apollo missions (see, e.g., [3]) to selecting landfill sites (see, e.g., [10]).

In our problem, we have an additional requirement—that all the values y_i must be positive. Thus, we must only compare solutions with $y_i > 0$ when selecting an alternative with the largest possible value of the weighted average.

In general, the weighted average approach often leads to reasonable solutions of the multi-criteria optimization problem. However, as we will show, in the presence of the additional positivity requirement, the weighted average approach is not fully satisfactory.

A practical multi-criteria optimization must take into account that measurements are not absolutely accurate. In many practical application of the multi-criterion optimization problem (in particular, in applications to optimal sensor placement), the values y_i come from measurements, and measurements are never absolutely accurate. The results $\widetilde{y}_i$ of the measurements are close to the actual (unknown) values y_i of the measured quantities, but they are not exactly equal to these values. If:

- we measure the values y_i with higher and higher accuracy and,
- based on the measurement results $\widetilde{y}_i$, we conclude that the alternative $y = (y_1, \ldots, y_n)$ is better than some other alternative $y' = (y'_1, \ldots, y'_n)$,

then we expect that the actual alternative y is indeed either better than y' or at least of the same quality as y'. Otherwise, if we do not make this assumption, we will not be able to make any meaningful conclusions based on real-life (approximate) measurements.

The above natural requirement is not always satisfied for weighted average. Let us show that for the weighted average, this "continuity" requirement is not satisfied even in the simplest case when we have only two criteria y_1 and y_2. Indeed, let $w_1 > 0$ and $w_2 > 0$ be the weights corresponding to these two criteria. Then, the resulting strict preference relation $\succ$ has the following properties:

- if $y_1 > 0$, $y_2 > 0$, $y_1' > 0$, and $y_2' > 0$, and $w_1 \cdot y_1' + w_2 \cdot y_2' > w_1 \cdot y_1 + w_2 \cdot y_2$, then
$$y' = (y_1', y_2') \succ y = (y_1, y_2); \tag{1}$$
- if $y_1 > 0$, $y_2 > 0$, and at least one of the values y_1' and y_2' is non-positive, then
$$y = (y_1, y_2) \succ y' = (y_1', y_2'). \tag{2}$$

Let us consider, for every $\varepsilon > 0$, the tuple $y'(\varepsilon) \stackrel{\text{def}}{=} \left(\varepsilon, 1 + \frac{w_1}{w_2}\right)$, with $y_1'(\varepsilon) = \varepsilon$ and $y_2'(\varepsilon) = 1 + \frac{w_1}{w_2}$, and also the comparison tuple $y = (1, 1)$. In this case, for every $\varepsilon > 0$, we have

$$w_1 \cdot y_1'(\varepsilon) + w_2 \cdot y_2'(\varepsilon) = w_1 \cdot \varepsilon + w_2 + w_2 \cdot \frac{w_1}{w_2} = w_1 \cdot (1 + \varepsilon) + w_2 \tag{3}$$

and

$$w_1 \cdot y_1 + w_2 \cdot y_2 = w_1 + w_2, \tag{4}$$

hence $y'(\varepsilon) \succ y$. However, in the limit $\varepsilon \to 0$, we have $y'(0) = \left(0, 1 + \frac{w_1}{w_2}\right)$, with $y_1'(0) = 0$ and thus, $y'(0) \prec y$.

Towards a more adequate approach to multi-criterion optimization. We want to be able to compare different alternatives.

Each alternative is characterized by a tuple of n values $y = (y_1, \ldots, y_n)$, and only alternatives for which all the values y_i are positive are allowed. Thus, from the mathematical viewpoint, the set of all alternatives is the set $(R^+)^n$ of all the tuples of positive numbers.

For each two alternatives y and y', we want to tell whether y is better than y' (we will denote it by $y \succ y'$ or $y' \prec y$), or y' is better than y ($y' \succ y$), or y and y' are equally good ($y' \sim y$). These relations must satisfy natural properties. For example, if y is better than y' and y' is better than y'', then y is better than y''. In other words, the relation $\succ$ must be transitive. Similarly, the relation $\sim$ must be transitive, symmetric, and reflexive ($y \sim y$), i.e., in mathematical terms, an *equivalence relation*.

So, we want to define a pair of relations $\succ$ and $\sim$ such that $\succ$ is transitive, $\sim$ is an equivalence relation, and for every y and y', one and only one of the following relations hold: $y \succ y'$, $y' \succ y$, or $y \sim y'$.

It is also reasonable to require that if each criterion is better, then the alternative is better as well, i.e., that if $y_i > y_i'$ for all i, then $y \succ y'$.

Comment. Pairs of relations of the above type can be alternatively characterized by a *pre-ordering* relation

$$y' \succeq y \Leftrightarrow (y' \succ y \vee y' \sim y). \tag{5}$$

This pre-ordering relation must be transitive and—in our case—total (i.e., for every y and y', we have $y \succeq y' \vee y' \succeq y$). Once we know the pre-ordering relation $\succeq$, we can reconstruct $\succ$ and $\sim$ as follows:

$$y' \succ y \Leftrightarrow (y' \succeq y \,\&\, y \not\succeq y'); \tag{6}$$

$$y' \sim y \Leftrightarrow (y' \succeq y \,\&\, y \succeq y'). \tag{7}$$

Scale invariance: motivation. In general, the quantities y_i describe completely different physical notions, measured in completely different units. In our meteorological case, some of these values are wind velocities measured in meters per second, or in kilometers per hour, or in miles per hour. Other values are elevations described in meters, in kilometers, or in feet, etc. Each of these quantities can be described in many different units. A priori, we do not know which units match each other, so it is reasonable to assume that the units used for measuring different quantities may not be exactly matched.

It is therefore reasonable to require that the relations $\succ$ and $\sim$ between the two alternatives $y = (y_1, \ldots, y_n)$ and $y' = (y_1', \ldots, y_n')$ do not change if we simply change the units in which we measure each of the corresponding n quantities.

Comment. The importance of such invariance is well known in measurements theory, starting with the pioneering work of S. S. Stevens [41]; see also the classical books [34] and [26] (especially Chap. 22), where this invariance is also called *meaningfulness*.

Scale invariance: towards a precise description. When we replace a unit in which we measure a certain quantity q by a new measuring unit which is $\lambda > 0$ times smaller, then the numerical values of this quantity increase by a factor of λ, i.e., $q \to \lambda \cdot q$. For example, 1 cm is $\lambda = 100$ times smaller than 1 m, so the length $q = 2$ m, when measured in cm, becomes $\lambda \cdot q = 2 \cdot 100 = 200$ cm.

Let λ_i denote the ratio of the old to the new units corresponding to the i-th quantity. Then, the quantity that had the value y_i in the old units will be described by a numerical value $\lambda_i \cdot y_i$ in the new units. Therefore, scale-invariance means that for all $y, y' \in (R^+)^n$ and for all $\lambda_i > 0$, we have

$$y' = (y_1', \ldots, y_n') \succ y = (y_1, \ldots, y_n) \Rightarrow (\lambda_1 \cdot y_1', \ldots, \lambda_n \cdot y_n') \succ (\lambda_1 \cdot y_1, \ldots, \lambda_n \cdot y_n)$$

and

$$y' = (y'_1, \ldots, y'_n) \sim y = (y_1, \ldots, y_n) \Rightarrow (\lambda_1 \cdot y'_1, \ldots, \lambda_n \cdot y'_n) \sim (\lambda_1 \cdot y_1, \ldots, \lambda_n \cdot y_n).$$

Comment. In general, in measurements, in addition to changing the unit, we can also change the starting point. However, for the differences y_i, the starting point is fixed by the fact that 0 corresponds to the threshold value. So, in our case, only changing a measuring unit (= scaling) makes sense.

Continuity. As we have mentioned in the previous section, we also want to require that the relations $\succ$ and $\sim$ are *continuous* in the following sense: if $y'(\varepsilon) \succeq y(\varepsilon)$ for every ε, then in the limit, when $y'(\varepsilon) \to y'(0)$ and $y(\varepsilon) \to y(0)$ (in the sense of normal convergence in R^n), we should have $y'(0) \succeq y(0)$.

The main result. Let us now describe our requirements in precise terms.

Definition 2. *By a* total pre-ordering relation *on a set* Y, *we mean a pair of a transitive relation* $\succ$ *and an equivalence relation* $\sim$ *for which, for every* $y, y' \in Y$, *one and only one of the following relations hold:* $y \succ y'$, $y' \succ y$, *or* $y \sim y'$.

Comment. We will denote $y \succeq y' \stackrel{\text{def}}{=} (y \succ y' \vee y \sim y')$.

Definition 3. *We say that a total pre-ordering is* non-trivial *if there exist* y *and* y' *for which* $y' \succ y$.

Comment. This definition excludes the trivial pre-ordering in which every two tuples are equivalent to each other.

Definition 4. *We say that a total pre-ordering relation on the set* $(R^+)^n$ *is:*

- monotonic *if* $y'_i > y_i$ *for all* i *implies* $y' \succ y$;
- scale-invariant *if for all* $\lambda_i > 0$:
- $(y'_1, \ldots, y'_n) \succ y = (y_1, \ldots, y_n)$ *implies*

$$(\lambda_1 \cdot y'_1, \ldots, \lambda_n \cdot y'_n) \succ (\lambda_1 \cdot y_1, \ldots, \lambda_n \cdot y_n), \tag{8}$$

  and
- $(y'_1, \ldots, y'_n) \sim y = (y_1, \ldots, y_n)$ *implies*

$$(\lambda_1 \cdot y'_1, \ldots, \lambda_n \cdot y'_n) \sim (\lambda_1 \cdot y_1, \ldots, \lambda_n \cdot y_n). \tag{9}$$

- continuous *if whenever we have a sequence* $y^{(k)}$ *of tuples for which* $y^{(k)} \succeq y'$ *for some tuple* y', *and the sequence* $y^{(k)}$ *tends to a limit* y, *then* $y \succeq y'$.

Theorem [20] *Every non-trivial monotonic scale-invariant continuous total pre-ordering relation on* $(R^+)^n$ *has the following form:*

$$y' = (y'_1, \ldots, y'_n) \succ y = (y_1, \ldots, y_n) \Leftrightarrow \prod_{i=1}^{n} (y'_i)^{\alpha_i} > \prod_{i=1}^{n} y_i^{\alpha_i}; \quad (10)$$

$$y' = (y'_1, \ldots, y'_n) \sim y = (y_1, \ldots, y_n) \Leftrightarrow \prod_{i=1}^{n} (y'_i)^{\alpha_i} = \prod_{i=1}^{n} y_i^{\alpha_i}, \quad (11)$$

for some constants $\alpha_i > 0$.

Comment. In other words, for every non-trivial monotonic scale-invariant continuous total pre-ordering relation on $(R^+)^n$, there exist values $\alpha_1 > 0, \ldots, \alpha_n > 0$ for which the above equivalence hold. Vice versa, for each set of values $\alpha_1 > 0, \ldots, \alpha_n > 0$, the above formulas define a monotonic scale-invariant continuous pre-ordering relation on $(R^+)^n$.

For reader's convenience, the proof of the main result is presented in an Appendix.

It is worth mentioning that the resulting relation coincides with Cobb-Douglas production (utility) function [4, 43] and with the asymmetric version (see, e.g., [36]) of the bargaining solution proposed by the Nobelist John Nash (see next section).

Applications. We have applied this approach to selecting a site for the Eddy tower that we built at Jornada Experimental Range, a study site in the northern Chihuahuan Desert; see, e.g., [17, 18]. In this applications, the parameters y_i have already been identified in the previous research; see, e.g., [2].

The values α_i were selected based on the information provided by experts, who supplied us with pairs of (approximately) equally good (or equally bad) designs y and y' with different combinations of the parameters y_i. Each resulting resulting condition $\prod_{i=1}^{n} y_i^{\alpha_i} = \prod_{i=1}^{n} (y'_i)^{\alpha_i}$ can be equivalently described, after taking logarithms of both sides, as a linear equation $\sum_{i=1}^{n} \alpha_i \cdot \ln(y_i) = \sum_{i=1}^{n} \alpha_i \cdot \ln(y'_i)$. By solving this system of linear equations, we found the values α_i that reflect the expert opinion on the efficiency of Eddy towers.

A similar symmetry-based approach was used to design a network of radiotelescopes [24].

Comment. The above equations determine α_i modulo a multiplicative constant: if we multiply all the values α_i by the same constant, the equations remain valid. To avoid this non-uniqueness, we used normalized values of α_i, i.e., values that satisfy the additional normalizing equation $\sum_{i=1}^{n} \alpha_i = 1$.

8 Group Decision Making

Need for group decision making. In many practical situations, several people are affected by the planned decision. In such situations, we need to take into account preferences of all the participating agents.

For each participant P_i, we can determine the utility $u_{ij} \stackrel{\text{def}}{=} u_i(A_j)$ of all the alternatives $A_1, \ldots, A_m$. How to transform these utilities into a reasonable group decision rule?

Nash's bargaining solution. The answer to this question was, in effect, provided by a future Nobelist John Nash who, in [30, 31], has shown that under reasonable assumptions like symmetry, independence from irrelevant alternatives, and *scale invariance* (i.e., invariance under replacing the original utility function $u_i(A)$ with an equivalent function $a \cdot u_i(A)$), the only group decision rule is selecting an alternative A for which the product

$$u(A) \stackrel{\text{def}}{=} \prod_{i=1}^{n} u_i(A)$$

is the largest possible; see also [27, 29].

Here, the utility functions must be scaled in such a way that the "status quo" situation $A^{(0)}$ is assigned the utility 0. This re-scaling can be achieved, e.g., by replacing the original utility values $u_i(A)$ with re-scaled values $u_i'(A) \stackrel{\text{def}}{=} u_i(A) - u_i(A^{(0)})$.

Multi-agent decision making under interval uncertainty. What if we do not know the exact values of utility, we only know intervals $[\underline{u}_i(A), \overline{u}_i(A)]$? In this case, the first idea is to find all A_0 which can be Nash-optimal, i.e., for which $\overline{u}(A_0) \geq \max\limits_{A} \underline{u}(A)$, where

$$\underline{u}(A) \stackrel{\text{def}}{=} \prod_{i=1}^{n} \underline{u}_i(A) \text{and } \overline{u}(A) \stackrel{\text{def}}{=} \prod_{i=1}^{n} \overline{u}_i(A).$$

If we want to select a single alternative, then we should maximize $u^{\text{equiv}}(A) \stackrel{\text{def}}{=} \prod\limits_{i=1}^{n} u_i^{\text{equiv}}(A)$, where $u_i^{\text{equiv}}(A)$ are values obtained by using Hurwicz optimism-pessimism criterion.

Comment. An interesting aspect of this problem is that sometimes, we have a conflict situation; this happens, for example, in security situations. In such situations, only partial results are known; see, e.g., [21].

9 Beyond Optimization

Need to go beyond optimization. While optimization problems are ubiquitous, sometimes, we need to go beyond optimization: e.g., we need to make sure that the system is *controllable* for all disturbances within a given range.

In control situations, the desired value z depends both on the variables the variables that we can select (control variables) $u = (u_1, \ldots, u_m)$ and on the variables $x = (x_1, \ldots, x_n)$ describing the changing state of the world: $z = f(x, u)$. For each control variable u_j, we know the range U_j within which we can select its value, and for each variable x_i, we know the range X_i of its possible values. We want to find a range Z for which, for every state of the world $x_i \in X_i$, we can get $z \in Z$ by selecting appropriate control values $u_j \in U_j$:

$$\forall x \, \exists u \, (z = f(x, u) \in Z).$$

Interval computations: reminder. Interval computations [28] can be viewed as a degenerate case of this control problem in which there are no controls at all. In this case:

- we know the intervals $X_1, \ldots, X_n$ containing $x_1, \ldots, x_n$;
- we know that a quantity z depends on x: $z = f(x)$;
- we want to find the range Z of possible values of z:

$$Z = \left[\min_{x \in X} f(x), \max_{x \in X} f(x)\right].$$

In logical terms, we want to make sure that $\forall x \, (z = f(x) \in Z)$.

Reformulation in logical terms—of modal intervals. In the general control case, we want to make sure that $\forall x_{\in X} \, \exists u_{\in U} \, (f(x, u) \in Z)$. There is a logical difference between intervals X and U: the property $f(x, u) \in Z$ must hold

- *for all* possible values $x_i \in X_i$, but
- *for some* values $u_j \in U_j$.

We can thus consider pairs of intervals and quantifiers (*modal intervals* [11]):

- each original interval X_i is a pair $\langle X_i, \forall \rangle$, while
- controlled interval is a pair $\langle U_j, \exists \rangle$.

We can then treat the resulting interval Z as the "range" defined over such modal intervals:

$$Z = f(\langle X_1, \forall \rangle, \ldots, \langle X_n, \forall \rangle, \langle U_1, \exists \rangle, \ldots, \langle U_m, \exists \rangle).$$

Even further beyond optimization. In more complex situations, we need to go beyond control. For example, in the presence of an adversary, we want to make a decision x such that:

- for every possible reaction y of an adversary,
- we will be able to make a next decision x' (depending on y)
- so that after every possible next decision y' of an adversary,
- the resulting state $s(x, y, x', y')$ will be in the desired set:

$$\forall y \, \exists x \, \forall y' \, (s(x, y, x', y') \in S).$$

In this case, we arrive at general quantifier classes described, e.g., in [39].

Acknowledgments This work was supported in part by the National Science Foundation grants HRD-0734825 and HRD-1242122 (Cyber-ShARE Center of Excellence) and DUE-0926721, by Grant 1 T36 GM078000-01 from the National Institutes of Health, and by a grant on F-transforms from the Office of Naval Research. The authors is greatly thankful to the anonymous referees for valuable suggestions.

A Proof of the Theorem

1°. Due to scale-invariance (9), for every $y_1, \ldots, y_n, y_1', \ldots, y_n'$, we can take $\lambda_i = \dfrac{1}{y_i}$ and conclude that

$$(y_1', \ldots, y_n') \sim (y_1, \ldots, y_n) \Leftrightarrow \left(\frac{y_1'}{y_1}, \ldots, \frac{y_n'}{y_n}\right) \sim (1, \ldots, 1). \tag{12}$$

Thus, to describe the equivalence relation $\sim$, it is sufficient to describe the set of all the vectors $z = (z_1, \ldots, z_n)$ for which $z \sim (1, \ldots, 1)$. Similarly,

$$(y_1', \ldots, y_n') \succ (y_1, \ldots, y_n) \Leftrightarrow \left(\frac{y_1'}{y_1}, \ldots, \frac{y_n'}{y_n}\right) \succ (1, \ldots, 1). \tag{13}$$

So, to describe the ordering relation $\succ$, it is sufficient to describe the set of all the vectors $z = (z_1, \ldots, z_n)$ for which $z \succ (1, \ldots, 1)$.

Alternatively, we can take $\lambda_i = \dfrac{1}{y_i'}$ and conclude that

$$(y_1', \ldots, y_n') \succ (y_1, \ldots, y_n) \Leftrightarrow (1, \ldots, 1) \succ \left(\frac{y_1}{y_1'}, \ldots, \frac{y_n}{y_n'}\right). \tag{14}$$

Thus, it is also sufficient to describe the set of all the vectors $z = (z_1, \ldots, z_n)$ for which $(1, \ldots, 1) \succ z$.

2°. The above equivalence involves division. To simplify the description, we can take into account that in the logarithmic space, division becomes a simple difference: $\ln\left(\dfrac{y_i'}{y_i}\right) = \ln(y_i') - \ln(y_i)$. To use this simplification, let us consider the logarithms

$Y_i \stackrel{\text{def}}{=} \ln(y_i)$ of different values. In terms of these logarithms, the original values can be reconstructed as $y_i = \exp(Y_i)$. In terms of these logarithms, we thus need to consider:

- the set $S_\sim$ of all the tuples $Z = (Z_1, \ldots, Z_n)$ for which

$$z = (\exp(Z_1), \ldots, \exp(Z_n)) \sim (1, \ldots, 1), \tag{15}$$

  and
- the set $S_\succ$ of all the tuples $Z = (Z_1, \ldots, Z_n)$ for which

$$z = (\exp(Z_1), \ldots, \exp(Z_n)) \succ (1, \ldots, 1). \tag{16}$$

We will also consider the set $S_\prec$ of all the tuples $Z = (Z_1, \ldots, Z_n)$ for which

$$(1, \ldots, 1) \succ z = (\exp(Z_1), \ldots, \exp(Z_n)). \tag{17}$$

Since the pre-ordering relation is total, for every tuple z,

- either $z \sim (1, \ldots, 1)$,
- or $z \succ (1, \ldots, 1)$,
- or $(1, \ldots, 1) \succ z$.

In particular, this is true for $z = (\exp(Z_1), \ldots, \exp(Z_n))$. Thus, for every tuple Z, either $Z \in S_\sim$ or $Z \in S_\succ$ or $Z \in S_\prec$.

3°. Let us prove that the set $S_\sim$ is closed under addition, i.e., that if the tuples $Z = (Z_1, \ldots, Z_n)$ and $Z' = (Z'_1, \ldots, Z'_n)$ belong to the set $S_\sim$, then their component-wise sum

$$Z + Z' = (Z_1 + Z'_1, \ldots, Z_n + Z'_n) \tag{18}$$

also belongs to the set $S_\sim$.

Indeed, by definition (15) of the set $S_\sim$, the condition $Z \in S_\sim$ means that

$$(\exp(Z_1), \ldots, \exp(Z_n)) \sim (1, \ldots, 1). \tag{19}$$

Using scale-invariance (9) with $\lambda_i = \exp(Z'_i)$, we conclude that

$$(\exp(Z_1) \cdot \exp(Z'_1), \ldots, \exp(Z_n) \cdot \exp(Z'_n)) \sim (\exp(Z'_1), \ldots, \exp(Z'_n)). \tag{20}$$

On the other hand, the condition $Z' \in S_\sim$ means that

$$(\exp(Z'_1), \ldots, \exp(Z'_n)) \sim (1, \ldots, 1). \tag{21}$$

Thus, due to transitivity of the equivalence relation $\sim$, we conclude that

$$(\exp(Z_1) \cdot \exp(Z'_1), \ldots, \exp(Z_n) \cdot \exp(Z'_n)) \sim (1, \ldots, 1). \tag{22}$$

Since for every i, we have $\exp(Z_i) \cdot \exp(Z_i') = \exp(Z_i + Z_i')$, we thus conclude that

$$(\exp(Z_1 + Z_1'), \ldots, \exp(Z_n + Z_n')) \sim (1, \ldots, 1). \tag{23}$$

By definition (15) of the set $S_\sim$, this means that the tuple $Z + Z'$ belongs to the set $S_\sim$.

4°. Similarly, we can prove that the set $S_\succ$ is closed under addition, i.e., that if the tuples $Z = (Z_1, \ldots, Z_n)$ and $Z' = (Z_1', \ldots, Z_n')$ belong to the set $S_\succ$, then their component-wise sum

$$Z + Z' = (Z_1 + Z_1', \ldots, Z_n + Z_n') \tag{24}$$

also belongs to the set $S_\succ$.

Indeed, by definition (16) of the set $S_\succ$, the condition $Z \in S_\succ$ means that

$$(\exp(Z_1), \ldots, \exp(Z_n)) \succ (1, \ldots, 1). \tag{25}$$

Using scale-invariance (8) with $\lambda_i = \exp(Z_i')$, we conclude that

$$(\exp(Z_1) \cdot \exp(Z_1'), \ldots, \exp(Z_n) \cdot \exp(Z_n')) \succ (\exp(Z_1'), \ldots, \exp(Z_n')). \tag{26}$$

On the other hand, the condition $Z' \in S_\succ$ means that

$$(\exp(Z_1'), \ldots, \exp(Z_n')) \succ (1, \ldots, 1). \tag{27}$$

Thus, due to transitivity of the strict preference relation $\succ$, we conclude that

$$(\exp(Z_1) \cdot \exp(Z_1'), \ldots, \exp(Z_n) \cdot \exp(Z_n')) \succ (1, \ldots, 1). \tag{28}$$

Since for every i, we have $\exp(Z_i) \cdot \exp(Z_i') = \exp(Z_i + Z_i')$, we thus conclude that

$$(\exp(Z_1 + Z_1'), \ldots, \exp(Z_n + Z_n')) \succ (1, \ldots, 1). \tag{29}$$

By definition (16) of the set $S_\succ$, this means that the tuple $Z + Z'$ belongs to the set $S_\succ$.

5°. A similar argument shows that the set $S_\prec$ is closed under addition, i.e., that if the tuples $Z = (Z_1, \ldots, Z_n)$ and $Z' = (Z_1', \ldots, Z_n')$ belong to the set $S_\prec$, then their component-wise sum

$$Z + Z' = (Z_1 + Z_1', \ldots, Z_n + Z_n') \tag{30}$$

also belongs to the set $S_\prec$.

6°. Let us now prove that the set $S_\sim$ is closed under the "unary minus" operation, i.e., that if $Z = (Z_1, \ldots, Z_n) \in S_\sim$, then $-Z \stackrel{\text{def}}{=} (-Z_1, \ldots, -Z_n)$ also belongs to $S_\sim$.

Indeed, $Z \in S_\sim$ means that

$$(\exp(Z_1), \ldots, \exp(Z_n)) \sim (1, \ldots, 1). \tag{31}$$

Using scale-invariance (9) with $\lambda_i = \exp(-Z_i) = \dfrac{1}{\exp(Z_i)}$, we conclude that

$$(1, \ldots, 1) \sim (\exp(-Z_1), \ldots, \exp(-Z_n)), \tag{32}$$

i.e., that $-Z \in S_\sim$.

7°. Let us prove that if $Z = (Z_1, \ldots, Z_n) \in S_\succ$, then $-Z \stackrel{\text{def}}{=} (-Z_1, \ldots, -Z_n)$ belongs to $S_\prec$.

Indeed, $Z \in S_\succ$ means that

$$(\exp(Z_1), \ldots, \exp(Z_n)) \succ (1, \ldots, 1). \tag{33}$$

Using scale-invariance (8) with $\lambda_i = \exp(-Z_i) = \dfrac{1}{\exp(Z_i)}$, we conclude that

$$(1, \ldots, 1) \succ (\exp(-Z_1), \ldots, \exp(-Z_n)), \tag{34}$$

i.e., that $-Z \in S_\prec$.

Similarly, we can show that if $Z \in S_\prec$, then $-Z \in S_\succ$.

8°. From Part 3 of this proof, it now follows that if $Z = (Z_1, \ldots, Z_n) \in S_\sim$, then $Z + Z \in S_\sim$, then that $Z + (Z + Z) \in S_\sim$, etc., i.e., that for every positive integer p, the tuple

$$p \cdot Z = (p \cdot Z_1, \ldots, p \cdot Z_n) \tag{35}$$

also belongs to the set $S_\sim$.

By using Part 6 of this proof, we can also conclude that this is true for negative integers p as well. Finally, by taking into account that the zero tuple $0 \stackrel{\text{def}}{=} (0, \ldots, 0)$ can be represented as $Z + (-Z)$, we conclude that $0 \cdot Z = 0$ also belongs to the set $S_\sim$.

Thus, if a tuple Z belongs to the set $S_\sim$, then for every integer p, the tuple $p \cdot Z$ also belongs to the set $S_\sim$.

9°. Similarly, from Parts 4 and 5 of this proof, it follows that

- if $Z = (Z_1, \ldots, Z_n) \in S_\succ$, then for every positive integer p, the tuple $p \cdot Z$ also belongs to the set $S_\succ$, and
- if $Z = (Z_1, \ldots, Z_n) \in S_\prec$, then for every positive integer p, the tuple $p \cdot Z$ also belongs to the set $S_\prec$.

10°. Let us prove that for every rational number $r = \dfrac{p}{q}$, where p is an integer and q is a positive integer, if a tuple Z belongs to the set $S_\sim$, then the tuple $r \cdot Z$ also belongs to the set $S_\sim$.

Indeed, according to Part 8, $Z \in S_\sim$ implies that $p \cdot Z \in S_\sim$.

According to Part 2, for the tuple $r \cdot Z$, we have either $r \cdot Z \in S_\sim$, or $r \cdot Z \in S_\succ$, or $r \cdot Z \in S_\prec$.

- If $r \cdot Z \in S_\succ$, then, by Part 9, we would get $p \cdot Z = q \cdot (r \cdot Z) \in S_\succ$, which contradicts our result that $p \cdot Z \in S_\sim$.
- Similarly, if $r \cdot Z \in S_\prec$, then, by Part 9, we would get $p \cdot Z = q \cdot (r \cdot Z) \in S_\prec$, which contradicts our result that $p \cdot Z \in S_\sim$.

Thus, the only remaining option is $r \cdot Z \in S_\sim$. The statement is proven.

11°. Let us now use continuity to prove that for every real number x, if a tuple Z belongs to the set $S_\sim$, then the tuple $x \cdot Z$ also belongs to the set $S_\sim$.

Indeed, a real number x can be represented as a limit of rational numbers: $r^{(k)} \to x$. According to Part 10, for every k, we have $r^{(k)} \cdot Z \in S_\sim$, i.e., the tuple

$$Z^{(k)} \stackrel{\text{def}}{=} (\exp(r^{(k)} \cdot Z_1), \ldots, \exp(r^{(k)} \cdot Z_n)) \sim (1, \ldots, 1). \tag{36}$$

In particular, this means that $Z^{(k)} \succeq (1, \ldots, 1)$. In the limit,

$$Z^{(k)} \to (\exp(x \cdot Z_1), \ldots, \exp(x \cdot Z_n)) \succeq (1, \ldots, 1). \tag{37}$$

By definition of the sets $S_\sim$ and $S_\succ$, this means that $x \cdot Z \in S_\sim$ or $x \cdot Z \in S_\succ$.

Similarly, for $-(x \cdot Z) = (-x) \cdot Z$, we conclude that $-x \cdot Z \in S_\sim$ or

$$(-x) \cdot Z \in S_\succ. \tag{38}$$

If we had $x \cdot Z \in S_\succ$, then by Part 7 we would get $(-x) \cdot Z \in S_\prec$, a contradiction. Thus, the case $x \cdot Z \in S_\succ$ is impossible, and we have $x \cdot Z \in S_\sim$. The statement is proven.

12°. According to Parts 3 and 11, the set $S_\sim$ is closed under addition and under multiplication by an arbitrary real number. Thus, if tuples $Z, \ldots, Z'$ belong to the set $S_\sim$, their arbitrary linear combination $x \cdot Z + \ldots + x' \cdot Z'$ also belongs to the set $S_\sim$. So, the set $S_\sim$ is a linear subspace of the n-dimensional space of all the tuples.

13°. The subspace $S_\sim$ cannot coincide with the entire n-dimensional space, because then the pre-ordering relation would be trivial. Thus, the dimension of this subspace must be less than or equal to $n - 1$. Let us show that the dimension of this subspace is $n - 1$.

Indeed, let us assume that the dimension is smaller than $n - 1$. Since the pre-ordering is non-trivial, there exist tuples $y = (y_1, \ldots, y_n)$ and $y' = (y'_1, \ldots, y'_n)$ for which $y \succ y'$ and thus, $Z = (Z_1, \ldots, Z_n) \in S_\succ$, where $Z_i = \ln\left(\dfrac{y_i}{y'_i}\right)$. From $Z \in S_\succ$, we conclude that $-Z \in S_\prec$.

Since the linear space $S_\sim$ is a less than $(n - 1)$-dimensional subspace of an n-dimensional linear space, there is a path connecting $Z \in S_\succ$ and $-Z \in S_\prec$ which avoids $S_\sim$. In mathematical terms, this path is a continuous mapping $\gamma : [0, 1] \to R^n$

for which $\gamma(0) = Z$ and $\gamma(1) = -Z$. Since this path avoids $S_\sim$, every point $\gamma(t)$ on this path belongs either to $S_\succ$ or to $S_\prec$.

Let $\overline{t}$ denote the supremum (least upper bound) of the set of all the values t for which $\gamma(t) \in S_\succ$. By definition of the supremum, there exists a sequence $t^{(k)} \to \overline{t}$ for which $\gamma\left(t^{(k)}\right) \in S_\succ$. Similarly to Part 11, we can use continuity to prove that in the limit, $\gamma\left(\overline{t}\right) \in S_\succ$ or $\gamma\left(\overline{t}\right) \in S_\sim$. Since the path avoids the set $S_\sim$, we thus get $\gamma\left(\overline{t}\right) \in S_\succ$.

Similarly, since $\gamma(1) \notin S_\succ$, there exists a sequence $t^{(k)} \downarrow \overline{t}$ for which $\gamma\left(t^{(k)}\right) \in S_\prec$. We can therefore conclude that in the limit, $\gamma\left(\overline{t}\right) \in S_\succ$ or $\gamma(\overline{t}) \in S_\sim$—a contradiction with our previous conclusion that $\gamma\left(\overline{t}\right) \in S_\succ$.

This contradiction shows that the linear space $S_\sim$ cannot have dimension smaller than $n - 1$ and thus, that this space have dimension $n - 1$.

14°. Every $(n - 1)$-dimensional linear subspace of an n-dimensional superspace separates the superspace into two half-spaces. Let us show that one of these half-spaces is $S_\succ$ and the other is $S_\prec$.

Indeed, if one of the subspaces contains two tuples Z and Z' for which $Z \in S_\succ$ and $Z' \in S_\prec$, then the line segment $\gamma(t) = t \cdot Z + (1 - t) \cdot Z'$ containing these two points also belongs to the same subspace, i.e., avoids the set $S_\sim$. Thus, similarly to Part 13, we would get a contradiction.

So, if one point from a half-space belongs to $S_\succ$, all other points from this subspace also belong to the set $S_\succ$. Similarly, if one point from a half-space belongs to $S_\prec$, all other points from this subspace also belong to the set $S_\prec$.

15°. Every $(n - 1)$-dimensional linear subspace of an n-dimensional space has the form

$$\alpha_1 \cdot Z_1 + \ldots + \alpha_n \cdot Z_n = 0 \tag{39}$$

for some real values α_i, and the corresponding half-spaces have the form

$$\alpha_1 \cdot Z_1 + \ldots + \alpha_n \cdot Z_n > 0 \tag{40}$$

and

$$\alpha_1 \cdot Z_1 + \ldots + \alpha_n \cdot Z_n < 0. \tag{41}$$

The set $S_\succ$ coincides with one of these subspaces. If it coincides with the set of all tuples Z for which $\alpha_1 \cdot Z_1 + \ldots + \alpha_n \cdot Z_n < 0$, then we can rewrite it as

$$(-\alpha_1) \cdot Z_1 + \ldots + (-\alpha_n) \cdot Z_n > 0, \tag{42}$$

i.e., as $\alpha'_1 \cdot Z_1 + \ldots + \alpha'_n \cdot Z_n > 0$ for $\alpha'_i = -\alpha_i$.

Thus, without losing generality, we can conclude that the set $S_\succ$ coincides with the set of all the tuples Z for which $\alpha_1 \cdot Z_1 + \ldots + \alpha_n \cdot Z_n > 0$. We have mentioned that

$$y' = (y'_1, \ldots, y'_n) \succ y = (y_1, \ldots, y_n) \Leftrightarrow (Z_1, \ldots, Z_n) \in S_\succ, \tag{43}$$

where $Z_i = \ln\left(\frac{y'_i}{y_i}\right)$. So,

$$y' \succ y \Leftrightarrow$$

$$\alpha_1 \cdot Z_1 + \ldots + \alpha_n \cdot Z_n = \alpha_1 \cdot \ln\left(\frac{y'_1}{y_1}\right) + \ldots + \alpha_n \cdot \ln\left(\frac{y'_n}{y_n}\right) > 0. \quad (44)$$

Since $\ln\left(\frac{y'_i}{y_i}\right) = \ln(y'_i) - \ln(y_i)$, the last inequality in the formula (44) is equivalent to

$$\alpha_1 \cdot \ln(y'_1) + \ldots + \alpha_n \cdot \ln(y'_n) > \alpha_1 \cdot \ln(y_1) + \ldots + \alpha_n \cdot \ln(y_n). \quad (45)$$

Let us take exp of both sides of the formula (45); then, due to the monotonicity of the exponential function, we get an equivalent inequality

$$\exp(\alpha_1 \cdot \ln(y'_1) + \ldots + \alpha_n \cdot \ln(y'_n)) > \exp(\alpha_1 \cdot \ln(y_1) + \ldots + \alpha_n \cdot \ln(y_n)). \quad (46)$$

Here,

$$\exp(\alpha_1 \cdot \ln(y'_1) + \ldots + \alpha_n \cdot \ln(y'_n)) = \exp(\alpha_1 \cdot \ln(y'_1)) \cdot \ldots \cdot \exp(\alpha_n \cdot \ln(y'_n)),$$

where for every i, $e^{\alpha_i \cdot z_i} = (e^{z_i})^{\alpha_i}$, with $z_i \stackrel{\text{def}}{=} \ln(y'_i)$, implies that

$$\exp(\alpha_i \cdot \ln(y'_i)) = (\exp(\ln(y'_i)))^{\alpha_i} = (y'_i)^{\alpha_i}, \quad (47)$$

so

$$\exp(\alpha_1 \cdot \ln(y'_1) + \ldots + \alpha_n \cdot \ln(y'_n)) = (y'_1)^{\alpha_1} \cdot \ldots \cdot (y'_n)^{\alpha_n} \quad (48)$$

and similarly,

$$\exp(\alpha_1 \cdot \ln(y_1) + \ldots + \alpha_n \cdot \ln(y_n)) = y_1^{\alpha_1} \cdot \ldots \cdot y_n^{\alpha_n}. \quad (49)$$

Thus, due to (44), (45), (46), (48) and (49), the condition $y' \succ y$ is equivalent to:

$$\prod_{i=1}^{n} y_i^{\alpha_i} > \prod_{i=1}^{n} (y'_i)^{\alpha_i}. \quad (50)$$

Similarly, we prove that

$$y_1, \ldots, y_n) \sim y' = (y'_1, \ldots, y'_n) \Leftrightarrow \prod_{i=1}^{n} y_i^{\alpha_i} = \prod_{i=1}^{n} (y'_i)^{\alpha_i}. \quad (51)$$

The condition $\alpha_i > 0$ follows from our assumption that the pre-ordering is monotonic.

The theorem is proven.

References

1. Aliev, R., Huseynov, O., Kreinovich, V.: Decision making under interval and fuzzy uncertainty: towards an operational approach. In: Proceedings of the Tenth International Conference on Application of Fuzzy Systems and Soft Computing ICAFS'2012, Lisbon, Portugal, 29–30 August 2012
2. Aubinet, M., Vesala, T., Papale, D. (eds.): Eddy Covariance - A Practical Guide to Measurement and Data Analysis. Springer, Dordrecht (2012)
3. Binder, A.B., Roberts, D. L.: Criteria for Lunar Site Selection, Report No. P-30, NASA Appollo Lunar Exploration Office and Illinois Institute of Technology Research Institute, Chicago, Illinois, January 1970
4. Cobb, C.W., Douglas, P.H.: A theory of production. Am. Econ. Rev. **18**(Supplement), 139–165 (1928)
5. de Finetti, B.: Theory of Probability, Wiley, New York, Vol. 1, 1974; Vol. 2, 1975
6. Ehrgott, M., Gandibleux, X. (eds.): Multiple Criteria Optimization: State of the Art Annotated Bibliographic Surveys. Springer, New York (2002)
7. Ferson, S., Kreinovich, V., Hajagos, J., Oberkampf, W., Ginzburg, L.: Experimental Uncertainty Estimation and Statistics for Data Having Interval Uncertainty, Sandia National Laboratories (2007), Publ. 2007–0939
8. Fishburn, P.C.: Utility Theory for Decision Making. Wiley, New York (1969)
9. Fishburn, P.C.: Nonlinear Preference and Utility Theory. Johns Hopkins University Press, Baltimore (1988)
10. Fountoulis, I., Mariokalos, D., Spyridonos, E., Andreakis, E.: Geological criteria and methodology for landfill sites selection. In: Proceedings of the 8th International Conference on Environmental Science and Technology, Greece, 8–10 September 2003, pp. 200–207
11. Gardeñes, E., et al.: Modal intervals. Reliable Comput. **7**, 77–111 (2001)
12. Giang, P.H.: Decision with dempster-shafer belief functions: decision under ignorance and sequential consistency. Int. J. Approximate Reasoning **53**, 38–53 (2012)
13. Guo, P., Tanaka, H.: Decision making with interval probabilities. Eur. J. Oper. Res. **203**, 444–454 (2010)
14. Guo, P., Wang, Y.: Eliciting dual interval probabilities from interval comparison matrices. Inf. Sci. **190**, 17–26 (2012)
15. Helversen B.V., Mata R.: Losing a dime with a satisfied mind: positive affect predicts less search in sequential decision making, Psychology and Aging, **27**, 825–839 (2012)
16. Hurwicz, L.: Optimality Criteria for Decision Making Under Ignorance, Cowles Commission Discussion Paper, Statistics, No. 370, 1951
17. Jaimes A., Salayandia L., Gallegos I.: New cyber infrastructure for studying land-atmosphere interactions using eddy covariance techniques. In: Abstracts of the 2010 Fall Meeting American Geophysical Union AGU'2010, San Francisco, California, 12–18 December 2010
18. Jaimes, A., Herrera, J., Gonzĺez, L., Ramìrez, G., Laney, C., Browning, D., Peters, D., Litvak, M., Tweedie, C.: Towards a multiscale approach to link climate, NEE and optical properties from a flux tower, robotic tram system that measures hyperspectral reflectance, phenocams, phenostations and a sensor network in desert a shrubland. In: Proceedings of the FLUXNET and Remote Sensing Open Workshop: Towards Upscaling Flux Information from Towers to the Globe FLUXNET-SpecNet'2011, Berkeley, California, 7–9 June 2011
19. Jaimes, A.: A cyber-Tool to Optimize Site Selection for Establishing an Eddy Covraince and Robotic Tram System at the Jornada Experimental Range. University of Texas, El Paso (2008)

20. Jaimes, A., Tweedie, C., Kreinovich, V., Ceberio, M.: Scale-invariant approach to multi-criterion optimization under uncertainty, with applications to optimal sensor placement, in particular, to sensor placement in environmental research. Int. J. Reliab. Saf. **6**(1–3), 188–203 (2012)
21. Kiekintveld, C., Kreinovich, V.: Efficient approximation for security games with interval uncertainty, In: Proceedings of the AAAI Spring Symposium on Game Theory for Security, Sustainability, and Health GTSSH'2012, Stanford, 26–28 March 2012
22. Kintisch, E.: Loss of carbon observatory highlights gaps in data. Science **323**, 1276–1277 (2009)
23. Kreinovich, V.: Decision making under interval uncertainty, In: Abstracts of the 15th GAMM - IMACS International Symposium on Scientific Computing, Computer Arithmetic, and Verified Numerical Computation SCAN'2012. Novosibirsk, Russia, September 2012
24. Kreinovich, V., Starks, S.A., Iourinski, D., Kosheleva, O., Finkelstein, A.: Open-ended configurations of radio telescopes: a geometrical analysis. Geombinatorics **13**(2), 79–85 (2003)
25. Kyburg, H.E., Smokler, H.E. (eds.): Studies in Subjective Probability. Wiley, New York (1964)
26. Luce R.D., Krantz D.H., Suppes P., Tversky A.: Foundations of measurement, vol. 3, Representation, Axiomatization, and Invariance, Academic Press, California (1990)
27. Luce, R.D., Raiffa, R.: Games and Decisions: Introduction and Critical Survey. Dover, New York (1989)
28. Moore, R.E., Kearfott, R.B., Cloud, M.J.: Introduction to Interval Analysis. SIAM, Pennsylvania (2009)
29. Myerson, R.B.: Game theory. Analysis of conflict, Harvard University Press, Cambridge (1991)
30. Nash, J.F.: The bargaining problem. Econometrica **28**, 155–162 (1950)
31. Nash, J.: Two-person cooperative games. Econometrica **21**, 128–140 (1953)
32. Nguyen, H.T., Kreinovich, V., Wu, B., Xiang G.: Computing Statistics under Interval and Fuzzy Uncertainty, Springer, Berlin (2012)
33. Nguyen, H.T., Kreinovich, V.: Applications of Continuous Mathematics to Computer Science. Kluwer, Dordrecht (1997)
34. Pfanzangl, J.: Theory of Measurement. Wiley, New York (1968)
35. Raiffa, H.: Decision Analysis. McGraw-Hill, Columbus (1997)
36. Roth, A.: Axiomatic Models of Bargaining. Springer, Berlin (1979)
37. Savage, L.J.: The Foundations of Statistics. Wiley, New York (1954)
38. Sawaragi, Y., Nakayama, H., Tanino, T.: Theory of Multiobjective Optimization. Academic Press, Orlando (1985)
39. Shary, S.P.: A new technique in systems analysis under interval uncertainty and ambiguity. Reliable Comput. **8**, 321–418 (2002)
40. Steuer, R.E.: Multiple Criteria Optimization: Theory, Computations, and Application. Wiley, New York (1986)
41. Stevens, S.S.: On the theory of scales of measurement. Science **103**, 677–680 (1946)
42. Troffaes, M.C.M.: Decision making under uncertainty using imprecise probabilities. Int. J. Approximate Reasoning **45**, 7–29 (2007)
43. Walsh, C.: Monetary Theory and Policy. MIT Press, Cambridge (2003)

Dealing with Imprecision in Consumer Theory: A New Approach to Fuzzy Utility Theory

David Gálvez Ruiz and José Luís Pino Mejías

Abstract This chapter presents a new approach to dealing with imprecision in the Classical Consumer Utility Theory based on the concept of Marginal Rate of Substitution (MRS) and using the concept of fuzzy sets and fuzzy numbers. The methodology developed applies imprecision to MRS, whereas previous studies placed the imprecision factor on final utility values and functions. The chapter considers fuzzy elements applied to MRS and uses the necessary formulations to obtain the results in Utility Theory. In this fuzzy environment, the final consumer decision problem is framed as a fuzzy nonlinear programming problem, maintaining the classical structure in which consumers maximize their fuzzy utility subject to budget constraints, and showing that the consumer optimum choice is a fuzzy set. The chapter will also address the problem of aggregation of utility functions in order to offer a multi-criteria approach.

Keywords New directions of decision analysis under uncertainty · Muti-criteria decision making

1 Introduction

The study of consumer decision-making from the classical perspective is based on the concept of utility. The first approach to this concept was created by Stanley Jevons, who thought that people consume according to the pain or pleasure they receive from the action. He called this utility [33].

D. Gálvez Ruiz (✉) · J. L. Pino Mejías
Department of Statistics and Operational Research, Faculty of Mathematics. University of Seville, C/ Tarfia s.n., Sevilla 41012, Spain
e-mail: davidgalvez@us.es

J. L. Pino Mejías
e-mail: jlpino@us.es

D. Gálvez Ruiz · J. L. Pino Mejías
Quantitative Methods on Evaluation Research Group, University of Seville, Sevilla, Spain

P. Guo and W. Pedrycz (eds.), *Human-Centric Decision-Making Models for Social Sciences*, Studies in Computational Intelligence 502,
DOI: 10.1007/978-3-642-39307-5_9,

This starting point led to several classical definitions of utility, summarized masterfully by the great Keynesian economist Joan Robinson, who stated that utility is just the characteristic of goods that makes people want to buy, and these people buy goods to enjoy their utility [32].

These definitions led classical economists to establish utility as a measurable cardinal concept. Thus, for early theorists, utility is considered as a reality born from psychological introspection and as such, its treatment would generate a directly measurable quantity inherent to the good consumed. In this extreme view, measuring the utility is equivalent to associating a real number (units of utility) to the feeling obtained from the consumption of goods. Thus, a good that provides 20 units of utility would be twice as desirable as another one that provides only 10. However, towards the end of the nineteenth century this concept would be progressively revised, as both a conceptual and an instrumental field.

The next step came with Pareto when the cardinal utility view was definitively rejected. Pareto, suggests that satisfaction need is based not only on the objective inner properties of things, but on the subjective aspects of needs as well [29]. This allowed for a redefined concept of the utility function as establishing only preference relations. Therefore, utility levels can be represented by any increasing function. In that sense, the utility function only needs to establish a preference relation order, so that a good that provides 20 units of utility would be only more desirable than another one that provides only 10, but not necessarily twice as desirable.

This conception led to the subjective conception of utility and made it possible to introduce parameters in utility functions in order to fit the perceived needs of the consumer. Therefore, the utility function for one consumer will reach a different value for the consumption of a combination of goods from the value reached by the utility function of another consumer for the same combination of goods, so that any interpersonal comparison is invalid.

All these developments led to the construction of a new model for classical consumer theory, whose main architect was Hicks, offering the following premise: "We must now begin the work of killing and reject all concepts with trace of quantitative value on utility, to replace them where they are, by notions entirely free of them" [21].

His whole argument is based on the definition of marginal rate of substitution (MRS). The MRS of x_1 on x_2 is the amount of x_2 which is sufficient to compensate the consumer for the loss of a marginal unit of x_1. In other words, the marginal rate of substitution shows the amount of product x_1 the consumer would exchange for an additional unit of product x_2, maintaining a constant level of utility. With this definition, Hicks manages to move the quantitative element of consumer subjectivity from utility to MRS, establishing a relation between them.

At this point, economists and other social scientists have added more complexity to the model in order to adjust it to different scenarios with uncertainty and/or imprecision [1, 15]. One of these scenarios is the internal uncertainty or imprecision involved in the judgements made by consumers and decision-makers. While external or general uncertainty is derived from environmental factors which are exogenous to the consumer, such as the weather, or the place where consumption will occur, internal uncertainty or imprecision arises from several factors inherent to the consumer. These factors break the classical assumption of preference and judgement certainty and produce a lack of exactness, affecting the utility valuations from consumers. This imprecision is always present in a consumer decision-making process: first, because people may not be well acquainted with the alternatives they are being asked to value, and cannot easily express a preference for different combinations of goods; and second, and specially relevant in this chapter, consumers may be truly uncertain about the utility of a combination of goods or concerning MRS, uncertain about their tradeoffs.

Nevertheless, far from the Hicksian statement, the treatment up to now of uncertainty and imprecision in Utility Theory has always relied on utility values given by consumers or on utility functions inferred from these values. In this sense, at the beginning of imprecision treatment in utility, several studies gave imprecision to utility from its crisp values ([34] lists some of these most important works), yet this conception was replaced in more recent studies, which deal directly with imprecise values inferred or given by the consumers [8, 25]; imprecise preferences [2, 10, 15, 27, 28, 35]; or imprecise utility functions [7, 24].

However, it is unrealistic to suppose that consumers are able to assign a number (value) for utility and simply deal with the incomplete information associated with utility. Through the alternative approach developed in this chapter, it is possible to initially ignore the utility valuation and determine how many units of product x_1 the consumer would exchange for an additional unit of product x_2 maintaining a constant level of utility. This is the concept of Marginal Rate of Substitution. The present methodology launches a new approach which applies imprecision to MRS, whereas previous studies placed the imprecision factor on final utilities in one way or another. The new methodology considers fuzzy elements applied to MRS and uses the necessary formulations to obtain the results in Utility Theory.

First, the marginal rate of substitution and its relation to utility will be presented. From the reasoning above, the uncertainty and imprecision element is placed on MRS instead of on utility values or functions using fuzzy sets. Then, from the fuzzy marginal rate of substitution, the methodology reaches the fuzzy consumer utility function by solving non-polynomial fuzzy partial differential equations through a new variation of the Buckley-Feuring method for the polynomial form. Classical consumer theory has always implemented this step inversely, that is, obtaining MRS from utility functions with the assumption that consumers can assess these utility values.

In this fuzzy environment, the final consumer decision problem is framed as a fuzzy nonlinear programming problem in which consumers maximize their fuzzy utility subject to budget constraints. The methodology deals with the solution in this new scenario, so that the classical economic approach to the final consumer decision is tackled in a fuzzy environment while maintaining the maximization problem structure. A Multi-criteria approach to the problem is also given, thus addressing the issue of aggregation of utility functions. All the steps outlined above have been applied to different commonly used consumer utility structures, such as Cobb-Douglas utilities or Constant Elasticity Substitution (CES) utilities.

2 Preliminaries

2.1 Fuzzy Numbers

Fuzzy sets were introduced by Zadeh [39–42] to deal with the imprecision and vagueness in numeric and qualitative values. From that point, considerable strides have been made in order to clarify the concept of fuzziness and identify properties and operations.

The most accepted definition for fuzzy sets works as follows:

If X is a collection of objects generally denoted by x, then, a fuzzy set $\tilde{A}$ over X is a set of pairs:

$$\tilde{A} = \{(x, \mu_{\tilde{A}}(x)) | x \in X\}, \tag{1}$$

where $\mu_{\tilde{A}}(x)$ is called the *membership function* (or membership degree) of x in $\tilde{A}$, and maps X in the pattern space M. The range of $\mu_{\tilde{A}}(x)$ is a subset of nonnegative real numbers with finite supremum.

From this definition, the fuzzy number concept is developed as a special case of fuzzy set in which:

1. $\sup \mu_{\tilde{A}}(x) = 1, \quad \inf \mu_{\tilde{A}}(x) = 0$
2. $\forall \alpha \in [0, 1], \quad \mu(\alpha) = \{x \in X | \mu_{\tilde{A}}(x) \geq \alpha\}$ is convex.
3. $\mu_{\tilde{A}}(x)$ is continuous.
4. $\exists \{x_0 \in X | \mu_{\tilde{A}}(x_0) = 1\}$.

Obviously, there will be as many fuzzy numbers types as possible membership functions. One of them is the triangular fuzzy number, identified by:

$$\mu_{\tilde{A}}(x) = \begin{cases} \dfrac{x - a_1}{a_2 - a_1}, & a_1 \leq x \leq a_2 \\[2ex] \dfrac{a_3 - x}{a_3 - a_2}, & a_2 \leq x \leq a_3 \\[2ex] 0 & \text{otherwise} \end{cases} \tag{2}$$

where a_2 reaches the $\sup \mu_{\tilde{A}}(x) = 1$, and a_1 and a_3 reach the $\inf \mu_{\tilde{A}}(x) = 0$.

Other ways to define fuzzy numbers have also been developed. This chapter uses properties from the Goetschel-Voxman definition [19] to solve fuzzy partial differential equations. From this definition, a fuzzy number is a fuzzy set defined by (1) with $\mu : X \rightarrow [0, 1]$ that satisfies the properties:

1. μ is upper semicontinuous.

2. $\mu(x) = 0$ outside of some interval $[c, d]$.
3. There are real numbers a and b, $c \leq a \leq b \leq d$ such that μ is increasing on $[c, a]$, decreasing on $[b, d]$ and $\mu(x) = 1$ for each $x \in [a, b]$.

This definition makes $a = b$ for triangular fuzzy numbers [corresponding to a_2 in (2)] and for other fuzzy numbers that have only one element with membership function equal to one.

For $0 \leq \alpha \leq 1$, it is possible to define:

$$\mu(\alpha) = \begin{cases} \{x \mid \mu(x) \geq \alpha\} & \text{if } 0 < \alpha \leq 1 \\ cl(supp \quad \mu) & \text{if } \alpha = 0 \end{cases} \tag{3}$$

Where $cl(supp \quad \mu)$ denotes the closure of support of μ and $\mu(\alpha)$ is called α-cut set. Then, $\tilde{A} = \{(x, \mu_{\tilde{A}}(x)), \quad \mu : X \rightarrow [0, 1])|x \in X\}$ is a fuzzy number if and only if:

- $\mu(\alpha)$ is a closed and bounded interval for each α, $0 < \alpha \leq 1$, and
- $\mu(1) \neq \emptyset$

With this characterization, it is possible to identify a fuzzy number with the parameterized triples $\{(a_1(\alpha), a_2(\alpha), \alpha) \mid 0 \leq \alpha \leq 1\}$, where $a_1(\alpha)$ denotes the left hand endpoint of $\mu(\alpha)$ and $a_2(\alpha)$ denotes the right hand endpoint for $a_1, a_2 \in \tilde{A}$ over X.

Suppose that the functions $a_1 : [0, 1] \rightarrow X$ and $a_2 : [0, 1] \rightarrow X$ satisfy the conditions:

1. a_1 is a bounded increasing function,
2. a_2 is a bounded decreasing function,
3. $a_1(1) \leq a_2(1)$,
4. For $0 < h \leq 1 : \lim_{\alpha \to h^-} a_1(\alpha) = a_1(h)$, and $\lim_{\alpha \to h^-} a_2(\alpha) = a_2(h)$,
5. $\lim_{\alpha \to 0^+} a_1(\alpha) = a_1(0)$, and $\lim_{\alpha \to 0^+} a_2(\alpha) = a_2(0)$.

If $\tilde{A} = \{(x, \mu_{\tilde{A}}(x))|x \in X\}$ is a fuzzy number identified with the parameterized triple $\{(a_1(\alpha), a_2(\alpha), \alpha) \mid 0 \leq \alpha \leq 1\}$, then the functions a_1 and a_2 satisfy conditions 1–5.

2.2 *Elements of the Consumer Decision Problem*

- First, it is necessary to identify the set of alternatives which can be adopted as solutions in the decision-making problem. In this case, this set of alternatives has infinite elements, such as pairs (x_1, x_2), with $x_1 \in S_1 \subset I_1 = (0, M_1]$, $x_2 \in S_2 \subset I_2 = (0, M_2]$ amounts of goods X_1 and X_2 respectively. X_1 and X_2 are perfectly divisible.[1]

[1] This is a general assumption in classic economic theory, but not restrictive in many cases, i.e. when producers are consumers of raw materials or goods like wood, steel, fuel, energy, water, oil, seeds ....

- Utility function and preference rules.
  The utility function is applied to each alternative (x_1, x_2) in order to evaluate the combination of goods X_1 and X_2. The utility function establishes a preference order as follows:
  - An alternative (x_1, x_2) is strictly preferred to (x_1', x_2'):
  $$(x_1, x_2) \succ (x_1', x_2')$$
  if the alternative (x_1, x_2) is better than (x_1', x_2') in the consumer decision-making problem, that is, $U(x_1, x_2) > U(x_1', x_2')$.
  - An alternative (x_1, x_2) is weakly preferred to (x_1', x_2'):
  $$(x_1, x_2) \succeq (x_1', x_2')$$
  if the alternative (x_1, x_2) is not worse than (x_1', x_2') in the consumer decision making problem, that is, $U(x_1, x_2) \geq U(x_1', x_2')$.
  - An alternative (x_1, x_2) is indifferent to (x_1', x_2'):
  $$(x_1, x_2) \sim (x_1', x_2')$$
  if the consumer gets the same satisfaction choosing either of them, that is:
  $$U(x_1, x_2) = U(x_1', x_2')$$
  If a set of alternatives is indifferent,it is said that they are in the same *indifference curve*, which shows the geometrical representation of the set:
  $$I(x_1', x_2') = \{(x_1, x_2) \in S_1 \times S_2 \setminus (x_1, x_2) \sim (x_1', x_2')\} \tag{4}$$

A generic utility function $U(x_1, x_2)$ can be used to order preferences as shown above if the following axioms are held[2]:

1. Comparability.
$$\forall (x_1, x_2), (x_1', x_2') \in S_1 \times S_2 : (x_1, x_2) \succeq (x_1', x_2'), \text{ or } (x_1', x_2') \succeq (x_1, x_2),$$
or both.

2. Transitivity.
$$\forall (x_1, x_2), (x_1', x_2'), (x_1'', x_2'') \in S_1 \times S_2 :$$
$$\text{if } (x_1, x_2) \succeq (x_1', x_2') \text{ and } (x_1', x_2') \succeq (x_1'', x_2''), \Longrightarrow (x_1, x_2) \succeq (x_1'', x_2'')$$

3. Consistency between indifference and weak preference.

[2] The first four axioms are generic in decision making.

$\forall(x_1, x_2), (x_1', x_2') \in S_1 \times S_2 : (x_1, x_2) \sim (x_1', x_2') \Longleftrightarrow (x_1, x_2) \succeq (x_1', x_2')$ and $(x_1', x_2') \succeq (x_1, x_2)$

4. Consistency between weak preference and strict preference. One alternative is preferred to another alternative only if:

$$\forall(x_1, x_2), (x_1', x_2') \in S_1 \times S_2 : \quad (x_1, x_2) \succ (x_1', x_2') \Longleftrightarrow \neg((x_1', x_2') \succeq (x_1, x_2))$$

5. $U(x_1, x_2)$ is monotone non-decreasing $\forall(x_1, x_2) \in S_1 \times S_2$. Also, it is generally assumed that Utility is a continuous function in $(x_1, x_2) \in I_1 \times I_2$ with partial D_{x_1}, D_{x_2} [26].
6. Both partial:

$$\frac{\partial U(x_1, x_2)}{\partial x_i} = D_{x_i}, \quad i = 1, 2, \quad \forall(x_1, x_2) \in S_1 \times S_2 \text{ are non-increasing}$$

functions for goods.

If these axioms are held, it is possible to define a utility function which assigns a number to the consequence of each alternative (x_1, x_2). This utility function is used to assign ordinal preferences, so that:

$$U(x_1, x_2) \geq U(x_1', x_2') \Longleftrightarrow (x_1, x_2) \succeq (x_1', x_2')$$

that is, a higher value reached in the utility function by one alternative over another one, implies that the former alternative is preferred to the latter.

The utility function will also be measurable if, in addition to the previous conditions, it satisfies that if:

$$U(x_1, x_2) - U(x_1', x_2') \geq U(x_1'', x_2'') - U(x_1''', x_2''')$$

then,

$$((x_1, x_2) \longleftarrow (x_1', x_2')) \succeq ((x_1'', x_2'') \longleftarrow (x_1''', x_2'''))$$

that is, a change from (x_1', x_2') to (x_1, x_2) is, at least as good as a change from (x_1''', x_2''') to (x_1'', x_2'').

3 Utility and MRS

The new approach presented here will show that it is possible (and more realistic) to maintain the Hicksian framework, initially ignoring the utility valuation and determining how many units of product X_1 the consumer would exchange for an

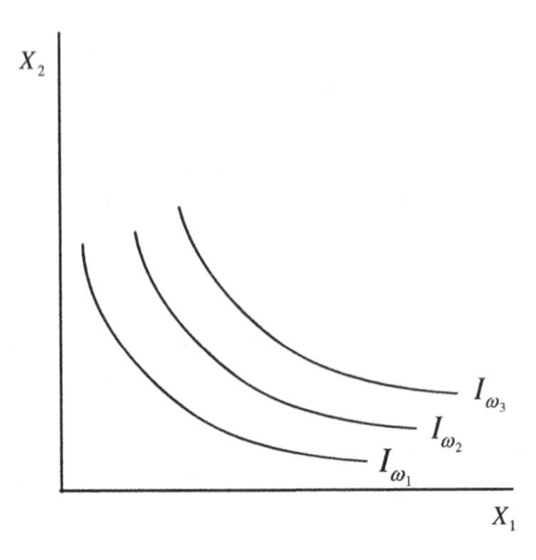

Fig. 1 Standard indifference curves for two goods

additional unit of product X_2 while maintaining a constant utility level, in other words, the marginal rate of substitution. Thus, let:

- x_i = Amount of good X_i.
- $U(x_1, x_2, ..., x_n)$ = Utility given by $(x_1, x_2, ..., x_n)$.
- Ω = Range of utility values.
- U^ω = Utility level $\omega \quad (U(\cdot) = \omega), \quad \omega \in \Omega$.

Following the classical economic approach, only two products are considered at the moment. In terms of utility function, the indifference curve described in (4) of these two goods is the set given by:

$$I(x_1', x_2') = \{(x_1, x_2) \in S_1 \times S_2 \setminus U(x_1, x_2) = U(x_1', x_2')\} \tag{5}$$

The MRS of x_2 over x_1 shows the amount of X_2 that the consumer is willing to exchange for an additional unit of X_1. It will be given by the indifference curve slope:

$$MRS_{21} = -\frac{dx_2}{dx_1} =| \text{ slope of indifference curve } |, \tag{6}$$

where the negative symbol is introduced in order to obtain the absolute value of MRS. This occurs because the slope is changing along the indifference curve: the lesser the amount of X_1 and greater the amount of X_2, the greater the valuation of marginal changes on x_1 in comparison to marginal changes on x_2. That implies a decrease in the absolute value of MRS_{21} while x_1 is increasing (Fig. 1).

Moreover, because utility is constant along the indifference curve, we have:

$$U(x_1, x_2) = U(x_1, x_2(x_1)) = C$$

$$\frac{\partial U}{\partial x_1} + \frac{\partial U}{\partial x_2}\frac{dx_2}{dx_1} = 0$$

$$-\frac{dx_2}{dx_1} = \frac{\partial U/\partial x_1}{\partial U/\partial x_2},$$

and from (6),

$$MRS_{21} = -\frac{dx_2}{dx_1} = \frac{\partial U/\partial x_1}{\partial U/\partial x_2} \tag{7}$$

If there is an explicit functional form of the utility function, then it is also possible to obtain the MRS_{21} in functional form from the expression above. In microeconomics the usual way to infer the utility function is through regressions from consumer valuations on utility levels or preferences[3]. From there, MRS_{21} may be obtained as a quotient of partial derivatives:

$$MRS_{21} = \frac{\partial U/\partial x_1}{\partial U/\partial x_2} = F(x_1, x_2; \boldsymbol{\beta}), \tag{8}$$

where $\boldsymbol{\beta}$ is a parameter vector determined by subjective preferences giving the inherent character of the consumer utility function as shown by the convexity level of indifference curves. That is, utility is a parametric function with parameters organized in $\boldsymbol{\beta}$. These parameters determine how the goods contribute to the utility while their values make the utility function inherent and different from one consumer to another[4].

With these foundations, this chapter introduces the inverse process, arriving at the utility function by solving the partial differential equations obtained through MRS.

---

[3] This issue is discussed extensively in microeconometrics. See, i.e. [26], or [6, 20]

[4] To illustrate this point, some examples of utility function are given:

- Perfect substitute goods:

$$U(x_1, x_2) = \beta_1 x_1 + \beta_2 x_2, \quad \boldsymbol{\beta} = (\beta_1, \beta_2)$$

- One good is bad:

$$U(x_1, x_2) = -\beta_1 x_1 + \beta_2 x_2, \quad \boldsymbol{\beta} = (\beta_1, \beta_2)$$

- Perfect complement goods:

$$U(x_1, x_2) = \min\{\beta_1 x_1, \beta_2 x_2\}, \quad \boldsymbol{\beta} = (\beta_1, \beta_2)$$

Different consumers will have different parameters values.

3.1 *Placing Imprecision on Marginal Rate of Substitution: The Fuzzy MRS*

As outlined above, it becomes more obvious to infer a function for MRS because marginal rate of substitution is more perceptive to valuations by consumers than final utilities are. However, this method is absent in the classical approach to consumer decision problems, perhaps because the step from MRS to utility function implies partial differential equations. To obtain the functional form of MRS described by (8), the well-known regression techniques used for consumer decision-making can also be applied. MRS obtained by means of consumers statements will only be an approximate value, considering the information available to consumers in an environment of internal uncertainty. For this reason, the treatment of MRS as a fuzzy concept is more appropriate:

$$\widetilde{MRS_{21}} = \frac{\partial \tilde{U}/\partial x_1}{\partial \tilde{U}/\partial x_2} = F(x_1, x_2; \tilde{\beta}), \tag{9}$$

where tilde means the fuzzy character of an element. x_r, $r = 1, 2$, $x_1 \in S_1 \subset I_1 = (0, M_1]$, $x_2 \in S_2 \subset I_2 = (0, M_2]$ are quantities of products X_1 and X_2. $\tilde{\beta} = (\tilde{\beta}_1, \tilde{\beta}_2, ..., \tilde{\beta}_k)$ is a triangular fuzzy parameter vector determined by the subjective preferences of each consumer, $\tilde{U}(x_1, x_2; \tilde{\beta})$ is the fuzzy utility of combination (x_1, x_2), that is positive and strictly increasing because X_1 and X_2 are treated as goods, and an increase in consumption implies an increase in utility. Finally, $F(x_1, x_2; \tilde{\beta})$ is the functional expression of MRS inferred from the consumer preferences by means of econometrics methods (for more information, see [20]). This means that, for example, if a carpenter is asked for how many units of type A wood would change for an additional unit of type B wood maintaining the same level of utility, it seems to be more realistic that he would say "around 6 units", or "six, more or less" than "I would change exactly six units", that is, an imprecise representation that gives a "close to real" marginal rate of substitution given the incomplete perceptions involved with valuations.

The fuzzy MRS will, therefore, be a fuzzy number with its associated membership functions. It is then appropriate to identify these fuzzy variables with a triangular membership function in order to reach the maximum membership value at values obtained from crisp consumer statements, and descend as we move away from those values[5].

[5] It is also reasonable to associate a quasi-Gaussian membership function to these variables. However, the triangular fuzzy numbers are especially handy in operations where they are involved. The fuzzy trapezoidal numbers could also represent the behavior of consumers. Therefore, the approach chosen in this paper is adaptable to any type of membership function by changing the conditions of differentiability given later.

4 From Fuzzy MRS to Fuzzy Utility: Solving Non-polynomial Fuzzy Partial Differential Equations (FPDE)

The final goal of this chapter is to reach a functional expression of $\tilde{U}$ from $\widetilde{MRS}$ by solving the fuzzy partial differential equation, and then to obtain the consumer decision in a fuzzy environment by maximizing the fuzzy utility subject to budget constraints. The chapter also addresses the problem of aggregation of utility function in order to offer a multi-criteria approach.

The key step to obtaining the final consumer decision in this fuzzy environment is to go from the fuzzy MRS as shown in (9) to the fuzzy utility expression:

$$\tilde{U}(x_1, x_2; \tilde{\beta})$$

solving the fuzzy partial differential equation.

For this purpose, this chapter makes use of the methodology developed for that purpose in [16, 17] originating from the previous work of Buckley and Feuring [3, 4].

The components of the fuzzy partial differential equation are:

- $x_r, \quad r = 1, 2, \quad x_1 \in S_1 \subset I_1 = (0, M_1], \quad x_2 \in S_2 \subset I_2 = (0, M_2]$ are quantities of products X_1 and X_2.
- $\tilde{\beta} = (\tilde{\beta}_1, \tilde{\beta}_2, ..., \tilde{\beta}_k)$, a triangular fuzzy parameter vector, reflects the particular characteristics of consumer preferences.
- $\mu_{\tilde{\beta}_j}(b)$ is the membership function of the element b in $\tilde{\beta}_j$. As noted, the marginal rate of substitution function is obtained from inferences made from the consumer responses about the exchange of products. The parameters inferred in this way are considered under uncertainty and imprecision. This imprecision gives them a fuzzy character, and they may be characterized as triangular fuzzy numbers. As such, it is preferable to consider the element of the fuzzy number $\tilde{\beta}_j$ with a higher value in the membership function, corresponding to the value of the parameter as previously approximated.
- $\mu_j(\alpha) = \{b \in \tilde{\beta}_j \mid \mu_{\tilde{\beta}_j}(b) \geq \alpha, \quad \alpha \in [0, 1]\}$ is the set called α-cut
  These sets are closed and bounded, so that it is possible to define, for a fuzzy number $\tilde{\beta}_j : \tilde{\beta}_j[\alpha] = [b_1(\alpha), b_2(\alpha)]$, where:
  - $b_1(\alpha)$ is the lower value b in which $\mu_{\tilde{\beta}_j}(b) \geq \alpha, \quad b \in \tilde{\beta}_j$.
  - $b_2(\alpha)$ is the higher value b in which $\mu_{\tilde{\beta}_j}(b) \geq \alpha, \quad b \in \tilde{\beta}_j$.
- $\tilde{U}(x_1, x_2, \tilde{\beta})$ is a positive and continuous fuzzy utility function with partial differentials D_{x_1}, D_{x_2}. The output of this function is the utility level given by $(x_1, x_2) \in S_1 \times S_2$. This function must also be strictly increasing in $(x_1, x_2) \in S_1 \times S_2$. Since we are dealing with goods, an increase in consumption implies an increase in utility levels.The fuzzy character shown by the tilde placed over U, is fixed by $\tilde{\beta}$.

- $\varphi(D_{x_1}, D_{x_2})$ is an expression with constant coefficients in (D_{x_1}, D_{x_2}) applied to $\tilde{U}(x_1, x_2, \tilde{\beta})$.
- $F(x_1, x_2, \tilde{\beta})$ is a continuous function in $(x_1, x_2) \in S_1 \times S_2$.

Following this notation, the specific fuzzy partial differential equation treated here has the following form:

$$\varphi(D_{x_1}, D_{x_2})\tilde{U}(x_1, x_2, \tilde{\beta}) = \frac{\partial\tilde{U}/\partial x_1}{\partial\tilde{U}/\partial x_2} = F(x_1, x_2, \tilde{\beta}). \tag{10}$$

4.1 The Buckley-Feuring Based Solution for Non-polynomial Fuzzy Partial Differential Equations

The new Buckley-Feuring (B-F) based solution developed for non-polynomial fuzzy partial differential equations uses a solution of the crisp partial differential equation described in (8):

$$\frac{\partial U/\partial x_1}{\partial U/\partial x_2} = F(x_1, x_2, \beta).$$

$$U(x_1, x_2) = G(x_1, x_2, \beta),$$

with G being continuous and monotone for all $(x_1, x_2) \in S_1 \times S_2$.

The next step is the fuzzification of G. This fuzzy expression will be the candidate solution for the fuzzy partial differential equation:

$$\tilde{Y}(x_1, x_2) = \tilde{G}(x_1, x_2, \tilde{\beta}),$$

Note that $\tilde{Y}$ is only the fuzzy representation of G, but as stated above, not necessarily the solution to the fuzzy partial differential equation. If it turns out that $\tilde{Y}(x_1, x_2)$ is a fuzzy number and it solves the Eq. (10), then it will be a solution and $\tilde{U}(x_1, x_2, \tilde{\beta}) = \tilde{Y}(x_1, x_2)$.

With this notation, it is possible to see that: $\tilde{Y}(x_1, x_2)[\alpha] = [y_1(x_1, x_2, \alpha), y_2(x_1, x_2, \alpha)]$, and $\tilde{F}(x_1, x_2, \tilde{\beta})[\alpha] = [f_1(x_1, x_2, \alpha), f_2(x_1, x_2, \alpha)]$, $\forall\alpha \in [0, 1]$.

and, by definition:

$$y_1(x_1, x_2, \alpha) = \min\{G(x_1, x_2, \beta), \quad \beta \in \tilde{\beta}[\alpha]\},$$
$$y_2(x_1, x_2, \alpha) = \max\{G(x_1, x_2, \beta), \quad \beta \in \tilde{\beta}[\alpha]\}$$

$$f_1(x_1, x_2, \alpha) = \min\{F(x_1, x_2, \beta), \quad \beta \in \tilde{\beta}[\alpha]\},$$
$$f_2(x_1, x_2, \alpha) = \max\{F(x_1, x_2, \beta), \quad \beta \in \tilde{\beta}[\alpha]\}$$

$$\forall x_1, x_2 \in S_1 \times S_2, \forall \alpha \in [0, 1]$$

If it is possible to apply the $\varphi(D_{x_1}, D_{x_2})$ operator to $y_i, i = 1, 2$, obtaining continuous expressions such that $\forall (x_1, x_2) \in S_1 \times S_2, \forall \alpha \in [0, 1]$:

- For $0 < h \leq 1 : \lim_{\alpha \to h^-} \varphi(D_{x_1}, D_{x_2}) y_1(x_1, x_2; \alpha) = \varphi(D_{x_1}, D_{x_2}) y_1(x_1, x_2; h)$, and $\lim_{\alpha \to h^-} \varphi(D_{x_1}, D_{x_2}) y_2(x_1, x_2; \alpha) = \varphi(D_{x_1}, D_{x_2}) y_2(x_1, x_2; h)$;
- $\lim_{\alpha \to 0^+} \varphi(D_{x_1}, D_{x_2}) y_1(x_1, x_2; \alpha) = \varphi(D_{x_1}, D_{x_2}) y_1(x_1, x_2; 0)$, and $\lim_{\alpha \to 0^+} \varphi(D_{x_1}, D_{x_2}) y_2(x_1, x_2; \alpha) = \varphi(D_{x_1}, D_{x_2}) y_2(x_1, x_2; 0)$,

then, it will be feasible to define the following expression for $\Gamma(x_1, x_2, \alpha)$:

$$\Gamma(x_1, x_2, \alpha) = [\Gamma_1(x_1, x_2, \alpha), \Gamma_2(x_1, x_2, \alpha)] \tag{11}$$

with: $\Gamma_1(x_1, x_2, \alpha) = \varphi(D_{x_1}, D_{x_2}) y_1(x_1, x_2, \alpha)$, $\Gamma_2(x_1, x_2, \alpha) = \varphi(D_{x_1}, D_{x_2}) y_2(x_1, x_2, \alpha)$

For a solution $\tilde{Y}$, it must be a fuzzy number for this one. If, for each pair $(x_1, x_2) \in S_1 \times S_2$, $\Gamma(x_1, x_2, \alpha)$ defines an α-cut for this one, then $\tilde{Y}(x_1, x_2)$ is **differentiable**, and we can write:

$$\varphi(D_{x_1}, D_{x_2}) \tilde{Y}(x_1, x_2)[\alpha] = \Gamma(x_1, x_2, \alpha), \quad \forall (x_1, x_2) \in S_1 \times S_2, \quad \forall \alpha \in [0, 1] \tag{12}$$

So that, it is necessary to test if $\Gamma(x_1, x_2, \alpha)$ really defines an α-cut for a fuzzy number and verifies the differentiability of $\tilde{Y}(x_1, x_2)$. For a triangular fuzzy number, the conditions are [19] :

1. $\varphi(D_{x_1}, D_{x_2}) y_1(x_1, x_2, \alpha)$ is an increasing function of α, for each $(x_1, x_2) \in S_1 \times S_2$.
2. $\varphi(D_{x_1}, D_{x_2}) y_2(x_1, x_2, \alpha)$ is a decreasing function of α, for each $(x_1, x_2) \in S_1 \times S_2$.
3. $\Gamma_1(x_1, x_2, 1) \leq \Gamma_2(x_1, x_2, 1)$ for each $(x_1, x_2) \in S_1 \times S_2$.

Once delimited the differentiability concept of $\tilde{Y}(x_1, x_2)$, it is possible to define the Buckley-Feuring based solution for this type of fuzzy partial differential equation. $\tilde{Y}(x_1, x_2)$ is a solution if the following conditions are satisfied:

1. $\tilde{Y}(x_1, x_2)$ is differentiable.
2. $\varphi(D_{x_1}, D_{x_2}) \tilde{Y}(x_1, x_2) = \tilde{F}(x_1, x_2, \tilde{\beta})$.

It is clear that if the differentiability conditions are held by the candidate to solution $\tilde{Y}(x_1, x_2)$, then it is a fuzzy number. To complete the conditions it is only necessary to test that:

$$\varphi(D_{x_1}, D_{x_2}) \tilde{Y}(x_1, x_2) = \tilde{F}(x_1, x_2, \tilde{\beta}) \tag{13}$$

or the equivalent conditions using (11) and (12) for all $(x_1, x_2) \in S_1 \times S_2$, and $\alpha \in [0, 1]$:

1. $\varphi(D_{x_1}, D_{x_2})y_1(x_1, x_2, \alpha) = f_1(x_1, x_2, \alpha)$.
2. $\varphi(D_{x_1}, D_{x_2})y_2(x_1, x_2, \alpha) = f_2(x_1, x_2, \alpha)$.

In this case $\tilde{Y}(x_1, x_2)$ corresponds to $\tilde{U}(x_1, x_2, \tilde{\beta})$.

4.2 Example: The Cobb Douglas-Utility Function

4.2.1 From MRS to Utility

In order to clarify the methodology developed above for obtaining fuzzy utility functions from the fuzzy marginal rate of substitution, an application to the Cobb-Douglas preference structure is shown as an example. The Cobb-Douglas utility function is one of the most used representation of preferences, generating demands in which the amount spent on each good is a constant proportion of income, so they are caused by preferences called "regular preferences".

These preference structures offer a very usual MRS given by the following expression:

$$\frac{\partial U/\partial x_1}{\partial U/\partial x_2} = \left(\frac{\lambda}{\gamma}\right)\frac{x_2}{x_1}, \qquad \beta = (\lambda, \gamma) \qquad \lambda, \gamma \in (0, 1),$$
$$\lambda + \gamma = 1, \quad \forall (x_1, x_2) \in S_1 \times S_2$$

Thus, if the consumer is willing to exchange quantities of one good for another at this rate, while maintaining the same level of utility, he will be describing a Cobb-Douglas marginal rate of substitution. However, under the basic hypothesis raised about imprecision, it would be more appropriate to treat this as a fuzzy expression due to a lack of acquaintance with the alternatives they are being asked to value, or to the uncertainty about the utility of a combination of goods. In this case, $\tilde{\beta} = (\tilde{\lambda}, \tilde{\gamma})$ and it may be written as:

$$\widetilde{MRS}_{21} = \frac{\partial \tilde{U}/\partial x_1}{\partial \tilde{U}/\partial x_2} = \left(\frac{\tilde{\lambda}}{\tilde{\gamma}}\right)\frac{x_2}{x_1}, \quad \tilde{\lambda}, \tilde{\gamma} \subseteq (0, 1) \text{ such as}$$
$$\forall l \in \tilde{\lambda}, \forall g \in \tilde{\gamma} : l + g = 1, \quad \forall (x_1, x_2) \in S_1 \times S_2$$

and, by definition, the fuzzy partial equation is given by (10):

$$\varphi(D_{x_1}, D_{x_2})U(x_1, x_2, \tilde{\beta}) = \frac{\partial \tilde{U}/\partial x_1}{\partial \tilde{U}/\partial x_2}, \quad \forall (x_1, x_2) \in S_1 \times S_2$$

A possible solution to this fuzzy partial differential equation in a crisp environment is the Cobb-Douglas utility function:

$$U(x_1, x_2; \beta) = x_1^{\lambda} x_2^{\gamma}, \quad \beta = (\lambda, \gamma), \quad \lambda, \gamma \in (0, 1), \quad \lambda + \gamma = 1,$$
$$\forall (x_1, x_2) \in S_1 \times S_2$$

Applying the fuzzification to λ and γ, they acquire a triangular fuzzy number form and $\tilde{\beta} = (\tilde{\lambda}, \tilde{\gamma})$, with $\tilde{\lambda}, \tilde{\gamma} \subseteq (0, 1)$ such as $\forall l \in \tilde{\lambda}, \forall g \in \tilde{\gamma} : l + g = 1$. While it can be tested that G holds all the conditions required regarding continuity, the solution candidate is the Cobb-Douglas fuzzy expression:

$$\tilde{Y}(x_1, x_2) = x_1^{\tilde{\lambda}} x_2^{\tilde{\gamma}}, \quad \tilde{\lambda}, \tilde{\gamma} \subseteq (0, 1) \text{ such as } \forall l \in \tilde{\lambda}, \forall g \in \tilde{\gamma} : l + g = 1,$$
$$\forall (x_1, x_2) \in S_1 \times S_2$$

The fuzzy parameters have membership functions with $\mu_{\tilde{\lambda}}(l)$ and $\mu_{\tilde{\gamma}}(g)$ respectively. From the α-cuts, it is possible to define:

$$\tilde{\lambda}[\alpha] = [l_1(\alpha), l_2(\alpha)], \quad \tilde{\gamma}[\alpha] = [g_1(\alpha), g_2(\alpha)]$$

And from these, for all $\alpha \in [0, 1]$:

$$\tilde{Y}(x_1, x_2)[\alpha] = [y_1(x_1, x_2, \alpha), y_2(x_1, x_2, \alpha)], \text{ and}$$
$$\tilde{F}(x_1, x_2, \tilde{\beta})[\alpha] = [f_1(x_1, x_2, \alpha), f_2(x_1, x_2, \alpha)],$$

where:

$$y_1(x_1, x_2, \alpha) = \min\{G(x_1, x_2, \beta), \quad \beta \in \tilde{\beta}[\alpha]\},$$
$$y_2(x_1, x_2, \alpha) = \max\{G(x_1, x_2, \beta), \quad \beta \in \tilde{\beta}[\alpha]\}$$

$$f_1(x_1, x_2, \alpha) = \min\{F(x_1, x_2, \beta), \quad \beta \in \tilde{\beta}[\alpha]\},$$
$$f_2(x_1, x_2, \alpha) = \max\{F(x_1, x_2, \beta), \quad \beta \in \tilde{\beta}[\alpha]\}$$

$$\forall \alpha \in [0, 1], \quad x_1 \in S_1, x_2 \in S_2.$$

At this point, it is necessary to clarify the relations between these α-cuts. Although two parameters are considered, the following relation $\forall l \in \tilde{\lambda}, \forall g \in \tilde{\gamma} : l + g = 1$ is imposed, so that, as a Cobb-Douglas utility, only parameters satisfying this condition can be considered. Therefore, once delimited $l_1(\alpha)$, $g_2(\alpha)$ will be assigned as

$1 - l_1(\alpha)$, and vice-versa if they are in the same expression of a Cobb-Douglas MRS or utility function. That is, it appears that $y_1(x_1, x_2, \alpha) = \min\{G(x_1, x_2, \beta), \quad \beta \in \tilde{\beta}[\alpha]\}$ must be expressed by:

$$x_1^{l_1(\alpha)} x_2^{g_1(\alpha)},$$

while $l_1(\alpha)$ is the minimum value of $l \in \tilde{\lambda}[\alpha]$ and $g_1(\alpha)$ is the minimum value of $g \in \tilde{\gamma}[\alpha]$. However, the expression above must be rejected since, as the Cobb-Douglas utility specifies, the sum of these values of parameters have to be equal to one, and only $g_2(\alpha) = 1 - l_1(\alpha)$ can be considered in the same expression, once $l_1(\alpha)$ has been delimited or vice-versa.

In the Cobb-Douglas function proposed, for all $(x_1, x_2) \in S_1 \times S_2$:

$$y_1(x_1, x_2; \alpha) = x_1^{l_1(\alpha)} x_2^{g_2(\alpha)}, \qquad l_1(\alpha), g_2(\alpha) \in (0, 1), \quad l_1(\alpha) + g_2(\alpha) = 1$$

$$y_2(x_1, x_2; \alpha) = x_1^{l_2(\alpha)} x_2^{g_1(\alpha)}, \qquad l_2(\alpha), g_1(\alpha) \in (0, 1), \quad l_2(\alpha) + g_1(\alpha) = 1$$

$$f_1(x_1, x_2; \alpha) = \left(\frac{l_1(\alpha)}{g_2(\alpha)}\right) \frac{x_2}{x_1}, \qquad l_1(\alpha), g_2(\alpha) \in (0, 1), \quad l_1(\alpha) + g_2(\alpha) = 1$$

$$f_2(x_1, x_2; \alpha) = \left(\frac{l_2(\alpha)}{g_1(\alpha)}\right) \frac{x_2}{x_1}, \qquad l_2(\alpha), g_1(\alpha) \in (0, 1), \quad l_2(\alpha) + g_1(\alpha) = 1$$

Since the expression holds continuity conditions, to test the differentiability from the expression $\Gamma(x_1, x_2, \alpha) = [\Gamma_1(x_1, x_2, \alpha), \Gamma_2(x_1, x_2, \alpha)]$ by verifying if $\Gamma(x_1, x_2, \alpha)$ defines an α-cut of a triangular fuzzy number for each pair (x_1, x_2), the following conditions must be satisfied:

1.

$$\Gamma_1(x_1, x_2, \alpha) = \varphi(D_{x_1}, D_{x_2}) y_1(x_1, x_2, \alpha) = \left(\frac{l_1(\alpha)}{g_2(\alpha)}\right) \frac{x_2}{x_1},$$
$$l_1(\alpha), g_2(\alpha) \in (0, 1)$$

is an increasing function of α, for each pair $(x_1, x_2) \in S_1 \times S_2$.
This occurs if $\varphi(D_{x_1}, D_{x_2}) y_1(x_1, x_2, \alpha)$ has a positive derivative on α. While $\forall l \in \tilde{\lambda}, \forall g \in \tilde{\gamma} : l + g = 1$, and $l_1(\alpha) \in \tilde{\lambda}$, and $g_2(\alpha) \in \tilde{\gamma}$, it is possible to make $g_2(\alpha) = 1 - l_1(\alpha)$, and:

$$\varphi(D_{x_1}, D_{x_2}) y_1(x_1, x_2; \alpha) = \left(\frac{l_1(\alpha)}{1 - l_1(\alpha)}\right) \frac{x_2}{x_1}, \qquad l_1(\alpha) \in (0, 1).$$

Testing the condition:

$$\frac{d}{d\alpha}\left(\frac{l_1(\alpha)}{1-l_1(\alpha)}\right)=\frac{l_1(\alpha)'(1-l_1(\alpha))+l_1(\alpha)'l_1(\alpha)}{(1-l_1(\alpha))^2}.$$

As $\tilde{\lambda}$ is a triangular fuzzy number and $l_1(\alpha)$ is defined from its α-cuts, $l_1(\alpha)$ satisfies the condition and is an increasing function with $l_1(\alpha)' > 0$. Since, its range is (0, 1), the numerator and denominator are positive expressions, so that:

$$\frac{d}{d\alpha}\left(\frac{l_1(\alpha)}{1-l_1(\alpha)}\right) > 0,$$

and the condition is satisfied.

2.

$$\Gamma_2(x_1, x_2, \alpha) = \varphi(D_{x_1}, D_{x_2})y_2(x_1, x_2, \alpha) = \left(\frac{l_2(\alpha)}{g_1(\alpha)}\right)\frac{x_2}{x_1},$$
$$l_2(\alpha), g_1(\alpha) \in (0, 1)$$

is a decreasing function of α, for each pair $(x_1, x_2) \in S_1 \times S_2$.
Again, this occurs if $\varphi(D_{x_1}, D_{x_2})y_2(x_1, x_2, \alpha)$ has a negative derivative on α, where it is possible to make $g_1(\alpha) = 1 - l_2(\alpha)$:

$$\frac{d}{d\alpha}\left(\frac{l_2(\alpha)}{1-l_2(\alpha)}\right)=\frac{l_2(\alpha)'(1-l_2(\alpha))+l_2(\alpha)'l_2(\alpha)}{(1-l_2(\alpha))^2}.$$

It is now possible to use $\tilde{\lambda}$ as a triangular fuzzy number and $l_2(\alpha)$ is defined from its α-cuts, so that, in analogy, $l_2(\alpha)$ satisfies the condition and is a decreasing function with $l_2(\alpha)' < 0$. Since, its range is (0, 1), the numerator is a negative expression and the denominator is a positive one, so that:

$$\frac{d}{d\alpha}\left(\frac{l_2(\alpha)}{1-l_2(\alpha)}\right) < 0,$$

and the condition is satisfied.

3. $\Gamma_1(x_1, x_2; 1) \leq \Gamma_2(x_1, x_2; 1), \quad \forall(x_1, x_2) \in S_1 \times S_2.$
In this case:

$$\left(\frac{l_1(1)}{g_2(1)}\right)\frac{x_2}{x_1} \leq \left(\frac{l_2(1)}{g_1(1)}\right)\frac{x_2}{x_1}, \quad \forall(x_1, x_2) \in S_1 \times S_2.$$

Again, $\forall l \in \tilde{\lambda}, \forall g \in \tilde{\gamma} : l + g = 1$, and while $l_1(1), l_2(1) \in \tilde{\lambda}$, and $g_1(1), g_2(1) \in \tilde{\gamma}$, it is possible to make $g_2(1) = 1 - l_1(1)$, $g_1(\alpha) = 1 - l_2(\alpha)$ and the analogous condition will be:

$$\left(\frac{l_1(1)}{1-l_1(1)}\right)\frac{x_2}{x_1} \le \left(\frac{l_2(1)}{1-l_2(1)}\right)\frac{x_2}{x_1}, \quad l_1(1), l_2(1) \in (0,1)$$
$$\forall (x_1, x_2) \in S_1 \times S_2$$

with x_1, x_2 as constants. It is necessary to test that:

$$\frac{l_1(1)}{1-l_1(1)} \le \frac{l_2(1)}{1-l_2(1)} \Leftrightarrow \frac{l_2(1)}{l_1(1)} \ge \frac{1-l_2(1)}{1-l_1(1)}.$$

Knowing that $l_1(1)$ and $l_2(1)$ are defined by a triangular fuzzy number, and:

$$l_2(1) \ge l_1(1) \Leftrightarrow \frac{l_2(1)}{l_1(1)} \ge 1 \text{ and } 1-l_2(1) \le 1-l_1(1) \Leftrightarrow \frac{1-l_2(1)}{1-l_1(1)} \le 1,$$

so that:

$$\frac{l_2(1)}{l_1(1)} \ge \frac{1-l_2(1)}{1-l_1(1)} \Leftrightarrow \frac{l_1(1)}{1-l_1(1)} \le \frac{l_2(1)}{1-l_2(1)},$$

and the third condition holds.

At this point, it is possible to say that

$$\tilde{Y}(x_1, x_2) = \tilde{G}(x_1, x_2, \tilde{\lambda}, \tilde{\gamma}) = x_1^{\tilde{\lambda}} x_2^{\tilde{\gamma}}$$

$$\tilde{\lambda}, \tilde{\gamma} \subseteq (0,1) \text{ such as } \forall l \in \tilde{\lambda}, \forall g \in \tilde{\gamma} : l + g = 1, \quad \forall (x_1, x_2) \in S_1 \times S_2$$

is differentiable and a good candidate for a solution. Still, there is one more necessary step: the solution must satisfy condition (13), that is

$$\varphi(D_{x_1}, D_{x_2})\tilde{Y}(x_1, x_2) = \tilde{F}(x_1, x_2, \tilde{\beta}), \qquad \forall (x_1, x_2) \in S_1 \times S_2$$

In our case:

$$\varphi(D_{x_1}, D_{x_2})\tilde{Y}(x_1, x_2) = \left(\frac{\tilde{\lambda}}{\tilde{\gamma}}\right)\frac{x_2}{x_1}, \text{ and}$$

$$\tilde{F}(x_1, x_2; \tilde{\beta}) = \left(\frac{\tilde{\lambda}}{\tilde{\gamma}}\right)\frac{x_2}{x_1}.$$

Thus, the Cobb-Douglas fuzzy utility function

$$\tilde{U}(x_1, x_2, \tilde{\lambda}, \tilde{\gamma}) = x_1^{\tilde{\lambda}} x_2^{\tilde{\gamma}},$$

$$\tilde{\lambda}, \tilde{\gamma} \subseteq (0,1) \text{ such as } \forall l \in \tilde{\lambda}, \forall g \in \tilde{\gamma} : l + g = 1, \quad \forall (x_1, x_2) \in S_1 \times S_2$$

is a solution for this non-polynomial fuzzy partial differential equation and the fuzzy utility function.

5 The Multi-Criteria Fuzzy Utility Function

A common scenario for the consumer is one in which it is hard to determine how many units of one product he would exchange for an additional unit of another product, maintaining a constant level of utility, as this decision would depend on several factors or criteria that influence the utility that the goods or services provide to the consumer. That is why in several cases, it is more desirable and more realistic to determine the fuzzy marginal rate of substitution in each of these criteria and then proceed to their aggregation in order to obtain an expression of total or final fuzzy utility. Through this method, a carpenter might more accurately determine the units of type A wood that he would exchange for an additional unit of type B wood if considered separately each of the criteria that influence his decision, such as: quality, appearance, malleability, ease of collection, etc..

Considering the reasoning above, it is important to adapt the previous analysis to a multi-criteria framework as described in [13], thus enabling a model that more closely reflects the real decision-making process faced by consumers.

The elements involved in this multi-criteria scenario are the same as shown above but include criteria and their utilities. Criteria or attributes are directly measurable objectives which serve as a reference for evaluating the suitability of the alternatives. Thus, criteria define the set $K \subset N = C_1, C_2, ..., C_k$, where k is the number of criteria taken into account in the problem. These criteria will reach values c_i in C_i criterion, and if it refers to an alternative or a combination of amount of goods (x_1, x_2), it will be $C_i(x_1, x_2)$, which shows the value reached on C_i criterion by the alternative (x_1, x_2). This value is called *consequence* in terms of the i-th criterion if the combination (x_1, x_2) is chosen. Therefore, the values of criteria $C_1, C_2, ..., C_k$ compose a k-dimensional consequence space in which, for a given alternative $(x_1, x_2) \in S_1 \times S_2$:

$$c(x_1, x_2) = (C_1(x_1, x_2), C_2(x_1, x_2), ..., C_k(x_1, x_2))$$

assigns the point that corresponds to the alternative (x_1, x_2) in the consequence space, denoted $\mathbf{C}^k$.

For a multi-criteria utility model of k criteria with a consequences space $\mathbf{C}^k$, it is possible to add uni-criterion utilities through an additive or multiplicative form if certain conditions of independence are satisfied. In particular, it is only possible to assign these aggregation forms to utility functions if the criteria are *mutually utility independent*.[6]

[6] Actually, the term *mutually utility independent* is used in expected utility to show the independence. Outside of expected utility, the correct term is *weak independent in difference*, although the same term often used in expected utility can also be used here.

Therefore, the multi-criteria consumer decision problem will deal with k marginal rates of substitution corresponding to each one of the k criteria involved:

$$\widetilde{MRS_{21}}^{i} = \frac{\partial \tilde{U}_i / \partial x_1}{\partial \tilde{U}_i / \partial x_2} = F_i(x_1, x_2; \tilde{\beta}_i), \quad i = 1, ..., k \tag{14}$$

As such, $\tilde{\beta}_i = (\tilde{\beta}_1, \tilde{\beta}_2, ..., \tilde{\beta}_k)$ is a triangular fuzzy parameter vector determined by the subjective preferences of each consumer in the i-th criterion, and it determines the inner characters of utility function for that consumer, while $\tilde{U}_i(x_1, x_2; \tilde{\beta}_i)$ is the fuzzy marginal utility of (x_1, x_2) considering only the i-th criterion with the same properties described in the sections above.

Once the fuzzy utility functions are obtained at each criterion by means of the methodology described above, the next step is the aggregation of the k utility functions of the k criteria, that is, to reach an expression $\tilde{U}(x_1, x_2; \tilde{\beta})$ from $\tilde{U}_i(x_1, x_2; \tilde{\beta}_i), \quad i = 1, ..., k$.

5.1 Mutual Utility Independence

As previously stated, mutual utility independence in criteria is needed in order to add uni-criterion utilities through an additive or multiplicative form.

If J is a subset of the criteria index set $K = 1, ..., k$, for each $i \in J$, the *preordering induced* by $(C_i)_{i \in J}$ is the one induced by $\preceq$ on $\prod_{i \notin J} C_i$ for each $i \in J$, the k attributes of K will be mutually utility independent if, for each subset $J \subseteq K$, the preordering induced by $(C_i)_{i \in J}$ on $\prod_{i \notin J} C_i$ is independent of $(C_i)_{i \in J}$.

That is, mutual utility independence implies that:

$$\text{If } (c_{i \in J}, c_{s \notin J})(c_{i \in J}, c_{t \notin j}) \succeq (c_{i \in J}, c_{l \notin J})(c_{i \in J}, c_{m \notin J}),$$

then, it follows that:

$$(c_{h \in J}, c_{s \notin J})(c_{h \in J}, c_{t \notin J}) \succeq (c_{h \in J}, c_{l \notin J})(c_{h \in J}, c_{m \notin J}) \text{ for some } c_h, h \in J.$$

In other words, mutual utility independence implies that for each subset J of criteria, $c_i, i \in J$ does not determine any relations over criteria out of J, that is, relations such as $c_a \succeq c_b, a, b \notin J$, only depends on c_a and c_b values.

A final approach to the condition of mutual utility independence ensures that the total fuzzy utility function acquires an additive or multiplicative form in the consumer decision-making problem, since the uni-criterion fuzzy utility functions have been obtained independently from marginal rates of substitution in a single criterion environment in which the consumer considers the other criteria as constant (*ceteris paribus*). In this sense, the order of the criteria values do not depend on the values of the other criteria once criteria have been properly scaled [23].

If mutual utility independence is satisfied, it is possible to add uni-criterion utilities by an additive or multiplicative form:

- A total utility function (Multi-criteria Utility) has an additive form if:

$$U(x_1, x_2; \beta) = \sum_{i=1}^{k} w_i U_i(x_1, x_2; \beta_i), \quad i = 1, ..., k \tag{15}$$

- A total utility function (Multi-criteria Utility) has a multiplicative form if:

$$1 + WU(x_1, x_2; \beta) = \prod_{i=1}^{k} [1 + Ww_i U_i(x_1, x_2; \beta_i)], \quad i = 1, ..., k \tag{16}$$

where, following the notation used in the single criterion treatment case, $U(x_1, x_2; \beta)$ is the multi-criteria or total utility function for goods x_1 and x_2, $U_i(x_1, x_2; \beta_i)$ is the uni-criterion utility function on the i-th criterion, w_i are weights of criteria, and they represent the relevance of the i-th criterion. Finally, W is a constant scale.

Although both the additive and multiplicative models require mutual utility independence, a total utility function in additive form must also satisfy an additional independence condition to be implemented. This is called *additive independence* between criteria, which makes it more restrictive than the multiplicative expression. Still, the multiplicative model is far from ideal in practical applications due to its many technical problems resulting from a more complex form.

The disadvantages of the multiplicative model implementation have led many researchers to analyze results obtained by applying the additive model in frameworks where the theoretical multiplicative model fits better. Works like [36] and [37], are good references in this area. In these papers, numerous simulations are carried out using the additive model in cases where the multiplicative model fits better in theory, showing how the errors introduced by the use of additive models were small for a wide range of multi-criteria decision problems. It was also found that these errors were introduced by the simplification of utility functions used to collect uni-criterion decision-maker preferences and not by the aggregation model alone. Given such results, it is possible to conclude that the multiplicative model can be replaced in many cases by an additive one, in order to obtain greater simplicity with a minimum loss of precision. However, these studies have also shown that the farther the real framework is from the required additive independence for an additive model, the more errors produced when the multiplicative model is replaced by an additive one [36, 37].

To avoid the need to replace a multiplicative aggregation model with an additive one, and to not be hindered by the errors caused by violation of the additive independence, it is necessary to satisfy additive conditions in the decision-making process. These conditions have been proposed by several authors from different perspectives and were established in 1960 by Gerard Debreu [9] from a topological point of view. In 1964 Duncan Luce and Tukey [12] established a set of axioms of additivity from an algebraic point of view, while Fishburn, in 1965 [14], derived the additivity conditions with a probabilistic approach. This last perspective has been the most commonly used to determine additivity in expected utility models.

However, this probabilistic approach is hardly appropriate here, since it is not based on expected utility theory. Therefore, this chapter will establish axioms to ensure an additive aggregation model of uni-criterion utility functions based on the work of Debreu and Luce and Tukey (see Appendix A and B for an explanation of Debreu and Luce and Tukey conditions for additivity). If the decision maker holds these axioms, there will be an additive model.

5.2 Additive Utility

This chapter considers that Luce-Tukey axioms, or the Debreu condition if more than two criteria are considered, are satisfied by the decision-maker. This consideration is not far from reality, while the Debreu condition is very consistent with consumer behavior, and multi-criteria decision-making problems usually have three or more criteria involved. The same can be said about the Luce-Tukey axioms, in which the ordering axiom is satisfied automatically because it corresponds with the multi-criteria decision-making axioms, the second and third axioms (solution to equations and cancelation) are feasibly held in the consumer decision-making problem or they can be imposed on the decision-maker in addition to the primary axioms to ensure consistency, while the archimedean axiom will be satisfied automatically if consequence space is continuous since preferences are evaluated by means of an increasing utility function, that is, the preference relation between, i.e. $C_1(x_1, x_2)$ and $C_2(x_1', x_2')$ is established by $U_1(x_1, x_2)$ and $U_2(x_1', x_2')$, with utility being an increasing function.

Failure to comply with these axioms would incur in a modeling error where an additive scheme is adopted, although, as mentioned above, the need for more information, and the model complexity often leads to the use of an additive model, given that results differ little in comparison with the higher level of complexity required for the multiplicative model. This use of additive aggregation is especially prevalent since the mid-twentieth century in the classical consumer theory which laid the foundation for this chapter.

Once considered the additive form of the multi-criteria utility function, the problem lies in obtaining the expression of the sum of fuzzy numbers, given that the uni-criterion fuzzy utility functions are fuzzy numbers with their membership functions and α-cuts as defined in a previous section.

If the additive nature of the total utility function is determined, it will be obtained through the sum of the utility functions corresponding to each criterion. Because these functions are fuzzy numbers, the total utility function is the result of a sum of finite fuzzy numbers, whose expression depends on the interaction between them.

In [11] there is a good explanation of the consequences of the interactions between fuzzy numbers as they affect the addition of those numbers. Again, since the uni-criterion fuzzy utility functions have been independently obtained from marginal rates of substitution in a single criterion environment and considering other criteria to be constant (*ceteris paribus*), there are not any kind of constraints that determine a relation of interactivity, that is, if $y_i, i = 1, ..., k$ are variables with values on $\tilde{U}_i(x_1, x_2; \tilde{\beta}_i), i = 1, ..., k$, then, $(y_1, y_2, ..., y_k)$ is a vector of variables with values on $\tilde{U}_1 \times \tilde{U}_2 \times \cdots \times \tilde{U}_k$, as y_i will have values on $\tilde{U}_i$ without any constraint imposed by values of y_h on $\tilde{U}_h, i \neq h$, with $i, h \in K$, and therefore the expression of the sum of two fuzzy utility functions is, in terms of α-cuts:

$$(\tilde{U}_i(x_1, x_2, \tilde{\beta}_i) + \tilde{U}_h(x_1, x_2, \tilde{\beta}_h))[\alpha] = \tilde{U}_i(x_1, x_2, \tilde{\beta}_i)[\alpha] + \tilde{U}_h(x_1, x_2, \tilde{\beta}_h)[\alpha] =$$

$$[u_{i1}(x_1, x_2; \alpha), u_{i2}(x_1, x_2; \alpha)] + [u_{h1}(x_1, x_2; \alpha), u_{h2}(x_1, x_2; \alpha)] =$$

$$[u_{i1}(x_1, x_2; \alpha) + u_{h1}(x_1, x_2; \alpha), u_{i2}(x_1, x_2; \alpha) + u_{h2}(x_1, x_2; \alpha)]$$

where, following the notation:

- $[u_{i1}(x_1, x_2; \alpha)$ is the lower value $U_i(x_1, x_2, \beta_i)$ in which $\mu(U_i(x_1, x_2, \beta_i)) \geq \alpha, \quad U_i(x_1, x_2, \beta_i) \in \tilde{U}_i(x_1, x_2, \tilde{\beta}_i)$.
- $[u_{i2}(x_1, x_2; \alpha)$ is the higher value $U_i(x_1, x_2, \beta_i)$ in which $\mu(U_i(x_1, x_2, \beta_i)) \geq \alpha, \quad U_i(x_1, x_2, \beta_i) \in \tilde{U}_i(x_1, x_2, \tilde{\beta}_i)$.
- $[u_{h1}(x_1, x_2; \alpha)$ is the lower value $U_h(x_1, x_2, \beta_h)$ in which $\mu(U_h(x_1, x_2, \beta_h)) \geq \alpha, \quad U_h(x_1, x_2, \beta_h) \in \tilde{U}_h(x_1, x_2, \tilde{\beta}_h)$.
- $[u_{h2}(x_1, x_2; \alpha)$ is the higher value $U_h(x_1, x_2, \beta_h)$ in which $\mu(U_h(x_1, x_2, \beta_h)) \geq \alpha, \quad U_h(x_1, x_2, \beta_h) \in \tilde{U}_h(x_1, x_2, \tilde{\beta}_h)$.

This expression is used to characterize the multi-criteria fuzzy utility function obtained through fuzzy marginal rates of substitution. In addition, the sum of non-interactive fuzzy numbers results in another fuzzy number, so that the multi-utility function is also a fuzzy number. [7]

5.3 Weights in the Fuzzy Multi-criteria Utility Function

At this point, we can summarize that the multi-criteria fuzzy utility function has additive form, and this addition is expressed as seen above for non-interactive fuzzy numbers. Thus, using the notation above:

$$\tilde{U}(x_1, x_2; \tilde{\beta}) = \sum_{i=1}^{k} \tilde{U}_i(x_1, x_2; \tilde{\beta}_i)$$

[7] The addition of more than two non-interactive fuzzy numbers is required to use the associative property and to link operations of addition if expressed in terms of membership function. That is, $\tilde{U}_1 + \tilde{U}_2 + \tilde{U}_3 = (\tilde{U}_1 + \tilde{U}_2) + \tilde{U}_3$, obtaining expressions of the sum of three fuzzy numbers. This procedure allows us to continue to link sums from the association adding k fuzzy numbers.

Nevertheless, it is easy to find decision scenarios in which criteria do not have the same importance for the consumer, that is, each criterion has a different weight according to this importance. If weights of criteria are specified, the total utility will be given by the fuzzy expression of Eq. (15):

$$\tilde{U}(x_1, x_2; \tilde{\beta}) = \sum_{i=1}^{k} w_i \tilde{U}_i(x_1, x_2; \tilde{\beta}_i)$$

Since this expression is more adequate in most cases, it is necessary to mention some of its implications:

- First, different interval scale based methods are used in multi-criteria utility theory to asses weighting in different aggregation rules [5, 38]. Since scaling these weights is a procedure based on decision-maker rationality, the method chosen must satisfy specific conditions in order to maintain the additive independence in utility. Because of the introduction of weights, the preference relations between two criteria are evaluated by $w_i U_i(x_1, x_2; \beta_i)$ instead of by utility function alone. That is, the relation between $C_i(x_1, x_2)$ and $C_h(x_1, x_2)$ is defined by $w_i U_i(x_1, x_2; \beta_i)$ and $w_h U_h(x_1, x_2; \beta_h)$. One of the conditions that is common for all such methods is that the sum of the criteria weights must be equal to one or differ from this by only a small quantity [38].
- Second, it is necessary to introduce operations such as:

  $$w_i \tilde{U}_i(x_1, x_2; \tilde{\beta}_i),$$

  where w_i is a positive real number and $\tilde{U}_i(x_1, x_2; \tilde{\beta}_i)$ is a fuzzy number.
  An introduction to these operations can be found in [22]. Continuing with the characterization of fuzzy numbers by α-cuts, the expression of these operations is:

  $$w_i \cdot \tilde{U}_i(x_1, x_2; \tilde{\beta}_i)[\alpha] = [w_i u_{i1}(x_1, x_2; \alpha), w_i u_{i2}(x_1, x_2; \alpha)].$$

  Therefore, weights can be included as an element in the fuzzy utility function and the fuzzy multi-criteria utility function can be characterized by α-cuts:

  $$\tilde{U}(x_1, x_2; \tilde{\beta})[\alpha] = [\sum_{i=1}^{k} w_i u_{i1}(x_1, x_2; \alpha), \sum_{i=1}^{k} w_i u_{i2}(x_1, x_2; \alpha)] \quad (17)$$

5.4 Example: The Cobb Douglas Multi-criteria Fuzzy Utility Function

5.4.1 Adding Uni-Criteria Fuzzy Utility Functions

If k criteria with Cobb-Douglas utilities are considered, the following k expressions can be obtained by applying the methodology developed:

$$\tilde{U}_i(x_1, x_2; \tilde{\beta}_i) = x_1^{\tilde{\lambda}_i} x_2^{\tilde{\gamma}_i}, \qquad i = 1, ..., k$$

$$\tilde{\beta}_i = (\tilde{\lambda}_i, \tilde{\gamma}_i), \quad \tilde{\lambda}_i, \tilde{\gamma}_i \subseteq (0, 1) \text{ such as } \forall l_i \in \tilde{\lambda}_i, \forall g_i \in \tilde{\gamma}_i : l_i + g_i = 1, \\ \forall (x_1, x_2) \in S_1 \times S_2$$

characterized by the following α-cuts for all $(x_1, x_2) \in S_1 \times S_2$:

$$\tilde{U}_i(x_1, x_2; \tilde{\beta}_i)[\alpha] = [u_{i1}(x_1, x_2; \alpha), u_{i2}(x_1, x_2; \alpha)] = [x_1^{l_{i1}(\alpha)} x_2^{g_{i2}(\alpha)}, x_1^{l_{i2}(\alpha)} x_2^{g_{i1}(\alpha)}],$$

$$l_{i1}(\alpha), g_{i1}(\alpha), l_{i2}(\alpha), g_{i2}(\alpha) \in (0, 1), \quad l_{i1}(\alpha) + g_{i2}(\alpha) = 1, \quad l_{i2}(\alpha) + g_{i1}(\alpha) = 1.$$

Multi-criteria or total utility will be given by (15). In this case:

$$\tilde{U}(x_1, x_2; \tilde{\beta}) = \sum_{i=1}^{k} w_i \tilde{U}_i(x_1, x_2; \tilde{\beta}_i) = \sum_{i=1}^{k} w_i x_1^{\tilde{\lambda}_i} x_2^{\tilde{\gamma}_i},$$

$$\tilde{\lambda}_i, \tilde{\gamma}_i \subseteq (0, 1) \text{ such as } \forall l_i \in \tilde{\lambda}_i, \forall g_i \in \tilde{\gamma}_i : l_i + g_i = 1, \quad \forall (x_1, x_2) \in S_1 \times S_2$$

characterized by these α-cuts:

$$\tilde{U}(x_1, x_2; \tilde{\beta})[\alpha] = [\sum_{i=1}^{k} w_i x_1^{l_{i1}(\alpha)} x_2^{g_{i2}(\alpha)}, \sum_{i=1}^{k} w_i x_1^{l_{i2}(\alpha)} x_2^{g_{i1}(\alpha)}]$$

$$l_{i1}(\alpha), g_{i1}(\alpha), l_{i2}(\alpha), g_{i2}(\alpha) \in (0, 1), \quad l_{i1}(\alpha) + g_{i2}(\alpha) = 1, \quad l_{i2}(\alpha) + g_{i1}(\alpha) = 1.$$

5.4.2 Adding Different Uni-Criteria Fuzzy Utility Functions

A multi-criteria utility could also result from an additive aggregation of criteria with different preference structures, that is, involving different fuzzy utility functions. Again, in this case, the generic form of the fuzzy utility will be given by (15), characterized by α-cuts as stated in (17).

An example can be provided obtaining a multi-criteria fuzzy utility function for two goods using a CES uni-criterion fuzzy utility function in one criterion (U_1) and a Cobb-Douglas fuzzy utility function in another one (U_2). The CES utility function (Constant Elasticity Substitution) is one of the most commonly used utility functions. It arises in direct analogy with the theory of production and it is widely accepted as the basis for the representation of reported utility for two normal and divisible goods. In this sense, it is especially recommended for consumption choices of raw materials, where the company behaves as a consumer. If there is no initial predisposition for either of the goods, the fuzzy CES MRS has the following form:

$$\frac{\partial \tilde{U}_1/\partial x_1}{\partial \tilde{U}_1/\partial x_2} = \left(\frac{x_1}{x_2}\right)^{\tilde{\beta}-1}, \qquad \tilde{\beta} \subseteq (0, 1].$$

It is then possible to prove that the CES fuzzy utility function provided as a solution for this fuzzy partial differential equation is the fuzzification of the crisp CES utility function [18], that is:

$$\tilde{U}_1(x_1, x_2; \tilde{\boldsymbol{\beta}}_1) = (x_1^{\tilde{\beta}} + x_2^{\tilde{\beta}})^{1/\tilde{\beta}}, \qquad \tilde{\boldsymbol{\beta}}_1 = (\tilde{\beta}) \;\; \tilde{\beta} \subseteq (0, 1], \quad \forall (x_1, x_2) \in S_1 \times S_2$$

with α-cuts:

$$u_{11}(x_1, x_2; \alpha) = (x_1^{b_1(\alpha)} + x_2^{b_1(\alpha)})^{1/b_1(\alpha)}, \qquad 0 < b_1(\alpha) \leq 1, \quad \forall (x_1, x_2) \in S_1 \times S_2$$

$$u_{12}(x_1, x_2; \alpha) = (x_1^{b_2(\alpha)} + x_2^{b_2(\alpha)})^{1/b_2(\alpha)}, \qquad 0 < b_2(\alpha) \leq 1, \quad \forall (x_1, x_2) \in S_1 \times S_2$$

Therefore[8]:

---

[8] If there exists an initial predisposition for any good, the fuzzy CES MRS would adopt this form:

$$\frac{\partial \tilde{U}_1/\partial x_1}{\partial \tilde{U}_1/\partial x_2} = \left(\frac{\tilde{\lambda}_1}{\tilde{\lambda}_2}\right)\left(\frac{x_1}{x_2}\right)^{\tilde{\beta}-1}, \qquad \tilde{\beta} \subseteq (0, 1], \quad \tilde{\lambda}_1, \tilde{\lambda}_2 \subseteq [0, \infty), \quad \forall (x_1, x_2) \in (S_1 \times S_2),$$

for which the fuzzification of the CES utility function:

$\tilde{U}_1(x_1, x_2; \tilde{\boldsymbol{\beta}}_1) = (\tilde{\lambda}_1 x_1^{\tilde{\beta}} + \tilde{\lambda}_2 x_2^{\tilde{\beta}})^{1/\tilde{\beta}}$, $\tilde{\boldsymbol{\beta}}_1 = (\tilde{\beta}, \tilde{\lambda}_1, \tilde{\lambda}_2)$ $\tilde{\beta} \subseteq (0, 1]$, $\tilde{\lambda}_1, \tilde{\lambda}_2 \subseteq [0, \infty)$, $\forall (x_1, x_2) \in (S_1 \times S_2)$is not a solution [18].

$$\tilde{U}(x_1, x_2; \tilde{\beta}) = \sum_{i=1}^{k=2} w_i \tilde{U}_i(x_1, x_2; \tilde{\beta}_i) = w_1(x_1^{\tilde{\beta}} + x_2^{\tilde{\beta}})^{1/\tilde{\beta}} + w_2 x_1^{\tilde{\lambda}} x_2^{\tilde{\gamma}},$$

$\tilde{\beta} \subseteq (0, 1]$, $\tilde{\lambda}, \tilde{\gamma} \subseteq (0, 1)$ such as $\forall l \in \tilde{\lambda}, \forall g \in \tilde{\gamma} : l+g = 1$, $\forall (x_1, x_2) \in S_1 \times S_2$

characterized by α-cuts as stated in (17), with:

$$\sum_{i=1}^{k} w_i u_{i1}(x_1, x_2; \alpha) = w_1(x_1^{b_1(\alpha)} + x_2^{b_1(\alpha)})^{1/b_1(\alpha)} + w_2 x_1^{l_1(\alpha)} x_2^{g_2(\alpha)}$$

$$\sum_{i=1}^{k} w_i u_{j2}(x_1, x_2; \alpha)] = w_1(x_1^{b_2(\alpha)} + x_2^{b_2(\alpha)})^{1/b_2(\alpha)} + w_2 x_1^{l_2(\alpha)} x_2^{g_1(\alpha)}$$

$$0 < b_1(\alpha) \leq 1, \quad 0 < b_2(\alpha) \leq 1, \quad l_1(\alpha), g_1(\alpha) \in (0, 1), \quad l_2(\alpha), g_2(\alpha) \in (0, 1),$$

$$l_1(\alpha) + g_2(\alpha) = 1, \quad l_2(\alpha) + g_1(\alpha) = 1, \quad \forall (x_1, x_2) \in S_1 \times S_2$$

6 Final Consumer Decision

In a fuzzy environment, the consumer decision problem is framed in fuzzy mathematical programming, with generic form:

$$\begin{aligned} \max_{x_1, x_2} \quad & \tilde{U}(x_1, x_2; \tilde{\beta}) \\ s.t. \quad & p_1 x_1 + p_2 x_2 \leq R \\ & x_r \in S_r \end{aligned} \tag{18}$$

where p_1 represents the price of good X_1, p_2 the price of good X_2 and R, the available income to the consumer.

An initial approach to the final consumer decision problem can be established from an economic perspective.

An intuitive analysis will produce the initial arguments. The parameters which are part of the utility function determine the convexity of the indifference curves. Thus, each value of these parameters has a particular convexity level associated with it. By introducing an element of uncertainty or imprecision and treating these parameters as fuzzy numbers with their respective membership functions, the range of possibilities is widened. In a context like this, there are as many forms of indifference curves as elements contained in the fuzzy number (do not forget that a fuzzy number is a fuzzy set with additional features), each one with its respective membership function.

That is, in a fuzzy environment, there are as many families of indifference curves as possible parameter values and, in each family of curves, the decision maker will

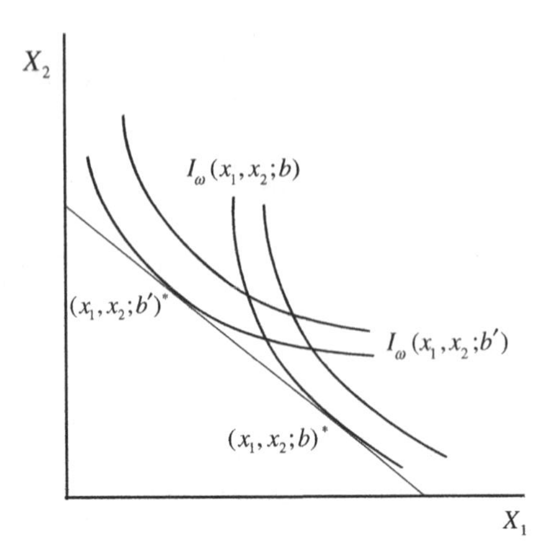

Fig. 2 Consumer equilibria in a fuzzy environment

be placed on the farthest curve but not exceeding the budget constraint. Thus, the solution to this problem will be a fuzzy set given by the tangent points of the budget constraint line with the different families of indifference curves associated with each possible value of the parameters. Each of these solutions has the membership function corresponding to the value of the parameter which generates the family of indifference curves associated to it. Figure 2 offers a demonstration of this explanation.

We can first consider a uni-dimensional parameter on the fuzzy utility function, $\tilde{\beta}$. The multi-dimensional parameter case will be tackled in the subsequent mathematical formalization. As the chart shows, for $b \in \tilde{\beta}$ and $b' \in \tilde{\beta}$, $b \neq b'$ two families of indifference curves are generated with different convexity levels: $I_\omega(x_1, x_2; b)$ and $I_\omega(x_1, x_2; b')$. Each family has a membership function equal to the membership function of the parameter which generates it, that is $\mu_{\tilde{\beta}}(b)$ and $\mu_{\tilde{\beta}}(b')$. As there exist as many families of indifference curves as elements in $\tilde{\beta}$, the consumer optimal decision set is the set of tangent points of each family of indifference curves with the budget constraint line. In the chart, optimums are given by $(x_1, x_2; b)^*$ and $(x_1, x_2; b')^*$, yet there will be as many points as families of indifference curves generated by the elements of $\tilde{\beta}$. Thus, the set of solutions will be a fuzzy set in which each solution included has membership function corresponding to the element of $\tilde{\beta}$ which generates the family of indifference curves containing the solution. With the elements shown in the chart above:

$$(x_1, x_2; b)^* \in (x_1, x_2; \tilde{\beta})^* \longrightarrow \mu_{(x_1, x_2; \tilde{\beta})^*}(x_1, x_2; b)^* = \mu_{\tilde{\beta}}(b)$$

$$(x_1, x_2; b')^* \in (x_1, x_2; \tilde{\beta})^* \longrightarrow \mu_{(x_1, x_2; \tilde{\beta})^*}(x_1, x_2; b')^* = \mu_{\tilde{\beta}}(b')$$

These conclusions can be reached by the general formulation of a fuzzy mathematical programming problem described in [31]:

$$\begin{aligned} \max_{x_1,x_2} \quad & \tilde{f}(\boldsymbol{x}; \tilde{\beta}) \\ s.t. \quad & \tilde{g_q}(\boldsymbol{x}; \tilde{\boldsymbol{a_q}})\tilde{R}_q\tilde{d}_q \ q \in \mathcal{M} \end{aligned} \tag{19}$$

where:

- $\tilde{\beta}$ is a multidimensional fuzzy set with membership function $\mu_{\tilde{\beta}} = (\mu_{\tilde{\beta}_1}(b_1), \mu_{\tilde{\beta}_2}(b_2), ..., \mu_{\tilde{\beta}_k}(b_k))$. However, it is necessary to combine these membership functions in one, reflecting a global membership of $(b_1, b_2, ..., b_k)$, with $b_1 \in \tilde{\beta}_1, b_2 \in \tilde{\beta}_2, ..., b_k \in \tilde{\beta}_k$. The solution can be a triangular norm which works as an aggregation operator:

$$\mu_{\tilde{\beta}}(\boldsymbol{b}) = T[\mu_{\tilde{\beta}_1}(b_1), \ldots, \mu_{\tilde{\beta}_k}(b_k))]$$

- $\tilde{\boldsymbol{a_q}}$ is a multidimensional fuzzy set with all coefficients of q-th constraint, $q \in \mathcal{M}$ (set of constraints) with membership function $\mu_{\tilde{\boldsymbol{a_q}}}$. As before, it is necessary to use a triangular norm as an aggregation operator to unify these membership functions.
- $\tilde{d}_q$ is a uni-dimensional fuzzy set with membership function $\mu_{\tilde{d}_q}$, $q \in \mathcal{M}$.
- $\tilde{R}_q$ is a fuzzy relation. This fuzzy relation is considered a fuzzy extension of R_q, and $\mu_{\tilde{R}_q} = \mu_{R_q}$. The membership function is given by:

$$\begin{aligned} &\mu_{\tilde{R}_q}(\tilde{g_q}(\boldsymbol{x}; \tilde{\boldsymbol{a_q}}), \tilde{d}_q) \\ &= T[\mu_{R_q}(u, v), T(\mu_{\tilde{g_q}(\boldsymbol{x};\tilde{\boldsymbol{a_q}})}(u), \mu_{\tilde{d}_q}(v))] \\ &= T[\mu_{\tilde{g_q}(\boldsymbol{x};\tilde{\boldsymbol{a_q}})}(u), \mu_{\tilde{d}_q}(v)] \quad | \, u R_q v \end{aligned}$$

  where T is a triangular norm.
- $\tilde{f}(\boldsymbol{x}; \tilde{\beta})$ is the objective value for an element of $\tilde{\beta}$. Its membership function will be determined by:

$$\mu_{\tilde{f}(\boldsymbol{x};\tilde{\beta})}(z) = \begin{cases} \mu_{\tilde{\beta}}(\beta) \mid \beta \in \tilde{\beta}, f = z & \text{if } f(\boldsymbol{x}; z)^{-1} \neq \emptyset \\ 0 & \text{otherwise} \end{cases}$$

- $\tilde{g_q}(\boldsymbol{x}; \tilde{\boldsymbol{a_q}})$ is the value of the q-th constraint for an element of $\tilde{\boldsymbol{a_q}}$. Its membership function $\mu_{\tilde{g_q}(\boldsymbol{x};\tilde{\boldsymbol{a_q}})}(z)$ is:

$$\mu_{\tilde{g_q}(\boldsymbol{x};\tilde{\boldsymbol{a_q}})}(z) = \begin{cases} \mu_{\tilde{\boldsymbol{a_q}}}(\boldsymbol{a_q}) \mid \boldsymbol{a_q} \in \tilde{\boldsymbol{a_q}}, g_q = z & \text{if } g_q(\boldsymbol{x}; z)^{-1} \neq \emptyset \\ 0 & \text{otherwise} \end{cases}$$

Now, it is possible to define the feasible set of the problem as the fuzzy subset $\tilde{X} \in \mathbb{R}^n$, with membership function:

$$\mu_{\tilde{X}}(\boldsymbol{x}) = A(\mu_{\tilde{R}_1}(\tilde{g_1}(\boldsymbol{x}; \tilde{\boldsymbol{a_1}}), \tilde{d}_1), ..., \mu_{\tilde{R}_m}(\tilde{g_m}(\boldsymbol{x}; \tilde{\boldsymbol{a_m}}), \tilde{d}_m),$$

where A is defined as an aggregation operator, such as a triangular norm. The elements of the α-cut are called α-feasible solutions. Moreover, the expression $\mu_{\tilde{R}_q}(\tilde{g_q}(\boldsymbol{x};\tilde{\boldsymbol{a_q}}),\tilde{d_q})$ can be understood as the level at which the qth constraint is satisfied.

However, in a consumer decision problem, there is not any fuzzy element in the constraints. That is, they adopt the following form: $g_q(\boldsymbol{x};\boldsymbol{a_q})R_q d_q$ and:

$$\mu_{\tilde{g_q}(\boldsymbol{x};\tilde{\boldsymbol{a_q}})} = \chi_{g_q(\boldsymbol{x};\boldsymbol{a_q})}, \quad \forall q \in \mathcal{M}$$

$$\mu_{\tilde{d_q}} = \chi_{d_q}$$

where χ represents the characteristic function[9] of $g_q(\boldsymbol{x};\boldsymbol{a_q})$, and therefore, there is no level at which a constraint is satisfied, and the qth constraint is either satisfied or not:

$$\mu_{\tilde{R}_q}(\tilde{g_q}(\boldsymbol{x};\tilde{\boldsymbol{a_q}}),\tilde{d_q}) = \begin{cases} 1 & \text{if } g_q(\boldsymbol{x};\boldsymbol{a_q})R_q d_q \\ 0 & \text{otherwise} \end{cases}$$

Applying the A T-norm:

$$\mu_{\tilde{X}}(\boldsymbol{x}) = A(\mu_{\tilde{R}_1}(\tilde{g_1}(\boldsymbol{x};\tilde{\boldsymbol{a_1}}),\tilde{d_1}), \ldots, \mu_{\tilde{R}_m}(\tilde{g_m}(\boldsymbol{x};\tilde{\boldsymbol{a_m}}),\tilde{d_m}) = \chi_X(\boldsymbol{x}).$$

Therefore, the feasible set of a fuzzy optimization problem with no fuzzy elements placed on the constraints is the same feasible set X as in the standard crisp optimization problem.

Once the feasible set is established, the next step is to define the optimal solution. At this point, many interpretations and possible approaches arise in fuzzy optimization. The broadest approach does not require any external assessments and defines the optimal solution set as the fuzzy set $\tilde{X}^*$ whose function membership is given by:

$$\mu_{\tilde{X}^*}(\boldsymbol{x}) = A_g(\mu_{\tilde{f}(\boldsymbol{x};\tilde{\beta})}(\boldsymbol{x}), \mu_{\tilde{X}}(\boldsymbol{x})),$$

where A_g represents an aggregation operator as a triangular norm. The elements of the α-cuts are called α-optimal solutions.

In a problem with non-fuzzy constraints:

[9] The characteristic function of a crisp set A that shows its membership:

$$\chi_A(x) = \begin{cases} 1 & \text{if } x \in A, \\ 0 & \text{otherwise} \end{cases}$$

$$\mu_{\tilde{X}*}(\boldsymbol{x}) = A_g(\mu_{\tilde{f}(\boldsymbol{x};\tilde{\beta})}(\boldsymbol{x}), \chi_X(\boldsymbol{x}))$$

$$\mu_{\tilde{X}*}(\boldsymbol{x}) = \begin{cases} \mu_{\tilde{f}(\boldsymbol{x};\tilde{\beta})}(\boldsymbol{x}) & \text{if } \boldsymbol{x} \in X \\ 0 & \text{otherwise} \end{cases}$$

Applying it to the consumer decision problem defined in (18):

$$\begin{aligned} \max_{x_1,x_2} \quad & \tilde{U}(x_1, x_2; \tilde{\beta}) \\ s.t. \quad & p_1x_1 + p_2x_2 \leq R \\ & x_r \in S_r \end{aligned}$$

the solution is given by:

$$\mu_{\tilde{X}*}(x_1, x_2) = \begin{cases} \mu_{\tilde{U}(x_1,x_2;\tilde{\beta})}(x_1, x_2) & \text{if } (x_1, x_2) \in X \\ \\ 0 & \text{otherwise} \end{cases}$$
$$= \begin{cases} \mu_{\tilde{\beta}}(\beta) \mid \beta \in \tilde{\beta}, \quad f(\boldsymbol{x}; \beta) = z & \text{if } f(\boldsymbol{x}; z)^{-1} \neq \emptyset \\ \\ 0 & \text{otherwise} \end{cases}$$

However, this optimal solution set contains the same elements as the feasible set, so that it provides little added information involving the imprecision of fuzzy numbers. In practice, there are several possibilities for redefining this set of solutions, including the two shown below.

One possibility is the one adopted in the graphical example explained above, developed through an economic and intuitive approach. It is based on prior knowledge of the fuzzy optimization problem and on a clear idea of what the problem constitutes, applying this previous knowledge to reduce the optimal solutions set. In this case, as noted before, that prior knowledge enables the highest utility levels to be achieved on the budget constraint line, making it possible to modify the feasible solutions set in (18) *a posteriori*, restricted to the items placed on the budget line:

$$\begin{aligned} \max_{x_1,x_2} \quad & \tilde{U}(x_1, x_2; \tilde{\beta}) \\ s.t. \quad & p_1x_1 + p_2x_2 = R \\ & x_r \in S_r \end{aligned}$$

In this approach, optimal solutions coincide with the solutions anticipated in the previous graphical analysis, thus the optimal solution set will be given by the fuzzy set with membership function expressed as:

$$\mu_{\tilde{X}*}(x_1, x_2) = \begin{cases} \mu_{\tilde{U}(x_1,x_2;\tilde{\beta})}(x_1, x_2) & \text{if } (x_1, x_2) \in X \\ 0 & \text{otherwise} \end{cases}$$

$$= \begin{cases} \mu_{\tilde{\beta}}(\beta) \mid \beta \in \tilde{\beta}, \quad f(\boldsymbol{x}; \beta) = z & \text{if } f(\boldsymbol{x}; z)^{-1} \neq \emptyset \\ 0 & \text{otherwise} \end{cases}$$

According to [31], a different method for dealing with this problems is to introduce an external and arbitrary value $\tilde{d_0}$ called *fuzzy objective* that provides a reference for evaluating values from $\tilde{f}(\boldsymbol{x}; \tilde{\beta})$ by fuzzy relation $\tilde{R}_0$, establishing a new constraint $\tilde{f}(\boldsymbol{x}; \tilde{\beta})\tilde{R}_0\tilde{d_0}$ that ensures values obtained will be over $\tilde{d_0}$. With these elements, the optimal set will be given by the fuzzy subset with membership function:

$$\mu_{\tilde{X}*}(\boldsymbol{x}) = A_g[\mu_{\tilde{R}_0}(\tilde{f}(\boldsymbol{x}; \tilde{\beta}), \tilde{d_0}), \mu_{\tilde{X}}(\boldsymbol{x})],$$

with A_g a triangular norm, and:

$$\mu_{\tilde{X}}(\boldsymbol{x}) = A(\mu_{\tilde{R}_1}(\tilde{g_1}(\boldsymbol{x}; \tilde{\boldsymbol{a_1}}), \tilde{d_1}), ..., \mu_{\tilde{R}_m}(\tilde{g_m}(\boldsymbol{x}; \tilde{\boldsymbol{a_m}}), \tilde{d_m}) = \chi_X(\boldsymbol{x})$$

for the consumer decision problem with crisp constraints.

Finally, for the problem treated, we have:

$$\mu_{\tilde{X}*}(\boldsymbol{x}) = A_g[\mu_{\tilde{R}_0}(\tilde{f}(\boldsymbol{x}; \tilde{\beta}), \tilde{d_0}), \chi_X(\boldsymbol{x})]$$

$$= \begin{cases} T[\mu_{\tilde{U}(x_1,x_2;\tilde{\beta})}(u), \mu_{\tilde{d_0}}(v)] \mid u \geq v & \text{if } \boldsymbol{x} \in X \\ 0 & \text{otherwise} \end{cases}$$

$$= \begin{cases} T[\mu_{\tilde{\beta}}(\beta), \mu_{\tilde{d_0}}(v)], \begin{array}{l} \beta \in \tilde{\beta} \\ U(x_1, x_2; \beta) = u \\ U^{-1}(x_1, x_2; u) \neq \emptyset \mid u \geq v \end{array} & \text{if } \boldsymbol{x} \in X, \\ 0 & \text{otherwise} \end{cases}$$

7 Conclusions

This chapter presents a mathematical approach to consumer utility theory which, instead of relying on utility values, addresses imprecision in consumer decision-making through the use of fuzzy MRS. The development of new tools, including non-polynomial fuzzy partial differential equation solutions, has been key to the implementation of this new approach. The methodology established has been created from the perspective of Classical Consumer Utility Theory, and can be applied in most of the cases in which the classical theory is usually applied, although some restrictions and limitations arise as a direct consequence of this treatment from a perspective reflecting classical theory.

It is also important to note that the methodology described in this chapter approaches consumer imprecision strictly as a mathematical modeling problem, and is not intended to serve as a functional decision-making support tool.

The main drawback encountered for the procedures introduced is that the mathematical formulations are for only two goods or services. As outlined in the methodology, dealing with the problem for more than two goods (or services) will require the introduction of fuzzy partial differential equation systems in order to obtain fuzzy utilities from fuzzy marginal rates of substitution. Although the introduction of these systems would clearly complicate the mathematical apparatus, and although the necessary procedures have yet to be developed, the problem statement and methodology given in this chapter would be perfectly adaptable and valid if it were possible to overcome those mathematical limitations.

One possible solution to this limitation is to divide the basket of consumer goods in pairs of comparable and interchangeable goods for the consumer, thus obtaining a close to real situation of consumption. An example situation for this method would be a carpenter facing a decision about the consumption of raw materials, deciding the amount of different types of wood to consume, for example type A, type B, type C, and type D. If wood types A and B are high-quality woods with similar characteristics, while wood types C and D are of a different, lower-quality, the consumer might divide his intake into two parts: high-quality wood, and other-quality wood, having two types of wood (two consumer goods) in each part. Modeling such a decision would thus involve two utility functions of two goods each, one for the consumption of high-quality wood and another for the consumption of other-quality wood. In this case, the solution can be addressed in two ways: the first arises if the consumer can divide his disposable income into the amounts allocated for the consumption of high-quality wood and other-quality wood, respectively, then the problem can be divided into two independent problems as shown in the chapter; the second scenario arises if it is not possible to divide the disposable income, in which case it will be necessary to add the utilities provided by both pairs of wood types and optimize the added utility subject to the common budget constraint. This approach, dividing the basket of consumer goods, is not as far removed from consumer decision-making as it may seem; as consumers face decisions about consumption, marginal rates of substitution between, for example, a personal cleanliness good and a concrete food item would not usually be considered. Marginal rates of substitution are rarely even considered between goods within the same "group", unless they are sufficiently interchangeable, as, for example, one kind of vegetable and one kind of fruit (both in "food group"). In this sense, if the process of splitting the basket of consumer goods is applied, there are many cases in which it is not easy to find real consumption situations in which more than two goods come out as interchangeable within the same group of goods as marginal rates of substitution are considered by the consumer. Therefore, the methodology developed in the chapter may be applicable to a large percentage of consumer decision-making situations.

Lastly, it is important to highlight that the model presented in this chapter is highly flexible and adaptable to a large number of scenarios. In this sense, it can be applied in cases in which non-triangular membership functions are considered, by simply adjusting the conditions involved in differentiability and subsequent sections. The methodology is also versatile when dealing with the imprecision or internal uncertainty placed on the budget constraints. Addressing the issue of uncertainty, the model is adaptable to situations in which consumers are truly uncertain about the prices of one good or both, or else about the final available income. This last approach requires fuzzy elements on the budget constraints, and the final decision problem has also been raised in a generic form which is applicable to this case and others cases with different fuzzy elements on the budget restrictions. Additionally, with the necessary adjustments, the model is also applicable to scenarios where there is dependence or interaction between criteria. However, the use of more complex tools, such as Choquet integrals, will be required to aggregate the criteria.

Appendix A. Additive Independence

Appendix A.1 Topological Approach by Debreu

If we assume a complete preference preordering $\succeq$ on $K = \prod_{i=1}^{k} C_i$ such that $\{C_i \in K \mid C_i \succeq C_h\}$ and $\{C_i \in K \mid C_i \preceq C_h\}$ are closed for all $C_h \in K$ (this will always occur if there is a finite number of criteria), it will be necessary to hold the following conditions for additive form on utility function:

1. The n factors of K are mutually utility independent.
2. More than two of them are essential.

If $K = \{C_1, ..., C_k\}$ is the set of criteria, their factors $K_1, ...K_n$ are separable spaces such that:

$$K = \prod_{f=1}^{n} K_f$$

Though it would be possible for one factor to contain more than one criterion, in this case, the uni-criterion utility functions are built for each one, so that, each factor (component of total utility) corresponds with each criterion.

In general, a factor K_f is essential if there exists a criterion C_i included in another factor $K_{f'}$ such that not all criteria of K_f are indifferent according to the preordering established by c_i.

As we said, in the case of consumer decision-making framework analyzed here, each factor corresponds to each criterion, therefore, as already shown, it is necessary to have as many summands as uni-criterion utility functions, that is, $n = k$, and $K_f = C_i$.

It is then possible to redefine the essential character of criteria: if c_i is the set of possible values for criteria C_i, and c_h represents the same for C_h, a criteria C_i is essential if there exists:

$$C_i(x_1, x_2), C_i(x_1', x_2') \in c_i, C_h(x_1, x_2) \in c_h \text{ such that}$$

$$(C_i(x_1, x_2), C_h(x_1, x_2)) \not\sim (C_i(x_1', x_2'), C_h(x_1, x_2)).$$

The biggest limitation of this condition may be that it requires at least three essential factors, which means that it can only be applied in decision-making frameworks with three or more criteria. Nevertheless, it is easily satisfied in any consumer decision-making problem by all criteria involved.

For cases in which only two criteria are taken into account, it will be more accurate to use the Luce and Tukey axioms.

Appendix A.2. Algebraic Approach by Luce and Tukey

Adapting Luce and Tukey conditions to the consumer decision problem: If C_1 and C_2 are the two criteria involved in the decision-making process, with: $c_1 = \{C_1(x_1, x_2), C_1(x_1', x_2'), C_1(x_1'', x_2''), \ldots\}$ and $c_2 = \{C_2(x_1, x_2), C_2(x_1', x_2'), C_2(x_1'', x_2''), \ldots\}$ the sets of values that can be reached by C_1 and C_2 respectively, $c_1 \times c_2$ is formed by pairs $(C_1(x_1, x_2), C_2(x_1, x_2)), (C_1(x_1, x_2), C_2(x_1', x_2')), (C_1(x_1', x_2'), C_2(x_1, x_2))$, etc. Considering the binary relation $\succeq$, criteria will be additive if the following axioms are satisfied:

I Ordering axiom: $\succeq$ is a weak order meeting the following axioms:

- Reflexivity:

$$(C_1(x_1, x_2), C_2(x_1, x_2)) \succeq (C_1(x_1, x_2), C_2(x_1, x_2)),$$
$$\forall C_1(x_1, x_2) \in c_1, C_2(x_1, x_2) \in c_2.$$

- Transitivity:

$$\text{If } (C_1(x_1, x_2), C_2(x_1, x_2)) \succeq (C_1(x_1', x_2'), C_2(x_1', x_2')),$$

$$\text{and } (C_1(x_1', x_2'), C_2(x_1', x_2')) \succeq (C_1(x_1'', x_2''), C_2(x_1'', x_2'')),$$

$$\text{then: } (C_1(x_1, x_2), C_2(x_1, x_2)) \succeq (C_1(x_1'', x_2''), C_2(x_1'', x_2'')).$$

- $\succeq$ is closed:

$$(C_1(x_1, x_2), C_2(x_1, x_2)) \succeq (C_1(x_1', x_2'), C_2(x_1', x_2'))$$

$$\text{or } (C_1(x_1, x_2), C_2(x_1, x_2)) \preceq (C_1(x_1', x_2'), C_2(x_1', x_2')), \text{ or both.}$$

- Definition:
  * $(C_1(x_1, x_2), C_2(x_1, x_2)) \sim (C_1(x_1', x_2'), C_2(x_1', x_2'))$ only if :

$$(C_1(x_1, x_2), C_2(x_1, x_2)) \succeq (C_1(x_1', x_2'), C_2(x_1', x_2')) \text{ and}$$

$$(C_1(x_1, x_2), C_2(x_1, x_2)) \preceq (C_1(x_1', x_2'), C_2(x_1', x_2')).$$

  * $(C_1(x_1, x_2), C_2(x_1, x_2)) \succ (C_1(x_1', x_2'), C_2(x_1', x_2'))$ only if:

$$\neg[(C_1(x_1', x_2'), C_2(x_1', x_2')) \succeq (C_1(x_1, x_2), C_2(x_1, x_2))].$$

II Solution to equations:

For each $C_1(x_1, x_2) \in c_1$ and $C_2(x_1, x_2), C_2(x_1', x_2') \in c_2$, the equation $(C_1(x_1^*, x_2^*), C_2(x_1, x_2)) = (C_1(x_1, x_2), C_2(x_1', x_2'))$ has a solution $C_1(x_1^*, x_2^*) \in c_1$ and,
For each $C_1(x_1, x_2), C_1(x_1', x_2') \in c_1$ and $C_2(x_1, x_2) \in c_2$, the equation $(C_1(x_1, x_2), C_2(x_1, x_2)) = (C_1(x_1', x_2'), C_2(x_1^*, x_2^*))$ has a solution $C_2(x_1^*, x_2^*) \in c_2$.

III Cancelation:

For all $C_1(x_1, x_2), C_1(x_1', x_2'), C_1(x_1'', x_2'') \in c_1$ and $C_2(x_1, x_2), C_2(x_1', x_2'), C_2(x_1'', x_2'') \in c_2$,

$$\text{if } (C_1(x_1, x_2), C_2(x_1'', x_2'')) \succeq (C_1(x_1'', x_2''), C_2(x_1', x_2'))$$

$$\text{and } (C_1(x_1'', x_2''), C_2(x_1, x_2)) \succeq (C_1(x_1', x_2'), C_2(x_1'', x_2'')),$$

$$\text{then: } (C_1(x_1, x_2), C_2(x_1, x_2)) \succeq (C_1(x_1', x_2'), C_2(x_1', x_2')).$$

IV Archimedean axiom: if $\{C_1(x_1, x_2)_i, C_2(x_1, x_2)_i\}$, $i = 0, 1, 2, \ldots$ is a non-trivial and increasing *dual standard sequence*, for each $C_1(x_1', x_2') \in c_1$, $C_2(x_1', x_2') \in c_2$, then, there exist two integers (positive or negative) m and n such that:

$$\begin{aligned}(C_1(x_1, x_2)_n, C_2(x_1, x_2)_n) &\succeq (C_1(x_1', x_2'), C_2(x_1', x_2')) \\ &\succeq (C_1(x_1, x_2)_m, C_2(x_1, x_2)_m).\end{aligned}$$

An infinite sequence $\{C_1(x_1, x_2)_t, C_2(x_1, x_2)_t\}$, $t = 0, 1, 2, \ldots$, with $C_1(x_1, x_2)_t \in c_1$, $C_2(x_1, x_2)_t \in c_2$ is a *dual standard sequence* (*dss*) when $(C_1(x_1, x_2)_m, C_2(x_1, x_2)_n) = (C_1(x_1, x_2)_p, C_2(x_1, x_2)_q)$ for $m+n = p+q$ for any m, n, p, q integer, positive, negative or null.
A *dss* will be trivial if $C_1(x_1, x_2)_t = C_1(x_1, x_2)_0$, or $C_2(x_1, x_2)_t = C_2(x_1, x_2)_0$.

References

1. Aliev, R., Pedrycz, W., et al.: Fuzzy logic-based generalized decision theory with imperfect information. Inf. Sci. **189**, 18–42 (2012)
2. Banerjee, A.: Fuzzy choice functions, revealed preference rationality. Fuzzy Sets Syst. **70**, 13–43 (1995)
3. Buckley, J.J., Feuring, T.: Introduction to fuzzy partial differential equations. Fuzzy Sets. Syst. **105**(2), 241–248 (1999)
4. Buckley, J.J., Feuring, T.: Fuzzy differential equations. Fuzzy Sets Syst. **110**, 43–54 (2000)
5. Chuoo, E.U., Shoner, B., Wedley, W.C.: Interpretation of criteria weights in multicriteria decision making. Comput. Ind. Eng. **37**, 527–541 (1999)
6. Colin, A., Trivedi, P.: Microeconometrics: Methods and Applications. Cambridge University Press, Cambridge (2005)
7. van Kooten, C., et al.: Preference uncertainty in non-market valuation: a fuzzy aproach. Am. J. Agric. Econ. **83**(3), 487–500 (2001)
8. Dean, P.-K., et al.: Product and cost estimation with fuzzy multi-attribute utility theory. Eng. Econ. **44**(4), 303–331 (1999)
9. Debreu, G.: Topological methods in cardinal utility theory. Math. Methods Soc. Sci. 1959, 16–26 (1960) (Cowles Foundation paper 156).
10. De Wilde, P.: Fuzzy utility and equilibria. IEEE Trans. Syst. Man Cybern. **34**(4), 1774–1785 (2004)
11. Dubois, D., Prade, H.: Additions of interactive fuzzy numbers. IEEE Trans. Autom. Control **26**(4), 926–936 (1981)
12. Duncan, R., Tukey, J.W.: Simultaneous conjoint measurement: a new type of fundamental measurement. J. Math. Psychol. **1**, 1–27 (1964)
13. Figueira, J., Greco, S., Ehrgott, M. (eds.): Multiple Criteria Decision Analysis. State of the Art Surveys. Springer, Berlin (2005)
14. Fishburn, P.: Independence in utility theory with whole product sets. Operat. Res. **13**(3), 28–45 (1965)
15. Fodor, J., De Baets, B., Perny, P. (eds.): Preferences and Decisions Under Incomplete Knowledge. Springer, Berlin (2000)
16. Gálvez, D. and Pino, J.L.: The extension of Buckley-Feuring solutions for non-polynomial fuzzy partial differential equations. Application to Microeconomics Utility Theory. In: Proceedings of NAFIPS (The 28th North American Fuzzy Information Processing Society Annual Conference). IEEE, 2009.
17. Gálvez, D. and Pino, J.L.: The extension of Buckley-Feuring solutions for non-polynomial fuzzy partial differential equations. Application to Microeconomics Utility Theory and consumer decision. In: Proceedings of Fuzz-IEEE (2009 IEEE International Conference on Fuzzy Systems). IEEE, 2009.
18. Gálvez, D.: Tratamiento de la imprecisión en la teoría de la utilidad del consumidor. Universidad de Sevilla. 2009.
19. Goetschel, R., Voxman, W.: Elementary fuzzy calculus. Fuzzy Sets Syst. **18**, 319–330 (1986)
20. Green, W.: Análisis Econométrico. Prentice hall, New Jersey (2002)
21. Hicks, JR: Value and Capital: An Inquiry into Some fundamental Principles of Economic Theory, pp. 18. Clarendon Press, Oxford (1939).
22. Kaufmann, A., Gupta, M.M.: Introduction to fuzzy arithmetic: theory and applications. Van Nostrand Reinhold, New York (1985)
23. Keeney, R.L., Raiffa, H.: Decisions with Multiple Objectives: Preferences and Value Trade-offs. Wiley, New York (1976)
24. Mathieu-Nicot, B.: Fuzzy expected utility. Fuzzy Sets Syst. **20**(2), 163–173 (1986)
25. Mesiar, R.: Fuzzy set approach to the utility, preference relations, and agregation operators. Eur. J. Operat. Res. **176**(1), 414–422 (2007)
26. Mora, J. J.: Introducción a la teoría del consumidor : de las preferencias a la estimación. Universidad ICESI (2002).

27. Nakamura, K.: Preference relations on a set of fuzzy utilities as a basis for decision making. Fuzzy Sets Syst. **20**(2), 147–162 (1986)
28. Orlovsky, S.A.: Decision making with a fuzzy preference relation. Fuzzy Sets Syst. **1**(3), 155–167 (1978)
29. Pareto, V.: Manual of Political Economy. Augustus M, Kelley, New York (1971)
30. Ponsard, C.: Fuzzy mathematical models in economics. Fuzzy Sets Syst. **28**(3), 273–283 (1988)
31. Ramík, J., Vlach, M.: Generalized Concavity in Fuzzy Optimization and Decision Analysis. Kluwer Academic Publishers, The Netherlands (2002)
32. Robinson, J.: Economics is a Serious Subject. W. Heffer and Sons, Cambridge (1932)
33. Rothbard, M.N.: History economic thought, vol. 1. Economic thought to Adam Smith, Union Editorial (1999)
34. Rommelfanger, H. J.: Decision Making in fuzzy environment. Ways for getting practical decision models. Paper on line at http://www.uni-frankfurt.de. University of Frankfurt (1999)
35. Salles, M.: Fuzzy utility. In: Barbera, S., et al. (eds.) Handbook of Utility Theory, vol. 1. Springer, Berlin (1999)
36. Stewart, T.J.: Simplified approaches for multi-criteria decision making under uncertainty. J. Multi-Criteria Decis. Anal. **4**(4), 246–258 (1995)
37. Stewart, T.J.: Robustness of additive value function methods in MCDM. J. Multi-Criteria Decis. Anal. **5**(4), 301–309 (1996)
38. Tenekedjiev, K., Nikolova, N.: Justification and numerical realization of the uniform method for finding point estimates of interval elicited scaling constants. Fuzzy Optim. Decis. Making **7**(2), 119–145 (2008)
39. Zadeh, L.A.: Fuzzy Sets. Inf. Control **8**, 338–353 (1965)
40. Zadeh, L.A.: The concept of Linguistic variables and its applications to approximate reasoning. Part I Inf. Sci. **8**(3), 199–249 (1975)
41. Zadeh, L.A.: The concept of Linguistic variables and its applications to approximate reasoning. Part II Inf. Sci. **8**(4), 301–357 (1975)
42. Zadeh, L.A.: The concept of Linguistic variables and its applications to approximate reasoning. Part III Inf. Sci. **9**(1), 43–80 (1975)

Decision Making Under Z-Information

R. A. Aliev and Lala M. Zeinalova

Abstract Rational decisions are based on information usually uncertain, imprecise and incomplete. The existing decision theories deal with three levels of generalization of decision making relevant information: numerical valuation, interval valuation and fuzzy number valuation. The classical decision theories, such as expected utility theory proposed by von Neumann and Morgenstern, and subjective expected utility theory proposed by Savage use the first level of generalization, i.e. numerical one. These approaches require that the objective probabilities or subjective probabilities and utility values be precisely known. But in real world in many cases it becomes impossible to determine the precise values of needed information. Interval analysis and classical fuzzy set theories have been applied in making decisions and many fruitful results have been achieved. But a problem is that in the mentioned above decision theories the reliability of the decision relevant information is not well taken into consideration. Prof. L. Zadeh introduced the concept of Z-numbers to describe the uncertain information which is more generalized notion closely related with confidence (reliability). Use of Z-information is more adequate and intuitively meaningful for formalizing information structure of a decision making problem. In this chapter we consider two approaches to decision making with Z-information. The first approach is based on reducing of Z-numbers to classical fuzzy numbers, and generalization of expected utility approach and use of Choquet integral with an integrant represented by Z-numbers. A fuzzy measure is calculated on a base of a given Z-information. The second approach is based on direct computation with Z-numbers. To illustrate a validity of suggested approaches to decision making with Z-information the numerical examples are used.

R. A. Aliev
Joint MBA Program (USA, Azerbaijan), Azadlyg ave. 20,Baku AZ1010, Azerbaijan
e-mail: raliev@asoa.edu.az

L. M. Zeinalova (✉)
Department of Computer-Aided Control Systems, Azerbaijan State Oil Academy,
Azadlyg ave. 20, Baku AZ1010 , Azerbaijan
e-mail: Zeynalova-69@mail.ru

P. Guo and W. Pedrycz (eds.), *Human-Centric Decision-Making Models for Social Sciences*, Studies in Computational Intelligence 502,
DOI: 10.1007/978-3-642-39307-5_10, © Springer-Verlag Berlin Heidelberg 2014

Keywords Decision making · Z-number · Choquet integral · Fuzzy utility

1 A Brief Review of Existing Decision Theories

For decision making the first axiomatic foundation of the utility paradigm was the expected utility (EU) theory of von Neumann and Morgenstern [34]. This model compares finite-outcome lotteries (alternatives) on the base of their utility values under conditions of precisely known utilities and probabilities of outcomes.

The assumptions of von Neumann and Morgenstern expected utility model stating that objective probabilities of events are known makes this model unsuitable for majority of real-world applications.

We have no representative experimental data or complete knowledge to determine objective probabilities. For such cases, Savage suggested a theory able to compare alternative actions on the base of a DM's experience or vision [32]. Savage's theory is based on a concept of subjective probability suggested by Ramsey [28] and de Finetti [13]. Subjective probability is DM's probabilistic belief concerning occurrence of an event and is assumed to be used by humans when no data on objective (actual) probabilities of outcomes is available. Savage's subjective expected utility (SEU) theory based on the use of subjective probabilities in the expected utility paradigm of von Neumann and Morgenstern instead of objective probabilities. SEU became a base of almost all the utility models for decision making under uncertainty.

Prospect Theory (PT) of Kahneman and Tversky [18, 33] is the one of the most famous theories in the new view on the utility concept. This theory is successful because it includes psychological aspects that form human behavior. Kahneman and Tversky uncovered series of features of human behavior in decision making and used them to construct their utility model.

Choquet Expected Utility (CEU) was suggested by Schmeidler [30] as a model with a new view on belief and representation of preferences in contrast to the SEU model. In CEU a belief is described by a capacity [10]—not necessarily additive measure.

Kahneman and Tversky, the authors of PT, suggested cumulative prospect theory (CPT) as a more advanced theory which can be applied, in contrast to PT, both for decisions under risk and uncertainty.

The principle of uncertainty aversion was formalized in form of an axiom by Gilboa and Schmeidler [17]. This axiom is one of the axioms underlying a famous utility model called Maximin Expected Utility (MMEU) [17]. According to the axiomatic basis of this model, there is a unique closed and convex set C of priors (probability measures) over states of nature and overall utility of an act is a minimum among all its expected utilities each obtained for prior one $P \in C$.

Ghirardato, Maccheroni and Marinacci suggested a generalization of MMEU [15] consisting in using all its underlying axioms except uncertainty aversion axiom. The obtained model is referred to as α – MMEU.

The main disadvantages of the MMEU are that in real problems it is difficult to strictly constrain the set of priors and various priors should not be considered equally relevant to a problem at hand. From the other side, in MMEU each act is evaluated on the base of only one prior. In order to cope with these problems Klibanoff et al. suggested a smooth ambiguity model as a more general way to formalize decision making under ambiguity than MMEU [19].

There are many different approaches for describing imprecision of probability relevant information. One of the approaches is the use of hierarchical imprecise models. These models capture the second-order uncertainty inherent in real problems. According to this approach an expert opinion on probability assessments is usually imprecise [2, 11, 35].

The existing decision theories are not developed for applications in fuzzy environment and consequently require more deterministic information. Development of new theories is now possible due to an increased computational power of information processing systems which allows for computations with imperfect information, particularly, imprecise and partially true information, which are much more complex than computations over numbers and probabilities. There is the series of fuzzy approaches to decision making like fuzzy AHP [14, 22], fuzzy TOPSIS [22, 37], fuzzy Expected Utility [9, 16, 23]. However, they are mainly fuzzy generalizations of the mathematical structures of the existing theories used with intent to account for vagueness, impreciseness and partial truth. Direct fuzzification of the existing theories often leads to inconsistency and loss of properties of the latter.

Approaches that are based on fuzzy description of the most part of a decision problem are lack of mathematical proof of an existence of a utility function. From the other side, many of the existing fuzzy approaches follow too simple models like EU model.

In [1] authors present a fuzzy-logic-based decision theory with imperfect information. This theory is developed for the framework of mix of fuzzy information and probabilistic information and is based on a fuzzy utility function represented as a fuzzy-valued Choquet integral.

On basis of analysis of the existing decision theories described above we may conclude that the existing decision models yielded good results, but nowadays there is a need in generation of more realistic decision models.

These approaches require that the objective probabilities or subjective probabilities and utility values be precisely known. But in real world in many cases it becomes impossible to determine the precise values of needed information. Interval analysis and classical fuzzy set theories have been applied in making decisions and many fruitful results have been achieved. But a problem is that in the mentioned above decision theories the reliability of the decision relevant information is not well taken into consideration.

In [21] Zadeh introduced the concept of Z-numbers to describe the uncertain information which is more generalized notion. A Z-number is an ordered pair of fuzzy numbers $(\tilde{A}, \tilde{R})$. Here $\tilde{A}$ is a value of some variable and $\tilde{R}$ represents an idea of certainty or other closely related concept such as sureness, confidence, reliability, strength of truth, or probability [29]. It should be noted that in everyday decision making most decisions are in the form of Z-numbers. Zadeh suggests operations for computation with Z-numbers, using the extension principle. In [29] author uses Z-numbers to provide information about an uncertain variable in the form of Z-valuations, assuming that this uncertain variable is random. In [29] author also offers an illustration of a Z-valuation, showing how to make decisions and answer questions. Also an alternative formulation is used for the information contained in the Z-valuations in terms of a Dempster-Shafer belief structure that made use of type-2 fuzzy sets. Simplified version of Z-valuation of decision relevant information is considered in [8]. In [7] authors considered a multi-criteria decision making problem using Z-numbers. For this purpose the Z-numbers are converted to classical fuzzy number and a priority weight of each alternative is determined.

In this chapter we consider two approaches to decision making with Z-information. The first approach is based on reducing of Z-numbers to classical fuzzy numbers, and generalization of expected utility approach and use of Choquet integral with an integrant represented by Z-numbers. A fuzzy measure is calculated on a base of a given Z-information. The second approach is based on direct computation with Z-numbers. To illustrate a validity of suggested approaches to decision making with Z-information the numerical examples are used.

The study is organized as follows. In Sect. 2 we present required preliminaries and cover some prerequisite material. In Sect. 3 we consider a generalization of Expected Utility Theory using Z-number. In Sect. 4 we present a method of Choquet integral based decision making using Z-information. In Sect. 5 we consider the second approach based on direct computation with Z-numbers applying it to the same problem. In Sect. 6 we cover application of the suggested method to a real-life business problem of hotel management using the suggested approaches. Concluding comments are included in Sect. 7.

2 Preliminaries

Definition 1 *Fuzzy sets* [4].

Let $\mathcal{X}$ be a classical set of objects, called the universe, whose generic elements are denoted x. Membership in a classical subset $\mathcal{A}$ of $\mathcal{X}$ is often viewed as a characteristic function $\mu_{\mathcal{A}}$ from $\mathcal{X}$ to $\{0, 1\}$ such that

$$\mu_{\mathcal{A}}(x) = \begin{cases} 1 & \textit{iff } x \in \mathcal{A} \\ 0 & \textit{iff } x \notin \mathcal{A} \end{cases}$$

where $\{0, 1\}$ is called a valuation set; 1 indicates membership while 0—non membership.

If the valuation set is allowed to be in the real interval [0,1], then $\mathcal{A}$ is called a fuzzy set, $\mu_{\mathcal{A}}$ is the grade of membership of x in A: $\mu_{\mathcal{A}}(x) : \mathcal{X} \to [0,1]$.

Let $\mathcal{E}^n$ be a space of all fuzzy subsets of $\mathcal{R}^n$. These subsets satisfy the conditions of normality, convexity, and are upper semicontinuous with compact support.

Definition 2 *A Z-number* [21].

A Z-number is an ordered pair of fuzzy numbers,$(\tilde{A}, \tilde{R})$. $\tilde{A}$-is a fuzzy restriction on the values which a real-valued uncertain variable is allowed to take. $\tilde{R}$ is a measure of reliability of the first component.

Example

(anticipated budget deficit, about three million USD, likely);

(price of oil in the near future, significantly over 50 dollars/barrel, very likely).

Denote Ω a universe of discourse and denote $\mathcal{F}$ a σ-algebra of subsets of Ω.

Definition 3 *Choquet integral* [6, 23, 25, 31, 39, 40]. Let $\varphi:\Omega \to R$ be a measurable real-valued function on Ω and $\eta:\mathcal{F} \to [0,1]$ be a non-additive measure defined over $\mathcal{F}$. The Choquet integral of φ with respect to η is defined as

$$\int_{\Omega} \varphi d\eta = \sum_{i=1}^{n} \left(\eta(B_{(i)}) - \eta(B_{(i-1)})\varphi(\omega_{(i)})\right) \tag{1a}$$

where index (i) implies that elements $\omega_i \in \Omega, i = 1, \ldots, n$ are permuted such that $\varphi(\omega_{(i)}) \geq \varphi(\omega_{(i+1)}), \varphi(\omega) = 0$ and $B_{(i)} = \{\omega_{(1)}, \ldots, \omega_{(i)}\} \subseteq \Omega$.

A value of fuzzy utility function for an action is determined as a fuzzy number-valued Choquet integral

$$\int_{\Omega} \tilde{\varphi} d\tilde{\eta} = \sum_{i=1}^{n} \left(\tilde{\eta}(B_{(i)}) - \tilde{\eta}(B_{(i-1)})\tilde{\varphi}(\omega_{(i)})\right) \tag{1b}$$

(i)means that utilities are ranked such that $\tilde{\varphi}(\omega_{(1)}) \geq \ldots \geq \tilde{\varphi}(\omega_{(n)}), \tilde{\varphi}(\omega) = 0$. Let $\tilde{\mathcal{F}}(\Omega) = \left\{\tilde{V} \left| \mu_{\tilde{V}} : \Omega \to [0,1]\right.\right\}$ be the class of all fuzzy subsets of Ω.

Definition 4 [1, 41]. A subclass $\tilde{\mathcal{F}}$ of $\tilde{\mathcal{F}}(\Omega)$ is called a fuzzy σ-algebra if it has the following properties:

(1) $\varnothing, \Omega \in \tilde{\mathcal{F}}$

(2) if $\tilde{V} \in \tilde{\mathcal{F}}$, then $\tilde{V}^c \in \tilde{\mathcal{F}}$

(3) if $\left\{\tilde{V}_n\right\} \subset \tilde{\mathcal{F}}$, then $\bigcup_{n=1}^{\infty} \tilde{V}_n \in \tilde{\mathcal{F}}$

Definition 5 *Fuzzy number-valued fuzzy measure* [1, 41]. A fuzzy number-valued fuzzy measure ((z) fuzzy measure) on $\tilde{\mathcal{F}}$ is a fuzzy number-valued fuzzy set function $\tilde{\eta}:\tilde{\mathcal{F}} \to \mathrm{E}^1$ with the properties:

(1) $\tilde{\eta}(\varnothing) = 0$;

(2) if $\tilde{V} \subset \tilde{W}$ then $\tilde{\eta}(\tilde{V}) \leq \tilde{\eta}(\tilde{W})$;

(3) if $\tilde{V}_1 \subset \tilde{V}_2 \subset \ldots, \tilde{V}_n \subset \ldots \in \tilde{\mathcal{F}}$, then $\tilde{\eta}(\bigcup_{n=1}^{\infty} \tilde{V}_n) = \lim\limits_{n\to\infty} \tilde{\eta}(\tilde{V}_n)$;

(4) if $\tilde{V}_1 \supset \tilde{V}_2 \supset \ldots, \tilde{V}_n \in \tilde{\mathcal{F}}$, and there exists n_0 such that $\tilde{\eta}(\tilde{V}_{n_0}) \neq \tilde{\infty}$, then $\tilde{\eta}(\bigcap_{n=1}^{\infty} \tilde{V}_n) = \lim\limits_{n\to\infty} \tilde{\eta}(\tilde{V}_n)$.

Definition 6 *Lower prevision* [2, 3, 12, 20, 24–26]. A coherent lower prevision is defined as a lower expectation functional from the set of gambles to the real numbers that satisfies some rationality criteria. This function is conjugate to another that is called a coherent upper prevision. When a coherent lower prevision coincides with its conjugate coherent upper prevision, we call it a linear prevision. An unconditional lower prevision $\underline{P}(B)$ is coherent if and only if it is the lower envelope of dominating linear previsions.

If the lower prevision $\underline{P}$ is represented as the lower envelope of a closed convex set P of linear previsions then

$$\underline{P} = \min\{P(B)\}, B \subset S \tag{2}$$

Lower prevision $\underline{P}$ is characterized by probability density function of each linear prevision in extreme points [36].

In particular case, when linear prevision is a probability measure the lower prevision is the lower envelope of multiple priors. In this work we use lower prevision as non-additive measure. So we can define η as $\underline{P}$.

Example [1]

Let Ω be a nonempty set and $\Omega = [0,1]$. Consider values of a fuzzy number-valued fuzzy measure $\tilde{\eta}$ for some fuzzy subsets $s_i \subset \Omega, i = \overline{1,3}$ and their unions. Then the corresponding values of the fuzzy number-valued fuzzy measure $\tilde{\eta}_{\tilde{P}l}$ can be as the triangular fuzzy numbers given in Table 1:

Definition 7 *Expected utility* [23, 27]

Let $P : S \to R$ be any probability measure on a set of states S such that $P(s) > 0$ for all $s \in S$. For each $s \in S$ define $v : \mathcal{X} \to \mathcal{R}$. Then

$$U(f) = \sum_{s \in S} P(s)v(f(s)) \tag{3}$$

where f is an act, $x = f(s)$ is an outcome, $v(f(s))$ is a utility in state s and $U(f)$ is the expected value of utility.

Table 1 The values of the fuzzy number-valued fuzzy measure $\tilde{\eta}_{\tilde{P}l}$

$\tilde{B} \subset S$	$\{s_1\}$	$\{s_2\}$	$\{s_3\}$	$\{s_1 \cup s_2\}$	$\{s_1 \cup s_3\}$	$\{s_2 \cup s_3\}$
$\tilde{\eta}_{\tilde{P}l}(\tilde{B})$	(0.3,0.4,0.4)	(0,0.1,0.1)	(0.3,0.5,0.5)	(0.3,0.5,0.5)	(0.6,0.9,0.9)	(0.3,0.6,0.6)

3 A Generalization of Expected Utility Theory Using Z-Number

A set of acts $f_1, f_2, \ldots, f_n$ with a number of possible utilities $\tilde{Z}_{v(f_i(s_1))}, \tilde{Z}_{v(f_i(s_2))}, \ldots, \tilde{Z}_{v(f_i(s_m))}$ in states $s_1, s_2, \ldots, s_m \in S$ and the corresponding state probabilities $\tilde{Z}_{P(s_1)}, \tilde{Z}_{P(s_2)}, \ldots, \tilde{Z}_{P(s_m)}$ are given and described by Z-numbers (Tables 2 and 3). Then we can determine the value of expected utility function for each act.

In payoff Table 2 $\tilde{R}_1$ is a confidence degree for the value of utility.

As decision maker usually is uncertain about first-order imprecise probabilities, we describe the probabilities of states of nature as Z-numbers (Table 2).

In Table 3 $\tilde{R}_2$ is a confidence degree for the value of probability of the state of nature.

Now using the Z-valuations of the values of utilities and probabilities of states of nature we can determine the values of expected utility for any act, represented as Z-numbers.

$$\begin{aligned}
\tilde{Z}_{U(f_1)} &= (\tilde{P}(s_1), \tilde{R}_2) \times (\tilde{v}(f_1(s_1)), \tilde{R}_1) + (\tilde{P}(s_2), \tilde{R}_2) \times (\tilde{v}(f_1(s_2)), \tilde{R}_1)+, \ldots, \\
&\quad + (\tilde{P}(s_m), \tilde{R}_2) \times (\tilde{v}(f_1(s_m)), \tilde{R}_1) \\
\tilde{Z}_{U(f_2)} &= (\tilde{P}(s_1), \tilde{R}_2) \times (\tilde{v}(f_2(s_1)), \tilde{R}_1) + (\tilde{P}(s_2), \tilde{R}_2) \times (\tilde{v}(f_2(s_2)), \tilde{R}_1)+, \ldots, \\
&\quad + (\tilde{P}(s_m), \tilde{R}_2) \times (\tilde{v}(f_2(s_m)), \tilde{R}_1) \\
\tilde{Z}_{U(f_n)} &= (\tilde{P}(s_1), \tilde{R}_2) \times (\tilde{v}(f_n(s_1)), \tilde{R}_1) + (\tilde{P}(s_2), \tilde{R}_2) \times (\tilde{v}(f_n(s_m)), \tilde{R}_1)+, \ldots, \\
&\quad + (\tilde{P}(s_m), \tilde{R}_2) \times (\tilde{v}(f_n(s_m)), \tilde{R}_1)
\end{aligned} \tag{4}$$

Now we have to choose the act with maximal expected utility i.e. the decision making problem in this case consists in the determination of an optimal action $f^* \in A$ as the following

$$\tilde{Z}_{U(f^*)} = \max_{f \in A}(\tilde{Z}_{U(f_1)}, \tilde{Z}_{U(f_2)}, \ldots, \tilde{Z}_{U(f_n)}) \tag{5}$$

Table 2 The payoff table with utilities as Z-numbers

	s_1	s_2	...	s_m
f_1	$(\tilde{v}(f_1(s_1)), \tilde{R}_1)$	$(\tilde{v}(f_1(s_2)), \tilde{R}_1)$	...	$(\tilde{v}(f_1(s_m)), \tilde{R}_1)$
f_2	$(\tilde{v}(f_2(s_1)), \tilde{R}_1)$	$(\tilde{v}(f_2(s_2)), \tilde{R}_1)$	...	$(\tilde{v}(f_2(s_m)), \tilde{R}_1)$
...	...	...	...	...
f_n	$(\tilde{v}(f_n(s_1)), \tilde{R}_1)$	$(\tilde{v}(f_n(s_2)), \tilde{R}_1)$		$(\tilde{v}(f_n(s_m)), \tilde{R}_1)$

Table 3 Probabilities of states as Z-numbers

$(\tilde{P}(s_1), \tilde{R}_2)$	$(\tilde{P}(s_2), \tilde{R}_2)$	...	$(\tilde{P}(s_m), \tilde{R}_2)$

Let outcomes $\tilde{Z}_{v(f_i(s_j))} = (\tilde{v}(f_i(s_j)), \tilde{R}_1)$ and the probabilities $\tilde{Z}_{P(s_j)} = (\tilde{P}(s_j), \tilde{R}_2)$ of the states $s_j \in S$, where $\tilde{R}_1 = \{(x_2, \mu_{\tilde{R}_1}(x)) : x_2 \in [0,1]\}$, $\tilde{R}_2 = \{(y_2, \mu_{\tilde{R}_2}(y)) : y_2 \in [0,1]\}$ are represented by trapezoidal and triangle fuzzy numbers respectively.

In this study it is assumed that it is given only NL-described reasonable knowledge about probability distribution over S. It means that a state s_j is assigned a linguistic probability $\tilde{P}_j$ that can be described by Z-number. Initial data for the problem are represented by given linguistic probabilities for $m-1$ states of nature whereas for one of the given states the probability is unknown. So at first it is required to obtain the unknown probability. To determinate an unknown probability of state $s_j - \tilde{Z}_{P(s_j)}$ on a base of given probabilities $\tilde{Z}_{P(s_1)}, \tilde{Z}_{P(s_2)}, \ldots, \tilde{Z}_{P(s_{j-1})}, \ldots, \tilde{Z}_{P(s_{j+1})}, \ldots, \tilde{Z}_{P(s_m)}$ we use the method suggested in [1]. In the framework of Computing with Words the problem of obtaining the unknown linguistic probability for state $\tilde{s}_j$ given linguistic probabilities of all other states is a problem of propagation of generalized constraints. Formally this problem is formulated as follows:

given

$$\tilde{P}(\tilde{s}_i) = \tilde{P}_i; \tilde{s}_i \in \varepsilon^n, \tilde{P}_i \in \varepsilon^1_{[0,1]}, i \in \{1, \ldots, j-1, j+1, \ldots, n\} \tag{6}$$

find unknown

$$\tilde{P}\left(\tilde{s}_j\right) = \tilde{P}_j, \tilde{P}_j \in \varepsilon^1_{[0,1]}$$

This problem reduces to a variation problem of constructing the membership function $\mu_{\tilde{P}_j}(\cdot)$ of an unknown fuzzy probability $\tilde{P}_j$ [1]:

$$\mu_{\tilde{P}_j}(p_j) = \sup_\rho \min_{i=\{1,\ldots,j-1,j+1,\ldots,n\}} (\mu_{\tilde{P}_i}(\int_S \mu_{\tilde{s}_i}(s)\rho(s)ds)) \tag{7}$$

subject to $\int_S \mu_{\tilde{s}_j}(s)\rho(s)ds = p_j, \int_S \rho(s)ds = 1$

Given the payoff table and the complete probability distribution we can evaluate the values of expected utility on base of (4). For this aim we use computation with Z-numbers which falls within the province of Computing with words. Computation with Z-information in this study is based on converting of Z-numbers to classical fuzzy numbers [8].

To convert the given Z-numbers on outcomes and probabilities first we determine the expected values of fuzzy numbers R_1 and R_2 describing reliability of variables of outcome and probability:

$$\alpha_1 = \frac{\int x\mu_{\tilde{R}_1}(x)dx}{\int \mu_{\tilde{R}_1}(x)dx}, \tag{8}$$

$$\alpha_2 = \frac{\int y\mu_{\tilde{R}_2}(y)dy}{\int \mu_{\tilde{R}_y}(y)dy} \tag{9}$$

Now we can represent the values of outcome and probability variables as trapezoidal fuzzy numbers: $\tilde{Z}^{\alpha_1}_{v_{s_j}(f_i(s))} = (\mathrm{a}_1, \mathrm{a}_2, \mathrm{a}_3, \mathrm{a}_4; \alpha_1)$, $\tilde{Z}^{\alpha_2}_{P(s_j)} = (c_1, c_2, c_3; \alpha_2)$. Then we convert this weighted Z-number to fuzzy number: $\tilde{Z}'_{v_{s_j}(f_i(s))} = (\sqrt{\alpha_1}\mathrm{a}_1, \sqrt{\alpha_1}\mathrm{a}_2, \sqrt{\alpha_1}\mathrm{a}_3, \sqrt{\alpha_1}\mathrm{a}_4; 1)$, $\tilde{Z}'_{P(s_j)} = (\sqrt{\alpha_2}c_1, \sqrt{\alpha_2}c_2, \sqrt{\alpha_2}c_3; 1)$

As we have an ordinary fuzzy numbers with trapezoidal and triangular membership functions then we can obtain fuzzy values of utility function $U(f(s))$ for each alternative by (4):

$$\tilde{Z}'_{U(f_i)} = \sum_{\substack{i=1 \\ j=1}}^{k,n} \tilde{Z}'_{v(f_i(s))} \times \tilde{Z}'_{P(s_j)} = (\sqrt{\alpha_1\alpha_2} \times (\sum_{i,j=1}^{k,n} \tilde{v}(f_i(s)) \times \tilde{P}(s_j)); 1) \quad (10)$$

An optimal action $f^* \in A$ is obtained in accordance with (5).

The value of Z-number for optimal utility function may be described as

$$\tilde{Z}_{U(f_i)} = (\tilde{Z}'_{U(f_i)}/\sqrt{\alpha_1\alpha_2}); \tilde{R}_3). \quad (11)$$

where $(\tilde{Z}'_{U(f_i)}/\sqrt{\alpha_1\alpha_2}); \tilde{R}_3)$ describes the reliability of the utility function. More preferable act is determined by ranking $\tilde{Z}_{U(f_i)}$ using ranking procedure given in Sect. 5.

4 Choquet Integral Based Decision Making Using Z-Information

Formally the problem is formulated as follows. Decision-making under Z-information can be considered as 4-tuple $(S, \tilde{Z}_X, \mathcal{A}, \succeq)$, where $S = \{s_1, s_2, \ldots, s_n\}$– *a space of mutually exclusive and exhaustive states of nature,* $\tilde{Z}_X$– a set of outcomes, described by Z-valuation. A is the set of actions that are functions $f : S \rightarrow \tilde{Z}_X$, $\succeq$ is the non-additive preference relation on the set of actions. In decision-making under uncertainty, a probability over S is imprecise. $\mathcal{F}_S$ is a σ—algebra of subsets B of S. Denote by $\mathcal{A}_0$ the set of all $\mathcal{F}_S$-measurable step-valued functions from S to X and denote $\mathcal{A}_C$ the constant actions in $\mathcal{A}_0$. Let A be a convex subset of X^S which includes $\mathcal{A}_C$. X can be considered as a subset of some linear space, and X^S can then be considered as a subspace of the linear space of all functions from S to the first linear space. The problem is to determine preferences among alternatives by means of a utility function.

The suggested decision-making methodology uses Choquet expected utility for description of preferences. The utility function used here is as follows

$$\tilde{Z}'_{U(f_i)} = \int_S Z'_{v(f_i(s))} dZ'_\eta \tag{12}$$

The decision making problem in this case consists in the determination of an optimal action $f^* \in A$ such that

$$\tilde{Z}'_{U(f_i^*)} = \max_{f \in A} \left\{ \int_S \tilde{Z}'_{v(f_i(s))} d\tilde{Z}'_\eta \right\} \tag{13}$$

As it was mentioned above the outcomes $\tilde{Z}_{v(f_i(s_j))} = (\tilde{v}(f_i(s_j)), \tilde{R}_1)$ and the probabilities $\tilde{Z}_{P(s_j)} = (\tilde{P}(s_j), \tilde{R}_2)$ of the states $s_j \in S$ where $\tilde{R}_1 = \{(x_2, \mu_{\tilde{R}_1}(x)) : x_2 \in [0,1]\}$ and $\tilde{R}_2 = \{(y_2, \mu_{\tilde{R}_2}(y)) : y_2 \in [0,1]\}$ are represented by trapezoidal and triangle fuzzy numbers. At first it is required to determine the unknown probability of state $s_j - \tilde{Z}_{P(s_j)}$ on a base of given probabilities $\tilde{Z}_{P(s_1)}, \tilde{Z}_{P(s_2)}, \ldots, \tilde{Z}_{P(s_{j-1})}, \ldots, \tilde{Z}_{P(s_m)}$ by formulas (6 and 7).

Given the payoff Table 2 and the complete probability distribution we can evaluate the values of Choquet integral on base of (12) [7, 8, 30, 38].

Given the complete probability distribution we construct measure as lower prevision.

The determination of a lower prevision Z'_η from linguistic probability distribution $\tilde{P}$ has a great role in the determination of the preferences in this model.

When the states of nature are just some elements, the measure is defined [1] as

$$\tilde{Z}'_{\eta_{\tilde{P}}}(H) = \bigcup_{\alpha \in (0,1]} a \cdot \left[\tilde{Z}'^{\alpha}_{\eta_{\tilde{P}_{left}}}(H), \tilde{Z}'^{\alpha}_{\eta_{\tilde{P}_{right}}}(H) \right], H \subset S = \{s_1, \ldots, s_m\} \tag{14}$$

where

$$\tilde{Z}'^{\alpha}_{\eta_{\tilde{P}}}(H) = inf \left\{ \sum_{s_j \in H} p(s_j), \ldots, p(s_m) \right\}, (p(s_1), \ldots, p(s_m)) \in P^\alpha$$

$$P^\alpha = \left\{ (p(s_1), \ldots, p(s_m)) \in P_1^\alpha \times \ldots \times P_m^\alpha | \sum_{j=1}^{m} p(s_j) = 1 \right\},$$

Here $P_1^\alpha, \ldots, P_m^\alpha$ are α-cuts of fuzzy probabilities $\tilde{P}_1, \ldots, \tilde{P}_m$, $p(s_1), \ldots, p(s_m)$ are basic probabilities for $\tilde{P}_1, \ldots, \tilde{P}_m$, $\times$ denotes the Cartesian product.

Now we can construct a fuzzy measure with triangle membership function from fuzzy set of possible probability distributions as its lower probability function (lower prevision) taking into consideration (14) and the method used in [1].

As we have an ordinary fuzzy numbers with trapezoidal and triangular membership functions then we can obtain the fuzzy values of utility function $\tilde{U}(f(s))$ for each alternative by (1b):

$$\tilde{Z}'_{U(f_i)} = \int\limits_S Z'_{v(f_i(s))} dZ'_\eta = \int\limits_S \sqrt{\alpha_1} \times \tilde{v}(f(s);1) d\tilde{Z}'_{\eta_{\tilde{P}}}$$
$$= \sum_{j=1}^{m} \left(\tilde{Z}'_{\eta_{\tilde{P}}}(B_{(j)}) - \tilde{Z}'_{\eta_{\tilde{P}}}(B_{(j-1)})\right) \tilde{Z}_{v(f_i(s))}$$
$$= \sum_{j=1}^{m} \left(\tilde{Z}'_{\eta_{\tilde{P}}}(B_{(j)}) - \tilde{Z}'_{\eta_{\tilde{P}}}(B_{(j-1)})\right)\left(\sqrt{\alpha_1} \times \tilde{v}(f_i(s);1)\right) \quad (15)$$

An optimal action $f^* \in A$ is obtained in accordance with (13) using ranking procedure given in Sect. 5.

5 Direct Computation with Z-Numbers

Let us now apply the second approach of direct computation with Z-numbers without transformation of Z-number into ordinal fuzzy number. A Z-number $(\tilde{A}, \tilde{R})$ can be interpreted as $\Pr ob(X\ is\ \tilde{A})$is $\tilde{R}$. This expresses that we do not know the true probability density over X, but have a constraint in form of a fuzzy subset $\tilde{P}$ of the space **P** of all probability densities over X. This restriction induces a fuzzy probability $\tilde{R}$. Let p be density function over X. The probability $\Pr ob(X is \tilde{A})$ (probability that $X is \tilde{A}$) is determined on the base of the definition of the probability of a fuzzy subset as

$$\Pr ob(X\ is\ \tilde{A}) = \int\limits_{-\infty}^{+\infty} \mu_A(x) p_X(x) dx.$$

Then the degree to which p satisfies the Z-valuation $\Pr ob(X\ is\ \tilde{A})$ is $\tilde{R}$ is $\mu_P(p) = \mu_R(\Pr ob(X \text{ is } \tilde{A}) \text{ is } \tilde{R})) = \mu_R(\int\limits_{-\infty}^{+\infty} \mu_A(x) p_X(x) dx)$.

Here p is taken as some a parametric distribution. The density function of a normal distribution is

$$p_X(x) = normpdf(x, m, \sigma) = \frac{1}{\sigma\sqrt{2\pi}} \exp\left(-\frac{(x-m)^2}{2\sigma^2}\right).$$

In this situation, for any m, σ we have

$$\Pr ob_{m,\sigma}(X \text{ is } \tilde{A}) = \int\limits_{-\infty}^{+\infty} \mu_A(x) p_{m,\sigma}(x) dx = \int_{-\infty}^{+\infty} \mu_A(x) \frac{1}{\sigma\sqrt{2\pi}} \exp\left(\frac{(x-m)^2}{2\sigma^2}\right) dx$$
$$= (trapmf(x, [a_1, a_2, a_3, a_4])^* normpdf(x, m, \sigma), -\inf, +\inf)$$

Then the space P of probability distributions will be the class of all normal distributions each uniquely defined by its parameters m, σ.

Let $X = (\tilde{A}_X, \tilde{R}_X)$ and $Y = (\tilde{A}_Y, \tilde{R}_Y)$ be two independent Z-numbers. Consider determination of $W = X + Y$. First, we need compute $\tilde{A}_X + \tilde{A}_Y$ using Zadeh's extension principle:

$$\mu_{(A_X+A_Y)}(w) = \sup_x(\mu_{A_X}(x) \wedge \mu_{A_Y}(w-x)), \quad \wedge = \min .$$

As the sum of random variables involves the convolution of the respective density functions we can construct $\tilde{P}_W$, the fuzzy subset of $\mathbf{P}$, associated with the random variable W. Recall that the convolution of density functions p_1 and p_2 is defined as the density function

$$p = p_1 \oplus p_2$$

such that

$$p(w) = \int_{-\infty}^{+\infty} p_1(x)p_2(w-x)dx = \int_{-\infty}^{+\infty} p_1(w-x)p_2(x)dx$$

One can then find the fuzzy subset $\tilde{P}_W$. For any $p_W \in \mathbf{P}$, one obtains

$$\mu_{P_W}(p_W) = \max_{p_U, p_V}[\mu_{P_X}(p_X) \wedge \mu_{P_Y}(p_Y)],$$

subject to

$$p_W = p_X \oplus p_Y,$$

that is,

$$p_W(w) = \int_{-\infty}^{+\infty} p_U(x)p_V(w-x)dx = \int_{-\infty}^{+\infty} p_U(w-x)p_V(x)dx.$$

Given $\mu_{P_X}(p_X) = \mu_{P_X}(m_X, \sigma_X)$ and $\mu_{P_Y}(p_Y) = \mu_{P_Y}(m_Y, \sigma_Y)$ as

$$\mu_{P_X}(m_X, \sigma_X) = \mu_{B_X}\left(\int_{-\infty}^{+\infty} \mu_{A_X}(x)\frac{1}{\sigma_X\sqrt{2\pi}}\exp\left(\frac{(x-m_X)^2}{2\sigma_X^2}\right)dx\right),$$

$$\mu_{P_Y}(m_Y, \sigma_Y) = \mu_{B_Y}\left(\int_{-\infty}^{+\infty} \mu_{A_Y}(x)\frac{1}{\sigma_Y\sqrt{2\pi}}\exp\left(\frac{(x-m_Y)^2}{2\sigma_Y^2}\right)dx\right)$$

one can define $\tilde{P}_W$ as follows

$$p_W = p_{m_X,\sigma_Y} \oplus p_{m_X,\sigma_Y},$$
$$p_W(w) = p_{m_W,\sigma_W} = normpdf[w, m_W, \sigma_W]$$
$$= (normpdf(x, m_X, \sigma_X) * normpdf(w - x, m_Y, \sigma_Y), -\inf, +\inf)$$
$$= \int_{-\infty}^{+\infty} \frac{1}{\sigma_X\sqrt{2\pi}} \exp\left(\frac{(x - m_X)^2}{2\sigma_X^2}\right) \frac{1}{\sigma_Y\sqrt{2\pi}} \exp\left(\frac{(w - x - m_Y)^2}{2\sigma_Y^2}\right) dx$$

where

$$m_W = m_X + m_Y \text{ and } \sigma_W = \sqrt{\sigma_X^2 + \sigma_Y^2},$$
$$\mu_{P_W}(p_W) = \sup(\mu_{P_X}(p_X) \wedge \mu_{P_Y}(p_Y))$$

subject to

$$p_W = p_{m_X,\sigma_Y} \oplus p_{m_Y,\sigma_Y}$$

$B_W$is found as follows.

$$\mu_{B_W}(b_W) = \sup(\mu_{\tilde{P}_W}(p_W))$$

subject to

$$b_W = \int_{-\infty}^{+\infty} p_W(w)\mu_{A_W}(w)dw$$

Let us now consider determination of $W = X \cdot Y$. $\tilde{A}_U \cdot \tilde{A}_V$ is defined by:

$$\mu_{(A_X \cdot A_Y)}(w) = \sup_x(\mu_{A_X}(x) \wedge \mu_{A_Y}(\frac{w}{x})), \quad \wedge = \min.$$

the probability density p_W associated with W is obtained as

$$p_W = p_{m_X,\sigma_X} \otimes p_{m_Y,\sigma_Y},$$

$$p_W(w) \quad = \quad p_{m_W,\sigma_W} \quad = \quad \int_{-\infty}^{+\infty} \frac{1}{\sigma_X\sqrt{2\pi}} \exp\left(\frac{(x-m_X)^2}{2\sigma_X^2}\right) \frac{1}{\sigma_Y\sqrt{2\pi}} \exp\left(\frac{(\frac{w}{x}-m_Y)^2}{2\sigma_Y^2}\right) dx$$

where

$$m_W = \frac{m_X m_Y}{\sigma_X \sigma_Y} + r,$$

and

$$\sigma_W = \frac{\sqrt{m_X^2\sigma_Y^2 + m_Y^2\sigma_X^2 + 2m_X m_Y \sigma_X \sigma_Y r + \sigma_X^2\sigma_Y^2 + \sigma_X^2\sigma_Y^2 r^2}}{\sigma_X\sigma_Y},$$

where r is correlation coefficient.

If X and Y are two independent random variables, then

$$m_W = \frac{m_X m_Y}{\sigma_X\sigma_Y}, \text{ and } \sigma_W = \frac{\sqrt{m_X^2\sigma_Y^2 + m_X^2\sigma_X^2 + \sigma_X^2\sigma_Y^2}}{\sigma_X\sigma_Y},$$

if take into account compatibility conditions $\sigma_X\sigma_Y = 1$.

The other steps are analogous to those of determination of $W = X + Y$.

Assume we compare two courses of action f_1 and f_2 [29]. We must select between these two based on the objective of getting the largest utilities. Assume we have information about the utilities associated with these two courses of action expressed in terms of Z-valuations. These are

$$\tilde{Z}_{U(f_1)} \text{ is } (\tilde{A}_1, \tilde{R}_1) \text{ if we select } f_1$$
$$\tilde{Z}_{U(f_2)} \text{ is } (\tilde{A}_2, \tilde{R}_2) \text{ if we select } f_2$$

A main problem is how to choose between these two alternatives, f_1 and f_2.

We have to represent the information obtained from Z-valuations on a space of probability distributions P. These two observations initiate the possibility distributions G_1 and G_2 over P. For any probability density $p_k \in P$, we have $G_i(p_k) = R_i(\int_S A_i(x)p_k(x)dx)$. We now have to select between these G_i. We obtain for each p_k its expected value $E_k = \int_S xp_k(x)dx$ and using this we can determine two possibility distributions F_1 and F_2 as

$$F_i = \bigcup_{p_k} \left\{ \frac{G_i(p_k)}{E_k} \right\}$$

The numerical value of each F_i is determined as

$$e_i = \sum_{p_k} E_k \cdot G_i(P_k)$$

and we choose the action with the largest value for e_i .

6 Applications

We consider the business problem under imprecise information described by Z-valuation. Suppose a hotel is considering the construction of an additional wing. The possibility of adding 30 (f_1), 40 (f_2) and 50 (f_3) rooms is evaluating. The success of the extension depends on a combination of local government legislation and competition in the field. There are three states of nature: positive legislation and low competition (s_1), positive legislation and strong competition (s_2), no legislation and low competition (s_3). Also we have the values anticipated payoffs (in percentage). The problem is to find how many rooms to build in order to maximize the return on investment. Z-valuation for the utilities of the each act taken at various states and probabilities on states are provided in Tables 4 and 5, respectively.

Here $\tilde{Z}_{v(f_i(s_j))} = (\tilde{v}(f_i(s_j)), \tilde{R}_1)$, where the outcomes are the trapezoidal fuzzy numbers and corresponding reliability is a triangular fuzzy number:

$\tilde{Z}_{v(f_1(s_1))} = (\tilde{v}(f_1(s_1)), \tilde{R}_1) =$ (high; likely) = [(7, 8, 9, 10;1), (0.6, 0.7, 0.8; 1)],

$\tilde{Z}_{v(f_1(s_2))} = (\tilde{v}(f_1(s_2)), \tilde{R}_1) =$ (below than high; likely) = [(6, 7, 8, 9;1), (0.6, 0.7, 0.8; 1)],

$\tilde{Z}_{v(f_1(s_3))} = (\tilde{v}(f_1(s_3)), \tilde{R}_1) =$ (medium; likely) = [(4, 5, 6, 7;1), (0.6, 0.7, 0.8; 1)],

$\tilde{Z}_{v(f_2(s_1))} = (\tilde{v}(f_2(s_1)), \tilde{R}_1) =$ (below than high; likely) = [(6, 7, 8, 9;1), (0.6, 0.7, 0.8; 1)],

$\tilde{Z}_{v(f_2(s_2))} = (\tilde{v}(f_2(s_2)), \tilde{R}_1) =$ (low; likely) = [(3, 4, 5, 6;1), (0.6, 0.7, 0.8; 1)],

$\tilde{Z}_{v(f_2(s_3))} = (\tilde{v}(f_2(s_3)), \tilde{R}_1) =$ (below than high; likely) = [(6, 7, 8, 9;1), (0.6, 0.7, 0.8; 1)],

$\tilde{Z}_{v(f_3(s_1))} = (\tilde{v}(f_3(s_1)), \tilde{R}_1) =$ (below than high; likely) = [(6, 7, 8, 9;1), (0.6, 0.7, 0.8; 1)],

$\tilde{Z}_{v(f_3(s_2))} = (\tilde{v}(f_3(s_2)), \tilde{R}_1) =$ (high; likely) = [(7, 8, 9, 10;1), (0.6, 0.7, 0.8; 1)],

$\tilde{Z}_{v(f_3(s_3))} = (\tilde{v}(f_3(s_3)), \tilde{R}_1) =$ (medium; likely) = [(4, 5, 6, 7;1), (0.6, 0.7, 0.8; 1)].

Let the probabilities for s_1 and s_2 be Z-numbers $\tilde{Z}_{P(s_j)} = (\tilde{P}(s_j)), \tilde{R}_2)$, where the probabilities and the corresponding reliability are the triangular fuzzy numbers:

$\tilde{Z}_{P(s_1)} = (\tilde{P}(s_1)), \tilde{R}_2) =$ (medium; quite sure) = [(0.25, 0.3, 0.35; 1), (0.8, 0.9, 1; 1)].

$\tilde{Z}_{P(s_2)} = (\tilde{P}(s_2)), \tilde{R}_2) =$ (more than medium; quite sure) = [(0.35, 0.4, 0.45; 1), (0.8, 0.9, 1; 1)].

In accordance with [1] we have calculated probability fors_3:

$\tilde{Z}_{P(s_3)} = (\tilde{P}(s_3)), \tilde{R}_2) =$ {low; quite sure} =[(0.2, 0.3, 0.4; 1), (0.8, 0.9, 1; 1)].

Table 4 The utility values of actions under various states

	s_1	s_2	s_3
f_1	(high; likely)	(below than high; likely)	(medium; likely)
f_2	(below than high; likely)	(low; likely)	(below than high; likely)
f_3	(below than high; likely)	(high; likely)	(medium; likely)

Table 5 The values of probabilities of states of nature

$s_1 =$ (medium; quite sure)	$s_2 =$ (more than medium; quite sure)	$s_3 =$ (low; quite sure)

Then we convert the value of fuzzy reliability into a numerical number based on (8 and 9):

$$\alpha_1 = \frac{\int x\mu_{\tilde{R}_1}(x)dx}{\int \mu_{\tilde{R}_1}(x)dx} = 0.7,$$

$$\alpha_2 = \frac{\int y\mu_{\tilde{R}_2}(y)dy}{\int \mu_{\tilde{R}_y}(y)dy} = 0.9$$

Given the complete fuzzy probability distribution $\tilde{P}(s_j), j = \overline{1,3}$, we add the weight of the reliability to the restriction and have the weighted Z-number for the outcomes and the probabilities:

$\tilde{Z}^{\alpha_1}_{v(f_1(s_1))} = (7, 8, 9, 10; 0.7), \tilde{Z}^{\alpha_1}_{v(f_1(s_2))} = (6, 7, 8, 9; 0.7), \tilde{Z}^{\alpha_1}_{v(f_1(s))} = (4, 5, 6, 7; 0.7),$
$\tilde{Z}^{\alpha_1}_{v(f_2(s_1))} = (6, 7, 8, 9; 0.7), \tilde{Z}^{\alpha_1}_{v(f_2(s_2))} = (3, 4, 5, 6; 0.7), \tilde{Z}^{\alpha_1}_{v(f_2(s))} = (6, 7, 8, 9; 0.7),$
$\tilde{Z}^{\alpha_1}_{v(f_3(s_1))} = (6, 7, 8, 9; 0.7), \tilde{Z}^{\alpha_1}_{v(f_3(s_2))} = (7, 8, 9, 10; 0.7), \tilde{Z}^{\alpha_1}_{v(f_3(s_3))} = (4, 5, 6, 7; 0.7),$
$\tilde{Z}^{\alpha_2}_{P(s_1)} = (0.25, 0.3, 0.35; 0.9), \tilde{Z}^{\alpha_2}_{P(s_2)} = (0.35, 0.4, 0.45; 0.9), \tilde{Z}^{\alpha_2}_{P(s_3)} = (0.2, 0.3, 0.4; 0.9).$

Now we convert the obtained weighted numbers to fuzzy numbers:

$$\begin{aligned}
\tilde{Z}'_{v(f_1(s_1))} &= (5.85, 6.69, 7.52, 8.36; 1),\\
\tilde{Z}'_{v(f_1(s_2))} &= (5.01, 5.85, 6.69, 7.52; 1),\\
\tilde{Z}'_{v(f_1(s_3))} &= (3.34, 4.18, 5, 01, 5, 85; 1),\\
\tilde{Z}'_{v(f_2(s_1))} &= (5.01, 5.85, 6.69, 7.52; 1),\\
\tilde{Z}'_{v(f_2(s_2))} &= (2.50, 3.34, 4.18, 5.01; 1),\\
\tilde{Z}'_{v(f_2(s_3))} &= (5.01, 5.85, 6.69, 7.52; 1),\\
\tilde{Z}'_{v(f_3(s_1))} &= (5.01, 5.85, 6.69, 7.52; 1),\\
\tilde{Z}'_{v(f_3(s_2))} &= (5.85, 6.69, 7.52, 8.36; 1),\\
\tilde{Z}'_{v(f_3(s_3))} &= 3.34, 4.18, 5.01, 5.85; 1).\\
\tilde{Z}'_{P(s_1)} &= (0.23, 0.28, 0.33; 1),\\
\tilde{Z}'_{P(s_2)} &= (0.33, 0.37, 0.42; 1),\\
\tilde{Z}'_{P(s_3)} &= (0.18, 0.28, 0.37; 1).
\end{aligned}$$

Given these data and following the proposed decision making method, we get the expected values of utility for acts f_1, f_2, f_3:

$$\begin{aligned}
\tilde{Z}'_{U(f_1)} &= \tilde{Z}'_{v(f_1(s_1))} * \tilde{Z}'_{P(s_1)} + \tilde{Z}'_{v(f_1(s_2))} * \tilde{Z}'_{P(s_2)} + \tilde{Z}'_{v(f_1(s_3))} * \tilde{Z}'_{P(s_3)}\\
&= (3.69, 4.32, 9.48, 12.42; 1),
\end{aligned}$$

$$\tilde{Z}'_{U(f_2)} = \tilde{Z}'_{v(f_2(s_1))} * \tilde{Z}'_{P(s_1)} + \tilde{Z}'_{v(f_2(s_2))} * \tilde{Z}'_{P(s_2)} + \tilde{Z}'_{v(f_2(s_3))} * \tilde{Z}_{P(s_3)}$$
$$= (2.97, 3.61, 8.49, 10.79; 1).$$
$$\tilde{Z}'_{U(f_3)} = \tilde{Z}'_{v(f_3(s_2))} * \tilde{Z}'_{P(s_1)} + \tilde{Z}'_{v(f_3(s_2))} * \tilde{Z}'_{P(s_2)} + \tilde{Z}'_{v(f_3(s_3))} * \tilde{Z}'_{P(s_3)}$$
$$= (3.77, 4.40, 10.08, 12.54; 1).$$

The value of Z-number for optimal utility function in accordance with (11) is determined as

$$\tilde{Z}_{U(f_1)} = (4.65, 5.45, 11.95, 15.65; 0.79, 0.89, 0.99),$$
$$\tilde{Z}_{U(f_2)} = (3.75, 4.55, 10.7, 13.6; 0.79, 0.89, 0.99),$$
$$\tilde{Z}_{U(f_3)} = (4.75, 5.55, 12.7, 15.8; 0.79, 0.89, 0.99).$$

As we have the Z-number valued utility functions then we can select between the preferences ranking them.

Ranking of fuzzy values of utilities gives a preference to the third alternative, i.e. $f_3 \succ f_1 \succ f_2$.

Given these data and following the proposed decision making method, we can obtain an overall utility as a fuzzy-valued Choquet integral:

$$\tilde{Z}'_{U(f_i)} = \tilde{Z}'_{\eta_{\tilde{P}}}(\{s_{(1)}\} - \tilde{Z}'_{\eta_{\tilde{P}}}\{s_{(0)}\}) * \tilde{Z}'_{v_{s(1)}(f_i(s))} + \tilde{Z}'_{\eta_{\tilde{P}}}(\{s_{(1)}, s_{(2)}\} - \tilde{Z}'_{\eta_{\tilde{P}}}\{s_{(1)}\})$$
$$* \tilde{Z}'_{v_{s(2)}(f_i(s))} + \tilde{Z}'_{\eta_{\tilde{P}}}(\{s_{(1)}, s_{(2)}, s_{(3)}\} - \tilde{Z}'_{\eta_{\tilde{P}}}\{s_{(1)}, s_{(2)}\}) * \tilde{Z}'_{v_{s(3)}(f_i(s))}$$

The states are ranked as:
For the first alternative $\tilde{Z}'_{v_{s_1}(f_1(s_1))} > \tilde{Z}'_{v_{s_2}(f_1(s_2))} > \tilde{Z}'_{v_{s_3}(f_1(s_3))}$,
For the second alternative $\tilde{Z}'_{v_{s_1}(f_1(s_1))} > \tilde{Z}'_{v_{s_3}(f_1(s_3))} > \tilde{Z}'_{v_{s_2}(f_1(s_2))}$,
For the third alternative $\tilde{Z}'_{v_{s_2}(f_1(s_2))} > \tilde{Z}'_{v_{s_1}(f_1(s_1))} > \tilde{Z}'_{v_{s_3}(f_1(s_3))}$.
The α-cuts of $\tilde{Z}'_{\tilde{\eta}_{\tilde{P}}}(\{s_1, s_2\}, \tilde{Z}'_{\tilde{\eta}_{\tilde{P}}}(\{s_1, s_3\}$ are found as the solutions of (14).
So we can determine the triangular fuzzy numbers

$$Z'_{\eta_{\tilde{P}}}(\{s_1, s_2\} = (0.62, 0.71, 0.71)$$
$$Z'_{\eta_{\tilde{P}}}(\{s_1, s_3\} = (0.57, 0.62, 0.62)$$

Given this, the values of the utility function for the alternatives are as follows:

$$\tilde{Z}'_{U(f_1)} = (-0.75, 5.61, 6.45, 11.79)$$
$$\tilde{Z}'_{U(f_2)} = (-1.99, 5.53, 6.36, 13.30)$$
$$\tilde{Z}'_{U(f_3)} = (-0.11, 6.29, 7.21, 12.66).$$

According to (11) we can determine the Z-number valued utility function for each alternative

$$\tilde{Z}_{U(f_1)} = (-0.94, 7.07, 8.13, 14.85; 0.79, 0.89, 0.99),$$
$$\tilde{Z}_{U(f_2)} = (-2.51, 6.97, 8.02; 0.79, 0.89, 0.99),$$
$$\tilde{Z}_{U(f_3)} = (-0.14, 7.93, 9.08; 0.79, 0.89, 0.99).$$

Ranking of Z-number valued utility functions gives a preference to the third alternative, i.e. $f_3 \succ f_1 \succ f_2$.

We applied the suggested in Sect. 5 approach to the same hotel management problem and get $f_3 \succ f_1 \succ f_2$.

7 Conclusion

We first analyzed the main existing decision making theories and concluded that in almost all these theories a reliability of the decision relevant information is not well taken into consideration. We then recalled the concept of Z-numbers introduced by Zadeh and showed how we can use a Z-valuation to make decisions. We investigated two approaches to decision making with Z-information. The first approach is based on reducing of Z-numbers to classical fuzzy numbers, and generalization of expected utility approach and use of Choquet integral with an integrant represented by Z-numbers. A fuzzy measure is calculated on a base of a given Z-information. The second approach is based on direct computation with Z-numbers. To illustrate a validity of suggested approaches to decision making with Z-information the numerical examples were used.

References

1. Aliev, R., Pedrycz, W., Fazlollahi, B., Huseynov, O., Alizadeh, A.: Fuzzy logic-based generalized decision theory with imperfect information. Inf. Sci. **189**, 18–42 (2012)
2. Aliev, R.A., Pedrycz, W., Huseynov, O.H., Zeinalova, L.M.: Decision making with second order information granules. In: Granular Computing and Intelligent Systems, pp. 327–374. Springer, Berlin (2011)
3. Aliev, R.A., Aliev, R.R.: Soft Computing and Its Applications. World Scientific Press, New Jersey, London, Singapore, Hong Kong (2001)
4. Aliev, R.A., Fazlollahi, B., Aliev, R.R.: Soft Computing and its Applications in Business and Economics. Springer, Berlin (2004)
5. Aliev, R.A., Jafarov, S.M., Gardashova, L.A., Zeinalova, L.M.: Principles of Decision Making and Control Under Uncertainty. Nargiz, Baku (1999)
6. Alo, R., Korvin, A., Modave, F.: Using Fuzzy Functions to Select an Optimal Action in Decision Theory. NAFIPS- FLINT, pp. 348–353 (2002)
7. Kang, B., Wei, D., Li, Y., Deng, Y.: Decision Making Using Z-numbers under Uncertain Environment. J. Inf. Comput. Sci. **8**(7), 2807–2814 (2012)
8. Kang, B., Wei, D., Li, Y., Deng, Y.: A method of converting Z-number to classical fuzzy number. J. Inf. Comput. Sci. **9**(3), 703–709 (2012)
9. Borisov, A.N., Alekseyev, A.V., Merkuryeva, G.V., Slyadz, N.N., Gluschkov, V.I.: Fuzzy information processing in decision making systems. Radio i Svyaz, Moscow (in Russian) (1989)

10. Choquet, G.: Theory of capacities. Annales de l'Institut Fourier **5**, 131–295 (1953)
11. Cooman, G.: Precision-imprecision equivalence in a broad class of imprecise hierarchical uncertainty models. J. Stat. Plan. Infer. **105**(1), 175–198 (2000)
12. Cooman, G., Troffaes, M., Miranda, E.: N-monotone lower previsions and lower integrals. J. Intell. Fuzzy Syst.: Appl. Eng. Technol. **16**(4), 253–263 (2005)
13. De Finetti, B.: Theory of Probability, vol. 1. John Wiley and Sons, New York (1974)
14. Enea, M., Piazza, T.: Project selection by constrained fuzzy AHP. Fuzzy Optim. Decis. Making **3**(1), 39–62 (2004)
15. Ghirardato, P., Maccheroni, F., Marinacci, M.: Differentiating ambiguity and ambiguity attitude. J. Econ. Theory **118**, 133–173 (2004)
16. Gil, M.A., Jain, P.: Comparison of experiments in statistical decision problems with fuzzy utilities. IEEE Trans. Syst. Man Cybern. **22**(4), 662–670 (1992)
17. Gilboa, I., Schmeidler, D.: Maximin Expected utility with a non-unique prior. J. Math. Econ. **18**, 141–153 (1989)
18. Kahneman, D., Tversky, A.: Prospect theory: an analysis of decision under uncertainty. Econometrica **47**, 263–291 (1979)
19. Klibanoff, P., Marinacci, M., Mukerji, S.: A smooth model of decision making under ambiguity. Econometrica **73**(6), 1849–1892 (2005)
20. Levi, I.: On indeterminate probabilities. J. Philos. **71**, 391–418 (1974)
21. Lotfi, A., Zadeh, : A note on a Z-number. Inf. Sci. **181**, 2923–2932 (2011)
22. Lu, J., Zhang, G., Ruan, D., Wu, F.: Multi-objective group decision making. In: Methods, Software and Applications with Fuzzy Set Techniques. Electrical and Computer Engineering Series, vol. 6, Imperial College Press, London (2007)
23. Mathieu Nicot, B.: Fuzzy expected utility. Fuzzy Sets Syst. **20**(2), 163–173 (1986)
24. Miranda, E., Zaffalon, M.: Conditional models: coherence and inference through sequences of joint mass functions. J. Stat. Plan. Infer. **140**(7), 1805–1833 (2010)
25. Modave, F., Iourinski, D.: Axiomatization of qualitative multicriteria decision making with the sugeno integral. In: Advances in Soft Computing. Intelligent Systems Design and Applications, Ed. Springer, pp. 77–86, May 2003
26. Nguyen, H.T., Walker, E.A.: A first course in Fuzzy Logic. CRC Press, Boca Raton (2000)
27. Anand, P., Pattanaik, P., Puppe, C. (eds.): The Handbook of Rational and Social Choice. Oxford Scholarship Online, Oxford University, Oxford (2009)
28. Ramsey, F.P.: Truth and probability. In: Braithwaite, R.B., Plumpton, F. (eds.) The Foundations of Mathematics and Other Logical Essays. K Paul, Trench, Truber and Co, London (1931)
29. Yager, Ronald R.: On Z-valuations using Zadeh's Z-numbers. Int. J. Intell. Syst. **27**, 259–278 (2012)
30. Schmeidler, D.: Subjective probability and expected utility without additivity. Econometrica **57**(3), 571–588 (1989)
31. Schmitt, E., Bombardier, V., Wendling, L.: Improving fuzzy rule classifier by extracting suitable features from capacities with respect to the choquet integral. IEEE Trans. Syst. Man Cybern. **38**(5),1195-1206 (2008)
32. Savage, L.J.: The Foundations of Statistics. Wiley, New York (1954)
33. Tversky, A., Kahneman, D.: Advances in prospect theory: cumulative representation of uncertainty. J. Risk Uncertainty **5**(4), 297–323 (1992)
34. Von Neumann, J., Morgenstern, O.: Theory of Games and Economic Behaviour. Princeton University Press, Princeton (1944)
35. Walley, P.: Statistical inferences based on a second-order possibility distribution. Int. J. Gen Syst. **9**, 337–383 (1997)
36. Walley, P.: Measures of Uncertainty in Expert Systems. Artif. Intell. **83**, 1–58 (1996)
37. Wang, Y.M., Elhag, T.M.S.: Fuzzy TOPSIS method based on alpha level sets with an application to bridge risk assessment. Expert Syst. Appl. **31**(2), 309–319 (2006)
38. Yager, R.R.: Uncertainty modeling and decision support. Reliab. Eng. Syst. Saf. **85**, 341–354 (2004)

39. Yang, R., Wang, Z., Heng, P-A.: Fuzzified Choquet integral with a fuzzy valued integrand and its application on temperature prediction. IEEE Trans. Syst. Man Cybern. Part B **38**(2), (2008)
40. Wang, Z., Yan, J.: Choquet integral and its applications: a survey. Academy of Mathematics and Systems Science, CAS, China, Aug 2006
41. Zhang, W.: Fuzzy number-valued fuzzy measure and fuzzy number-valued fuzzy integral on the fuzzy set. Fuzzy Sets Syst. **49**, 357–376 (1992)

Approximations of One-dimensional Expected Utility Integral of Alternatives Described with Linearly-Interpolated *p*-Boxes

N. D. Nikolova, S. Ivanova and K. Tenekedjiev

Abstract In the process of quantitative decision making, the bounded rationality of real individuals leads to elicitation of interval estimates of probabilities and utilities. This fact is in contrast to some of the axioms of rational choice, hence the decision analysis under bounded rationality is called fuzzy-rational decision analysis. Fuzzy-rationality in probabilities leads to the construction of x-ribbon and p-ribbon distribution functions. This interpretation of uncertainty prohibits the application of expected utility unless ribbon functions were approximated by classical ones. This task is handled using decision criteria Q under strict uncertainty—Wald, maximax, Hurwicz$_{\alpha}$, Laplace—which are based on the pessimism-optimism attitude of the decision maker. This chapter discusses the case when the ribbon functions are linearly interpolated on the elicited interval nodes. Then the approximation of those functions using a Q criterion is put into algorithms. It is demonstrated how the approximation is linked to the rationale of each Q criterion, which in three of the cases is linked to the utilities of the prizes. The numerical example demonstrates the ideas of each Q criterion in the approximation of ribbon functions and in calculating the Q-expected utility of the lottery.

Keywords Quantitative decision making · Rational choice · Interval estimates · Bounded rationality · Fuzzy-rationality · Decision criteria under strict uncertainty · Expected utility

N. D. Nikolova (✉) · S. Ivanova · K. Tenekedjiev
Nikola Vaptsarov Naval Academy, 73 V. Drumev Str, 9026 Varna , Bulgaria
e-mail: natalianik@gmail.com

S. Ivanova
e-mail: sneji_di@abv.bg

K. Tenekedjiev
e-mail: Kiril.Tenekedjiev@fulbrightmail.org

P. Guo and W. Pedrycz (eds.), *Human-Centric Decision-Making Models for Social Sciences*, Studies in Computational Intelligence 502,
DOI: 10.1007/978-3-642-39307-5_11,

1 Introduction

The main objective of classical quantitative decision analysis is to find the balance between the preferences of the decision maker (DM), her risk preferences, and the uncertainties she faces in a particular decision situation. Utility theory offers *lotteries* as a model of uncertain alternatives [1]. They consist of a full set of disjoint events (states) each associated with a holistic consequence for the DM, called a *prize*. Lotteries are categorized according to the cardinality of the set of lotteries L and the set of prizes X. Ordinary lotteries apply to discrete sets of prizes. If the prize is a random variable that is probabilistically described by a distribution function $F(.)$, then generalized lotteries of either I, II or III type apply [2]. In problems with generalized lotteries of I type, in particular, the set of lotteries is discrete, whereas the set of prizes is continuous.

The uncertainty in the lottery model is measured by (usually subjectively elicited) *probabilities*, whereas the preferences over the prizes are measured by (always subjectively elicited) *utilities*. The works [3–5] argued that the choice of an action of the rational individual must be related to her subjective belief. Then her subjective probabilities might be deducted from her preference over the alternatives. For that reason they considered as useless the idea to divide decision theory in two parts—one that refers to the value system of the DM, and another that refers to the subjective description of uncertainty. Savage [4] introduced axioms as sufficient conditions for existence of both a utility function over the consequences, and a probability function over the events. De Finetti [6], De Groot [7], and Villigas [8] on the other hand, considered that even though it was possible to separate the description of uncertainty from the model of DM's value system, the subjective probability is related to the willingness to enter bets, even if though hypothetical. A detailed overview to the issues of axiomatization in quantitative decision analysis and a comparison of views may be found in [9].

In case the rationality of preference was ensured (via s certain axiom set), it is possible to construct the utility function $u(.)$ over the prizes, such that the more preferred the consequence the higher the utility. The seminal work [10] offers a thorough insight into the process of construction of utility functions for the case of one-dimensional (1-D) and multi-dimensional prizes. The main paradigm of utility theory is that all kind of lotteries should be ranked in descending order of their *expected utilities*, which is the utility of prizes, weighted by their probabilities [1].

In the ideal risky situation, the DM assigns unique probability measures to the chance of receiving a prize (this is the classical risky lottery) [11]. However, real DMs can only define subjective probabilities in an interval form. As discussed in [12], if all necessary probabilities and numerical characteristics of probability distributions that would allow the complete description of uncertainty in a decision problem is only known to belong to a given multi-dimensional *credal set M*, then those probabilities are called *imprecise, indefinite, interval, confidence*, etc. Empirical evidence and studies show that individuals would rather define interval than precise estimates (of utilities and probabilities) [13]. In any case, the elicitation of interval probabilities is a widely discussed issue. Approaches to handle it based on pair-wise comparisons of events using linear and quadratic programming, on modeling lower and upper probabilities via heuristical processes, or via multinormal data to obey invariance principles have been discussed [14–16]. Still, a great deal of details are yet to be investigated in the area of interval probabilities.

As claimed in [17], in the context of imprecise probabilities, the DM feels more comfortable with lotteries than with events. Knetsch [18] offers empirical proof in favor of the suggestion that the certainty equivalent of a lottery (the price of the lottery) is a wide interval, whose lower bound is the maximal reasonable price one would buy the lottery for, whereas its upper bound is the lowest price one would sell it. For that reason, the DM identifies the certainty equivalent's lower bounds of several simple lotteries (not in L), called *lower previsions* [19]. The general assumption is either that the DM is risk neutral [12] or that the utility function over the certainty equivalents has been constructed and each "lottery–lower prevision" pair sets a linear constraints to the set M, which in this case turns out to be convex. The lower prevision of a lottery's derivative is taken for the *upper prevision* of that same lottery [20]. The work [21] introduces a complex z-dimensional optimization procedure to construct an imprecise CDF (cumulative distribution function) based on z imprecise nodes. Those nodes are subjectively elicited in a dialog with the DM [22]. The set M may in any case be defined and constructed using an arbitrary method.

As a result of interval probabilities, utility theory assumptions are disobeyed, and partially non-transitive preferences are observed. For that reason, in [23], real DMs are referred to as fuzzy-rational—FRDM. Then the alternatives are represented as *fuzzy-rational lotteries* where the chance of receiving each prize is quantified by interval probability measures [22].

Since fuzzy-rational DMs only partially quantify uncertainty, then ranking fuzzy-rational lotteries is a problem of mixed type, and generalizes decisions under risk and under strict uncertainty. Here, it is necessary to choose the method Q under strict uncertainty to be applied, such as Wald, maximax,(Hurwicz$_\alpha$, Savage and Laplace criteria (see [24–27] regarding Q criteria under strict uncertainty). However, none of the Q criteria obeys the minimal rationality requirements of choice [28]. Yet they offer an approach to approximating fuzzy-rational lotteries by classical risky ones, which can then be ranked according to the expected utility criterion. Then the Q-expected utility criterion to rank fuzzy-rational lotteries is defined. These procedures benefit from the existing mathematical homology between the descriptions of the triples "event from a probability field–interval subjective probability–point estimate probability" and "object from an universe–degree of membership to an intuitionistic fuzzy set—degree of membership to a (classical) fuzzy set" [29] (see [30] regarding description of intuitionistic fuzzy sets). That allows transforming interval probabilities into point estimates using the operators that transform an intuitionistic fuzzy degree of membership into classical fuzzy degree of membership (see [31, 32] for further reading on intuitionistic fuzzy degree of membership).

This chapter focuses on the case of 1-D GL-I, where the underlying quantity X takes values in the interval $[x_{min}; x_{max}]$, the uncertainty is described by a 1-D CDF, $F(.)$, and the utility over the values of X is measured by $u(.)$. The expected utility of each uncertain prospect may then be calculated by the Stieltjes integral $E(u|F) = \int_{x_{min}}^{x_{max}} u(x)dFx$. However, a common way to construct $F(.)$ is via linear interpolation on subjectively interval-elicited nodes. The CDF constructed on interval-elicited nodes is called *ribbon CDF* [22]. It complies with the concept of p-boxes [33]. Depending on the parameter with an interval estimate, there are *x-ribbon* CDF and *p-ribbon* CDF. This chapter offers several approximations of the expected utility integral in the described setup by transforming the interval-elicited nodes according to certain assumptions. Since the task of approximating ribbon functions is one under strict uncertainty, those assumptions happen to be the pessimism and optimism of the FRDM. The general criterion Q will be the Hurwicz$_\alpha$ criterion (for $\alpha \in [0; 1]$), which is a pessimistic-optimistic decision rule. Special cases of the Hurwicz$_\alpha$ criterion are the Wald (pessimistic criterion with $\alpha = 1$) and maximax (optimistic criterion with $\alpha = 0$ criteria. Another criterion that is to be discussed is the Laplace criterion, which assumes uniform distribution of values within the uncertainty interval of the node. Of specific interest are the cases when the utility function is non-monotonic, which challenges the application of the Q methods, as some of them are substantially linked to the utilities of the prizes. Algorithms to solve the approximation task under those criteria for the case of x-ribbon and p-ribbon functions will be outlined. An example will demonstrate the importance of decision modeling and detailed analysis of available subjective information for the final decision.

In what follows, Sect. 2 offers a discussion on the ways to construct probability distributions based on interval estimates. In Sect. 3, the general case of ranking fuzzy-rational GL-I is presented. Two different interpretations are given in Sects. 4 and 5 depending on the type of ribbon function that describes the uncertainty in the lotteries. Section 6 offers an economically-oriented example, where the discussed procedures are applied.

2 Constructing Distributions on Interval Probabilities

Assume that a decision problem encapsulates several uncertain prospects, associated with a given 1-D quantity X (random variable), whose realizations $x \in (-\infty; +\infty)$. In an ideal case, the uncertainty in X would be entirely measured by a known 1-D *classical CDF*, $F(.)$:

$$F(x) = \text{CDF}(x) = P(X \leq x), \forall x \in (-\infty; +\infty) \quad (1)$$

In a fuzzy-rational setup, the uncertainty in X is partially quantified by a 1-D *ribbon CDF*—$F^R(.)$—that entirely lies between two 1-D classical CDF, called *lower* and *upper distributional bounds* $F^d(.)$ and $F^u(.)$[22]:

$$F^d(x) \leq F^R(x) \leq F^u(x), \forall x \in (-\infty; +\infty). \quad (2)$$

$$F^d(x) \leq F^u(x). \quad (3)$$

It often happens so that $F^R(.)$ is interpolated or approximated on nodes that are intervally elicited on the prize (i.e. the uncertainty intervals are on the quantile, that is the abscissa x). Then $F^R(.)$ is called *x-ribbon CDF*—$F^{xR}(.)$— with *lower* and *upper x-bounds* $F^{xd}(.)$ and $F^{xu}(.)$. This representation is in accordance with the concept of p-boxes [33]. Assume there are $z > 1$ number of elicited quantiles of $F^{xR}(.)$, which obey the following conditions:

$$\{(x_l^d; x_l^u; F_l) | l = 1, 2, \ldots, z\}, \tag{4}$$

$$\left| \begin{array}{l} x_1^d \leq x_2^d \leq \ldots \leq x_z^d, \\ x_1^u \leq x_2^u \leq \ldots \leq x_z^u, \\ x_l^d \leq x_l^u, l = 2, 3, \ldots, z-1, \\ x_1^d = x_1^u, x_z^d = x_z^u, \\ 0 = F_1 < F_2 < \ldots < F_z = 1. \end{array} \right. \tag{5}$$

Then a convenient way to assign the x-bounds is by linear interpolation on the lower and upper ends of the uncertainty intervals of those nodes, such that for all $x \in (-\infty; +\infty)$:

$$F^{xd}(x) = \begin{cases} 0, & \text{for } x < x_1^d, \\ F_l, & \text{for } x_l^d = x < x_{l+1}^d, l = 1, 2, \ldots, z-1, \\ F_l + \frac{(x - x_l^d)(F_{l+1} - F)}{x_{l+1}^d - x_l^d}, & \text{for } x_l^d < x < x_{l+1}^d, l = 1, 2, \ldots, z-1, \\ 1, & \text{for } x_z^d \leq x; \end{cases} \tag{6}$$

$$F^{xu}(x) = \begin{cases} 0, & \text{for } x < x_1^u, \\ F_l, & \text{for } x_l^u = x < x_{l+1}^u, l = 1, 2, \ldots, z-1, \\ F_l + \frac{(x - x_l^u)(F_{l+1} - F)}{x_{l+1}^d - x_l^d}, & \text{for } x_l^u < x < x_{l+1}^u, l = 1, 2, \ldots, z-1, \\ 1, & \text{for } x_z^u \leq x; \end{cases} \tag{7}$$

$$F^{xd}(x) \leq F^{xR}(x) \leq F^{xu}(x). \tag{8}$$

In may also happen so that $F^R(.)$ is interpolated or approximated on nodes that are intervally elicited on the probability value (i.e. whose uncertainty interval is on the quantile index, that is the ordinate p). Then $F^R(.)$ is called *p-ribbon CDF*—$F^{pR}(.)$—with *lower* and $F^{pd}(.)$ and $F^{pu}(.)$. Assume there are $z > 1$ number of elicited quantile indices of $F^{pR}(.)$, which obey the following conditions:

$$\{(x_l; F_l^d; F_l^u) | l = 1, 2, \ldots, z\}, \tag{9}$$

$$\left| \begin{array}{l} x_1 < x_2 < \ldots < x_z, \\ 0 = F_1^d \leq F_2^d \leq \ldots \leq F_z^d = 1, \\ 0 = F_1^u \leq F_2^u \leq \ldots \leq F_z^u = 1, \\ F_l^d \leq F_l^u, l = 2, 3, \ldots, z-1. \end{array} \right. \tag{10}$$

Then a convenient way to assign the p-bounds is by linear interpolation on the lower and upper ends of the uncertainty intervals of those nodes, such that for all $x \in (-\infty; +\infty)$:

$$F^{pd}(x) = \begin{cases} 0 & \text{for } x < x_1, \\ F_l^d & \text{for } x_l = x < x_{l+1}, \quad l = 1, 2, \ldots, z-1, \\ F_l^d + \frac{(x-x_l)(F_{l+1}^d - F_l^d)}{x_{l+1}-x_l} & \text{for } x_l < x < x_{l+1}, \quad l = 1, 2, \ldots, z-1, \\ 1 & \text{for } x_z \leq x; \end{cases} \tag{11}$$

$$F^{pu}(x) = \begin{cases} 0 & \text{for } x < x_1, \\ F_l^u & \text{for } x_l = x < x_{l+1}, \quad l = 1, 2, \ldots, z-1, \\ F_l^u + \frac{(x-x_l)(F_{l+1}^u - F_l^u)}{x_{l+1}-x_l} & \text{for } x_l < x < x_{l+1}, \quad l = 1, 2, \ldots, z-1, \\ 1 & \text{for } x_z \leq x; \end{cases} \tag{12}$$

$$F^{pd}(x) \leq F^{pR}(x) \leq F^{pu}(x). \tag{13}$$

3 Ranking 1-D Fuzzy-Rational GL-I: The General Case

Assume there are q alternatives that generate prizes from a piece-wise continuous set X according to continuous or mixed probability laws. Such alternatives may be represented as GL-I [5]. A 1-D GL-I with a 1-D ribbon CDF is referred to as a *1-D fuzzy-rational GL-I* [22] and takes the form:

$$g_i^{fr} = \langle F_i^R(x); x \rangle, i = 1, 2, \ldots, q. \tag{14}$$

Here, $F_i^R(.)$ has lower and upper distributional bounds $F_i^d(.)$ and $F_i^u(.)$. Ranking g_i^{fr} according to expected utility is then brought down to the following two stages:

(1) Using a Q criterion under strict uncertainty, each $F_i^R(.)$ is approximated by a 1-D classical CDF $F_i^Q(.)$, such that for all $x \in (-\infty; +\infty)$

$$F_i^d(x) \leq F_i^Q(x) \leq F_i^u(x), i = 1, 2, \ldots, q. \tag{15}$$

In that way each g_i^{fr} is approximated by a 1-D classical-risky GL-I, called *Q-generalized* (*1-D Q-GL-I*):

$$g_i^Q = \langle F_i^Q(x); x \rangle. \tag{16}$$

(2) The alternatives are ranked in descending order of the expected utilities of g_i^Q, which is calculated as a Stieltjes integral with an integrating function $F_i^Q(.)$:

$$E_i^Q(u|F_i^R) = \int_{-\infty}^{+\infty} u(x)dF_i^Q(x). \tag{17}$$

These three steps represent the *Q-expected utility* criterion to rank 1-D fuzzy-rational GL-I. A summary of decision criteria Q under strict uncertainty is offered in the *Appendix* to the chapter. For some of the Q criteria, the approximation of $F_i^R(.)$ by $F_i^Q(.)$ relies strongly on the 1-D utility function $u(.)$.

4 Ranking 1-D x-Fuzzy-Rational GL-I

If the uncertainty in a 1-D fuzzy-rational GL-I is described by an x-ribbon CDF, then the *1-D x-fuzzy-rational GL-I* takes the form:

$$g_i^{xfr} = \langle F_i^{xR}(x); x\rangle, i = 1, 2, \ldots, q. \tag{18}$$

The following steps to calculate the Q-expected utility of g_i^{xfr} are needed:

(1) Using a Q criterion under strict uncertainty, $F_i^{xR}(.)$ is piece-wise linearly approximated by a 1-D classical CDF $F_i^{xQ}(.)$ as in (21), with nodes that obey the conditions (19)–(20):

$$\left\{(x_l^{Q,(i)}; F_l^{(i)})|l = 1, 2, \ldots, z_i\right\}, \tag{19}$$

$$\left| \begin{array}{l} x_1^{Q,(i)} \leq x_2^{Q,(i)} \leq \ldots \leq x_{z_i}^{Q,(i)}, \\ x_l^{d,(i)} \leq x_l^{Q,(i)} \leq x_l^{u,(i)}, l = 2, 3, \ldots, z_i - 1, \\ x_1^{Q,(i)} = x_1^{d,(i)} = x_1^{u,(i)}, x_{zi}^{Q,(i)} = x_{zi}^{d,(i)} = x_{zi}^{u,(i)}. \end{array} \right. \tag{20}$$

$$F_i^{xQ}(x) = \begin{cases} 0 & \text{for } x < x_1^{Q,(i)}, \\ F_l^{(i)} & \text{for } x_l^{Q,(i)} = x < x_{l+1}^{Q,(i)}, l = 1, 2, \ldots, z_i - 1, \\ F_l^{(i)} + \frac{(x - x_l^{Q,(i)})(F_{l+1}^{(i)} - F_l^{(i)})}{x_{l+1}^{Q,(i)} - x_l^{Q,(i)}} & \text{for } x_l^{Q,(i)} < x < x_{l+1}^{Q,(i)}, l = 1, 2, \ldots, z_i - 1, \\ 1 & \text{for } x_z^{Q,(i)} \leq x. \end{cases} \tag{21}$$

In that way, g_i^{xfr} is approximated by a 1-D classical risky GL-I, called *xQ-generalized (1-D xQ-GL-I)*,

$$g_i^{xQ} = \langle F_i^{xQ}(x); x\rangle. \tag{22}$$

(2) The Q-expected utility of g_i^{xfr} is the expected utility of g_i^{xQ}, calculated using Riemann integral under the first summation symbol:

$$E_i^{xQ}(u|F_i^{xR}) = \int_{x_1^{Q,(i)}}^{x_{z_i}^{Q,(i)}} u(x)dF_i^{xQ}(x)$$

$$= \sum_{\substack{l=1 \\ x_{l+1}^{Q,(i)} > x_l^{Q,(i)}}}^{z_i-1} \frac{F_{l+1}^{(i)}-F_l^{(i)}}{x_{l+1}^{Q,(i)}-x_l^{Q,(i)}} \int_{x_l^{Q,(i)}}^{x_{l+1}^{Q,(i)}} u(x)dx + \sum_{\substack{l=1 \\ x_{l+1}^{Q,(i)} = x_l^{Q,(i)}}}^{z_i-1} (F_{l+1}^{(i)} - F_l^{(i)})u(x_l^{Q,(i)}). \tag{23}$$

This procedure is called the *xQ-expected utility.*

The calculation of the xQ-expected utility of the ith fuzzy-rational GL-I is brought down to the estimation of the inner quantiles $x_l^{Q,(i)}$, $l = 2, 3, \ldots, z_i - 1$, of the classical CDF in g_i^{xQ}. Task 1 generalizes this problem.

Task 1: Calculating the xQ-expected utility of a fuzzy-rational GL-I
Given:

- criterion under strict uncertainty Q ;
- 1-D utility function $u(.)$;
- number of approximating nodes $z_i > 1$;
- quantile indices $F_l^{(i)}, l = 1, 2, \ldots, z_i$, such that

$$0 = F_1^{(i)} \le F_2^{(i)} \le \ldots \le F_{z_i-1}^{(i)} \le F_{z_i}^{(i)} = 1; \tag{24}$$

- lower quantile bounds $x_l^{d,(i)}, l = 1, 2, \ldots, z_i$, such that

$$x_1^{d,(i)} \le x_2^{d,(i)} \le \ldots \le x_{z_i-1}^{d,(i)} \le x_{z_i}^{d,(i)}; \tag{25}$$

- upper quantile bounds $x_l^{u,(i)}, l = 1, 2, \ldots, z_i$, such that

$$x_1^{d,(i)} = x_1^{u,(i)} \le x_2^{u,(i)} \le \ldots \le x_{z_i-1}^{u,(i)} \le x_{z_i}^{u,(i)} = x_{z_i}^{d,(i)}, \tag{26}$$

$$x_l^{d,(i)} \le x_l^{u,(i)}, l = 2, 3, \ldots, z_i - 1; \tag{27}$$

- end quantiles

$$x_1^{Q,(i)} = x_1^{d,(i)} = x_1^{u,(i)}, \tag{28}$$

$$x_{z_i}^{Q,(i)} = x_{z_i}^{d,(i)} = x_{z_i}^{u,(i)}. \tag{29}$$

Find:

- inner quantiles $x_l^{Q,(i)}$, $l = 2, 3, \ldots, z_i - 1$, such that

$$x_1^{Q,(i)} \leq x_2^{Q,(i)} \leq \ldots \leq x_{z_i-1}^{Q,(i)} \leq x_{z_i}^{Q,(i)}, \tag{30}$$

$$x_l^{d,(i)} \leq x_l^{Q,(i)} \leq x_l^{u,(i)}. \tag{31}$$

Task 2 would be given different solutions depending on the Q criterion

4.1 Approximating x-Ribbon CDF Using the Wald Criterion ($Q = W$)

The Wald decision criterion under strict uncertainty assumes that the worst outcome always occurs (see Appendix 1). Its interpretation for 1-D x-fuzzy-rational GL-I implies to choose the quantiles $x_l^{W,(i)}$, $l = 2, 3, \ldots, z_i - 1$, so that to minimize the xW-expected utility of the lottery [34]:

$$\begin{aligned} E_i^{xW}(u|F_i^{xR}) &= \int_{x_1^{W,(i)}}^{x_{z_i}^{W,(i)}} u(x)dF_i^{xW}(x) = \sum_{l=1}^{z_i-1} \int_{x_l^{W,(i)}}^{x_{l+1}^{W,(i)}} u(x)dF_i^{xW}(x) \\ &= \sum_{\substack{l=1 \\ x_{l+1}^{W,(i)} > x_l^{W,(i)}}}^{z_i-1} \frac{F_{l+1}^{(i)} - F_l^{(i)}}{x_{l+1}^{W,(i)} - x_l^{W,(i)}} \int_{x_l^{W,(i)}}^{x_{l+1}^{W,(i)}} u(x)dx + \sum_{\substack{l=1 \\ x_{l+1}^{W,(i)} = x_l^{W,(i)}}}^{z_i-1} (F_{l+1}^{(i)} - F_l^{(i)})u(x_l^{W,(i)}) \\ &= \sum_{l=1}^{z_i-1} (F_{l+1}^{(i)} - F_l^{(i)}) I_l^{xW,(i)}. \end{aligned} \tag{32}$$

$$I_l^{xW,(i)} = \begin{cases} \frac{1}{x_{l+1}^{W,(i)} - x_l^{W,(i)}} \int_{x_l^{W,(i)}}^{x_{l+1}^{W,(i)}} u(x)dx & \text{for } x_{l+1}^{W,(i)} > x_l^{W,(i)}, \\ u(x_l^{W,(i)}) & \text{for } x_{l+1}^{W,(i)} = x_l^{W,(i)}. \end{cases}, \quad l = 1, 2, 3, \ldots, z_i - 1 \tag{33}$$

The variables $I_l^{xW,(i)}$ physically coincide with the expected utilities of hypothetical 1-D classical risky GL-I $g_l^{h,xW,(i)} = \langle F_l^{h,xW,(i)}(x); x \rangle$, where the 1-D classical CDF $F_l^{h,xW,(i)}(.)$ are linearly interpolated on two nodes $(x_l^{W,(i)}; 0)$ and $(x_{l+1}^{W,(i)}; 1)$. The 1-D classical CDF $F_l^{h,xW,(i)}(.)$, the hypothetical 1-D classical-risky GL-I $g_l^{h,xW,(i)}$ and their expected utilities $I_l^{xW,(i)}$ are unknown until the quantiles $x_l^{W,(i)}$,

$l = 2, 3, \ldots, z_i - 1$ were defined, so that to obey the conditions

$$\left| \begin{array}{ll} x_l^{d,(i)} - x_l^{W,(i)} \le 0, l = 2, 3, \ldots, z_i - 1, \\ x_l^{W,(i)} - x_l^{u,(i)} \le 0, l = 2, 3, \ldots, z_i - 1, \\ x_l^{W,(i)} - x_{l+1}^{W,(i)} \le 0, l = 2, 3, \ldots, z_i - 2, \\ x_1^{d,(i)} - x_2^{W,(i)} \le 0, \\ x_{z_i-1}^{W,(i)} - x_{z_i}^{d,(i)} \le 0. \end{array} \right. \tag{34}$$

The so-defined $(z_i - 2)$-dimensional non-linear optimization task with $3z_i - 5$ linear constraints can be redefined in a task of lower dimension provided that the following properties were considered:

(a) Since the weight coefficients $F_{l+1}^{(i)} - F_l^{(i)}$ of the variables $I_l^{xW,(i)}$ in $E_i^{xW}(u|F_i^{xR})$ are known and nonnegative, then the required quantile estimates should be defined so that to minimize the quantities $I_l^{xW,(i)}$;
(b) Let all quantiles, but the lth, be fixed at a certain level, where $l \in \{2, 3, \ldots, z_i - 1\}$. Assume that for the lth quantile

$$x_l^{W,(i)} \in [max\{x_l^{d,(i)},\ x_{l-1}^{W,(i)}\}; min\{x_l^{u,(i)},\ x_{l+1}^{W,(i)}\}]. \tag{35}$$

Then the change in $x_l^{W,(i)}$ only influences $I_{l-1}^{xW,(i)}$ and $I_l^{xW,(i)}$;

(c) Let for some $l \in \{2, 3, \ldots, z_i - 1\}$ the utility $u(.)$ be:

- monotonically increasing in the interval $x \in [x_l^{d,(i)}; x_l^{u,(i)}]$;
- bounded above by $u(x_l^{d,(i)})$ in the interval $x \in [x_{l-1}^{d,(i)}; x_l^{d,(i)}]$;
- bounded below by $u(x_l^{u,(i)})$ in the interval $x \in [x_l^{u,(i)}; x_{l+1}^{u,(i)}]$.

Then $I_{l-1}^{xW,(i)}$ and $I_l^{xW,(i)}$ are monotonically increasing functions of $x_l^{W,(i)}$;

(d) Let for some $l \in \{2, 3, \ldots, z_i - 1\}$, the utility function $u(.)$ be:

- monotonically decreasing in the interval $x \in [x_l^{d,(i)}; x_l^{u,(i)}]$;
- bounded below by $u(x_l^{d,(i)})$ in the interval $x \in [x_{l-1}^{d,(i)}; x_l^{d,(i)}]$;
- bounded above by $u(x_l^{u,(i)})$ in the interval $x \in [x_l^{u,(i)}; x_{l+1}^{u,(i)}]$.

Then $I_{l-1}^{xW,(i)}$ and $I_l^{xW,(i)}$ are monotonically decreasing functions of $x_l^{W,(i)}$;

(e) Let for some $l \in \{2, 3, \ldots, z_i - 1\}$ the utility function $u(.)$ be a constant in the interval $x \in [x_{l-1}^{d,(i)}; x_{l+1}^{u,(i)}]$ then $I_{l-1}^{xW,(i)}$ and $I_l^{xW,(i)}$ do not depend on changes in $x_l^{W,(i)}$.

When reducing the dimensionality of the optimization task, it is convenient to assign the quantiles $x_l^{W,(i)}, l = 1, 2, \ldots, z_i$, to 5 disjoint sets: "*known*", "*arbitrary*", "*left prone*", "*right prone*" and "*optimizing*", according to Algorithm 1.

Algorithm 1. Classification of the quantiles of an x-ribbon CDF according to Wald

1. All quantiles are marked as *optimizing*.
2. From left to right (for $l = 2, 3, \ldots, z_i - 1$) all *optimizing* quantiles, whose lower and upper limits coincide (i.e. $x_l^{d,(i)} = x_l^{u,(i)}$), are marked as *known*. The following assignment is made: $x_l^{W,(i)} = x_l^{d,(i)}$;
3. From left to right (for $l = 2, 3, \ldots, z_i - 1$) all *optimizing* quantiles that obey property *e* (i.e. $I_{l-1}^{xW,(i)}$ and $I_l^{xW,(i)}$ do not depend on changes in $x_l^{W,(i)}$), are marked as *arbitrary*;
4. From left to right (for $l = 2, 3, \ldots, z_i - 1$) all *optimizing* quantiles that obey property *c* (i.e. $I_{l-1}^{xW,(i)}$ and $I_l^{xW,(i)}$ are monotonically increasing functions of $x_l^{W,(i)}$,) are marked as *left prone*;
5. From right to left (for $l = z_i - 1, z_i - 2, \ldots, 3, 2$), all *optimizing* quantiles that obey property *d* (i.e. $I_{l-1}^{xW,(i)}$ and $I_l^{xW,(i)}$ are monotonically decreasing functions of $x_l^{W,(i)}$,) are marked as *right prone*;
6. From left to right (for $l = 2, 3, \ldots, z_i - 1$) all *arbitrary* quantiles, whose left neighbor is *known*, *left prone* or *optimizing*, is marked as *left prone*;
7. From left to right (for $l = 2, 3, \ldots, z_i - 1$) all *arbitrary* quantiles, whose right neighbor is *known*, *right prone* or *optimizing*, are marked as *right prone*;
8. From left to right (for $l = 2, 3, \ldots, z_i - 1$) all *arbitrary* quantiles, whose left neighbor ll and right neighbor lr do not overlap (i.e. $x_{ll}^{u,(i)} \leq x_{lr}^{d,(i)}$), are marked as *left prone*, and both neighbors are marked as *known*. The following assignments are made: $x_{ll}^{W,(i)} = x_{ll}^{u,(i)}$, $x_{ll}^{d,(i)} = x_{ll}^{u,(i)}$, $x_t^{d,(i)} = \max\{x_t^{d,(i)}; x_{ll}^{W,(i)}\}$, $t = ll+1, ll+2, \ldots, z_i - 1$, $x_{lr}^{W,(i)} = x_{lr}^{d,(i)}$, $x_{lr}^{u,(i)} = x_{lr}^{d,(i)}$ and $x_t^{u,(i)} = \min\{x_t^{u,(i)}; x_{lr}^{W,(i)}\}$, $t = 2, 3, \ldots, lr - 1$;
9. From left to right (for $l = 2, 3, \ldots, z_i - 1$) all quantiles, whose lower and upper limits coincide (i.e. $x_l^{d,(i)} = x_l^{u,(i)}$), are marked as *known*. The following assignment is made: $x_l^{W,(i)} = x_l^{d,(i)}$;
10. From left to right (for $l = 2, 3, \ldots, z_i - 1$) the first quantiles from the *arbitrary* group, whose left neighbor ll and right neighbor lr overlap (i.e. $x_{ll}^{u,(i)} > x_{lr}^{d,(i)}$), are marked as *optimizing*, and the other quantiles in the group are marked as *left prone*;
11. From left to right (for $l = 2, 3, \ldots, z_i - 1$) all *right prone* quantiles, which do not overlap with their right *left prone* neighbor (i.e. $x_l^{u,(i)} \leq x_{l+1}^{d,(i)}$), are marked as *known*, together with their right neighbors. The following assignments are made: $x_l^{W,(i)} = x_l^{u,(i)}$, $x_l^{d,(i)} = x_l^{u,(i)}$, $x_{l+1}^{W,(i)} = x_{l+1}^{d,(i)}$, and $x_{l+1}^{u,(i)} = x_{l+1}^{d,(i)}$;

12. From left to right (for $l = 2, 3, \ldots, z_i - 1$) all *right prone* quantiles, which do not overlap with their right *left prone* neighbor (i.e. $x_l^{u,(i)} > x_{l+1}^{d,(i)}$), are marked as *optimizing*;
13. From left to right (for $l = 2, 3, \ldots, z_i - 1$) all *left prone* quantiles, whose left neighbor is *known* or which do not overlap with their left neighbor (i.e. $x_l^{d,(i)} \geq x_{l-1}^{u,(i)}$), are marked as *known*. The following assignments are made: $x_l^{W,(i)} = x_l^{d,(i)}$ and $x_l^{u,(i)} = x_l^{d,(i)}$;
14. From right to left (for $l = z_i - 1, z_i - 2, \ldots, 3, 2$) all *right prone* quantiles, whose right neighbor is *known*, or which do not overlap with their right neighbor (i.e. $x_l^{u,(i)} \leq x_{l+1}^{d,(i)}$), are marked as *known*. The following assignments are made: $x_l^{W,(i)} = x_l^{u,(i)}$ and $x_l^{d,(i)} = x_l^{u,(i)}$;
15. If at least one quantile has been marked as *known* in steps 13 and 14, then go to step 13;
16. If at least one *optimizing* quantile has been marked as *arbitrary*, *left prone* or *right prone* in steps from 3 to 5, then go to step 3, otherwise—the end.

As a result of Algorithm 1:

- there are no *arbitrary* quantiles;
- if there are no *optimizing* quantiles, then there are only *known* quantiles;
- if there are *optimizing* quantiles, then there are no *right prone* quantiles with right *left prone* neighbors.
- the lower and upper bounds of all *known* quantiles coincide with a fixed values;
- all quantile bounds obey the initial conditions.

Let N be the cardinality of the set of *optimizing* quantiles. If $N = 0$, then all quantiles have been found and Task 1 was solved. If $N > 0$, then the function $E_i^{xW}(u|F_i^{xW})$ may be calculated only on selected permissible values of the *optimizing* quantilies according to Algorithm 2.

Algorithm 2. Calculating the maximal expected utility in the case of x-ribbon CDF after the quantiles have been categorized

1. From left to right (for $l = 2, 3, \ldots, z_i - 1$), all *optimizing* quantiles are set to coincide with the chosen values of the *optimizing* quantiles $x_l^{W,(i)}$, such that $x_l^{d,(i)} \leq x_l^{W,(i)} \leq x_l^{u,(i)}$;
2. From left to right (for $l = 2, 3, \ldots, z_i - 1$) all *left prone* quantiles are set to coincide with what is greater between their lower bound and the left neighbor: $x_l^{W,(i)} = max\{x_l^{d,(i)}, x_{l-1}^{W,(i)}\}$;
3. From right to left (for $l = z_i - 1, z_i - 2, \ldots, 3, 2$) all *right prone* quantiles are set to coincide with what is smaller between their upper bound and the right neighbor: $x_l^{W,(i)} = min\{x_l^{u,(i)}, x_{l+1}^{W,(i)}\}$;
4. The values $I_l^{xW,(i)}$, $l = 1, 2, 3, \ldots, z_i - 1$ are calculated using the defined $x_l^{W,(i)}$;
5. $E_i^{xW}(u|F_i^{xR})$ is calculated using the $I_l^{xW,(i)}$ defined in step 4.

It is again necessary to optimize $E_i^{xW}(u|F_i^{xR}$, but the dimensionality N of this task does not exceed (and is usually lower than) $z_i - 2$. From the $(3z_i - 5)$ number of linear constraints only those that include an *optimizing* quantile are analyzed. This solves Task 1 for $Q = W$ in the general case of an arbitrary utility function.

Assume that the utility function $u(.)$ is monotonically increasing in the interval $[x_1^{d,(i)}; x_{z_i}^{u,(i)}]$, so that for $x_a \in [x_1^{d,(i)}; x_{z_i}^{u,(i)}]$ and $x_b \in [x_1^{d,(i)}; x_{z_i}^{u,(i)}]$

$$if x_a > x_b, \text{then } u(x_a) \geq u(x_b). \tag{36}$$

Then all unknown quantiles $x_l^{W,(i)}$, for $l = 2, 3, \ldots, z_i - 1$ would obey property c in the general case (i.e. $I_{l-1}^{xW,(i)}$ and $I_l^{xW,(i)}$ are monotonically increasing functions of $x_l^{W,(i)}$). Then, in order to minimize $I_l^{xW,(i)}$, all quantiles would be set to their lower limits, which are the smallest values that obey the linear constraint:

$$x_l^{W,(i)} = x_l^{d,(i)}, l = 2, 3, \ldots, z_i - 1. \tag{37}$$

This solves Task 1 for $Q = W$ in the case of monotonically increasing utility function.

If the utility function $u(.)$ is monotonically decreasing in the interval $[x_1^{d,(i)}; x_{z_i}^{u,(i)}]$, then for $x_a \in [x_1^{d,(i)}; x_{z_i}^{u,(i)}]$ and $x_b \in [x_1^{d,(i)}; x_{z_i}^{u,(i)}]$

$$if x_a > x_b, \text{then } u(x_a) \leq u(x_b). \tag{38}$$

Then all unknown quantiles $x_l^{W,(i)}$, for $l = 2, 3, \ldots, z_i - 1$ would obey property d in the general case (i.e. $I_{l-1}^{xW,(i)}$ and $I_l^{xW,(i)}$ are monotonically decreasing functions of $x_l^{W,(i)}$). Then, in order to minimize $I_l^{xW,(i)}$, all quantiles would be set to their upper limits, which are the greatest values that obey the linear constraints:

$$x_l^{W,(i)} = x_l^{u,(i)}, l = 2, 3, \ldots, z_i - 1. \tag{39}$$

This solves Task 1 for $Q = W$ in the case of a monotonically decreasing utility function.

4.2 *Approximating x-Ribbon CDF Using the Maximax Criterion* ($Q = \neg W$)

The rationale behind the maximax criterion is opposite to that of the Wald criterion and the required quantiles $x_l^{\neg W,(i)}$, $l = 2, 3, \ldots, z_i - 1$ (i.e. the solution of Task 1) may be identified using the algorithms from Sect. 4.1 using the following substitution:

$$u(x) = -u(x), \forall x \in (-\infty; +\infty). \tag{40}$$

4.3 Approximating x-Ribbon CDF Using the Hurwicz$_\alpha$ Criterion $(Q = H)_\alpha$

The Hurwicz$_\alpha$ decision criterion under strict uncertainty assumes that the choice of an alternative should be guided by a numerical pessimism index $\alpha \in [0;\ 1]$ that is a weighted sum of the worst and the best outcome one can get from that alternative, and which measured the pessimism of the DM [25]. The application of that concept for 1-D x-fuzzy-rational GL-I (i.e. the solution of Task 1) implies to choose the quantiles $x_l^{H_\alpha,(i)}$, $l = 2, 3, \ldots, z_i - 1$ as weighted measures of the quantiles $x_l^{W,(i)}$ and $x_l^{\neg W,(i)}$:

$$x_l^{H_\alpha,(i)} = \alpha\, x_l^{W,(i)} + (1-\alpha)\, x_l^{\neg W,(i)}. \tag{41}$$

4.4 Approximating x-Ribbon CDF Using the Laplace Criterion $(Q = L)$

The approximation of a 1-D x-ribbon CDF using the Laplace criterion were initially described in [22]. The values of the required quantiles $x_l^{L,(i)}$, $l = 2, 3, \ldots, z_i - 1$, do not depend on the utility function. According to Laplace's insufficient reasoning principle, if no information is available for the quantiles (i.e. all quantiles are with maximum width, that is $x_l^{d,(i)} = x_1^{d,(i)} = x_1^{u,(i)}$, $x_l^{u,(i)} = x_{z_i}^{d,(i)} = x_{z_i}^{u,(i)}$, $l = 2, 3, \ldots, z_i - 1$), then the distribution must be uniform in the interval $[x_1^{d,(i)};\ x_{z_i}^{d,(i)}]$. Let the quantile with the $F_l^{(i)}$ index of this uniform distribution be called *quantile of complete ignorance* $x_l^{aL,(i)}$:

$$x_l^{aL,(i)} = x_1^{d,(i)} + (x_{z_i}^{d,(i)} - x_1^{d,(i)}) F_l^{(i)}, l = 2, 3, \ldots, z_i - 1 \tag{42}$$

Let $h_l^{x,(i)}$ be the homothety of the maximum uncertainty interval under strict uncertainty of the lth quantile $[x_1^{d,(i)};\ x_{z_i}^{d,(i)}]$ into the actual uncertainty interval $[x_l^{d,(i)};\ x_l^{u,(i)}]$ of that same quantile. Then according to [22] the required quantile $x_l^{L,(i)}$ will be the image of the quantile of complete ignorance $x_l^{aL,(i)}$ at the homothety $h_l^{x,(i)}$, which solves Task 1 for $Q = L$:

$$x_l^{L,(i)} = x_l^{d,(i)} + (x_l^{u,(i)} - x_l^{d,(i)}) \frac{x_l^{aL,(i)} - x_1^{d,(i)}}{x_{z_i}^{d,(i)} - x_1^{d,(i)}} = x_l^{d,(i)} + (x_l^{u,(i)} - x_l^{d,(i)}) F_l^{(i)}. \tag{43}$$

5 Ranking 1-D *p*-Fuzzy-Rational GL-I

If the uncertainty in a 1-D fuzzy-rational GL-I is described by an *p*-ribbon CDF, then the *1-D p- fuzzy-rational GL-I* takes the form

$$g_i^{pfr} = \langle F_i^{pR}(x); x\rangle, i = 1, 2, \ldots, q. \tag{44}$$

The following steps to calculate the *Q*-expected utility of g_i^{pfr} may be applied:

(1) Using a *Q* criterion under strict uncertainty, $F_i^{pR}(.)$ is piece-wise linearly approximated by a 1-D classical CDF $F_i^{pQ}(.)$ according to (47) with nodes as in (45) and (46):

$$\{(x_l^{(i)}; F_l^{Q,(i)})|l = 1, 2, \ldots, z_i\}, \tag{45}$$

$$\left|\begin{array}{l} 0 = F_1^{Q,(i)} \leq F_2^{Q,(i)} \leq \ldots \leq F_{z_i}^{Q,(i)} = 1, \\ F_l^{d,(i)} \leq F_l^{Q,(i)} \leq F_l^{u,(i)}, l = 2, 3, \ldots, z_i - 1. \end{array}\right. \tag{46}$$

$$F_i^{pR}(x) = \begin{cases} 0 & \text{for } x < x_1^{(i)}, \\ F_l^{Q,(i)} & \text{for } x_l^{(i)} = x < x_{l+1}^{(i)},\ l = 1,\ 2,\ \ldots,\ z_i - 1, \\ F_l^{Q,(i)} + \frac{(x - x_l^{(i)})(F_{l+1}^{Q,(i)} - F_l^{Q,(i)})}{x_{l+1}^{(i)} - x_l^{(i)}} & \text{for } x_l^{(i)} < x < x_{l+1}^{(i)},\ l = 1,\ 2,\ \ldots,\ z_i - 1, \\ 1 & \text{for } x_{z_i}^{(i)} \leq x. \end{cases} \tag{47}$$

In that way, g_i^{pfr} is approximated by a 1-D classical risky GL-I, called *pQ-generalized (1-D pQ-GL-I)*,

$$g_i^{pQ} = \langle F_i^{pQ}(x); x\rangle. \tag{48}$$

(2) The *Q*-expected utility of g_i^{pfr} is calculated as the expected utility of g_i^{pQ}, using Riemann integral under the first summation symbol:

$$\begin{aligned} E_i^{pQ}(u|F_i^{pR}) &= \int_{x_1^{(i)}}^{x_{z_i}^{(i)}} u(x) dF_i^{pQ}(x) \\ &= \sum_{\substack{l=1 \\ x_{l+1} > x_l}}^{z_i - 1} \frac{F_{l+1}^{Q,(i)} - F_l^{Q,(i)}}{x_{l+1}^{(i)} - x_l^{(i)}} \int_{x_l^{(i)}}^{x_{l+1}^{(i)}} u(x) dx + \sum_{\substack{l=1 \\ x_{l+1} = x_l^{(i)}}}^{z_i - 1} (F_{l+1}^{Q,(i)} - F_l^{Q,(i)}) u(x_l^{(i)}). \end{aligned} \tag{49}$$

This procedure is called *pQ-expected utility*. The calculation of the pQ-expected utility of the ith fuzzy-rational GL-I is brought down to the estimation of the inner quantile indices $F_l^{Q,(i)}, l = 2, 3, \ldots, z_i - 1$, of the classical CDF in the pQ-GL-I g_i^{pQ}. Task 2 generalizes this problem.

Task 2

Given:

- criterion under strict uncertainty Q;
- 1-D utility function $u(.)$;
- number of approximating nodes $z_i > 1$;
- quantiles $x_l^{(i)}, l = 1, 2, \ldots, z_i$, such that

$$x_1^{(i)} \leq x_2^{(i)} \leq \ldots \leq x_{z_i}^{(i)}; \tag{50}$$

- lower quantile index bounds $F_l^{d,(i)}, l = 1, 2, \ldots, z_i$, such that

$$0 = F_1^{d,(i)} \leq F_2^{d,(i)} \leq \ldots \leq F_{z_i-1}^{d,(i)} \leq F_{z_i}^{d,(i)} = 1; \tag{51}$$

- upper quantile index bounds $F_l^{u,(i)}, l = 1, 2, \ldots, z_i$, such that

$$0 = F_1^{u,(i)} \leq F_2^{u,(i)} \leq \ldots \leq F_{z_i-1}^{u,(i)} \leq F_{z_i}^{u,(i)} = 1; \tag{52}$$

- end quantile indices

$$F_1^{Q,(i)} = F_1^{d,(i)} = F_1^{u,(i)} = 0, \tag{53}$$

$$F_{z_i}^{Q,(i)} = F_{z_i}^{d,(i)} = F_{z_i}^{u,(i)} = 1. \tag{54}$$

Find:

- inner quantile indices $F_l^{Q,(i)}, l = 2, 3, \ldots, z_i - 1$, such that

$$F_2^{Q,(i)} \leq F_3^{Q,(i)} \leq \ldots \leq F_{z_i-2}^{Q,(i)} \leq F_{z_i-1}^{Q,(i)}, \tag{55}$$

$$F_l^{d,(i)} \leq F_l^{Q,(i)} \leq F_l^{u,(i)}. \tag{56}$$

Task 2 will be given different solutions depending on the Q criterion.

5.1 Approximating p-Ribbon CDF Using the Wald Criterion ($Q = W$)

The application of the Wald decision criterion under strict uncertainty in the case of a 1-D p-fuzzy-rational GL-I implies to choose the quantile indices $F_l^{W,(i)}, l = 2, 3, \ldots, z_i - 1$, so that to minimize the pW-expected utility of the lottery:

$$
\begin{aligned}
E_i^{pW}(u|F_i^{pR}) &= \int\limits_{x_1^{(i)}}^{x_{z_i}^{(i)}} u(x)dF_i^{pW}(x) = \sum_{l=1}^{z_i-1} \int\limits_{x_l^{(i)}}^{x_{l+1}^{(i)}} u(x)dF_i^{pW}(x) \\
&= \sum_{\substack{l=1 \\ x_{l+1}^{(i)} > x_l^{(i)}}}^{z_i-1} \frac{F_{l+1}^{W,(i)} - F_l^{W,(i)}}{x_{l+1}^{(i)} - x_l^{(i)}} \int\limits_{x_l^{(i)}}^{x_{l+1}^{(i)}} u(x)dx + \sum_{\substack{l=1 \\ x_{l+1}^{(i)} = x_l^{(i)}}}^{z_i-1} (F_{l+1}^{W,(i)} - F_l^{W,(i)})u(x_l^{(i)}) \\
&= I_{z_i-1}^{p,(i)} + \sum_{l=2}^{z_i-1} F_l^{W,(i)}(I_{l-1}^{p,(i)} - I_l^{p,(i)}).
\end{aligned}
\tag{57}
$$

$$
I_l^{p,(i)} = \begin{cases} \dfrac{1}{x_{l+1}^{(i)} - x_l^{(i)}} \displaystyle\int\limits_{x_l^{(i)}}^{x_{l+1}^{(i)}} u(x)dx & \text{for } x_{l+1}^{(i)} > x_l^{(i)}, \\ u(x_l^{(i)}) & \text{for } x_{l+1}^{(i)} = x_l^{(i)}. \end{cases} \quad , l = 1, 2, 3, \ldots, z_i - 1
\tag{58}
$$

The variables $I_l^{p,(i)}$ physically coincide with the expected utilities of hypothetical 1-D classical risky GL-I $g_l^{h,p,(i)} = \langle F_l^{h,p,(i)}(x); x \rangle$, where the 1-D classical CDF $F_l^{h,p,(i)}(.)$ are linearly interpolated on two nodes $(x_l^{(i)}; 0)$ and $(x_{l+1}^{(i)}; 1)$. That is why $I_l^{p,(i)}$ will be referred to as *auxiliary expected utilities*.

After calculating $I_l^{p,(i)}, l = 1, 2, 3, \ldots, z_i - 1$, the required quantile indices $F_l^{W,(i)}, l = 2, 3, \ldots, z_i - 1$, may be identified by solving the following linear programming task: *Minimize the linear function* $\sum_{l=2}^{z_i-1} F_l^{W,(i)}(I_{l-1}^{p,(i)} - I_l^{p,(i)})$ *on* $F_2^{W,(i)}, F_3^{W,(i)}, \ldots, F_{z_i-1}^{W,(i)}$ *provided the following* $3z_i - 5$ *linear constraints hold*:

$$
\left|
\begin{array}{l}
F_l^{d,(i)} - F_l^{W,(i)} \le 0, l = 2, 3, \ldots, z_i - 1, \\
F_l^{W,(i)} - F_l^{u,(i)} \le 0, l = 2, 3, ..., z_i - 1, \\
F_l^{W,(i)} - F_{l+1}^{W,(i)} \le 0, l = 2, 3, ..., z_i - 2, \\
-F_2^{W,(i)} \le 0, \\
F_{z-1}^{W,(i)} - 1 \le 0.
\end{array}
\right.
\tag{59}
$$

A solution of that task may be found by the MATLAB function *linprog* that applies twice the Dantzig-Orden-Wolfe generalized simplex method [35]. That solves Task 2 for $Q = W$ in the general case of an arbitrary utility function.

In a special case (e.g. when the utility function $u(.)$ is monotonically increasing in the interval $[x_1^{(i)}; x_{z_i}^{(i)}]$ so that (36) applies for $x_a \in [x_1^{(i)}; x_{z_i}^{(i)}]$ and, $x_b \in [x_1^{(i)}; x_{z_i}^{(i)}]$) the following might hold for g_i^{pfr}

$$I_{l-1}^{p,(i)} \leq I_l^{p,(i)}, l = 2, 3, \ldots, z_i - 1. \tag{60}$$

Since the coefficients of the linear function $\sum_{l=2}^{z_i-1} F_l^{W,(i)}(I_{l-1}^{p,(i)} - I_l^{p,(i)})$ are entirely non-positive, then the minimum would be identified only for the greatest values of the unknown variables that obey the linear constraints, i.e.

$$F_l^{W,(i)} = F_l^{u,(i)}, l = 2, 3, \ldots, z_i-1. \tag{61}$$

That solves Task 2 for $Q = W$ in the case of increasing auxiliary expected utilities.

In another special case (e.g. when $u(.)$ is monotonically decreasing in the interval $[x_1^{(i)}; x_{z_i}^{(i)}]$ so that (38) applies for $x_a \in [x_1^{(i)}; x_{z_i}^{(i)}]$ and, $x_b \in [x_1^{(i)}; x_{z_i}^{(i)}]$) the following might hold for g_i^{pfr}

$$I_{l-1}^{p,(i)} \geq I_l^{p,(i)}, l = 2, 3, \ldots, z_i - 1. \tag{62}$$

Since the coefficients of the linear function $\sum_{l=2}^{z_i-1} F_l^{W,(i)}(I_{l-1}^{p,(i)} - I_l^{p,(i)})$ are entirely nonnegative, then the minimum would be identified only for the smallest values of the unknown variables that obey the linear constraints, i.e.

$$F_l^{W,(i)} = F_l^{d,(i)}, l = 2, 3, \ldots, z_i - 1. \tag{63}$$

That solves Task 2 for $Q = W$ in the case of decreasing auxiliary expected utilities.

5.2 Approximating p-Ribbon CDF Using the Maximax Criterion ($Q = \neg W$)

The required quantile indices $F_l^{\neg W,(i)}, l = 2, 3, \ldots, z_i - 1$, may be identified using the algorithms from Sect. 6.1 with the substitution (64), which solves Task 2 for $Q = \neg W$:

$$u(x) = -u(x), \forall x \in (-\infty; +\infty) \tag{64}$$

5.3 Approximating p-Ribbon CDF Using the Hurwicz$_\alpha$ Criterion ($Q = H_\alpha$)

The application of the Hurwicz$_\alpha$ decision criterion under strict uncertainty in the case of a 1-D p-fuzzy-rational GL-I implies to choose the quantile indices $F_l^{H_\alpha,(i)}, l = 2, 3, \ldots, z_i - 1$ as weighted measures of the quantile indices $F_l^{W,(i)}$ and $F_l^{\neg W,(i)}$, which solves Task 2 for $Q = H_\alpha$:

$$F_l^{H_\alpha,(i)} = \alpha F_l^{W,(i)} + (1-\alpha) F_l^{\neg W,(i)}, l = 2, 3, \ldots, z_i - 1. \tag{65}$$

5.4 Approximating p-Ribbon CDF Using the Laplace Criterion ($Q = L$)

The approximation of the 1-D p-ribbon CDF using Laplace criterion were initially described in [22]. The values of the required quantile indices $F_l^{L,(i)}, l = 2, 3, \ldots, z_i - 1$, do not depend on the utility function. According to Laplace's insufficient reasoning principle, if no information is available for the quantile indices (i.e. $F_l^{d,(i)} = 0$, $F_l^{u,(i)} = 1, l = 2, 3, \ldots, z_i - 1$), then the distribution must be uniform in the interval $[x_1^{(i)}; x_{z_i}^{(i)}]$. Let the quantile index of the quantile $x_l^{(i)}$ of this uniform distribution be called *quantile index of the complete ignorance* and be denoted as $F_l^{aL,(i)}$:

$$F_l^{aL,(i)} = \frac{x_l^{(i)} - x_1^{(i)}}{x_{z_i}^{(i)} - x_1^{(i)}}, l = 2, 3, \ldots, z_i - 1. \tag{66}$$

Let $h_l^{p,(i)}$ be the homothety of the maximal uncertainty interval under strict uncertainty of the lth quantile index [0; 1] into the actual uncertainty interval $[F_l^{d,(i)}; F_l^{u,(i)}]$ of that same quantile index. Then according to [22] the required quantile index $F_l^{L,(i)}$ will be the image of the quantile index of complete ignorance $F_l^{aL,(i)}$ at the homothety $h_l^{p,(i)}$, which solves Task 2 for $Q = L$:

$$F_l^{L,(i)} = F_l^{d,(i)} + (F_l^{u,(i)} - F_l^{d,(i)}) F_l^{aL,(i)} = F_l^{d,(i)} + (F_l^{u,(i)} - F_l^{d,(i)}) \frac{x_l^{(i)} - x_1^{(i)}}{x_{z_i}^{(i)} - x_1^{(i)}}. \tag{67}$$

6 Examples of Approximating Ribbon CDF Under Non-monotonic Utility

6.1 Approximating x-Ribbon CDFs

6.1.1 Setup

Consider a 1-D random variable X that represents the profit rate of a small enterprise in a competitive business area. It takes values (in thousand US dollars) in the interval [30; 42]. A 1-D x-ribbon CDF $F^{xR}(.)$ is constructed over the values of X. It is assigned by 9 inner nodes elicited by the FRDM, which obey (5): $(x_1^d = x_1^u = 30; F_1 = 0)$, $(x_2^d = 31; x_2^u = 32; F_2 = 0.1)$, $(x_3^d = 32; x_3^u = 33; F_3 = 0.2)$, $(x_4^d = 32.8; x_4^u = 33.5; F_4 = 0.3)$, $(x_5^d = 33; x_5^u = 34.5; F_5 = 0.4)$, $(x_6^d = 34; x_6^u = 36; F_6 = 0.5)$, $(x_7^d = 35.5; x_7^u = 37; F_7 = 0.6)$, $(x_8^d = 36; x_8^u = 37.5; F_8 = 0.7)$, $(x_9^d = 37; x_9^u = 40; F_9 = 0.8)$, $(x_{10}^d = 40; x_{10}^u = 41; F_{10} = 0.9)$, $(x_{11}^d = x_{11}^u = 42; F_{11} = 1)$. The so-constructed x-ribbon $F^{xR}(.)$ defines the x-fuzzy-rational GL-I $g^{xfr} = \langle F^{xR}(x); x\rangle$.

Assume, for the sake of the example, that the FRDM has non-monotonic preferences over the possible profit rates, resulting from the competitive profile of the business sector. In such a setup, a small company with a rather low profit rate would not manage to gain a sufficient market share. An eventual increase of the profit over a certain threshold, on the other hand, would put the enterprise on the spotlight of larger companies, i.e. of severe competition and eventual take-overs. The non-monotonic utility function in the interval [30; 42] has been linearly interpolation on eleven elicited inner nodes as follows: $u(30) = 0$, $u(31) = 0.06$, $u(32) = 0.09$, $u(33) = 0.15$, $u(34) = 0.3$, $u(35) = 0.55$, $u(36) = 0.7$, $u(37) = 0.6$, $u(38) = 0.4$, $u(39) = 0.2$, $u(40) = 0.15$, $u(41) = 0.1$, $u(42) = 0$ (see third graphics from below on Fig. 1).

The task is to approximate $F^{xR}(x)$ using the Wald, maximax, Hurwicz$_\alpha$, and Laplace criteria and then calculate the $L-$, $W-$, $\neg W-$ and the H_α-expected utility of g^{xfr}.

6.1.2 Approximation of $F^{xR}(x)$ Using the Wald Criterion

From (28) and (29) it follows that $x_1^W = 30$, $x_{11}^W = 42$. After applying Algorithm 1, six quantiles are identified as *known*: $x_2^W = 31$, $x_3^W = 32$, $x_4^W = 32.8$, $x_5^W = 33$, $x_6^W = 34$, and $x_{10}^W = 41$. The other three quantiles are *optimizing* and their estimates should be in the intervals $35.5 \le x_7^W \le 37$, $36 \le x_8^W \le 37.5$, $37 \le x_9^W \le 40$.

For example, according to Algorithm 2, the permissible combination $x_7^W = 36.5$, $x_8^W = 37$, $x_9^W = 39$ generates a xW-expected utility of the lottery of $E^{xW}(u|F^{xP})$=0.2413. The minimal possible value of the xW-expected utility $E^{xW}(u|F^{xP})$ is possible under the following values of the *optimizing* quantiles:

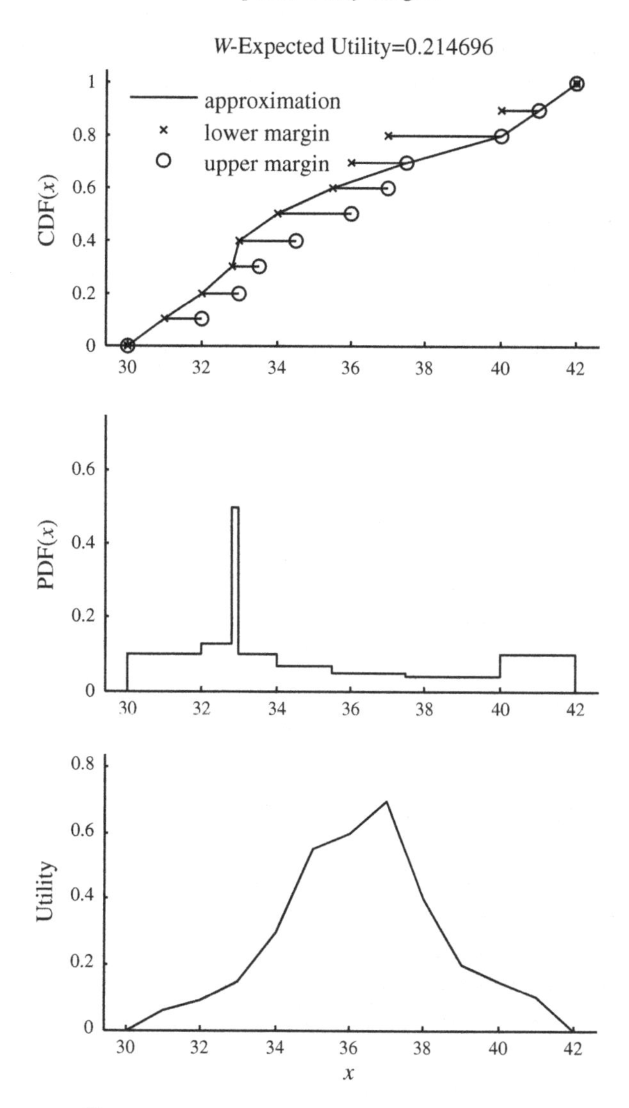

Fig. 1 Graphics of $F^{xW}(x)$, density (PDF) and utility function $u(x)$ of the FRDM over the values of X in the interval $[30; 42]$

$x_7^W = 35.5$, $x_8^W = 37.5$, $x_9^W = 40$. The optimization is performed using three-dimensional scanning method with a step of 1/10 of the interval's width.

Therefore, $F^{xR}(.)$ is approximated by $F^{xW}(.)$ on the nodes $(x_1^W = 30; F_1 = 0)$, $(x_2^W = 31; F_2 = 0.1)$, $(x_3^W = 32; F_3 = 0.2)$, $(x_4^W = 32.8; F_4 = 0.3)$, $(x_5^W = 33; F_5 = 0.4)$, $(x_6^W = 34; F_6 = 0.5)$, $(x_7^W = 35.5; F_7 = 0.6)$, $(x_8^W = 37.5; F_8 = 0.7)$, $(x_9^W = 40; F_9 = 0.8)$, $(x_{10}^W = 41; F_{10} = 0.9)$, $(x_{11}^W = 42; F_{11} = 1)$. Then g^{xfr} is approximated by the 1-D xW-GL-I $g^{xW} = \langle F^{xW}(x); x\rangle$. Following (23), the expected utility of the xW-GL-I is $E_i^{xW}(u|F_i^{xR}) = 0.215$, which is also the xW-expected utility of g^{xfr}. The graphics of $F^{xW}(x)$ and its density (probability density function—PDF) are depicted in Fig. 1.

6.1.3 Approximation of $F^{xR}(x)$ Using the Maximax Criterion

From (28) and (29) it follows that $x_1^{\neg W} = 30$, $x_{11}^{\neg W} = 42$. After applying Algorithm 1 with the substitution (40), two quantiles are identified as *known*: $x_2^{\neg W} = 32$, $x_{10}^{\neg W} = 40$. Four of the other quantiles are marked as *right prone* and their estimates should be in the intervals $32 \leq x_3^{\neg W} \leq 33$, $32.8 \leq x_4^{\neg W} \leq 33.5$, $33 \leq x_5^{\neg W} \leq 34.5$, $34 \leq x_6^{\neg W} \leq 36$. The other three quantiles are marked as *optimizing* and their estimates should be in the intervals $35.5 \leq x_7^{\neg W} \leq 37$, $36 \leq x_8^{\neg W} \leq 37.5$, $37 \leq x_9^{\neg W} \leq 40$.

For example, at the permissible combination $x_7^{\neg W} = 36$, $x_8^{\neg W} = 36.5$, $x_9^{\neg W} = 38$, the *right prone* quantiles will be set at $x_3^{\neg W} = 33$, $x_4^{\neg W} = 33.5$, $x_5^{\neg W} = 34.5$, and $x_6^{\neg W} = 36$. According to Algorithm 2 and (40), the $x\neg W$-expected utility of the lottery is $E^{x\neg W}(u|F^{xP}) = 0.336$. The maximal possible value of the $x\neg W$-expected utility $E^{x\neg W}(u|F^{xP})$ is possible under the following values of the *optimizing* quantiles: $x_7^{\neg W} = 37$, $x_8^{\neg W} = 37$, $x_9^{\neg W} = 37$. At those values, the estimates of the *right prone* quantiles are again $x_3^{\neg W} = 33$, $x_4^{\neg W} = 33.5$, $x_5^{\neg W} = 34.5$, $x_6^{\neg W} = 36$. The optimization was performed using three-dimensional scanning method with a step of 1/10 of the interval's width.

Therefore, $F^{xR}(.)$ is approximated by $F^{x\neg W}(.)$ on the nodes ($x_1^{\neg W} = 30$; $F_1 = 0$), ($x_2^{\neg W} = 32$; $F_2 = 0.1$), ($x_3^{\neg W} = 33$; $F_3 = 0.2$), ($x_4^{\neg W} = 33.5$; $F_4 = 0.3$), ($x_5^{\neg W} = 34.5$; $F_5 = 0.4$), ($x_6^{\neg W} = 36$; $F_6 = 0.5$), ($x_7^{\neg W} = 37$; $F_7 = 0.6$), ($x_8^{\neg W} = 37$; $F_8 = 0.7$), ($x_9^{\neg W} = 37$; $F_9 = 0.8$), ($x_{10}^{\neg W} = 40$; $F_{10} = 0.9$), ($x_{11}^{\neg W} = 42$; $F_{11} = 1$). Then g^{xfr} is approximated by the 1-D $x\neg$W-GL-I $g^{x\neg W} = \langle F^{x\neg W}(x);\ x\rangle$. Following (23), the expected utility of $x\neg$W-GL-I is $E_i^{x\neg W}(u|F_i^{xR}) = 0.370$, which is also the $x\neg W$-expected utility of g^{xfr}. The graphics of $F^{x\neg W}(.)$ and its density are depicted in Fig. 2.

6.1.4 Approximation of $F^{xR}(x)$ Using the Hurwicz$_\alpha$ Criterion

Let $\alpha = 0.7$. From (28) and (29) it follows that $x_1^{H_{0.7}} = 30$, $x_{11}^{H_{0.7}} = 42$. According to (41), the required quantiles depend on x_l^W and $x_l^{\neg W}$, already calculated above. Here $x_2^{H_{0.7}} = 0.7\,x_2^W + (1-0.7)\,x_2^{\neg W,(i)} = 0.7 \times 31 + 0.3 \times 32 = 31.3$, and in the same fashion $x_3^{H_{0.7}} = 32.3$, $x_4^{H_{0.7}} = 33.01$, $x_5^{H_{0.7}} = 33.45$, $x_6^{H_{0.7}} = 34.6$, $x_7^{H_{0.7}} = 35.95$, $x_8^{H_{0.7}} = 37.35$, $x_9^{H_{0.7}} = 39.1$, $x_{10}^{H_{0.7}} = 40.7$.

Therefore, $F^{xR}(.)$ is approximated by $F^{xH_{0.7}}(.)$ on the nodes ($x_1^{H_{0.7}} = 30$; $F_1 = 0$), ($x_2^{H_{0.7}} = 31.3$; $F_2 = 0.1$), ($x_3^{H_{0.7}} = 32.3$; $F_3 = 0.2$), ($x_4^{H_{0.7}} = 33.01$; $F_4 =$

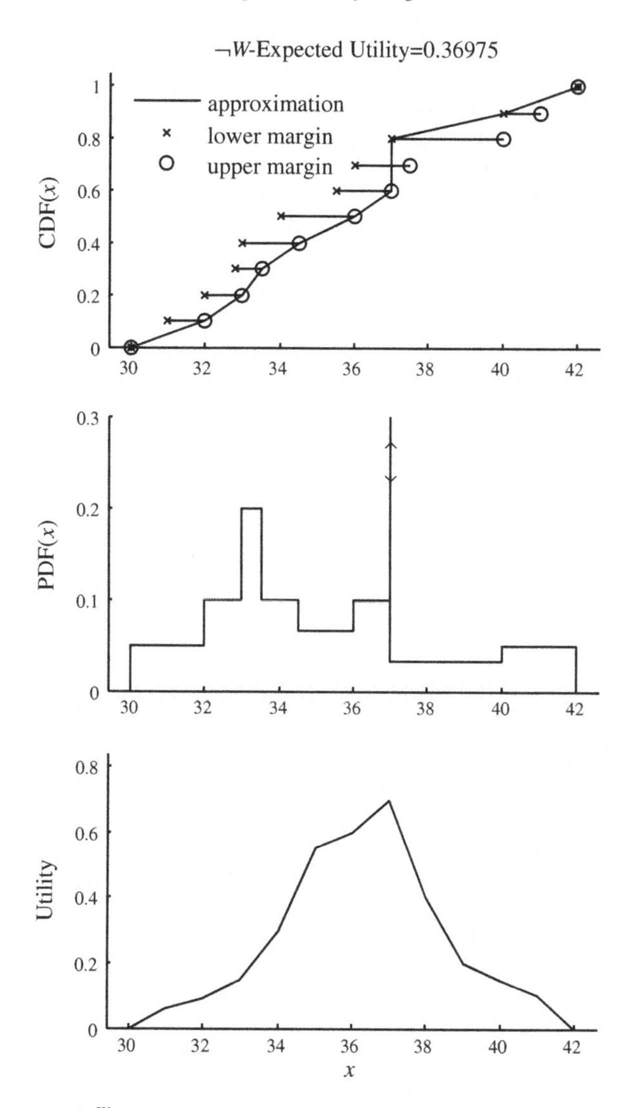

Fig. 2 Graphics of $F^{x \neg W}(.)$, density (PDF) and utility function $u(x)$ of the FRDM over the values of X in the interval [30; 42]

0.3), $(x_5^{H_{0.7}} = 33.45;\ F_5 = 0.4)$, $(x_6^{H_{0.7}} = 34.6;\ F_6 = 0.5)$, $(x_7^{H_{0.7}} = 35.95;\ F_7 = 0.6)$, $(x_8^{H_{0.7}} = 37.35;\ F_8 = 0.7)$, $(x_9^{H_{0.7}} = 39.1;\ F_9 = 0.8)$, $(x_{10}^{H_{0.7}} = 40.7;\ F_{10} = 0.9)$, $(x_{11}^{H_{0.7}} = 42;\ F_{11} = 1)$. Then g^{xfr} is approximated by the 1-D $xH_{0.7} -$ GL-I $g^{xH_{0.7}} = \langle F^{xH_{0.7}}x); x\rangle$. Following (23), the expected utility of $xH_{0.7} -$ GL-I is $E_i^{xH_{0.7}}(u|F_i^{xR}) = 0.2542$, which is also the $xH_{0.7}$-expected utility of g^{xfr}. The graphics of $F^{xH_{0.7}}(.)$ and its density are depicted in Fig. 3.

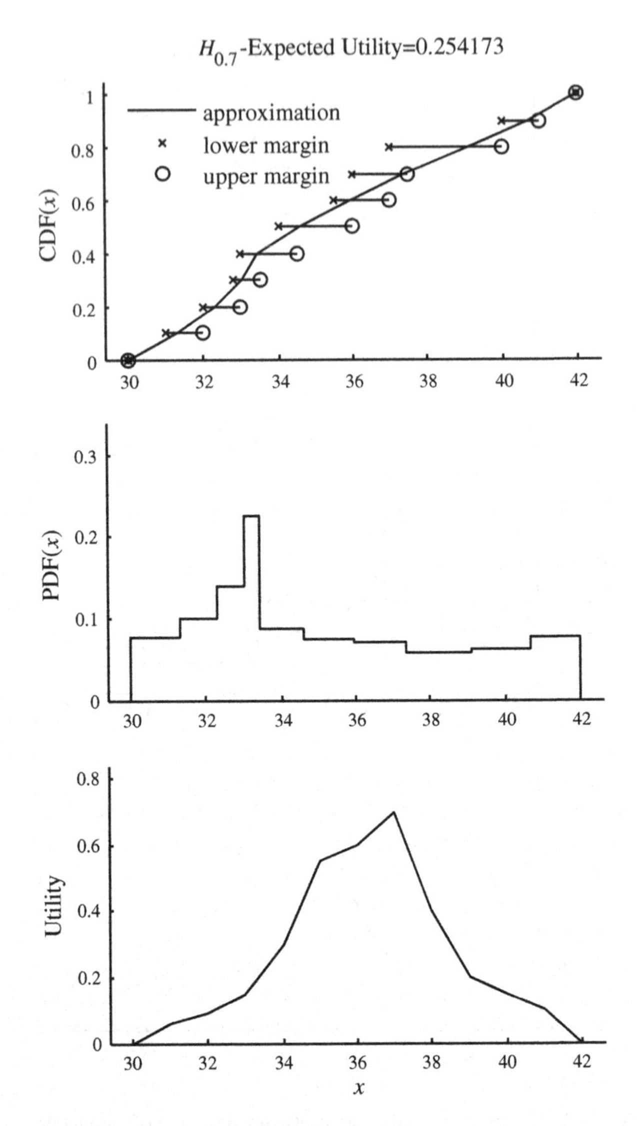

Fig. 3 Graphics of $F^{xH_{0.7}}(.)$, density (PDF) and utility function $u(x)$ of the FRDM over the values of X in the interval [30; 42]

6.1.5 Approximation of $F^{x^R}(x)$ Using the Laplace Criterion

From (28) and (29) it follows that $x_1^L = 30, x_{11}^L = 42$. According to (43) the required quantiles are: $x_2^L = x_2^d + (x_2^u - x_2^d)F_2 = 31 + (32 - 31) \times 0.1 = 31.1$, and in the same fashion $x_3^L = 32.2, x_4^L = 33.01, x_5^L = 33.6, x_6^L = 35, x_7^L = 36.4, x_8^L = 37.05, x_9^L = 39.4, x_{10}^L = 40.9$.

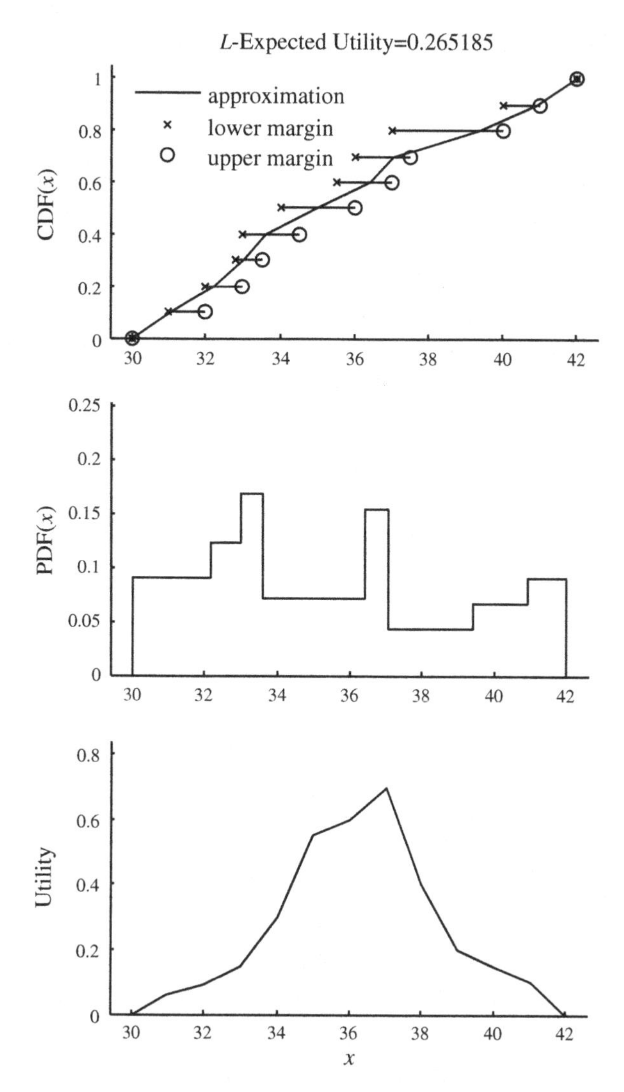

Fig. 4 Graphics of $F^{xL}(x)$, density (PDF) and utility function $u(x)$ of the FRDM over the values of X in the interval [30; 42]

Therefore $F^{xR}(.)$ is approximated by $F^{xL}(.)$ on the nodes ($x_1^L = 30$; $F_1 = 0$), ($x_2^L = 31.1$; $F_2 = 0.1$), ($x_3^L = 32.2$; $F_3 = 0.2$), ($x_4^L = 33.01$; $F_4 = 0.3$), ($x_5^L = 33.6$; $F_5 = 0.4$), ($x_6^L = 35$; $F_6 = 0.5$), ($x_7^L = 36.4$; $F_7 = 0.6$), ($x_8^L = 37.05$; $F_8 = 0.7$), ($x_9^L = 39.4$; $F_9 = 0.8$), ($x_{10}^L = 40.9$; $F_{10} = 0.9$), ($x_{11}^L = 42$; $F_{11} = 1$). Then g^{xfr} is approximated by the 1-D *xL*-GL-I $g^{xL} = \langle F^{xL}(x); x \rangle$. Following (23), the expected utility of the *xL*-GL-I is $E_i^{xL}(u|F_i^{xR}) = 0.265$, which is also the *xL*-expected utility of g^{xfr}. The graphics of $F^{xL}(x)$ and its density are depicted in Fig. 4.

6.2 Approximating p-Ribbon CDFs

6.2.1 Setup

Consider the same 1-D random variable X as in Sect. 5.2.1. A 1-D p-ribbon CDF $F^{pR}(.)$ is constructed over the values of X. It is assigned by 7 inner nodes elicited by the FRDM, which obey (10): ($x_1 = 30$; $F_1^d = F_1^u = 0$), ($x_2 = 31.5$; $F_2^d = 0.07$, $F_2^u = 0.13$), ($x_3 = 33$; $F_3^d = 0.23$, $F_3^u = 0.31$), ($x_4 = 34.5$; $F_4^d = 0.41$, $F_4^u = 0.51$), ($x_5 = 36$; $F_5^d = 0.51$, $F_5^u = 0.65$), ($x_6 = 37.5$; $F_6^d = 0.68$, $F_6^u = 0.80$), ($x_7 = 39$; $F_7^d = 0.76$, $F_7^u = 0.88$), ($x_8 = 40.5$; $F_8^d = 0.85$, $F_8^u = 0.95$), ($x_9 = 42$; $F_9^d = F_9^u = 1$). The so-defined p-ribbon $F^{pR}(.)$ defines the p-fuzzy-rational GL-I $g^{pfr} = \langle F^{pR}(x); x\rangle$.

The utility function is the same as the one from Sect. 6.1.1 (see the third graph in Fig. 5).

The task is to approximate $F^{pR}(x)$ using the Wald, maximax, the Hurwicz$_\alpha$, and the Laplace, criteria and then calculate the L-, W-, $\neg W$-, and H_α-expected utility of g^{pfr}.

6.2.2 Approximation of $F^{pR}(x)$ Using the Wald Criterion

From (53) and (54) it follows that $F_1^W = 0$, $F_9^W = 1$. The linear programming task, defined in Sect. 5.1 with constraints in (59) is solved using *linprog*. It generates the following results: $F_2^W = 0.13$, $F_3^W = 0.31$, $F_4^W = 0.51$, $F_5^W = 0.65$, $F_6^W = 0.68$, $F_7^W = 0.76$, $F_8^W = 0.85$.

Therefore, $F^{pR}(.)$ is approximated by $F^{pW}(.)$ on the nodes ($x_1 = 30$; $F_1^W = 0$), ($x_2 = 31.5$; $F_2^W = 0.13$), ($x_3 = 33$; $F_3^W = 0.31$), ($x_4 = 34.5$; $F_4^W = 0.51$), ($x_5 = 36$; $F_5^W = 0.65$), ($x_6 = 37.5$; $F_6^W = 0.68$), ($x_7 = 39$; $F_7^W = 0.76$), ($x_8 = 40.5$; $F_8^W = 0.85$), ($x_9 = 42$; $F_9^W = 1$). Then g^{pfr} is approximated by the 1-D pW-GL-I $g^{pW} = \langle F^{pW}(x); x\rangle$. Following (49), the expected utility of the pW-GL-I is $E_i^{pW}(u|F_i^{pW}) = 0.2286$, which is also the pW-expected utility of g^{pfr}. The graphics of $F^{pW}(x)$ and its density are depicted in Fig. 5.

6.2.3 Approximation of $F^{pR}(x)$ Using the Maximax Criterion

From (53) and (54) it follows that $F_1^{\neg W} = 0$, $F_9^{\neg W} = 1$. The linear programming task, defined in Sect. 5.2 and (64) with constraints in (59), is solved using *linprog*. It generates the following results: $F_2^{\neg W} = 0.07$, $F_3^{\neg W} = 0.23$, $F_4^{\neg W} = 0.41$, $F_5^{\neg W} = 0.51$, $F_6^{\neg W} = 0.80$, $F_7^{\neg W} = 0.88$, $F_8^{\neg W} = 0.95$.

Therefore, $F^{pR}(.)$ is approximated by $F^{p\neg W}(.)$ on the nodes ($x_1 = 30$; $F_1^{\neg W} = 0$), ($x_2 = 31.5$; $F_2^{\neg W} = 0.07$), ($x_3 = 33$; $F_3^{\neg W} = 0.23$), ($x_4 = 34.5$; $F_4^{\neg W} = 0.41$), ($x_5 = 36$; $F_5^{\neg W} = 0.51$), ($x_6 = 37.5$; $F_6^{\neg W} = 0.80$), ($x_7 = 39$; $F_7^{\neg W} = 0.88$),

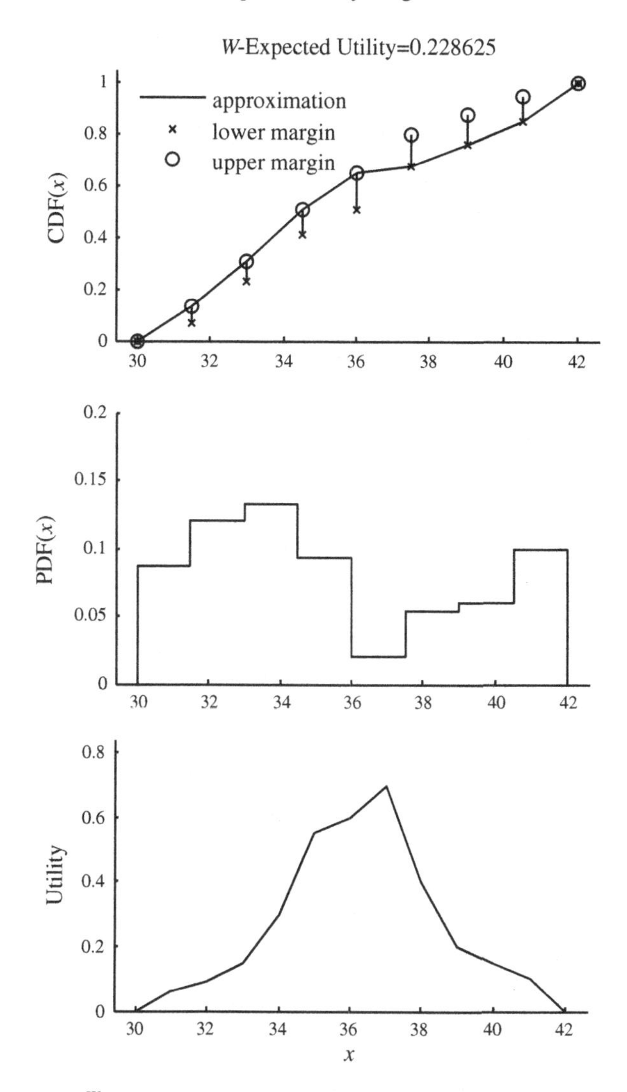

Fig. 5 Graphics of $F^{pW}(x)$, density (PDF) and utility function $u(x)$ of the FRDM over the values of X in the interval [30; 42]

$(x_8 = 40.5;\ F_8^{\neg W} = 0.95)$, $(x_9 = 42;\ F_9^{\neg W} = 1)$. Then g^{pfr} is approximated by the 1-D $p\neg W$-GL-I $g^{p\neg W} = \langle F^{p\neg W}(x); x\rangle$. Following (49), the expected utility of the $p\neg$W-GL-I is $E_i^{p\neg W}(u|F_i^{p\neg W}) = 0.3532$, which is also the $p\neg W$-expected utility of g^{pfr}. The graphics of $F^{p\neg W}(x)$ and its density are depicted in Fig. 6.

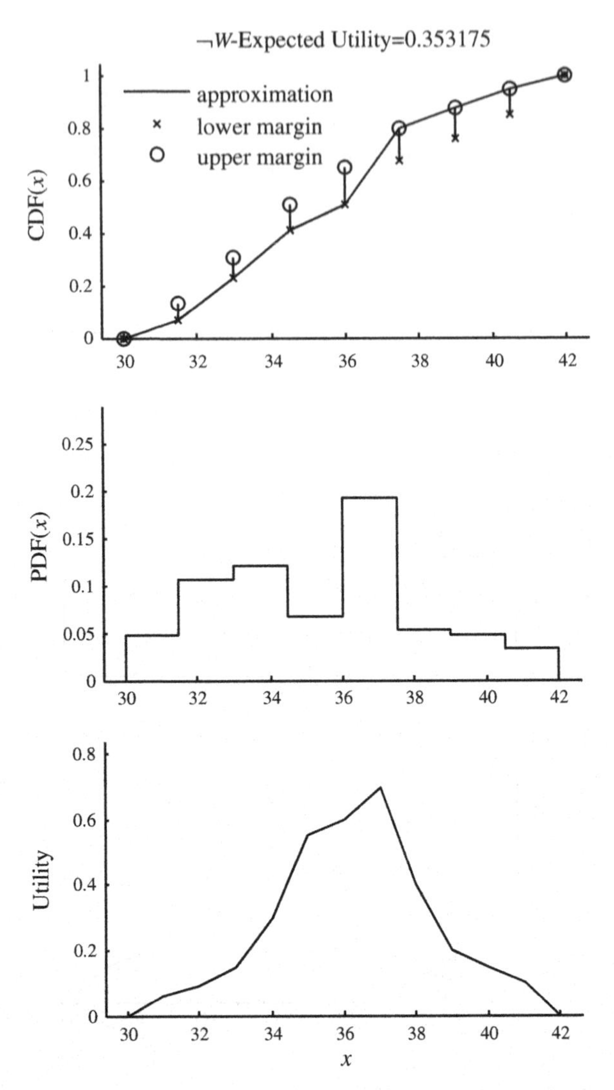

Fig. 6 Graphics of $F^{p\neg W}(.)$, density (PDF) and utility function $u(x)$ of the FRDM over the values of X in the interval [30; 42]

6.2.4 Approximation of $F^{pR}(x)$ Using the Hurwicz$_\alpha$ Criterion

Let $\alpha = 0.7$. From (53) and (54) it follows that $F_1^{H_{0.7}} = 0$, $F_9^{H_{0.7}} = 1$. According to (65), the required quantiles depend on F_l^W and $F_l^{\neg W}$. Here $F_2^{H_{0.7}} = 0.7F_2^W + (1-0.7)F_2^{\neg W,(i)} = 0.7 \times 0.13 + 0.3 \times 0.07 = 0.112$, and in the same fashion $F_3^{H_{0.7}} = 0.286$, $F_4^{H_{0.7}} = 0.48$, $F_5^{H_{0.7}} = 0.608$, $F_6^{H_{0.7}} = 0.716$, $F_7^{H_{0.7}} = 0.796$, $F_8^{H_{0.7}} = 0.88$.

Therefore, $F^{pR}(.)$ is approximated by $F^{pH_{0.7}}(.)$ on the nodes ($x_1 = 30$; $F_1^{H_{0.7}} = 0$), ($x_2 = 31.5$; $F_2^{H_{0.7}} = 0.112$), ($x_3 = 33$; $F_3^{H_{0.7}} = 0.286$), ($x_4 = 34.5$; $F_4^{H_{0.7}} = 0.48$), ($x_5 = 36$; $F_5^{H_{0.7}} = 0.608$), ($x_6 = 37.5$; $F_6^{H_{0.7}} = 0.716$), ($x_7 = 39$; $F_7^{H_{0.7}} = 0.796$), ($x_8 = 40.5$; $F_8^{H_{0.7}} = 0.88$), ($x_9 = 42$; $F_9^{H_{0.7}} = 1$). Then g^{pfr} is approximated by the 1-D $pH_{0.7}$-GL-I $g^{pH_{0.7}} = \langle F^{pH_{0.7}}(x); x\rangle$. Following (49), the expected utility of the $pH_{0.7}$-GL-I is $E_i^{pH_{0.7}}(u|F_i^{pR}) = 0.266$, which is also the $pH_{0.7}$-expected utility of g^{pfr}. The graphics of $F^{pH_{0.7}}(.)$ and its density are depicted in Fig. 7.

6.2.5 Approximation of $F^{pR}(x)$ Using the Laplace Criterion

From (53) and (54) it follows that $F_1^L = 0$, $F_9^L = 1$. According to (67) the required quantile indices are: $F_2^L = F_2^d + (F_2^u - F_2^d)\frac{x_2 - x_1}{x_9 - x_1} = 0.07 + (0.13 - 0.07) \times [(31.5 - 30)/(42 - 0)] = 0.0775$, and in the same fashion $F_3^L = 0.25$, $F_4^L = 0.4475$, $F_5^L = 0.58$, $F_6^L = 0.755$, $F_7^L = 0.85$, $F_8^L = 0.9375$.

Therefore $F^{pR}(.)$ is approximated by $F^{pL}(.)$ on the nodes ($x_1 = 30$; $F_1^L = 0$), ($x_2 = 31.5$; $F_2^L = 0.0775$), ($x_3 = 33$; $F_3^L = 0.25$), ($x_4 = 34.5$; $F_4^L = 0.4475$), ($x_5 = 36$; $F_5^L = 0.58$), ($x_6 = 37.5$; $F_6^L = 0.755$), ($x_7 = 39$; $F_7^L = 0.85$), ($x_8 = 40.5$; $F_8^L = 0.9375$), ($x_9 = 42$; $F_9^L = 1$). Then g^{pfr} is approximated by the 1-D pL-GL-I $g^{pL} = \langle F^{pL}(x); x\rangle$. Following (49), the expected utility of the pL-GL-I is $E_i^{pL}(u|F_i^{pR}) = 0.3126$, which is also the pL-expected utility of g^{pfr}. The graphics of $F^{pL}(x)$ and its density are depicted in Fig. 8.

7 Conclusions

This chapter discussed the influence of fuzzy-rationality on the calculation of the expected utility integral of fuzzy-rational GL-I. It stemed from the interval form of the nodes for interpolation/approximation of the ribbon CDF. The 1-D GL-I envisaged probability distributions of the prize. Due to the use of ribbon functions, those lotteries transformed into fuzzy-rational ones of either x-fuzzy-rational or p-fuzzy-rational type. A two-step ranking procedure was formalized in both cases. It envisaged approximation of the ribbon functions by classical, partially linear ones in order to apply expected utility.

The two-step scheme was further extended into tasks, related to the application of the criteria Q under strict uncertainty to transform the interval nodes of the ribbon functions into point estimates. This process aimed at complying with the pessimism-optimism rationale behind the specific Q criterion, namely Wald, maximax, Hurwicz$_\alpha$, and Laplace criteria. Several algorithms were elaborated for that purpose. The numerical example further demonstrated the application of those algorithms in an economic setup. It showed the different values the expected utility of a

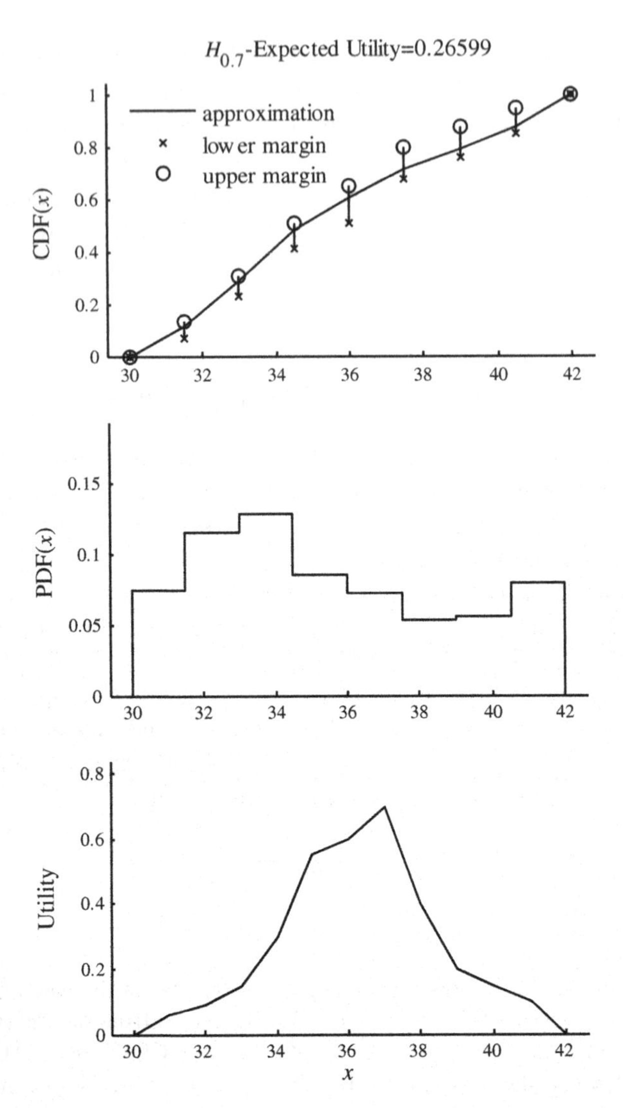

Fig. 7 Graphics of $F^{pH_{0.7}}(.)$, density (PDF) and utility function $u(x)$ of the FRDM over the values of Xin the interval [30; 42]

given 1-D GL-I might get depending on the pessimism-optimism philosophy of the criteria under strict uncertainty. Generating a final decision in the example was not performed, as there was only a single lottery to analyze. In a larger study, however, the usage of a specific Q-expected utility would lead to different ranking of lotteries. All calculation and visualization procedures in the chapter were performed with the help of original software written in MATLAB R2011b, which is available free of charge upon request from the authors.

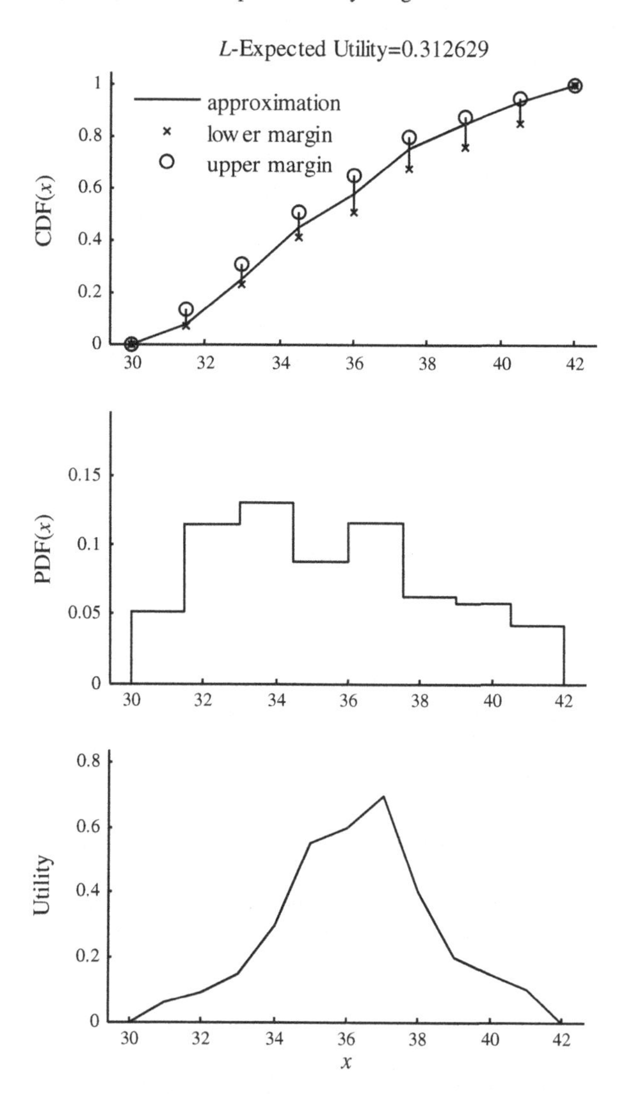

Fig. 8 Graphics of $F^{p^L}(x)$, density (PDF) and utility function $u(x)$ of the FRDM over the values of X in the interval [30; 42]

Appendix 1: Decision Criteria Under Strict Uncertainty

A lottery is a generic model of an uncertain decision alternative. Let's assume that the DM has to rank n uncertain alternatives according to preference, which can result in r different holistic consequences x_j called prizes, indexed according to the preferences of the DM, x_1 being the most preferred, and x_r being the most unwanted one. The consequences of the ith alternative will be x_j if the event $\theta_{i,j}$ occurs, where

$\theta_{i,1}, \theta_{i,2}, \ldots, \theta_{i,r}$, $i = 1, 2, \ldots, n$, called *states*, form a full set of disjoint events. An ordinary lottery may be denoted as (A1) with the conditions in (A2):

$$l_i = \langle \theta_{i,1}, x_1; \theta_{i,2}, x_2; \ldots; \theta_{i,r}, x_r \rangle, \tag{A1}$$

$$\begin{aligned} & x_1 \succsim x_2 \succsim \ldots \succsim x_r \\ & \theta_{i,j} \succsim_l \varnothing, \\ & \theta_{i,1} \cup \theta_{i,2} \cup \ldots \cup \theta_{i,r} = \Theta, \\ & \theta_{i,j} \cap \theta_{i,k} = \varnothing \text{ for } j \neq k. \end{aligned} \tag{A2}$$

Here, $\varnothing$ denotes the null event, Θ denotes the certain event, $\succsim_l$ denotes the binary relation "at least as likely as" defined over events, and $\succsim$ denotes the binary relation "at least as preferred as", defined over prizes or lotteries.

Assume the DM has constructed a utility function $u(.)$ over the consequences x_j, such that it measures her relative preferences in the sense [36]:

$$u(x_j) \geq u(x_k) \text{ iff } \quad x_j \succsim x_k. \tag{A3}$$

Unfortunately, the elicitation of utilities from real DMs results in uncertainty intervals, which is another demonstration of fuzzy rationality. For reasons of simplicity, it is assumed here that the values of the utility function are point estimates of some kind of their uncertainty intervals. It is also assumed that those correctly reflect the preferences of the DM.

If the only thing the DM knows about the uncertainty in the problem is which states are possible and which are not, then the decision is said to be under *strict uncertainty*. Let $b(.)$ be a discrete Boolean function defined over $\theta_{i,j}$ and with range {'*t*' , '*f*'}, where '*t*' stands for '*true*' and is assigned when the state is possible, whereas '*f*' stands for '*false*' and is assigned when the state is impossible:

$$b(\theta_{i,j}) = \begin{cases} 't', & \text{for } \theta_{i,j} \succ_l \varnothing \\ 'f', & \text{for } \theta_{i,j} \sim_l \varnothing \end{cases}. \tag{A4}$$

Here $\sim_l$ and $\succ_l$ denote respectively the binary relations "equally likely to" and "more likely than" defined over events. In the strict uncertainty case, for each state $\theta_{i,j}$ (for $i = 1, 2, \ldots, n$ and $j = 1, 2, \ldots, r$) the DM can define the value $b(\theta_{i,j})$ subject to the condition:

$$b(\theta_{i,1}) \vee b(\theta_{i,2}) \vee \ldots \vee b(\theta_{i,r}) = 't'. \tag{A5}$$

In (A5), $\vee$ is the Boolean operator "and". The lottery (A1) with the known function $b(.)$ can be better represented as

$$l_i = \langle \langle \theta_{i,1}, b(\theta_{i,1}) \rangle, x_1; \langle \theta_{i,2}, b(\theta_{i,2}) \rangle, x_2; \ldots; \langle \theta_{i,r}, b(\theta_{i,r}) \rangle, x_r \rangle. \tag{A6}$$

The lottery (A6) subject to (A2) and (A5) can be called *a strictly uncertain lottery*. There are criteria to rank strictly uncertain lotteries. Four of the most widespread are the Savage, Laplace, Wald and Hurwicz$_\alpha$ criteria.

Savage's minimax criterion [24] constructs the so-called regret table, and recommends the alternative which minimizes the maximal regret. Its idea will not be used in this chapter.

Laplace's criterion [37] is based on the principle of insufficient reasoning, which says that if no information is available regarding a set of random events, then they might be assumed equally probable. In this way the probabilities are known and the resulting problem under risk must be solved by the expected utility criterion, which here degenerates to the mean utility of all possible states:

$$c_i = \frac{\sum\limits_{\substack{j=1\\ b(\theta_{i,j})='t'}}^{r} u(x_j)}{\sum\limits_{\substack{j=1\\ b(\theta_{i,j})='t'}}^{r} 1}. \tag{A7}$$

Then the recommended choice is l_k for which the mean utility is maximal.

Wald's criterion of maximin return suggests ranking actions in descending order of their worst outcomes [26]. The security level denotes the worst possible outcome from the ith alternative

$$s_i = \min_{\substack{j=1\\ b(\theta_{i,j})='t'}}^{r} \{u(x_j)\}. \tag{A8}$$

Then the recommended choice is l_k for which the security level is maximal. This criterion is suitable for extreme pessimists.

If the DM is an extreme optimist, she can use the maximax criterion, which suggests ranking actions in descending order of their best outcomes. The optimism level o_i denotes the best possible outcome from the ith alternative

$$o_i = \max_{\substack{j=1\\ b(\theta_{i,j})='t'}}^{r} \{u(x_j)\}. \tag{A9}$$

Then the recommended choice is l_k for which the optimism level is maximal. Despite the symmetry between (A8) and (A9), the maximax criterion (A9) is not used in practice, whereas Wald's criterion is often preferred as a decision tool.

Hurwicz argues that people usually do not express such extremes of pessimism or optimism as the previous criteria suggested [25]. He introduced the optimism-pessimism index $\alpha \in [0; 1]$ to weight the security level and the optimism index for each alternative in the form

$$h_i^\alpha = \alpha s_i + (1 - \alpha) o_i. \tag{A10}$$

Hurwicz suggested to rank alternatives in descending order of h_i^α, which can be called the Hurwicz$_\alpha$ strict uncertainty criterion. Then the recommended choice is l_k for which h_i^α is maximal.

The optimism-pessimism index α is a measure of people's pessimism. It is specific to each DM and applies to all decision situations. In order to elicit α, the DM can be offered the choice between: a) lottery l_1, giving consequences with utility 1, v, and 0 respectively at states $\theta_{1,1}$, $\theta_{1,2}$, and $\theta_{1,3}$, where $b(\theta_{1,1}) = 't'$, $b(\theta_{1,2}) = 'f'$ and $b(\theta_{1,3}) = 't'$; b) lottery l_2 giving consequences with utility 1, v, and 0 respectively at states $\theta_{2,1}$, $\theta_{2,2}$, and $\theta_{2,3}$, where $b(\theta_{1,1}) = 'f'$, $b(\theta_{1,2}) = 't'$, and $b(\theta_{1,3}) = 'f'$. The value of v varies until the DM becomes indifferent between the lotteries, at which point $\alpha = 1 - v$.

The rationality of each criterion can be assessed against a set of reasonable properties of the decisions generated by that criterion [36]. However, analysis shows that any decision criterion under strict uncertainty does not (and will not) possess this set of properties of choice and thus is (and will be) irrational. One possible explanation is that problems, where the DM knows nothing about the uncertainty, do not actually exist.

References

1. Von Neumann, J., Morgenstern, O.: Theory of Games and Economic Behavior, 2nd edn. Princeton University Press, Princeton (1947)
2. Tenekedjiev, K.: Decision problems and their place among operations research. Autom. Inform. **XXVIII**(1), 7–10 (2004)
3. Ramsay, F.P.: Truth and Probability, The Logical Foundations of Mathematics and Other Essays. Kegan Paul, London (1931) (Reprinted: Kyburg, H.E. Jr., Smolker, H.E. (eds.). (1964). Studies in Subjective Probability. Wiley, 61–92)
4. Savage, L.J.: The Foundations of Statistics, 1st edn. Wiley, New York (1954)
5. Pratt, J.W., Raiffa, H., Schlaifer, R.: Introduction to Statistical Decision Theory. MIT Press, Cambridge (1995)
6. De Finetti, B.: La prevision: ses lois logiques, ses sorces subjectives. Annales de l'Institut Henri Poincare **7**, 1–68 (1937) (Translated in Kyburg, H.E. Jr., Smokler, H.E. (eds.): Studies in Subjective Probability, pp. 93–158. Wiley, New York (1964))
7. De Groot, M.H.: Optimal Statistical Decisions. McGraw-Hill, New York (1970)
8. Villigas, C.: On qualitative probabilityσ-algebras. Ann. Math. Stat. **35**, 1787–1796 (1964)
9. French, S., Insua, D.R.: Statistical Decision Theory. Arnold, London (2000)
10. Keeney, R.L., Raiffa, H.: Decisions with Multiple Objectives: Preference and Value Tradeoffs. Cambridge University Press, Cambridge (1993)
11. Bernstein, P.L.: Against the Gods—The Remarkable Story of Risk. Wiley, New York (1996)
12. De Cooman, G., Zaffalon, M.: Updating beliefs with incomplete observations. Artif. Intell. **159**, 75–125 (2004)
13. Cano, A., Moral, S.: Using probability trees to compute marginals with imprecise probabilities. Int. J. Approximate Reasoning **29**, 1–46 (2002)

14. Guo, P., Wang, Y.: Eliciting dual interval probabilities from interval comparison matrices. Inf. Sci. **190**, 17–26 (2012)
15. Guo, P., Tanaka, H.: Decision making with interval probabilities. Eur. J. Oper. Res. **203**, 444–454 (2010)
16. Walley, P.: Inferences from multinomial data: learning about a bag of marbles. J. Royal Stat. Soc. B, **58**, 3–57 (1996)
17. Miranda, E., de Cooman, G., Couso, I.: Lower previsions induced by multi-valued mappings. J. Stat. Plann. Infer. **133**, 173–197 (2005)
18. Knetsch, J.: The endowment effect and evidence of nonreversible indifference curves. In: Kahneman, D., Tversky, A. (eds.) Choices, Values and Frames, pp. 171–179. Cambridge University Press, Cambridge (2000)
19. Walley, P.: Statistical Reasoning with Imprecise Probabilities. Chapman and Hall, London (1991)
20. Troffaes, M.C.M.: Decision making under uncertainty using imprecise probabilities. Int. J. Approximate Reasoning **45**, 7–29 (2007)
21. Kozine, I., Utkin, L.: Constructing imprecise probability distributions. Int. J. General Syst. **34**(4), 401–408 (2005)
22. Tenekedjiev, K., Nikolova, N.D., Toneva, D.: Laplace expected utility criterion for ranking fuzzy rational generalized lotteries of I type. Cybern. Inf. Technol. **6**(3), 93–109 (2006)
23. Nikolova, N.D., Shulus, A., Toneva, D., Tenekedjiev, K.: Fuzzy rationality in quantitative decision analysis. J. Adv. Comput. Intell. Intell. Inf. **9**(1), 65–69 (2005)
24. Yager, R.R.: Decision making using minimization of regret. Int. J. Approximate Reasoning **36**, 109–128 (2004)
25. Yager, R.R.: Generalizing variance to allow the inclusion of decision attitude in decision making under uncertainty. Int. J. Approximate Reasoning **42**, 137–158 (2006)
26. Fabrycky, W.J., Thuesen, G.J., Verma, D.: Economic Decision Analysis. Prentice Hall, Englewood Cliffs (1998)
27. Hackett, G., Luffrum, P.: Business Decision Analysis—An Active Learning Approach. Blackwell, New York (1999)
28. Milnor, J.: Games against nature. In: Thrall, R., Combs R.D. (eds.) Decision Processes, pp. 49–59. Wiley, Chichester (1954)
29. Szmidt, E., Kacprzyk, J.: Probability of intuitionistic fuzzy events and their application in decision making. In: Proc. EUSFLAT-ESTYLF joint conference, September 22–25, Palma de Majorka, Spain, pp. 457–460 (1999)
30. Atanassov, K.: Intuitionistic Fuzzy Sets. Springer, Heidelberg (1999)
31. Atanassov, K.: Review and New Results on Intuitionistic Fuzzy Sets, Preprint IM-MFAIS, vol. 1–88. Sofia (1988)
32. Atanassov, K.: Four New Operators on Intuitionistic Fuzzy Sets, Preprint IM-MFAIS, vol. 4–89. Sofia (1989)
33. Utkin, L.V. (2007). Risk analysis under partial prior information and non-monotone utility functions. Int. J. Inf. Technol. Decision Making. **6**(4), 625–647 (2007)
34. Nikolova, N.D.: Three criteria to rank x-fuzzy-rational generalized lotteries of I type. Cybern. Inf. Technol. **7**(1), 3–20 (2007)
35. Dantzig, G.B., Orden, A., Wolfe, Ph: The generalized simplex method for minimizing a linear form under linear inequality restraints. Pacific J. Math. **5**(2), 183–195 (1955)
36. French, S.: Decision Theory: An Introduction to the Mathematics of Rationality. Ellis Horwood, Chichester (1993)
37. Rapoport, A.: Decision Theory and Decision Behaviour-Normative and Descriptive Approaches. USA, Kluwer Academic Publishers (1989)

Human-Centric Cognitive Decision Support System for Ill-Structured Problems

Tasneem Memon, Jie Lu and Farookh Khadeer Hussain

Abstract The solutions to ill-structured decision problems greatly rely upon the intuition and cognitive abilities of a decision maker because of the vague nature of such problems. To provide decision support for these problems, a decision support system (DSS) must be able to support a user's cognitive abilities, as well as facilitate seamless communication of knowledge and cognition between itself and the user. This study develops a cognitive decision support system (CDSS) based on human-centric *semantic de-biased associations (SDA) model* to improve ill-structured decision support. The SDA model improves ill-structured decision support by refining a user's cognition through reducing or eliminating bias and providing the user with *validated* domain knowledge. The use of semantics in the SDA model facilitates the natural representation of the user's cognition, thus making the transfer of knowledge/cognition between the user and system a natural and effortless process. The potential of *semantically defined cognition* for effective ill-structured decision support is discussed from a human-centric perspective. The effectiveness of the approach is demonstrated with a case study in the domain of *sales*.

Keywords Human-centric decision systems · Cognitive decision support · Cognitive biases

T. Memon · J. Lu (✉) · F. K. Hussain
The Decision Systems and e-Service Intelligence (DeSI) Laboratory, The Centre for Quantum Computation and Intelligent Systems (QCIS), School of Software, Faculty of Engineering and Information Technology, University of Technology, Sydney, NSW, Australia
e-mail: Tasneem.Memon@uts.edu.au

J. Lu
e-mail: Jie.Lu@uts.edu.au

F. K. Hussain
e-mail: Farookh.Hussain@uts.edu.au

P. Guo and W. Pedrycz (eds.), *Human-Centric Decision-Making Models for Social Sciences*, Studies in Computational Intelligence 502,
DOI: 10.1007/978-3-642-39307-5_12,

1 Introduction

Over the years, *decision support systems (DSS)* have proven to be very successful in providing decision support for *well-structured* decision problems. The major challenge for DSS, however, has been to provide adequate support for *ill-structured*[1] decision problems. This shortcoming of DSS arises because the solution to an *ill-structured* decision problem relies mainly upon the *cognitive abilities* of a decision maker rather than standard optimized procedures. The current DSS do not provide support for *cognition* because of the inherent limitations of current computer systems.

Information storage and processing in a computer system differs significantly from the *knowledge* and *cognitive abilities* of the human mind. Likewise, facilities of the human mind such as *knowledge*, *perception*, *intuition* and *natural language* are unfamiliar to the computer system's processing system which is based purely on formal logic. A computer system's skills, such as *storing huge amounts of information* and *fast and accurate processing of data*, are nearly impossible for the human mind to achieve. Although this difference between humans and computers is complementary in many applications, such as simulation software, multimedia content management, and image processing, it becomes a barrier in decision making, especially for *ill-structured* problems.

During the course of decision making, a decision maker evaluates the situation with the help of his *knowledge* and *cognitive maps* (also called *mental models*[2]) [35, 52]. By contrast, the information provided by a DSS is in the form of *quantitative results*, such as *reports*, *forms*, *documents* and *graphs*. Therefore, a decision maker is required to constantly *internalize*[3] the DSS output into *mental models* to acquire a better perspective on the situation; and to convert his *mental models* into appropriate *information extracting queries*, to extract the required information from DSS. During this process of conversion, a great deal of *knowledge* is lost which could otherwise prove fundamental to reach optimal decision.

To prevent this loss of *knowledge*, it is essential to enable DSS to: (a) extract *knowledge* from the decision maker in a *natural* way, such as through *natural language*; (b) store it in a format similar to that of the *knowledge* in human mind, such as *mental models*; (c) output the required *knowledge* in a format *instinctive* to human perception, such as *semantic diagrams* of decision alternatives.

This chapter discusses how integrating *human-centric techniques* in a DSS can prevent the *loss of knowledge*. The *human-centric* techniques can enable the DSS to seamlessly transfer *knowledge* and *cognition* to and from a decision maker and store it without conversion, to provide improved support for *ill-structured* decision problems. A *semantic de-biased associations (SDA) model* is introduced here, which facilitates the transfer of *knowledge*, to and from the user (decision maker), in a

[1] The description of *ill-structured decision problems* is given in Sect. 2.

[2] For the scope of this chapter, we shall refer to them as *mental models*.

[3] Please refer to Nonaka's [36] *knowledge spiral model* for the definition of *internalization*.

human-centric format. The SDA model allows user to input *knowledge* in *natural language,* eliminating the need to convert *knowledge* in structured data format; thus preventing the *loss of knowledge.* The *knowledge* in SDA model is stored in *semantic* representation of *mental models*, which keeps the original format of the *knowledge* intact. The output is provided in graphical representation of *mental models.* The graphical representation assists user to intuitively understand the situation in an efficient manner, saving valuable decision making time. The SDA model and its architecture are described in Sect. 4.

The rest of the chapter is organized as follows. Section 2 describes the types of decision problems, followed by a discussion on decision support systems. Section 3 discusses the role of *cognition* and *human-centric approaches* in ill-structured decision support, as well as proposes the conceptual model of a *human-centric cognitive DSS.* Section 4 describes the *semantic de-biased associations (SDA) model,* its structure, and its effectiveness in providing support for *ill-structured* decision problems. Section 5 presents a case study to demonstrate the performance of the SDA model, followed by conclusions and future work.

2 Decision Problems and Decision Support Systems

The success of a DSS relies greatly upon the type of decision problem it is handling. Decision problems can be classified into two categories according to the characteristics they possess. This section describes the types of decision problems. A brief overview of DSS and Intelligent DSS is then presented, followed by an illustration of the *conceptual model of human-centric cognitive DSS.*

2.1 Types of Decision Problems

A decision problem can be divided into two types: *well-structured* and *ill-structured* [47].

2.1.1 Well-Structured Decision Problems

Well-structured decision problems are those problems for which *the existing state, goal state* and *constraint parameters* are well defined [16]. These problems have a single correct solution [46]. Since the initial and final states of such problems are known, procedures, rules and policies can be devised to solve them. A *well-structured* problem will have the same result every time it is solved. Problems such as finding the value of x in a quadratic equation, the procedure of fixing an overheating car engine, or calculating the distance of a star from our solar system, are examples of *well-structured* problems.

2.1.2 Ill-Structured Decision Problems

Ill-structured problems, on the other hand, are those in which the *existing state* is vague, and the *goal state* is unclear and non-quantitative; thus, a course of action cannot be devised to reach the goal state [47]. *Ill-structured* problems do not have a single correct solution; that is, there can be more than one appropriate solution depending upon the current circumstances of the system. According to Simon [45], the category of ill-structured problems is a residual category of *well-structured* problems; that is, all the problems which are not *well-structured* can be called *ill-structured* problems. Every *ill-structured* problem is unique in terms of *existing situation, goal, objective, desired outcome*, and the *process* necessary to reach an optimal solution. Examples of *ill-structured* problems are: devising a strategy to increase profit for a product, creating efficient traffic infrastructure for a population of two million, or selecting a suitable candidate for an executive position.

2.2 Decision Support Systems

The term *decision making* refers to the process of making the choice for further course of action based upon certain criteria, circumstances and available information. The computerized systems that support decision makers in this process are referred to as *decision support systems (DSS)*. A DSS is "an interactive computer-based system or subsystem intended to help decision makers use communications technologies, data, documents, knowledge and/or models to identify and solve problems, complete decision process tasks, and make decisions" [39].

Since its conception in 1965, the DSS field has evolved in several directions based on the technologies introduced over time. These include the *model-driven*[4] or *model-oriented* DSS [2, 13, 44], *data-driven DSS*[5] [40], *communication-driven DSS*[6] [7], *Group Decision Support Systems* [38], *document-driven DSS*[7] [48] and *web-based DSS*.[8]

---

[4] A model-driven DSS emphasizes access to and manipulation of financial, optimization and/or simulation models.

[5] A data-driven DSS emphasizes access to and manipulation of a time-series of internal company data and sometimes external and real-time data.

[6] Communication-driven DSS use network and communication technologies to facilitate decision-relevant collaboration and communication. Tools used include groupware, video conferencing, and computer-based bulletin boards.

[7] A document-driven DSS uses computer storage and processing technologies to provide document retrieval and analysis. Documents may include scanned documents, hypertext documents, images, sounds and video.

[8] A web-based DSS is simply a system which is implemented using web-based technologies. A web-based DSS can be any type of DSS, such as a communication-driven, model-driven or data-driven system.

Apart from these categories, a classification which has proven to be the most promising in supporting decision making for *ill-structured problems*, is *intelligent DSS (IDSS)*. These systems have also been called *suggestion DSS* [2]. The term was coined by Clyde Holsapple in 1977. The idea of a DSS intelligently synthesizing *decision alternatives* and supporting a decision maker's *cognitive abilities* appealed to many researchers; thus the work on merging AI techniques with DSS to produce *IDSS* started. This field has not matured as rapidly as was anticipated; however, it is proving itself to be very promising for *ill-structured decision support*.

2.3 Intelligent Decision Support for Ill-Structured Problems

A DSS that is designed to behave *intelligently* through incorporating an *artificial intelligent* technique is called an *intelligent decision support system (IDSS)*. The early IDSS were *rule-based* decision support systems, also known as *expert systems* [12]. However, the IDSS evolved in different directions over time, based on advances in the field of *artificial intelligence (AI)*. AI techniques, such as *neural networks, fuzzy logic*, and*genetic algorithms*, were incorporated into the DSS to achieve intelligent support for *ill-structured* decision problems [12].

In 1991, Nonaka wrote a paper with the title "The Knowledge-creating Company", in which he identified *knowledge* as the most valued asset, and the key factor, for business organisations to gain competitive advantage. *Knowledge* can comprise *skills, ideas, observations, experiences* and *intuition* [34, 36]. This research paper was arguably the starting point for IDSS research to evolve in the revolutionary direction of *knowledge warehouse-based DSS (KWDSS)*. Knowledge warehouse (KW) technology is still in its infancy, and there is as yet no standard storage structure proposed for it.

Following the studies conducted in psychology and organisational behaviour, which recognized the significance of the *cognitive abilities* of decision makers in solving complex decision problems [14], DSS researchers employed techniques to incorporate *cognition* into DSS [8, 43, 53]. These systems have been named *cognitive decision support systems (CDSS)*. The CDSS use *mental models* as the means to store a decision maker's *cognition* [35].

Storing *cognition* in the system is a step towards achieving the goal of preventing the *loss of knowledge*. However, merely storing cognition is not sufficient to develop effective DSS for ill-structured problems. It is essential to employ a *human-centric* interface to facilitate the seamless transfer of *knowledge* between the decision maker and the *cognition storage* in the DSS. The techniques, such as *natural language processing, semantics*, and *graphical representation of decision alternatives*, can be used to accomplish the task knowledge transfer.

This chapter proposes a conceptual model of the *semantics-driven cognitive decision support system (SCDSS)*, which incorporates *human-centric techniques* to reduce biases and support decision maker's cognition. (Fig. 2). The overview of SCDSS conceptual model will be presented in Sect. 3.3. But prior to that, following is an account of the role of *cognition* in *ill-structured* decision support, and the challenges of employing it effectively in DSS, followed by a discussion on the role of the *human-centric approach* in *ill-structured* decision support.

3 Cognition and Human-Centric Computing in Ill-Structured Decision Support

As discussed in the previous section, *cognition based DSS* can provide better support for a decision maker's thinking process and knowledge creation than conventional DSS. We discuss below how cognition can be incorporated into a DSS, following which the role that *human-centric computing* (HC^2) techniques can play in providing an appropriate interface for the transfer of human-computer cognition is discussed.

3.1 Cognitive Mapping in Ill-Structured Decision Support

The interpretation of the objective world in the human mind is highly subjective. It can be constructed *socially*, but a large part of the mental representation of the world is *personally* constructed. This mental representation, which is constructed upon personal values and experiences, is stored in the human mind in the form of *cognitive maps* or *mental models*.

The theory of *mental models* was proposed by Kenneth Craik [9], which states that the mind develops models of real world, and utilizes them to deal with the present and the future situations in a better way. Later, based on this theory, Johnson-Laird [23] emphasized that to make inferences, mind relies upon intuition and mental models rather than formal logic (see [22] for a comprehensive history of mental models). Especially in complex situations with imprecise information, *mental models* become the key factor to make sense out of the insufficient information [52]. They assist us in transforming the *tacit knowledge* into *explicit knowledge*, so as to communicate it to others. They have proved to be the fundamental analysis tool, and critical success factor, in complex and *ill-structured decision situations* [29], p. 182; [33].

Mental model representation has been used extensively in DSS to deal with *ill-structured decision problems* [8, 35]. *Mental models* assist decision makers to understand and formulate *ill-structured decision problems*, and to *select* the relevant parts of the *problem space* [53]. The computerized *mental model representation* allows the system to collect and store decision makers' *cognition* in natural format, which prevents *loss of knowledge*. Being able to retain the *cognition* of decision makers in the system implies that while decision makers may leave the organisation, their *knowledge* and *cognition* remains within the organisation for future use.

There is nevertheless a downside to the use of *mental models*. During the construction of mental *models* through an experience, *biases* are introduced into the *mental models* based on the prevailing conditions. Biases are inclinations towards or against certain ideas or entities which often lead to poor decision making (see Chen and Lee [8, 10] and Korte [26] for a detailed account of biases and their effects on decision results). However, the decision makers are bound to use their biased mental models because of the limited or ambiguous knowledge available in complex and ill-structured decision situations. Cognitive biases and their effects on ill-structured decision support are described below.

Cognitive Biases

The human mind forms *mental models* during an experience and, depending on the circumstances surrounding that experience, introduces *biases* in them [18]. While mental models are the fundamental tool for solving ill-structured decision problems, the inherent biases become a barrier in achieving optimal solutions [5, 21, 27, 30]. Following is a brief discussion on the impact of biases in ill-structured decision making.

Cognitive biases refer to the inclinations towards or against certain ideas or entities that are generated within mental models during everyday experiences [18, 25]. Once formed, bias can consistently influence our judgement, and our future perceptions of the world and happenings in it [3, 19]. Bias is considered to be "a negative consequence of adopting heuristics" as it "entices decision makers away from making optimal decisions" [11], p. 760. For a comprehensive taxonomical classification of decision biases, refer to Arnott [4]. The presence of cognitive bias may result in enormous damage, especially in business and medical domains; thus it is essential to mitigate all such bias [6, 15, 26, 30].

Croskerry [10] has outlined thirty biases in the clinical domain. Das and Teng [11] have grouped the biases involved in *strategic decision making* into four categories, which are:

i. *Prior hypotheses and focusing on limited targets:* The decision makers choose to rely on their previous beliefs, or rely heavily on one piece of information, and may ignore information that indicates otherwise. The cognitive biases *anchoring, attentional bias*, and *focusing effect* come under this category.
ii. *Exposure to limited alternatives:* Problems are broken down into smaller constructs and fewer alternatives are considered, relying more upon intuition. The cognitive bias *availability* comes under this category.
iii. *Insensitivity to outcome probabilities:* Subjective judgement is used rather than rational probability. The cognitive bias *neglect of probability* comes under this category.
iv. *Illusion of manageability:* Over optimism, overestimation of the control level. The cognitive bias *illusion of control* comes under this category.

Cognitive biases hinder the achievement of optimal decision solutions, which results in enormous loss in such fields as *business* and *clinical medicine* [10, 26]. Thus, it is essential to devise techniques to reduce or eliminate bias, in order to reach better solutions to decision problems [5, 11, 26, 31].

This chapter addresses four biases, which are: *availability, framing, contextual* and *group biases*. Following is a brief description of these biases.

Availability Bias. Availability bias is the phenomenon of considering only those *alternatives* that are easily recalled from memory [49]. To reduce this bias, a CDSS must be able to assist a user in bringing maximum relevant knowledge from *long-term memory* to *working memory* [7].

Framing Bias. Framing bias refers to the inclination created by the way a decision situation is presented [50]. Framing bias can be reduced or eliminated by making a decision maker draw, or study, causal maps regarding a decision situation [20], or by providing them with adequate *contextual information* about a decision situation [51].

Contextual Bias. Contextual bias is a *false permanent impression* about a decision alternative being successful, or otherwise depending on the environment it was generated in. For example, a user may find *presales services* very useful for *increasing sales* in the case of "e-reader", and may develop a mental model based on this experience as:

Pre-sales service --- increases ---> Sales

However, this mental model may not be applicable in an alternative situation, such as for the sale of the product "books". It is essential to have the historical contextual information about this mental model stored with the model, so that a user can delineate where it may or may not be applicable.

Group Bias. Group bias refers to the biases of a team or group with similar backgrounds, jobs, experiences, training, values and goals [24, 41, 42]. Group biases can be removed by allowing users from different backgrounds review one another's experiences (recommendations) [24].

3.2 Human-Centric Approach in Ill-Structured Decision Support

The *concepts* in the human mind about the real world are highly *abstract*. We humans tend to associate these *concepts* with certain words or expressions which do not describe the *concepts* in their entirety. Nevertheless, our mind interprets the *abstract concepts* behind those words and expressions correctly. This ability to make sense out of *abstract* or *vague* information essentially becomes the key to dealing with complex and unclear situations, and to reaching suitable solutions to such problems [48].

In contrast to the human ability to perceive, intuitively recognize, understand and communicate, a computer system works on a pre-defined set of instructions, each of

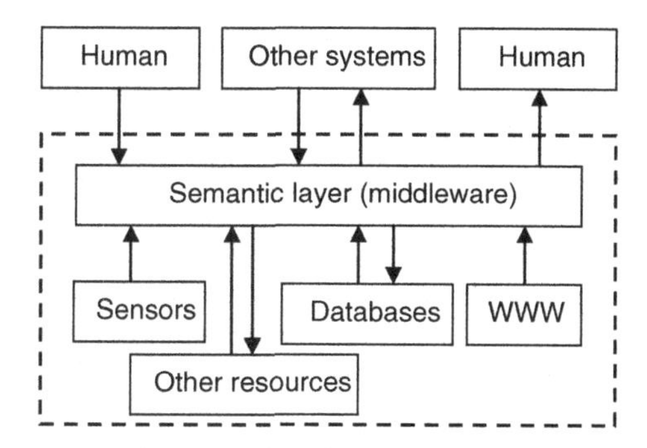

Fig. 1 Architecture of human-centric system. Taken from [37]

which is designed to perform a certain task. Computer systems can only recognize or understand the information for which they are programmed. There are obvious inadequacies in computer systems when it comes to *intelligence*, and it may be a long time before machines are enabled with human-like *intelligence*. Nevertheless, they can be programmed to interface with humans in a more human-like, intuitive and natural way, so that *knowledge* can be seamlessly transferred between humans and computers. This is where HC^2 comes in. The basic idea behind HC^2 is to make computer interfaces *natural, intuitive* and somewhat *intelligent* for humans. The architecture of HC^2 is given in Fig. 1.

Since this chapter focuses on incorporating *cognition* into DSS through *mental model representation*, HC^2 techniques that are suitable for creating seamless communication between human cognition and *mental model representation* will be discussed. These techniques are *natural language processing (NLP)* and *semantics*. Following is an overview of these techniques.

3.2.1 Natural Language Processing

The very basic form of communication among individuals is *natural language*, such as English. A natural language is a rich set of linguistic constructs, such as *the parts of speech*. Humans comprehend the meaning behind the *language* not only by these constructs, that is, *words*, but also by their interrelationships and their placement in the text [12]. A word may represent different contexts when used in different places in a sentence or with different words. For instance, adding one word, "to", to the word "ran" in the following sentences signifies two different meanings:

(a) She ran *to* the hotel.
(b) She ran the hotel.

The inherent nature of computer instructions, on the other hand, is predefined. Each command performs exactly the same task each time it is run. This nature of computer systems has some drawbacks when it comes to communication between the user and the computer.

- Only a fraction of *information* or *knowledge* can be transferred to the system due to inherent limitations in processing and *understanding* the information; hence the *loss of knowledge.*
- The user must learn proper syntax before being able to communicate with the system
- The user must convert their ideas, queries, or information from an *intuitive* format to a system-defined format before they can be presented to the system. This fact implies that the focus of the user will be more on the communication format, rather than the actual task.

As discussed in Sect. 3.1, it is essential to prevent the *loss of knowledge* during the communication between a decision maker and the system in order to provide better support for ill-structured decision making. A NLP equipped user interface allows the user to input their *knowledge* and *cognition* in the *intuitive* manner, as well as *query* the system in *natural language*, without having to convert their thoughts into *computer system* syntax. In this way, the decision maker will be able to focus more on the problem at hand, rather than on the method of interaction with the system. Following is a brief overview of NLP.

The goal of NLP is to investigate the factors behind the comprehension of complex *natural language* constructs by humans, and explore the ways to implement them into computer systems [12]. NLP addresses the issue of how to make machines understand a *word, sentence, paragraph*, and eventually a *document* in their entirety.

To enable machines to *understand* natural language, it is first essential to recognize the factors which assist humans to correctly comprehend the meaning behind a *word, sentence, paragraph* or whole *document*. According to Feldman [12], humans process language on seven levels of *understanding* to extract the context behind it: *phonological, morphological, lexical, syntactic, semantic, discourse* and *pragmatic.* Following is a brief description of these levels.

- **Phonological level.** This level refers to the pronunciation of words and is important for speech recognition systems.
- **Morphological level.** This level deals with the *morphemes.* A morpheme is the smallest meaningful unit of a language. Prefixes, suffixes and root words are examples of morpheme.
- **Lexical level.** This level deals with the dictionary meanings of words and their analysis in terms of *parts of speech.*
- **Syntactic level.** Syntax deals with the meaning conveyed by the structure of a sentence. This level helps to identify a word's meaning and grammatical classification (part of speech) depending on its location in the sentence, even when its actual meaning is unknown.
- **Semantic level.** This level refers to the meaning of the words themselves, within the context of the sentence in which they appear. For example, the word "draw" by itself may bring to mind "to draw a picture". When it used in "*Draw* the drapes please", however, it signifies the entirely different concept of "opening/closing the curtains".

- **Discourse level.** This level determines the structure and the role of a particular part of the text in terms of the structure of the whole document. For example, if a NLP tool knows the standard structure of a research paper, it may determine the nature of a piece of text in the document, based on certain characteristics, such as *location;* that is to say, it can sense whether a piece of text is the abstract, introduction, results or conclusion of a research paper [12].
- **Pragmatic level.** This refers to the external or general knowledge (or *common sense*), that we use to extract the meaning from a document or conversation.

Among these levels, *semantics* has grown to be a field in its own right in computational research. Following is an overview of *semantics.*

3.2.2 Semantics

The study of *semantics* deals with the *meaning* or *context* of words in a language. The subject of *semantics* stems from philosophy and has been under scrutiny for thousands of years. Ancient philosophers such as Plato and Aristotle discussed the topic, as well as labelling the relationship between a *word* and its *meaning.* Plato described this relationship as *naturalist*, stating that the *meaning* of a *word* is derived from the *sound* it makes. Conversely, Aristotle called this relationship *conventionalist*, opining that the relationship between the *meaning* and *sound* of a word is absolutely arbitrary.

There are many facets of semantics and its classification, but for the scope of this chapter, we shall describe semantics in terms of *linguistic semantics* and *non-linguistic semantics. Linguistic semantics* is the study of *meaning* behind spoken or written language. *Non-linguistic semantics*, on the other hand, is the study of *meaning* behind non-verbal expressions and body language. Similarly, in computer systems, *semantics* have been applied at both *linguistic* and *non-linguistic* levels. Before examining *linguistic semantics*, we will briefly describe *non-linguistic semantics* in computer systems.

(a) Non-Linguistic Semantics

As stated above, non-linguistic semantics deals with the *meaning* and *intention* behind non-verbal communication such as gestures and body language. The *semantics* behind non-linguistic communication, such as physical and emotional gestures, when interpreted correctly, can be of high significance in understanding the intentions of a speaker. It can provide the information that linguistic communication may be unable to express. The *virtual doctor system* (VDS) by [14] is an example of such system. VDS extracts the *physical* and *emotional* information from a patient by the use of cameras and other sensory equipment. The system has two ontologies that store the *semantics* of the various *physical* and *mental* gestures of the patient, which, along with *linguistic* input by the patient, assist in diagnosing the patient. This system shows that non-linguistic semantics increases a machine's understanding of a human's intentions during communication.

(b) Linguistic Semantics

The linguistic semantics in computers are applied in three categories: *logical language semantics, programming language semantics* and *natural language semantics. Logical language semantics* represents the meaning (truth values) of certain propositions in terms of set-theoretic models. *Programming language semantics* defines the meaning of programming language commands, that is, which command will perform what task. *Natural language semantics* deals with the *meaning (context or intention of the speaker)* behind the *natural language* constructs.

As discussed in Sect. 2, storing *cognition* is a step towards improving ill-structured decision support. To store *cognition* in the system, *mental model representation* is used. The NLP interface facilitates the transfer of knowledge from the decision maker's mind to the system and stores it in the *mental model representation.* An improvement over *mental model representation* is to integrate *semantics*, so that the *context* and *meaning* behind the *cognition* can be stored along with the *cognition* itself.

From the discussion in previous sections, it can be concluded that to support *ill-structured* decision problems, four aspects need to be considered:

(a) Enable the DSS to provide *cognitive* support to the decision maker by storing *cognition* in the DSS in an intuitive format;
(b) Prevent the *loss of knowledge* during communication between the decision maker and the system by providing a *human-centric* interface to facilitate the seamless transfer of *knowledge*;
(c) Mitigate the bias from the *cognition* stored in the DSS.
(d) Incorporate *semantics* in the *mental model representation* to transfer the *context* of the *knowledge* along with the *knowledge* itself.

In the next section, we will discuss the *semantic de-biased associations (SDA) model*, which incorporates the *human-centric* techniques of NLP and semantics, to enable us to lay a foundation for the transfer and storage of *cognition* in DSS.

3.3 The Conceptual Model of Semantics-Driven Cognitive Decision Support System (SCDSS)

We have proposed a *human-centric* CDSS, namely *semantics-driven cognitive decision support system (SCDSS).* The conceptual model of SCDSS is illustrated in Fig. 2. The system is designed to support a decision maker's cognition during decision making process by the following process. A decision maker develops ideas, perceptions, skills and intuition over time through experiences, which generate *mental models.* SCDSS proposes to apply NLP techniques to retrieve these mental models in an intuitive manner, and pass then to the de-biasing mechanism. The de-biasing techniques allow to eliminate any biases that have been introduced by the personal preferences

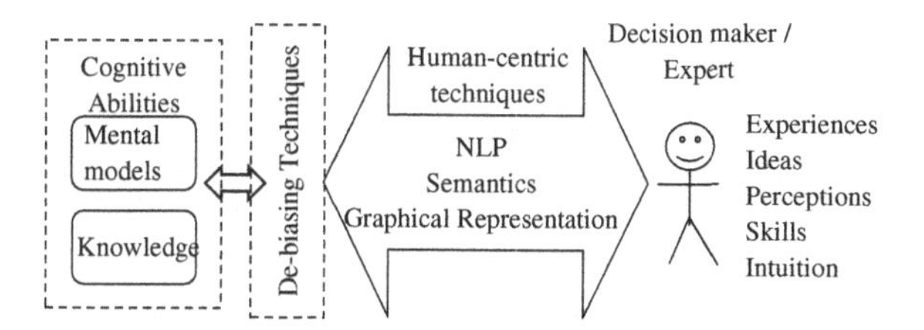

Fig. 2 The conceptual model of semantics-driven cognitive decision support system (SCDSS)

or inclinations of the decision maker. The de-biased *mental models* and their corresponding *knowledge* can then be stored in the system. Later at the time of decision making, the required de-biased *mental models* can be retrieved from the storage and presented to the decision maker, thus supporting them to make better informed decisions.

The focus of this paper is, however, a component of the SCDSS, which is *semantic de-biased associations (SDA) model.* The SDA model [32] is proposes an improvement over the conventional *mental model representation* technique, by incorporating *semantics* in it. This approach has certain advantages in terms of de-biasing and accurate representation of human cognition in a *human-centric* format. The following section describes the SDA model and its benefits in detail.

4 Semantic De-Biased Association Model

The SDA model is designed to store *cognition* in human-like format and to provide an interface which allows a decision maker to communicate with the system in an *intuitive* way. This enables the decision maker to concentrate on the decision problem, rather than on the format of the communication.

The SDA model allows the storage of cognition in a *natural* format by incorporating *semantics* in the conventional *mental model representation* (Fig. 4). Following is an overview of the SDA model architecture, illustrating the *semantic mental model representation.*

4.1 The Architecture of the SDA Model

The architecture comprises three main components: *cognitive knowledge base (CKB)*, the *knowledge management layer*, and the *user interface* (Fig. 3). The core component of the system is CKB, which stores *cognition* through *concepts, associations*, and *cases*, described as follows:

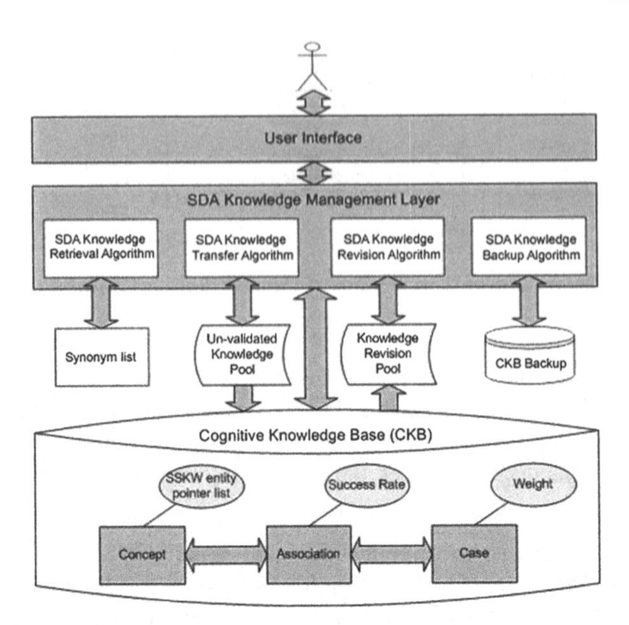

Fig. 3 Architecture of the SDA model [32]

- ***Concepts*** are the notions within a *problem domain* (such as *loss of sales*), in the form of a single word, such as "sales", or a collection of words, such as "bad delivery". Each *concept* is assigned a URI for unique identification.
- ***Associations*** are *semantic triples*, which are made up of two *concepts* and a *relationship* between them. Associations show the type, effect and purpose of the relationship between the two attached concepts. An *association* provides the foundation for attaching de-biasing information, and each *association* has many *cases* or none attached to it (*cases* are described below). An *association* is assigned a *success rate*, which is an average of its success for all the attached *cases*.
- ***Cases*** are the past decision problems attached to *associations*. These *cases* show past decision situations, and the role of the *association* in their solution. Every *case* has a *weight* for each *association* used in solving the *case*. The *weight* shows the degree of importance of the corresponding *association* in solving the *case*.

The CKB is managed by the *SDA knowledge management layer*, which consists of four algorithms. The *SDA knowledge retrieval algorithm* is designed to fetch knowledge from the CKB according to the user query. The algorithm uses the *synonym list* to change the verbs used in the user query in the *association* names in the CKB. The *SDA knowledge transfer algorithm* helps users to transfer their personal experiences and mental models to the CKB. These experiences are transferred to the *un-validated knowledge pool*, until they are verified by other users. Once the knowledge is verified by the verification process defined by the *knowledge transfer algorithm*, it is forwarded to the CKB. The *SDA knowledge revision algorithm* is used to update, delete or refine the old or out-dated knowledge in the CKB. The *SDA knowledge backup algorithm* backs up the knowledge, which has been proven to be useful for the problem domain, to the CKB backup.

At the time of decision making, the retrieval of particular mental models is carried out by identifying *concepts* and *associations* in the user's query, and by fetching related mental models from the CKB, which makes the process fast and reliable, retrieving only relevant and precise knowledge.
System users have unique login identities, with different privileges. Users are divided into three categories: *domain experts, managers* and *employees.*

- **Domain experts** are specialists in the field who have more than 10 years' successful decision making experience;
- **Managers** are executives who are qualified for the decision maker's job but who do not have as much experience as *domain experts;*
- **Employees** are subordinates of the *managers*, who have vast experience of the organisation and its market. Employees are authorized by *managers* to use the system.

4.2 Human-Centric Approach in the SDA Model

Following is an overview of the *human-centric* approaches used in the SDA model and their advantages in preventing the *loss of knowledge* and improvement of *ill-structured decision support.*

4.2.1 Human-Centric Cognition Storage Using Semantics

Figure 4 illustrates how a *semantic* mental model relationship is defined in the SDA model, compared to the conventional mental model relationship. A conventional causal relationship is an undefined directed link from one concept to another (Fig. 4a); whereas in the SDA model, a relationship is *semantically* defined and *labelled* (Fig. 4b). This *semantic relationship*, along with the two attached *concepts*,

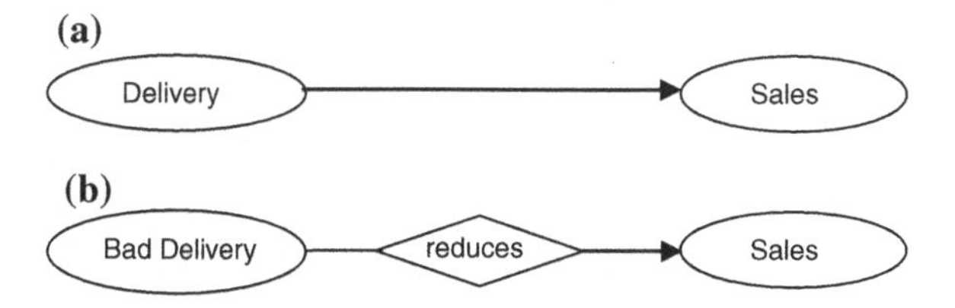

Fig. 4 **a** Relationship in conventional mental model representation; **b** Relationship (association) in proposed SDA mental model representation

forms a *semantic triple*, such as "bad delivery *reduces* sales". The purpose of applying *semantics* is twofold: (a) it allows the storage of *cognition* in a natural format, which matches the mental model representation in the human mind; (b) it allows storage of the context behind the *cognition*, preventing the *loss of knowledge*. The *semantics* provides a basis for the assignation of distinctive *contextual information (cases)* and performance measuring parameters *(weight* and *success rate)* to the stored *mental models (associations)*. These elements, i.e. *cases, weights* and *success rate*, provide objective information about a particular *association* to the decision maker, thereby assisting to reduce bias. Moreover, labelling a relationship makes it possible to perform an exhaustive search on the knowledge (mental models) in a fast and reliable manner. To the best of our knowledge, this is the first effort to represent mental models *semantically*, in the form of *triples*. The next section describes the *human-centric interface* in detail.

4.2.2 Seamless Transfer of Cognition Through NLP

As can be seen from the architecture of the SDA model, the *cognition* of a decision maker can be stored in a *natural* manner with the help of *semantic mental model representation*. The SDA model provides a *human-centric interface* for communication between the system storage and the decision maker, which allows the input of *knowledge* in *natural language* format. Figure 5 gives an example that illustrates the effortless process of transferring *cognition* from a user's mind to the CKB.

Figure 5 illustrates the process of the *seamless transfer of cognition* provided by the SDA model. A salesman discovers that when he provides *pre-sales services* to a customer, the customer is more likely to buy the product. This experience of the salesman forms the *mental model* in his mind, which he can describe in *natural language* as "pre-sales service increases sales". The *human-centric* user interface of the SDA model allows this salesman to transfer this mental model as it is into the system. Thus, the salesman enters "pre-sales service *increases* sales". The system

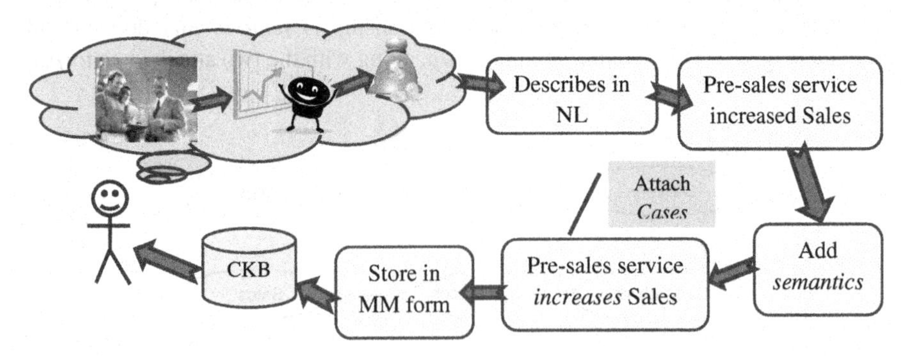

Fig. 5 Seamless transfer of cognition

asks the salesman for any previous experience which has led him to conclude that "pre-sales service *increases* sales". The salesman can choose whether he wants to give this information to the system. The system stores this mental model as *semantically-defined cognition.*

4.2.3 Intuitive Graphical Representation as Output

It is said that a picture is worth a thousand words. When it comes to complex situations, a pictorial representation makes more sense to the human mind, more quickly, than reading lengthy reports defining the situation. Studies in decision making show that not only do the pictorial representations of a situation make a user quickly understand the situation, but also that they assist in avoiding bias [20]. Accordingly, the SDA model provides the results of a decision query in the form of a graphical representation of the stored *cognition* (Fig. 8). The graphical output assists the user to observe a situation clearly without missing any aspects of that situation, which may not be the case with text-only output. The graphics help the user locate the root of the problem quickly and easily.

4.3 Bias Mitigation in SDA

The SDA model mitigates four biases: *availability, framing,* and *contextual*, and *group biases.* The process of mitigating these biases is given below.

4.3.1 Availability Bias

The SDA model allows *concepts* to have multiple *associations* among themselves for multiple problem domains, such as *loss-of-sales* and *new-product-launch.* Users (*domain experts, managers* and *employees*) can record their experiences about different problem domains in the form of *concepts* and *associations.* These experiences are merged according to the *concepts*, with the help of *concept* identifiers (URIs), forming a comprehensive mental model about the problem domain. Note that mental models are not separated according to the problem domain; rather, they are connected to one another according to user experiences, thus creating a larger, all-inclusive mental model of the *application domain.* Since there is a wide range of experiences stored in the CKB the decision maker is presented with an extensive choice of possible decision alternatives. This helps the decision maker to recall previous experiences regarding all possible decision alternatives presented by the system, which would not have been remembered otherwise due to the *availability bias.* As a result, the *availability bias* is reduced, and in some case eliminated, with the help of the SDA-based CDSS.

4.3.2 Framing Bias

Cases are added to an *association* to add *contextual information* related to the decision situations in which the *association (knowledge)* was used to reach the solution. *Cases* present different perspectives on the particular *association*, such as where and whether the *association* has been successful in the past, and in which circumstances it has failed to assist in reaching an optimal solution. By providing the user with all the perspectives, positive and negative, on the *association*, the proposed system helps reduce or eliminate *framing bias.*

4.3.3 Contextual Bias

The importance or usefulness of an *association* for a particular *case* is measured by a *weight* assigned to the *case* regarding that *association.* A higher *weight* indicates the importance of the *association* in the attached *case (decision situation)*, or its success, whereas, a lower *weight* shows that the *association* was not particularly useful in solving a certain decision problem. Having a *weight* assigned to the *case* helps to identify the decision situations in which the *association* was most or least useful, which in turn assists in determining where the *association* can be applied successfully in the future to obtain optimal results. This helps to reduce or eliminate *contextual bias.* Different categories of users assist with the elimination of *group biases*, which tend to be similar among a homogeneous group of people due to their similar backgrounds, jobs, experiences, training, values and goals [41, 42]. By allowing users from different backgrounds to participate in the knowledge sharing and validation, such biases can be reduced to a great extent.

4.3.4 Group Bias

Group bias is the collective bias of a team or group with similar backgrounds, jobs, experiences, skill sets and goals [24, 42]. This bias can be mitigated by allowing people from different groups to review the experiences (cognition) of people from other groups [24]. Keeping this in mind, users of the SDA model are divided into three categories according to their background, experiences, skills, goals and position in the organisation: namely, *domain experts, managers*, and *employees.* The users of the SDA-based system are fundamental to the *knowledge validation process.* They cause the group bias present in the knowledge input to be neutralized, thereby mitigating the group bias.

4.4 The SDA Knowledge Cycle

A decision maker starts with some *experience* already acquired through training, study or observation. This *experience* generates *mental models.* These mental models are extracted by allowing the user to define these mental models in natural lan-

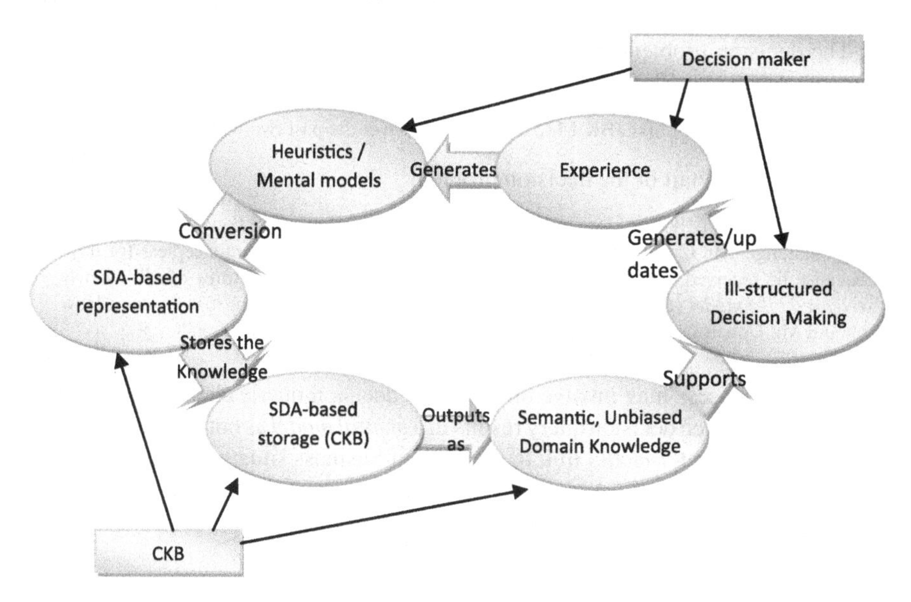

Fig. 6 SDA knowledge cycle

guage. Following the *validation process*, the mental models are sent to the *cognitive knowledge base* (*CKB*). The mental models are *de-biased* during the *definition* and *validation* process, and *semantics* are added to them. They are stored in the CKB in the natural format of semantic triples. Later, when a user (decision maker) queries the system regarding a decision situation, the system fetches the *semantically* defined, *de-biased knowledge (cognition)* from the CKB, and presents it to the user, to support ill-structured decision making. The decision maker, while using this knowledge, may learn or synthesize new knowledge from the mental models presented by the system, or discover a new pattern during the process. This will augment or update his previous knowledge, adding to his experience. Upon discovering a new pattern, or synthesizing new knowledge, the decision maker may decide to add this newly-acquired mental model to the system, thus continuing the *knowledge cycle* (Fig. 6).

5 Case Study

To test out the effectiveness of SDA model in supporting ill-structured decision maker, a DSS based on the SDA model is developed, called *semantics-driven cognitive decision support system (SCDSS)*. In this section, the process of solving an ill-structured decisionproblem with the help of SCDSS is illustrated through an example. The decision problem is taken from Niu et al [35] and is based on a fictitious company, *Adventure Works*.[9] The decision query is as follows:

> Why have the sales of bike (BK-M82S-38) dropped over the past two weeks?

Following is the detail of the decision problem:

> *Adventure Works (AW)* has dominated the market for over one year; however, they face a big challenge. Mr. Cobarol, the chief executive officer of *AW*, has been sleepless for days because he received very bad news from the marketing department: *sales of AW's newly released bike model (BK-M82S-38) have dropped over 40% over the past two weeks*. How should Mr. Cobarol respond appropriately and reverse this difficult situation?

The decision process may involve one or more decision queries being made by the decision maker. Each decision query results in a *mental model* as output. The *concepts* of the resultant *mental model* link to the current business situation regarding those *concepts*, whereas the *associations* link to their corresponding *cases, success rate*, and *weights* in the problem domain.

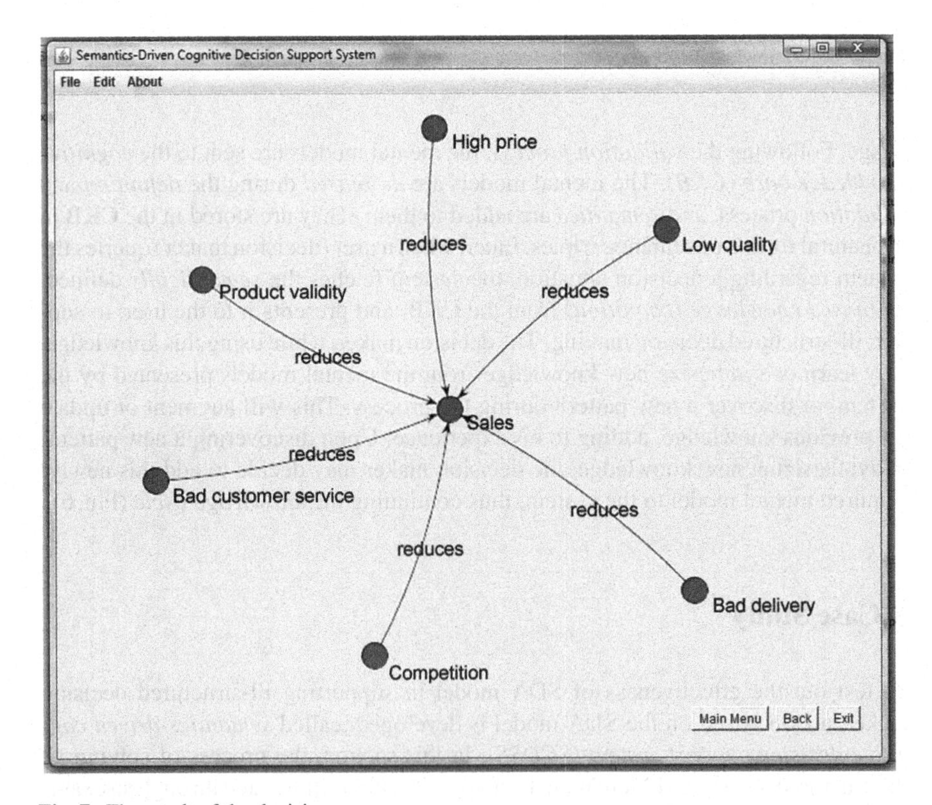

Fig. 7 The result of the decision query

[9] AdventureWorks is a sample database which comes with SQL Server installation [1]

As the first step, the system presents the user with three options: (1) Add knowledge to the system; (2) Query the system about a decision situation; (3) Ask strategic questions from other users. To query the system about a decision problem, the user will have to choose the second option: "Query the system about a decision process". This option allows the user to input the decision query. For this case study, the user inputs the following query.

Why have the sales of our product BK-M82S-38 been dropping for the past two weeks?

The system then fetches the corresponding mental model, along with the current state of the object(s) in the query from the backend, and displays the output in graphical format (Fig. 7). Clicking on a concept will open a corresponding summarized report on the business object in question. For example, clicking on the *bad delivery* will open the delivery status of the product *BK-M82S-38*. Clicking on an association, such as *bad delivery reduces sales*, will show the list of *cases* in which the *association* has been used in the past, as well as the *success rate* and *weight of* the *association* (Fig. 8). The decision maker can then compare the environmental parameters of the situations in which the association (decision alternative) has been successful with the current decision situation and base his choice on the comparison.

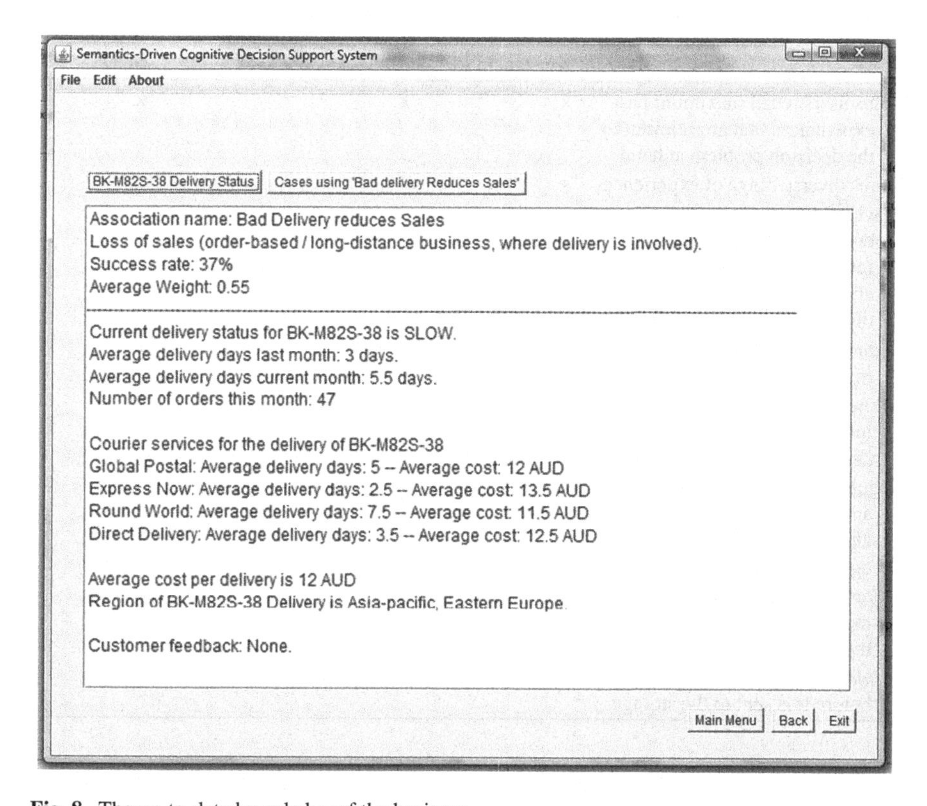

Fig. 8 The up-to-date knowledge of the business

5.1 Comparison

The solution to the decision problem stated above is also achieved with FACETS, the CDSS proposed by Li Niu [35]. FACETS focuses on improving the *situation awareness* of a decision maker. Given this decision problem, the decision maker inputs their observation of the situation in the CDSS. According to the input, the system fetches the mental model along with the data from the *data warehouse (DW)*. However, due to the conventional *mental model representation* in FACETS, which stores *concepts* connected with *undefined links*, the system takes four *decision cycles* to produce a comprehensive mental model, relevant to the situation. The SDA-based SCDSS, on the other hand, provides an improved*mental model representation*, which stores *concepts* connected with *semantically defined links.* The *semantics* incorporated in these links allows for an *efficient* and *accurate* searching, which help SCDSS to extract the relevant mental models in the first decision cycle. Thus the SCDSS offers improved performance in terms of efficiency and accuracy of the produced results. Table 1 summarizes the advantages of SDA-based SCDSS over the previous CDSS.

Table 1 Default simulation settings

Criteria	Yadav & Khazanchi [53]	Chen & Lee [8]	Li Niu at al. [35]	SDA-based SCDSS
Helps users recall maximum past experiences that are relevant to the decision problem at hand	x	x	–	x
Stores diverse range of experiences	x	x	x	x
Provides the functionality to continuously input and integrate mental models, allowing progressive evolution of the stored knowledge	–	–	–	x
Reduces/removes biases from the mental models	–	x	–	x
Allows to store the contextual information behind every *decision alternative* generated	–	–	–	x
Regularly measures the usefulness and accuracy of each decision alternative; and discard the *out-dated* ones	–	–	–	x
Stores the *cognition* in a meaningful way by incorporating *semantics*	–	–	–	x
Validates each *decision alternative* before it is sent to the storage	–	–	–	x

6 Conclusions

This chapter discusses the role of human-centric techniques in improving *ill-structured* decision support. A *cognition-driven* model, namely, the *semantic de-biased associations (SDA) model*, is described. The SDA model employs three human-centric techniques: NLP, semantics and the graphical representation of cognition. The NLP allows the seamless transfer of *cognition* from the decision maker to the system. The semantics assist in storing *cognition* in a human-centric format, along with the *context* behind the *cognition/knowledge*. The graphical representation is used to output the stored *cognition*, to help the user comprehend the decision situation quickly, as well as to identify the root of the problem. With the help of these human-centric techniques, the SDA model prevents the *loss of knowledge* during the communication between the decision maker and the DSS. The SDA model helps to save decision making time of users by fetching only precise and relevant knowledge about a decision situation through the use of *semantics*. The SDA model can effectively assist decision makers in the complex domains of *business* and *medicine*, where the situation information is limited or vague, and *time* and the *precision of knowledge* are the most critical success factors.

References

1. AdventureWorks Database' 2005, AdventureWorks Database, Microsoft (2011) http://msftdbprodsamples.codeplex.com/
2. Alter, S.L.: Decision Support Systems: Current Practice and Continuing Challange, 1st edn. Addison-Wesley, Mass (1980)
3. Ariely, D.: Predictably Irrational: The Hidden Forces That Shape Our Decisions, 1st edn. HarperCollins, Canada (2008)
4. Arnott, D.: A Taxonomy of Decision Biases (Technical report. No. 2002/01) Decision support systems laboratory. Monash University, School lof Information Management and Systems, Melbourne (2002)
5. Barnes, J.H.: Cognitive biases and their impact on strategic planning. Strateg. Manage. J. **5**(2), 129–137 (1984)
6. Bhandari, G., Hassanein, K., Deaves, R.: Debiasing investors with decision support systems: an experimental investigation. Decis. Support Syst. **46**(1), 399–410 (2008)
7. Chandrasekaran, B.: Designing decision support systems to help avoid biases and make robust decisions, with examples from army planning. Paper presented to the proceedings of army science conference 2008
8. Chen, J.Q., Lee, S.M.: An exploratory cognitive DSS for strategic decision making. Decis. Support Syst. **36**(2), 147–160 (2003)
9. Craik, K.: The Nature of Explanation, 1st updated edition edn. Cambridge University Press, Cambridge (1943)
10. Croskerry, P.: Achieving quality in clinical decision making: cognitive strategies and detection of bias. Acad. Emer. Med. **9**(11), 1184–1204 (2002)
11. Das, T.K., Teng, B.S.: Cognitive biases and strategic decision processes: an integrative perspective. J. Manage. Stud. **36**(6), 757–778 (1999)
12. Feldman, S.: NLP meets the jabberwocky: natural language processing in information (1999) Retrieval, http://www.scism.lsbu.ac.uk/inmandw/ir/jaberwocky.htm

13. Ferguson, R.L., Jones, C.H.: A computer aided decision system. Manage. Sci. **15**(10), 550–562 (1969)
14. Fujita, H., Hakura, J., Kurematsu, M.: Virtual doctor system (VDS): medical decision reasoning based on physical and mental ontologies. Paper presented to the proceedings of the 23rd international conference on industrial engineering and other applications of applied intelligent systems—Volume Part III. Cordoba, Spain, 2010
15. Gorini, A., Pravettoni, G.: An overview on cognitive aspects implicated in medical decisions. European J. Intern. Med. **22**(6), 547–553 (2011)
16. Greeno, J.G.: Natures of problem-solving abilities. In: Handbook of Learning and Cognitive Processes: V, pp. 239–270. Human information, Lawrence Erlbaum, Oxford (1978)
17. Grief, I.: Computer-supported cooperative work: A book of readings, vol. 1, 1st edn. Morgan Kaufmann Publishers, Inc., San Mateo (1988)
18. Hicks, E.P., Kluemper, G.T.: Heuristic reasoning and cognitive biases. Am. J. Orthod. and Dentofac. Orthop. **139**(3), 297–304 (2011)
19. Hilbert, M.: Toward a synthesis of cognitive biases: how noisy information processing can bias human decision making. Psychol. Bull. **138**, 211–237 (2012)
20. Hodgkinson, G.P., Bown, N.J., Maule, A.J., Glaister, K.W., Pearman, A.D.: Breaking the frame: an analysis of strategic cognition and decision making under uncertainty. Strateg. Manage. J. **20**(10), 977–985 (1999)
21. Janser, M.J.: Army War College. Senior Service College Fellowship, P., Center for, S. and International, S: Cognitive biases in military decision making. Army War College, U.S. (2007)
22. Johnson-Laird, P.: The history of mental models. In: Manktelow, K., Chung, M.C. (eds.) Psychology of Reasoning: Theoretical And Historical Perspectives, p. 382. Psychology Press, New York (2004)
23. Johnson-Laird, P.N.: Mental models: Towards a Cognitive Science of language, inference, and Consciousness. Harvard University Press, Cambridge (1983)
24. Kahneman, D., Lovallo, D., Sibony, O.: The big idea: before you make that big decision. Harvard Bus. Rev. **89**, 50–60 (2011)
25. Kahneman, D., Tversky, A.: Subjective probability: a judgment of representativeness. Cogn. Psychol. **3**(3), 430–454 (1972)
26. Korte, R.F.: Biases in decision making and implications for human resource development. Adv. Develop. Hum. Resources **5**(4), 440–457 (2003)
27. Krause, T.R., Hidley, J.H.: Protecting your decision making from cognitive bias. In: Taking the Lead in Patient Safety, pp. 149–176. John Wiley & Sons, Inc., New York (2008)
28. Liddy, E.D.: Enhanced text retrieval using natural language processing. Bull. Am. Soc. Inform. Sci. Technol. **24**(4), 14–16 (1998)
29. Lucas, H.C.: Computer-Based Information Systems in Organization. Science Research Associates, Chicago (1973)
30. Lucchiari, C., Pravettoni, G.: Cognitive balanced model: a conceptual scheme of diagnostic decision making. J. Eval. Clin. Pract. **18**(1), 82–88 (2012)
31. Maqsood, T., Finegan, A.D., Walker, D.H.T.: Biases and heuristics in judgment and decision making: the dark side of tacit knowledge. Issues Inform. Sci. Inform. Technol. **1**, 295–301 (2004)
32. Memon, T., Lu, J., Hussain, F. K.: Semantic de-biased associations (SDA) model to improve ill-structured decision support. In: Huang, T., Zeng, Z., Li, C., Leung, C. (eds.), Neural Information Processing. Lecture notes in computer science, vol. 7664, Springer berlin heidelberg, 483–490 (2012)
33. Mintzberg, H.: The Nature of Managerial Work. Harper and Row, New York (1973)
34. Nemati, H.R., Steiger, D.M., Iyer, L.S., Herschel, R.T.: Knowledge warehouse: an architectural integration of knowledge management, decision support, artificial intelligence and data warehousing. Decis. Support Syst. **33**, 143–161 (2002)
35. Niu, L., Lu, J., Zhang, G.: Cognition-driven decision support for business intelligence, SpringerLink, Heidelberg (2009)
36. Nonaka, I.: The knowledge-creating company. Harvard Bus. Rev. **69**, 96–104 (1991)

37. Pedrycz, W., Gomide, F.: Fuzzy Systems Engineering: Towards human-centric computing, John Wiley & Sons, Inc., United states (2007)
38. Pervan, G., Arnott, D., Dodson, G.: Trends in the study of group support systems, In: Proceedings of Group Decision and Negotiation 2005. INFORMS/University of Vienna, Vienna, 2005
39. Power, D.J.: A Brief History of Decision Support Systems. DSS, DSSResources.com, viewed Oct 2, 2009 http://dss.cba.uni.edu/dss/dsshistory.html (1999). Version 2.8, 31 May 2003
40. Power, D.J.: Decision support systems: A historical overview. In: Handbook on Decision Support Systems 1, vol. 1. Springer Berlin, Heidelberg (2008)
41. Rentsch, J.R., Hall, R.J.: Members of great teams think alike: A model of team effectiveness and schema similarity among team members. In: Beyerlein, M.M., Johnson, D.A. (eds.), Advances in interdisciplinary studies of work teams: Theories of self-managing work teams, vol. 1, Elsevier Science/JAI Press, (1994)pp. 223–261.
42. Rentsch, J.R., Klimoski, R.J.: Why do 'great minds' think alike?: antecedents of team member schema agreement. J. Organ. Beh. **22**(2), 107–120 (2001)
43. Schwenk, C.R.: The cognitive perspective on strategic decision making. J. Manage. Stud. **25**(1), 41–55 (1988)
44. Scottmorton, M.S.: Management Decision Systems—Computer-Based Support for Decision Making—Morton, MSS. Sloan Manage. Rev. **12**(3), 103–104 (1971)
45. Simon, H.A.: The structure of ill structured problems. Artif. Intell. **4**(3–4), 181–201 (1973)
46. Simon, H.A.: Information-processing theory of human problem solving. In: Estes, W.K. (ed.) Handbook of Learning and cognitive Process, p. 318. Lawrence Erlbaum, HIllsdale (1978)
47. Simon, H.A., Newell, A.: Heuristic problem solving: the next advance in operations research. Oper. Res. **6**(1), 1 (1958)
48. Swanson, E.B., Culnan, M.J.: Document-based systems for management planing and control: a classification, survey, and assessment. MIS Q. **2**(4), 31–46 (1978)
49. Tversky, A., Kahneman, D.: Judgment under uncertainty: heuristics and biases. Science **185**(4157), 1124–1131 (1974)
50. Tversky, A., Kahneman, D.: The framing of decisions and the psychology of choice. Science **211**(4481), 453–458 (1981)
51. Ubel, P.A., Smith, D.M., Zikmund-Fisher, B.J., Derry, H.A., McClure, J., Stark, A., Wiese, C., Greene, S., Jankovic, A., Fagerlin, A.: Testing whether decision aids introduce cognitive biases: results of a randomized trial. Patient Educ. Couns. **80**(2), 158–163 (2010)
52. Vosniadou, S.: Mental models in conceptual development. In: Magnani, L., Nersessian, N. (eds.) Model-Based Reasoning: Science, Technology, Values. Springer, New York (2002)
53. Yadav, S.B., Khazanchi, D.: Subjective understanding in strategic decision making: an information systems perspective. Decis. Support Syst. **8**(1), 55–71 (1992)

Decision-Making Under Conditions of Multiple Values and Variation in Conditions of Risk and Uncertainty

Ewa Roszkowska and Tom R. Burns

Abstract Empirical research shows that humans face many kinds of uncertainties, responding in different ways to the variations in situational knowledge. The standard approach to risk, based largely on rational choice conceptualization, fails to sufficiently take into account the diverse social and psychological contexts of uncertainty and risk. The article addresses this challenge, drawing on sociological game theory (SGT) in describing and analyzing risk and uncertainty and relating the theory's conceptualization of judgment and choice to a particular procedure of multi-criteria decision-making uncertainty, namely the TOPSIS approach. Part I of the article addresses complex risk decision-making, considering the universal features of an actor's or decision-maker's perspective: a model or belief structure, value complex, action repertoire, and judgment complex (with its algorithms for making judgments and choices). Although these features are universal, they are particularized in any given institutional or sociocultural context. This part of the article utilizes SGT to consider decision-making under conditions of risk and uncertainty, taking into account social and psychological contextual factors. Part II of the article takes up an established method, TOPSIS with Belief Structure (BS), for dealing with multi-criteria decision-making under conditions of uncertainty. One aim of this exercise is to identify correspondences between the SGT universal architecture and the operative components of the TOPSIS method. We expose, for instance, the different value components or diverse judgment algorithms in the TOPSIS procedure. One of the benefits of such an exercise is to suggest ways to link different decision methods and procedures in a comparative light. It deepens our empirical base and understanding

E. Roszkowska (✉)
Faculty of Economy and Management, University of Bialystok, Warszawska 63, 15-062 Bialystok, Poland
e-mail: erosz@o2.pl

T. R. Burns
Woods Institute for Environment and Energy, Stanford University, California and Department of Sociology, University of Uppsala, Box 821, 75108 Uppsala, Sweden
e-mail: tom.burns@soc.uu.se

P. Guo and W. Pedrycz (eds.), *Human-Centric Decision-Making Models for Social Sciences*, Studies in Computational Intelligence 502,
DOI: 10.1007/978-3-642-39307-5_13,

of values, models, action repertoires, and judgment structures (and their algorithms). The effort here is, of course, a limited one.

Keywords Complex decision making · Multiple values · Risk · Uncertainty · Judgment algorithms · Fuzzy judgment · TOPSIS method

1 Introduction

Risk has become understood among many researchers as a type of measured uncertainty concerning possibilities that entail a loss, damage, or other undesirable outcome. Some risk researchers distinguish sharply between risk and uncertainty, to stress the different degrees of knowledge facing decision-makers and the boundedness of risk calculations. Empirical research shows that humans face many kinds of uncertainty and respond in different ways to the variation in situational knowledge. The standard approach to risk, based largely on rational choice theory, fails to sufficiently take into account the diverse social and psychological contexts of uncertainty and risk. This article addresses this challenge, offering a new theoretical perspective on analyzing risk and uncertainty and relating a general theory of judgment and choice to a particular procedure of multi-criteria decision-making under uncertainty, namely the TOPSIS approach.

Part I of this article considers several key aspects of decision-making under risk and uncertainly, suggesting that risk and uncertainty judgments can be contextualized applying a few key concepts and principles of Sociological Game Theory (SGT). The analysis implies a variety of risk conceptions and models highly dependent on social and psychological contexts. According to SGT, social and psychological contexts can be specified by, among other things, the system of values of decision-makers; the degree of integration (commensurability) of values and likelihood estimates; the level and types of knowledge they have access to in the decision situation; and the particular judgment algorithms decision-makers are predisposed to use in a given context.

Part II applies the TOPSIS procedure of multi-criteria decision-making under conditions of uncertainty. One aim of this exercise is to show the application of a multi-criteria decision-making to a complex decision situation and at the same time to identify correspondences between the SGT universal architecture and the components of the TOPSIS method. This exercise exposes, for instance, the different value components and diverse judgment algorithms in the TOPSIS procedure. One of the benefits of such an exercise is to suggest ways to link different decision methods and procedures in a comparative light. It deepens our understanding of values, models, action repertoires, and judgment structures (and their algorithms) and their role in decision-making under conditions of uncertainty and risk.

2 Part I.Decision-Making Under Conditions of Risk and Uncertainty

2.1 From Risk to Uncertainty Back to Risk Again

The standard distinction between risk and uncertainty—and many do not accept this distinction—is misleading. Risk has become understood in decision and related sciences as a type of measured uncertainty concerning possibilities that denote a loss, damage, or other undesirable outcome.[1] Risk conceptions have diffused throughout modern society. From the 1700s on, the conceptions were applied to gambling problems. Applications spread until today they are found in business, government, law and judicial systems, engineering, new technologies, environment, climate change, among others [39].

The rational foundation for decision making under risk according to expected utility rules was formulated by John von Neumann and Oskar Morgenstern (1944) in their work *Theory of Games and Economic Behaviour* [48]. Their formulation has been modified and developed in diverse ways: expected utility hypothesis, state-preference approach to uncertainly, weighted expected utility, or non-linear expected utility, non-additive expected utility, prospect theory, among others. Some economists dispute the distinction between Knightian risk and uncertainty, arguing that they are one and the same thing. In Knightian uncertainty, the problem is that the agent *does not* assign probabilities, and not that she actually *cannot*, i.e. that uncertainty is really a problem of "knowledge" of the relevant probabilities, not of their "existence". Some economists also claim that there are actually no probabilities out there to be "known" because probabilities are really only "beliefs". Only very rarely are probabilities known with certainty. Therefore, the clear cases of "risk" (known probabilities) seem to be idealized textbook cases. Real life cases are characterized by uncertainty that does not come with exact probabilities. Hence, most of decisions are decisions "under uncertainty". Decisions "under risk", this does not mean exactly that these decisions are made under conditions of completely known probabilities. Rather, it means that researchers (and practitioners) have chosen to simplify their description of these decision problems by treating them as cases of known probabilities. That means that Knightian "uncertainty" may be the only relevant form of randomness for economics. Knightian "risk" is only possible in some specific and controlled scenarios when the alternatives are clear and experiments can be repeated; but this conception does not apply in many economic decision-making situations in the "real world", where the situations are usually unique and unprecedented, and the alternatives are not really all known or understood. In such situations, mathematical probability assignments usually cannot be made.

---

[1] In decision theory, lack of knowledge is divided into the two major categories '*risk*' and '*uncertainty*' [31], p.20 Chap. 7 where "*risk*" refers to situations where the decision-maker can assign mathematical probabilities to the randomness which he is faced with, where and "*uncertainty*" refers to situations when this randomness "cannot" be expressed in terms of specific mathematical probabilities.

2.2 Background: Simon, Tversky and Kahneman Revisited

Simon [43] did more than give us the idea of "bounded rationality." He stressed that humans typically operate with multiple values, "a vector of values". One should not assume, he proposed, commensurability of these values and the reducibility of multiple values to a common metric. He stressed also the role of decision procedures or algorithms to deal with complexity and uncertainty. He anticipated meta-level processes in judgment, in particular, the significance of meta-level judgment and innovation in decision-making. Simon believed that agents are faced with uncertainty about the future as well as costs in acquiring information. These factors limit the extent to which agents can make a fully rational decision and leaves them with "bounded rationality", they must make decisions by "satisficing," or choosing an option which will leave them satisfied "enough" but which might not be optimal. The term bounded rationality designated a type of rational choice theory that takes into account the cognitive and knowledge limitations of decision-makers. Bounded rationality has become an established notion in behavioral economics. Behavioral economics, which systematically links psychology to economics, emerged as a challenge to some parts of conventional economics, in particular standard expected utility or rational choice theory. It can be traced back to [30, 43, 45] Tversky and Kahneman (1985), [18, 44]. The work demonstrated that rational economic man, the all-seeing, all knowing figure on which much of contemporary economics had been constructed, was a purely fictional character. Faced with even simple sets of options to choose from, human beings make decisions that are inconsistent, suboptimal, and, sometimes, plain irrational. Rather than thinking things through logically, they rely on misleading rules of thumb and they leap to inappropriate conclusions. Moreover, they are heavily influenced by how the choices are presented to them and, sometimes, respond on the basis of completely irrelevant information ([13], pp. 30–34).

Daniel Kahneman and Amos Tversky, like Simon, recognized the multi-value character of human judgment and decision. Of particular importance in their "prospect theory" were the differences between positive values and negative values. They identified and analyzed many such asymmetries. In contrast to standard expected utility theory, they rejected the straightforward application of probabilities in their models, stressing that probabilities or likelihoods enter into the judgment calculus with "weights".

Both theories stressed and illustrated their differences with *expected utility theory*. While their work is widely recognized, and they became established icons of contemporary social science (Simon and Kahneman both won Nobel prizes in economics), they never really won the war. Rational choice as well as expected utility theory continues to dominate mainstream economics and a considerable part of political science and sociology. One major reason for this is that the Simon, and Tversky and Kahneman after him, did not conceptualize socially embedded processes, that is contextualize their models in social relations, institutions, and normative orders. At the same time, they failed to provide the grounds for a comprehensive new choice theory. In part, SGT is intended to fill these gaps. Drawing on Simon, this article addresses challenges such as the following:

- multi-value problems and dilemmas
- algorithms and meta-processes in judgment and decision processes
- framing, structuring and re-designing of choice components.

Simon's originality was in identifying the complexity of values, exploring the role of decision algorithms, and emphasizing human creativity and adaptability. Like Simon, Tversky and Kahneman saw judgment/decision-making as a process. In general, they explored and developed cognitive aspects of judgment and choice. "Framing" of the choice situation and "reference points" were critical concepts in their theory.

Their treatment of probability or likelihood was also particularly innovative, and contrasted sharply with expected utility theory. They rejected the practice of introducing "probability" directly into the calculus of judgment and choice (this was a brilliant insight since "probabilities" themselves were not value terms at the same level or possessing the same qualities as value (or utility). By transforming probabilities into weights (which were adjusted linearly), they solved the problem of a potential incommensurability between likelihood and values, and recognized that people weigh different likelihood estimates according to their risk avoidance or risk proneness.

Although Simon as well as Tversky and Kahneman made occasional references to "norms" and the social situation, neither of the approaches particularly recognized or stressed such social conditions of judgment and choice. Social institutional and role concepts as well as "sacrality" and human passions were basically alien ideas. Such social science concepts are part and parcel of the SGT approach.

Uncertainty and the heuristics of likelihood judgments. Tversky and Kahneman departed from expected utility theory especially in the area of uncertainty. They did not accept that people operate with well-defined, well-structured probability measures. And, even in cases where they might do so, they do not treat the probabilities on the same level or as directly integratable with values. Rather, likelihoods are weighted—typically reflecting other judgments such as risk-adverseness or risk-proneness. There is a great variety of likelihood beliefs and "models". Some are only fragments; others are relatively complete and may even correspond to well-known probability distributions.

2.3 *Social and Psychological Context of Risk and Uncertainly*

In a sociological perspective, uncertainty is a more fundamental concept than risk. Empirical research shows that humans face many types of uncertainty. The standard approach to risk, based largely on rational choice conceptualization, fails to take into account the diverse social and psychological contexts of uncertainty. That is, the social context of uncertainty and risk conceptualization and analysis is often ignored or neglected. As we point out below, one must consider not only how people establish order, predictability and certainty but also conditions where human agents are unable to do so, or their constructions fail. To the extent that order is established and maintained—and actions and interactions and their consequences are patterned (and frequencies can be identified and analyzed), many uncertainties are minimized. Stable patterns or institutional routines—the basis for some degree of certainty (and predictability)—may be established and maintained. Under these conditions, some of the more standard models of risk can be employed, for instance, where risk is measurable.

$$\textit{Risk} = \textit{(impact or assessment of a hazard, loss, or undesirable outcome)} \\ \text{x}\ \textit{(probability of a hazard, loss, undesirable outcome)} \quad (1)$$

Such a conception, however, decontextualizes many key social as well as physical factors (and shares a good deal of the weaknesses of rational choice theory [12]. Social contextualization, as suggest below, implies at least a variety of different risk conceptions and models ranging from qualitative ones to quantitative ones.

Also, meta-processes, as Tversky and Kahneman have demonstrated, operate to determine not only the values (or ordering) of different hazards but also a "revision" of the assessments of likelihood estimates, depending, for instance, on how risk-prone or risk-averse one is.

SGT can deal descriptively and analytically with varied and complex judgment and choice situations. Here we are extending that work to the realm of risk and uncertainty modelling. We make use of socially embedded multi-value judgment theory, a key part of SGT, to formulate a unified theory that deals with multiple values and different qualities and degrees of risk. A major feature of the approach is that it models judgment and decision in specified institutional and normative contexts. Such contexts define and activate actors' ROLES, with their VALUE vectors, MODELS of reality (actions, outcomes, likelihoods): Information in the form of particular rules, about action-outcome linkages, about likelihoods, ACT (pragmatics, rules about action possibilities, constraints), and J(udgment). In SGT, social and psychological context appears explicitly in:

(1) the particular values and value conceptions of actors;
(2) the level and types of knowledge they have;
(3) the degree of integration (commensurability) of values and likelihood estimates;
(4) the particular meta-processes and meta-judgments that frame and regulate risk judgment processes.

2.4 *Sociological Game Theory: SGT*

2.4.1 Classical Theory as a Variant of SGT

Sociological Game Theory (SGT) is an extension and generalization of classical game theory using the mathematical theory of rules and rule complexes [4, 7, 10, 11, 21, 22]. Those tools can formalize in a uniform way social theory concepts such as norm, value, belief, role, social relationship, and institution as well game. SGT is a cultural institutional approach to game conceptualization and analysis [1–3]; also see [35, 36, 40]. A well-specified game in the context or situation S_t at time t, G(t), is an interaction situation where the participating actors have defined roles and role relationships formalized as mathematical objects or complexes of rules.[2] The role complex includes, among other things:

- particular beliefs or rules that frame and define the reality of relevant interaction situations;
- norms and values relating, respectively, to what to do and what not to do and what is good or bad;
- repertoires of strategies, programs, and routines;
- a judgment complex to organize the determination of decisions and actions in the game.

SGT has identified and analyzed several types of judgment modalities, for instance: routine or habitual, normative, and instrumental modalities (see later discussion). The rule complex(es) of a game in a particular social context guide and regulate the participants in their actions and interactions at the same time that in "open games" the players may restructure and transform the game itself and, thereby, the conditions of their actions and interactions.

2.4.2 The General Approach of SGT

In the SGT approach, a well-specified game at time t is a particular multi-agent interaction situation where the participating players have defined roles and role relationships. Given a situation S_t in context t (time, space, social environment), a general *game structure* is represented as a particular rule complex $G(t)$ [7, 21]. The $G(t)$ complex includes as subcomplexes of rules the players' social roles vis-à-vis one another along with other relevant norms and rules, R. Suppose that a group or collective I = {1, ..., m} of players is involved in a game $G(t)$. $ROLE(i, t, G)$ denotes player i's role complex in $G(t)$ (we drop the "G" indexing of the role). The game structure $G(t)$ consists then of a configuration of two or more roles together with R, some general

[2] Rules and rule systems are key concepts in the new institutionalism [6, 24, 33–35, 38, 42], evolutionary sociology [5, 41], and ethnomethodology [20] and are closely related to important work in philosophy on "language games" [47] and linguistics [16, 17, 37] as well as work in mathematics and computer science [7, 8, 21, 22] among others.

norms (and rule complexes) of the game:

$$G(t) = [ROLE(1,t), ROLE(2,t), \ldots, ROLE(k,t); R], \quad (2)$$

R contains rules which describe and regulate the game such as the "rules of the game", general norms, practical rules and meta-rules, indicating, for instance, how seriously or strict the roles and rules of the game are to be implemented, and possibly of other rules specifying ways to adapt or to adjust the rule complexes to particular situations. In this way, the SGT approach enables the construction of multi-level models of the judgment and choice mechanisms on the basis of which actors address dilemmas, conflicts, problems of integration, and complex judgments. Meta-rules define connections, priorities, and other relationships among rules and rule complexes.

The main components of a generalized game are the general rules for the players and the roles for the different participants in the game (see Fig. 1). Each role consists of the following:

- a complex $MODEL(i, t)$ which describes the players' "situational view", providing the perspective on, and basis for understanding of the reality of an action or interaction situation. It consists of a complex of rules representing players' beliefs about themselves, their environment, action and interaction conditions, and action-consequence associations, which frames and defines the situational reality, key conditions, causal mechanisms, and possible scenarios of the interaction situation.
- a complex $VALUE(i, t)$ consisting of the players' values, goals and commitments. In this set there are rules assigning values to things and deeds, determining what is "good", "bad", "acceptable", "unacceptable".
- a complex $ACT(i, t)$ which includes acts, routines, programs, strategies which can be used by the players in order to respond or to deal with problems and challenges in the context of a choice situation; action enablers and constraints have normative as well as pragmatic aspects.
- $J(i, t)$, to organize the determination of decisions and actions in relation to other agents in situation S_t. The judgment complex—typically, an algorithm—consists of rules which enable the agent i to come to conclusions about truth, validity, value, or choice of strategic action(s) in the given situation. In general, judgment is a process of operation on objects. The types of objects on which judgments can operate are: values, norms, beliefs, data, and strategies as well as other rules and rule complexes. Also there are different kinds of outputs or *conclusions* of judgment operations such as evaluations, beliefs, data, programs, procedures, and other rules and rule complexes.

In general, $MODEL(i, t)$, $VALUE(i, t)$, $ACT(i, t)$, and $J(i, t)$ are complexes of actor i's rules which are activated in situation S and at moment of time $t \in T$.

The article focuses on the judgment and choice behavior of a single social actor in a situation S, which may or may not be a game. An actor's basis for making judgments and decisions are specified in her role(s) in the action context. In preparing or taking a decision, the actor compares her options with respect to her relevant norms and values for the situation in question. Multiple values and interests are often taken into

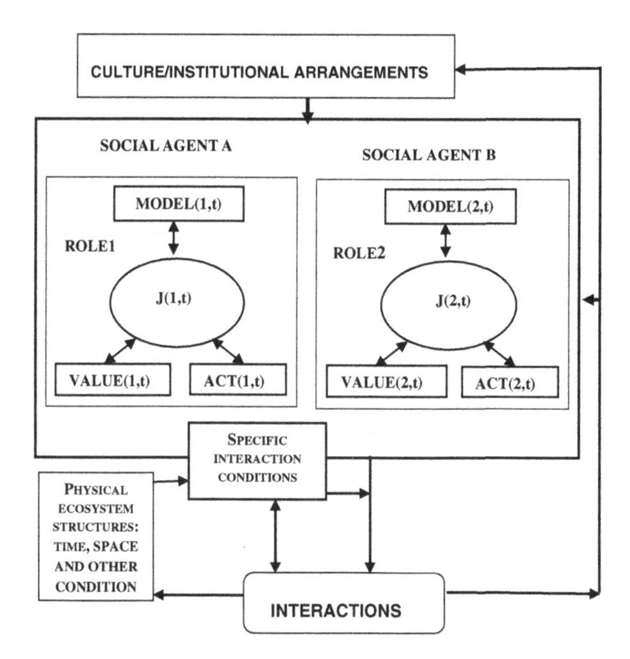

Fig. 1 Two role model of interaction embedded in cultural-institutional and ecological contexts. *Source* [4]

consideration. Action and interaction are multi-dimensional and open to differing interpretations and evaluation processes. The focus may be on, for instance [4, 11]:

- on the outcomes of the action ("consequentialism" or "instrumental rationality");
- compliance with a norm or law prescribing particular action(s) ("duty theory");
- the emotional qualities of the action ("feel good theory");
- the expressive qualities of the action (action oriented to communication and the reaction of others as in "dramaturgy");
- habitual or routine action;
- or combinations of these.

In an instrumental modality, for instance, the value of acts derives from assessments or evaluative judgments of action outcomes or "payoffs," whereas the value of action in the case of normative modality derives from judgments of the qualities of the action itself (including possibly the intentionality of the player). Note the operational differences between normative and instrumental modalities, particularly in open interaction situations where the players construct their actions. In the case of normative modality, the player constructs an action which corresponds to prescribed properties or qualities. In the case of instrumental modality where the players are supposed to accomplish an outcome or state of the world with prescribed features, they must find or construct an action that produces the prescribed effects—the properties of the action itself may be left unspecified. They require a cognitive model which links actions to outcomes, or allows them to specify such linkages.

What determines modality? From what has been stated earlier, the short answer is apparent: a player's role or particular rules in the role indicate which consequences one should attend to, thus indicating the appropriate modality or basis for determining action in a given situation. There may be practical constraints on such a determination, however. Situational constraints may be such that the player i is unable to determine the action on the basis of outcomes or qualities of the action. Information is lacking or there is substantial difficulty in assessing available information, or there are substantial constraints on acquiring information, so the actor would be unable to operate with, for instance, an instrumental modality (paying attention to outcomes), but may resort to acting "as if" utilizing a dramaturgical-communicative modality. Or, she makes use of rules of thumb, "standard operating procedures" or habits that have in the past led to appropriate outcomes. Habitual or routine type modalities entail executing a program, script, or procedure without deliberation or reflection, or weighing of alternatives. Such modalities are analytically distinguishable from consequentialist and normative modalities, where the actor makes evaluative judgments as well as calculations in the course of their activities. People often utilize the habitual modality for reason of efficiency—it requires much less situational data and time; information and operational costs are low in comparison to full-fledged instrumental or normative modalities. Sometimes no choice is available. Or, there is less risk involved in following orthodox routines. It is less risky in the sense that doing the normal is less likely to be questioned or criticized afterwards than doing something highly promising but full of uncertainties.
Role incumbents focus on specific qualia in particular contexts, because, among other things: such behavior is prescribed by their roles, and institutionalized in the form of routines; or the actor has no time or computational capability to deal with other qualia. The modalities of judgment and action differ in a business enterprise; market setting, community, family setting, competitive sport situation, interaction situations involving players who are enemies. In such ways, SGT encompasses a variety of judgment and choice modalities observable in social life.

2.4.3 Social Relationships, Contextualized Framing and Judgment Calculus

In the SGT perspective, human agents and their interactions are embedded in their social relations and institutional arrangements, and these conditions frame their choice components.

The framing of choice situations is a *multi-level process* (Fig. 2). Our earlier work has shown how people's social relationships frame their operative models (beliefs, perceptions), value complexes and evaluations, judgments and decisions, and patterns of interaction and potential equilibria [10].

Frame 1. In any given situation *S*, the initial framing is the *social definition of the situation*—what kind of choice or interaction situation is this; which institution or rule regime applies?

Frame 2. The institutional arrangement or normative order frames the particular social relationships, the roles of the actors and the norms they are to follow.

Frame 3. The components of roles—for instance, the belief models, the value complexes, and the judgment functions—also frame actors' dispositions and behavior in the situation.

So, the social relationships in a game G where actors have defined roles and normative rules provide the value, model, and judgment frames of the players vis-à-vis one another.

In sum, the normative or institutional context and actors' role relationships are usually the most important factors in framing choice situations. Note that these contexts of framing elicit value complexes and reference points, for example, differences between exchanging with a friend as opposed to engaging in a pure market exchange. In gift exchange or gift-giving, there are appropriate norms and reference points having to do with reciprocity and the nature of the particular social relationship in which the actors are involved. Issues of giving too much or giving too little are some of the considerations and meanings of the situation. For instance, someone in a high position might give too little—that is, the gift is below a socially established expectation or reference line for someone in a high position (Even in straightforward market exchanges there are reference points and complex judgments [9, 11]).

2.5 Context Dependent Risk Situations: Multi-level Risk Judgment

Risk[3] entails a composite judgment about the likelihood of damage or a potential negative impact from an action or event in terms of some characteristic value associated with the action or event. That is, a particular algorithm relates hazard value judgments and likelihood judgments to one another through multiplication (see Eq. 1).

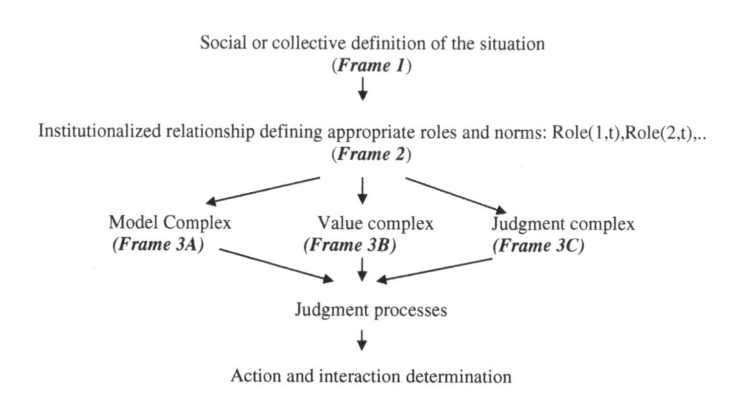

Fig. 2 The social framing process. *Source* The author

[3] Risk is a concept with multiple definitions and differences of approach, ranging from "an unwanted event which may or may not occur to "the statistical expected value of undesirable or unwanted events which may or may not occur" [14].

Other algorithms can be more complex and can take into account the fact that valuation and likelihood estimates may not be integratable in such terms. (In SGT, a variety of empirically meaningful algorithms can be used to relate hazard assessments and likelihood estimates depending on the social definition of situations as well norms and values specified by institutionalized relationships. For instance, the most common model involves a combinatorial algorithm (see Eq. 1) which simply "multiplies" cost/impact measure by likelihood (probability in some cases) to get an *expected loss).*

Human agents do not operate with a single conception or scale of risk. The variety of their conceptions and models reflects the complexity of their environment, particularly their social environment. Their judgments of likelihoods (likelihood estimates) are part of streams of judgment that make up complex, composite judgments and action determinations. For instance, actors in a given situation S consider possible gains, potential losses and likelihood estimates of these. The judgments may be combined in different ways depending on the algorithm(s) which the agents utilize—and which is a function of the social and psychological context in which the agents operate.

Social and psychological context. People make decisions not only on the basis of well-deliberated calculations, but also by intuition and routine. At the same time, since the actors follow different norms or values, risk means different things to them that is, they perceive and assess risk in different ways depending on the context in which they operate or are engage. As argued here, decision-making and risk-taking depend on actor's social and psychological context. In other words, the culture and institutional context must be taking into account in any analysis, since they influence the decision-making processes to a greater or lesser extent.

We can distinguish different contexts of decision-making with respect to the decision- maker and her particular situation. The models generated by SGT theory for the different contexts typically imply very different performance properties. We have proposed earlier that actors' social context (group, organization) and role relationships frame their judgment and choice situations. Part of this framing may entail operating with a common metric, e.g. money or "survival rates at intensive care units of a hospital," etc. Even if "distinctions" are made between costs and benefits (in common monetary terms), the net benefit (or "net cost") might be calculable, and the similar type of judgment algorithm can be utilized as in the purely uni-value models.

Some consequences are strictly distinguished and at the same time socially defined as non-commensurable. Many instances of cost-benefit analysis are of this character; hence, the development of cost-benefit methods which do not necessarily reduce costs and benefits to a single metric. Costs might be in money terms (Euros) and benefits are formulated in terms of reduction in mortality (or increased longevity) in the field of health. (Of course, these benefits have their monetary aspects, but the health care policymakers and physicians emphasize that people's lives and quality of life should not be reduced to monetary terms). Thus, the socially based framing of the situation is all important [19]. The question is whether a single value applies (or, alternatively, multiple values reducible to a common metric); or multiple, non-commensurable values apply in relation to a political, social, or moral question. Other dimensions of importance relate to the complexity of the judgment situation. The complexity of action-outcome structures makes for more or less difficult decision settings. But the question of complexity is very much a function of the social framing of the choice situation—that is, for instance, the extent to which an actor or actor(s) identify multiple, relevant and salient consequences of different actions and the extent to which these are or are not reducible to a single value, for example, monetary value.

Framing of the judgment and choice situation. In the conceptualization of context dependent choice situations, we distinguish models in terms of their degree of quantification and degree of commensurability or integratability. We can assume in some cases deterministic models, that is, the actor knows (or believes she knows) all of the consequences of her actions (or joint actions if with others). One of our main principles is that actors in the face of multiple, non-commensurable values, piecemeal beliefs and likelihood estimates, operate with algorithms enabling them to make composite judgments and decisions. While there are clear patterns, we stress the highly context dependent nature of the models: for instance, not only the level of certainty and the complexity of the situations with which they must deal.

Likelihood assessments. With respect to likelihood estimates: even if there are well-defined probabilities (based, e.g., on frequencies), these are evaluated (following Tversky and Kahneman) in a manner similar to the evaluation of "benefits" and "costs". Such evaluations of likelihood are expressed as, for instance, uncertainty acceptance or uncertainty averseness. The situations are characterized by particular organizations, group conditions, roles, norms; the stress is typically on "certainty", "security"—but in many situations, not. We speak of risk aversion or certainty demands, on the one hand; or, risk proneness or acceptance, on the other.

The likelihood estimates are necessarily systematic and coherent as in standard expected utility theory. People may operate, however, with very rough and piecemeal schemes, based on piecing together an array of likelihood estimates from different sources and belief systems. In a given social and psychological context, beliefs, pieces of information and hints from others, heuristics, and guesswork, etc provide a rough framework (or more precisely, a messy patchwork) which can be represented and analyzed in SGT.

Judgments and action determinations. The SGT conceptualization of decision-making implies that there are distinct ways in which evaluative judgments are combined with likelihood judgments. Different modalities for making such composite judgments generate varying decisions among alternatives, depending on the social context, key components of value and likelihood judgments, and ways in which these are combined through the judgment modalities as a basis for making decisions (one of the standard modalities consists of the usually expected value or utility format using a well-defined value or utility function combined with a probability distribution over outcomes). The approach is applicable to multi-value (multi-criteria) decision-making with various forms of likelihood judgments as a part of complex decision-making processes.

Discussion. The SGT conceptualization leads to the formulation of several distinct context-dependent decision-making models, distinguishable in terms of the properties of elementary judgment and judgment modalities. One model can represents classical decision theory with full quantification and commensurability of evaluative judgments and probability estimates and where a maximization of value principle applies in making decisions. Other models can entail judgments which are quantified but not commensurable and integratable and where actors apply algorithms or rule complexes for multi-criteria decision-making. Some models can involves non-quantification and non-commensurability but partial ordering of value judgments on diverse dimensions and fuzzy judgments about likelihoods and where actors apply an algorithm or rule complex that operates with partial orderings and fuzzy judgments to reach decisions. Yet other models concern a non-reducible "package" which is either a constructed partial ideal or comparable to an established ideal; decision-making is based on a "close enough" fuzzy judgment.

By taking into account the social context, SGT enables the construction of several distinct context-dependent models. In other words, it is a generative theory. This includes a purely meta-judgment model associated with high uncertainty about values (and desirable or undesirable conditions) at stake in the situation as well as uncertainty about the likelihood of different situations or events. Such high uncertainty may be the result of straightforward ignorance in the situation; or, even in the case that there is considerable information (e.g., about one's own values), they are judged unreliable, not a solid basis on which to make judgments about critical hazards in the situation and their likelihoods. Consequently, the actor falls back on deeper (or in our language, meta-judgments).

Our social and social psychological approach provides a point of departure for, among other things, identifying and understanding the role of human institutions in reducing uncertainty (and uncertainties). Institutions frame and define social action and interaction situations, specifying people's relationships and roles as well as appropriate norms. Given elaborate institutional arrangements, actors can know much of what is going on, what to expect. They can simulate and predict many action and interaction processes and outcomes in the situation. They can relate their behavior to one another, even in strategic games.

Institutional rules which specify situations are, however, never complete. In part, this is because the situations in which an institutional arrangement is applied vary and change over time. Thus, there are always residual uncertainties. Often, the actors involved apply pragmatic or heuristic rules, reducing residual uncertainty in a certain sense, "filling in" gaps. In some cases, when they fail to reduce uncertainty (or they choose to ignore what they know) they may simply act—risking that the action and its interactions fail to turn out as desired or as expected. When situations deviate greatly from what has been "normal," unanticipated and unintended outcomes are likely. Patterns or what have been more or less stable frequencies are violated. There may be no frequency measures. Of course, people may have (subjective) beliefs about likelihoods even in highly unstable or dynamic, evolving situations.

Situational anomalies are of particular interest to us. These arise because of changes in key dimensions of the situation—models and expectations no longer fit. Actions have unexpected outcomes, possibly very contrary to what is intended or expected. Such anomalies set the stage for actors to attempt to restructure the institutional arrangement to accomplish greater order, predictability and effectiveness.

In addition to the gaps and fuzziness of institutional arrangements in defining interaction situations, they often intentionally define special fields or arenas, which are intentionally "open" or "underspecified", settings where designated actors are allowed (or expected to) make judgments (possibly after negotiating) and determine actions and interactions. The actors characterize these social judgments and choice situations in terms of "alternatives", "action opportunities" (either given or constructed) and ranges of possible consequences. Actors' level of knowledge about the action opportunities and consequences, e.g. symptoms and diagnoses as well as prognoses, may of course vary considerably.

Our contextual approach makes it apparent that real-life situations (complex, dynamic contexts) are unlike the highly stylized "gambling situations." The latter are greatly simplified as well as de-contextualized from actual situations (although the actors participating have, of course, their lives (debts, obligations, etc) outside the gambling setting. Gambling houses then must be seen as particular types of institutionalized settings, which not only simplify the choice situation but also de-contextualize it as much as possible (in part, so as to trap their "customers").

3 Part II. Application of SGT to the Topsis Multi-Criteria Decision Procedure

3.1 Introduction

One of the main SGT principles is that a decision maker in the face of multiple, non-commensurable values, piecemeal beliefs and likelihood estimates, operate with particular algorithms enabling them to make composite judgments and decisions. In this part of the paper we consider situations where a decision-maker, in making her choices, utilizes a complex algorithm drawing upon her beliefs and rules concerning the likelihood of events and the uncertain associations between actions and consequences. Judgments of likelihood are combined with judgments of value and make up what we refer to as composite risk judgments, utilizing judgment algorithms. This analysis relates to the expanding fields of multi-criteria decision-making.

A variety of multi-criteria decision making (MCDM) methods as well as fuzzy MCDM methods [32] have been developed in relation to complex risk judgments. In general, multi-criteria decision-making researchers have formulated a variety of quantitative methods to compare, select, or rank multiple alternatives distinguished on multiple criteria. The procedures help decision-makers to formulate their preferences, to rank priorities, and to apply them to a particular decision context. The aim is to provide support to the decision-maker in the process of making a choice among different complex options. There are three steps in utilizing techniques involving analysis of alternatives: determining the relevant criteria and alternatives, attaching measures to the relative value or importance of the criteria, and assessing the alternatives on the basis of multiple criteria and determining a ranking among them. MCDM methods use crisp data, but such data are often inadequate for modeling many complex decision situations characterized by risk. Uncertainty prevails in risk judgment due to incomplete or non-obtainable information. A decision-maker is not always able to estimate the consequences of the option in a precise way, but, nonetheless, she is able to approximate or can use assessments in natural language terms instead of exact numerical values. Since the human risk judgments are often vague, the application of fuzzy concepts in risk assessment has come to be seen as highly relevant. The classical MCDM methods are extended to decision-making problems with interval data, fuzzy data or natural language terms which are represented usually by fuzzy numbers [15, 26, 27]. In risk assessment however, it is more appropriate to express decision-maker opinions by a series of linguistic variables in terms of degrees of belief. For example, the risk criterion of some alternative may be evaluated by the decision-maker as 70 % sure that the risk criterion is "highly risky" and 30 % sure that it is "medium risky", which can be expressed as follows: {(very high, 0.7), (medium, 0.3)}.

3.2 *The TOPSIS Procedure of Multi-Criteria Decision-Making Under Conditions of Uncertainty*

The*Technique for Order of Preference by Similarity to Ideal Solution* (TOPSIS) is a multi-criteria decision method [25]. It is based on the notion that a preferred or selected alternative should have the shortest distance from a positive ideal point and the longest distance from a negative or undesirable point. The method enables a comparison of a set of multi-criteria alternatives where weights are assigned for each criterion and an algorithm is used to rank order and choose among the multi-criteria alternatives.

The basic principle of this method is that the selected alternative should have the shortest distance from a Positive Ideal Solution (PIS) and the farthest distance from Negative Ideal Solution (NIS). PIS is the solution that maximises the benefit criteria and minimizes the cost criteria, while NIS is an alternative, which maximises the cost criteria and minimizes the benefit criteria. Extensions of TOPSIS were also developed, aiming at the adaptation of the original procedure to interval data or fuzzy conditions [15, 26, 27]. The TOPSIS method has thus been applied in risk analysis [29], as we do here, using the model to analyze multi-criteria risk situations. Thus, we suggest that TOPSIS can be usefully formulated and interpreted in the perspective of SGT.

The Model or Belief Structure (BS) of TOPSIS was developed to deal with MCDM problems under conditions of uncertainty [49–51] and later developed in relation to fuzzy conditions [52]. The BS model with degrees of belief is used to represent the performance of an option on a criterion. Suppose a selected criterion is assessed by a complete set of standards with evaluation grades represented by $H = \{H_1, \ldots, H_s, \ldots, H_n\}$, where H_s is the s-th evaluation grade with numerical value or utility functions $0 \leq u(H_s) \leq 1$, $s = 1, 2, \ldots, n$. A given assessment for criterion C is represented as the BS model in the following way:

$$S(C) = \{f(H_s, \beta_s) : \text{ where } s = 1, 2 \ldots, n\}. \tag{3}$$

We say that an assessment S is complete if $\sum_{s=1}^{n} \beta_s = 1$ and incomplete if $\sum_{s=1}^{n} \beta_s < 1$. The similarity $\tilde{s}_{ij}$ judgment between two grades is represented by numerical value or utility functions $u(H_i)$, and $u(H_j)$ is calculated as:

$$\tilde{s}_{ij}(H_i, H_j) = 1 - \left|u(H_i) - u(H_j)\right| \quad i, j = 1, 2 \ldots, n \tag{4}$$

The BS model consisting of n evaluation grades can be also described as a vector

$$B = (\beta_1, \ldots, \beta_s, \ldots, \beta_n) \tag{5}$$

Suppose now that we have two BS models, S_1 and S_2, with the corresponding vectors B_1 and B_2, respectively. The comparison between two such BS models, S_1 and S_2, can be transformed into the distance measure between two vectors B_1 and B_2. The distance between S_1 and S_2 is defined as:

$$d_{BS}(S_1, S_2) = d_{BS}(B_1, B_2) = \left(\frac{1}{2}(B_1 - B_2)\tilde{S}(B_1 - B_2)^T\right)^{\frac{1}{2}} \tag{6}$$

where $\tilde{S} = \left[\tilde{s}_{ij}\right]$ is a similarity matrix, which describes the differences among the evaluation grades $H_s, s = 1, 2, \ldots, n$.

The TOPSIS procedure for multi-criteria risk assessment. Suppose that there are m alternatives $A_i, i = 1, 2, \ldots, m$, that is, these options are part of an action or repertoire ACT. Each alternative is evaluated by the C_j risk criteria $j = 1, 2, \ldots, r$. To represent uncertainty, the BS model defined by (3) is applied to describe the risk judgments of decision-makers.

We assume that the assessment for risk criterion C_j on alternative A_i is represented as the BS model:

$$S_{ij} = \{f(H_s, \beta_s) : s = 1, 2 \ldots, n\}_{ij}, \text{where } i = 1, 2, \ldots., m; \; j = 1, 2, \ldots r \tag{7}$$

where:

- $H = \{H_1, \ldots, H_s, \ldots, H_n\}$ is a set of standards with evaluation risk grades,
- $u(H_s)$ value or utility function s-th risk grades $0 \leq u(H_s) \leq 1, s = 1, 2, \ldots, n$.

It means (7) that, in the decision-maker's judgment, alternative A_i is assessed to have an evaluation grade H_s with degree of belief β_s regarding criterion $C_j, s = 1, 2, \ldots, n$.

The risk analysis based on TOPSIS is illustrated in the following steps:

Step 1. *Structuring the MCDM problem.* Formulate the risk analysis problem by first constructing a MCDM decision matrix

$$M = \left[S_{ij}\right]_{mxr} \tag{8}$$

where $S_{ij} = \{f(H_s, \beta_s) : s = 1, 2 \ldots, n\}_{ij}$ is a BS model.

Let $w = [w_1, \ldots, w_r]$ be the weight vector, where w_j is the weight of criterion C_j. We have $\sum_{k=1}^{r} w_k = 1$.

Without loss of generality, it is assumed that every BS model is complete. If incompleteness obtains, the BS model has to be normalized.

Step 2. *Determination of Ideal Points (Positive and Negative).* Determine the Positive Ideal Points (PIBS) A^+ and Negative Ideal Point (NIBS) A^- respectively as:

$$A^+ = \left\{S_1^+, \ldots, S_r^+\right\}, \tag{9}$$

$$A^- = \left\{S_1^-, \ldots, S_r^-\right\} \tag{10}$$

where S_j^+ and S_j^- are BS models.

Step 3. *Calculate the "Distance" of Each Alternative from the Ideal Points,* PIBS (A^+) and NIBS (A^-). For each alternative A_i, the distance measure can be calculated using the following formula:

$$D_i^+ = \sqrt{\sum_{j=1}^{r} w_j d_{BS}(S_{ij}, S_j^+)} \tag{11}$$

$$D_i^- = \sqrt{\sum_{j=1}^{r} w_j d_{BS}(S_{ij}, S_j^-)} \tag{12}$$

where w_j is the weight of criterion C_j, $d_{BS}(S_{ij}, S_j^+)$ and $d_{BS}(S_{ij}, S_j^-)$ are the belief distance measures between the two BS models, S_{ij} and $S^+(S^-)$.

Step 4. *Judgment Algorithm.* Calculate the relative closeness of alternatives to the PIBS and NIBS. A closeness coefficient of the ith alternative (with closeness coefficient R_i) is defined so as to rank all possible alternatives. R_i represents the distances to PIBS $\left(A^+\right)$ and NIBS $\left(A^-\right)$, simultaneously and is calculated as:

$$R_i = \frac{D_i^-}{D_i^- + D_i^+} \tag{13}$$

where $0 \leq R_i \leq 1, \quad i = 1, 2, ..., r$.

Step 5. *Rank ordering algorithm.* Rank all alternatives according to descending R_i.

Step 6. *Selection rule or procedure.* Select the highest ranked alternative; alternatively, select the first alternative j that has a closeness coefficient R_j which is "satisficing". If there are several options that satisfice,[4] then one uses additional criteria to distinguish and rank order the options. Or, possibly, uses a coin-flipping algorithm to make the selection.

In risk situations, the risk evaluation grades are typically fuzzy. The set of fuzzy risk evaluation grades may be either triangular or trapezoidal fuzzy sets or their combinations. The Model or Belief Structures presented here can be extended to Fuzzy Belief Structures (FBS); these have been successfully applied in the areas of Fuzzy MCDM. Then, the evaluation grades as well the utilities of evaluation grades are representable by triangular or trapezoidal fuzzy numbers [23, 52]. Such risk analysis deal with incompleteness, ignorance, and vagueness in complex uncertain choice situations.

In the table below we indicate the correspondences between the universal SGT categories and the key components of the TOPSIS procedure (Table 1).

Numerical Example. In order to illustrate the method a numerical example is provided. Suppose a MCDM problem has three alternatives A_1, A_2 and A_3, and four risk criteria C_1, C_2, C_3, C_4. The decision-maker makes a judgment using a BS model for each alternative on each criterion. Suppose there are four evaluation risk grades $\{H_1, H_2, H_3, H_4\}$ ={"very high", "high", "medium", "low"}. Regarding C_1, if the decision-maker is 80% sure that alternative A_1 is very high risk and 20% sure that A_2 is high risk, the BS model should be expressed as $\{(H_1; 0.8), (H_2; 0.2), (H_3; 0), (H_4; 0)\}$ or (0.8, 0.2, 0,0). In this way, all judgments of the decision-maker are summarized and are collected in Table 2.

The relative importance of the four risk criteria is specified as $w = [0.3, 0.3, 0.2, 0.2]$. Suppose the utilities of the evaluation grades are: $u(H_1) = 1$ ("very high"), $u(H_2) = 0.7$ ("high"), $u(H_3) = 0.4$("medium"), and $u(H4) = 0$ ("low"). Then, the similarity matrix can be generated by (4) as:

$$\tilde{S} = \begin{bmatrix} 1 & 0.7 & 0.4 & 0 \\ 0.7 & 1 & 0.7 & 0.3 \\ 0.4 & 0.7 & 1 & 0.6 \\ 0 & 0.3 & 0.6 & 1 \end{bmatrix}$$

The NIBS has the form $\{1, 0, 0, 0\}$ and PIBS $\{0, 0, 0, 1\}$. The separation measures D^+ and D^- an be calculated using (11) and (12). Finally, one calculates the relative closeness R to the ideal solution for each alternative A using (13), and ranks the preference order in terms of the values of R_i . The results are shown in Tables 3 and 4.

[4] This derives from the concept of "satisficing" introduced by [43]. Elsewhere [10], we formulate a satisficing algorithm which compares the characteristics of an option to a vector of values specifying standards and determined "sufficient similarity" and, therefore satisfied.

Table 1 Links between SGT general model and the TOPSIS procedure for risk assessment

SGT components of the general theoretical model	The TOPSIS procedure for risk assessment
Model Complex describes the players' "situational view" and understanding of the action or interaction situation. Model contains beliefs about the situation Likelihood or assessments in the Model complex are generated from different belief systems and other sources in a given social and psychological context	*The complex decision situation formalized in:* The belief structure (BS) (as well as the belief decision matrix based on BSs) make up the complex decision situation. The BS model with degrees of belief represents the anticipated or expected performance of an alternative on diverse criterion
Value Complex consisting of players' values, goals and commitments. Assessments and evaluations are generated on the basis of appropriate or prescribed values and preferences in the social and psychological context	*Value and normative views of risk situation are formalized:* the weight vector of criteria; S-evaluation grades; concept of ideal points (positive, negative) and $u(H_S)$ utility functions s-th risk grades
ACT complex of potential options, typically defined or prescribed for the decision or interaction situation	Repertoire of m options or alternatives $A_i, i = 1, 2, 3, \ldots, m$
Single or/and composite judgment may be combined in different ways depending on the algorithm(s) which the player utilize; it will be a function of the social and psychological context, the value system(s), and social relationships, often in a context of vagueness or ambiguity	TOPSIS procedure entails several judgment algorithms such as: calculation of the distance measures of an option from PIBS (and NIBS); calculation of the relative distance to the PIBS (and NIBS); rank ordering of alternatives; application of a choice rule or criterion

Source The authors

In this case of risk assessment, the ranking of three alternative structures, as indicated in Table 4, is A_1, A_2, A_3, so, the alternative A_3 is the preferred option with respect to minimizing risk.

Table 2 Judgments from decision-maker

A_i	Criteria			
	C_1	C_2	C_3	C_4
A_1	(0.8, 0.2, 0,0)	(0, 0.6, 0.4, 0)	(0, 0, 0.7, 0.3)	(0.8, 0.2, 0, 0)
A_2	(0.5, 0.5, 0, 0)	(0, 0, 08., 0.2)	(0.2, 0.8, 0, 0)	(0.5, 0.5, 0, 0)
A_3	(0, 0.8, 0.2, 0)	(0.3, 0.7, 0, 0)	(0.4, 0.6, 0, 0)	(0, 0.8, 0.2, 0)

Source By the authors

Table 3 Belief distance from NIBS and PIBS

A_i	NIBS /PIBS	C_1	C_2	C_3	C_4
A_1	A^-	0.012	0.348	0.636	0.700
	A^+	0.892	0.508	0.196	0.100
A_2	A^-	0.075	0.616	0.192	0.604
	A^+	0.775	0.644	0.256	0.324
A_3	A^-	0.312	0.147	0.108	0.420
	A^+	0.592	0.727	0.748	0.064

Table 4 Separation measure

A_i	D_i^-	D_i^+	R_i	Rank
A_1	0.612	0.692	0,530	3
A_2	0.605	0.736	0.540	2
A_3	0.493	0.747	0.610	1

4 Summarizing Remarks

This article has considered complex risk decision-making, using SGT as a point of departure. SGT considers the universal features of an actor's or decision-maker's system: a model or belief structure, value complex, action repertoire and judgment complex (with its algorithms for making judgments and choices). Although these features are universal, they are particularized in any given institutional or socio-cultural context.

The article considered an established method, TOPSIS, for dealing with multi-criteria decision-making under conditions of uncertainty. One aim of this exercise was to identify correspondences between the SGT universal architecture and the components of the TOPSIS method. We could expose, for instance, the different value components as well as diverse judgment algorithms encompassed by the TOPSIS procedure.

One of the benefits of such an exercise is to suggest ways to link different decision methods and procedures in a comparative light. It deepens our empirical base and understanding of values, models, action repertoires, and judgment structures (and their algorithms). Of course, our effort here has been a modest one. In future papers, we expect to consider and compare a number of MCDMs and FMCDMs.

Acknowledgments This work was partially supported by the grant from Polish National Science Center (DEC-2011/03/B/HS4/03857).

References

1. Burns, T.R.: Models of social and market exchange: toward a sociological theory of games and human interaction. In: Calhoun, C., Meyer, M.W., Scott, W.R. (eds.) Structures of Power and Constraint: Essays in Honor of Peter M. Blau. Cambridge University Press, New York (1990)
2. Burns, T.R.: Two conceptions of human agency: rational choice theory and the social theory of action. In: Sztompka, P. (ed.) Human Agency and the Reorientation of Social Theory. Gordon and Breach, Amsterdam (1994)
3. Burns, T.R., Baumgartner, T., DeVille, P.: Man, Decision and Society. Gordon and Breach, London (1985)
4. Burns, T.R., Caldas, J.C., Roszkowska, E.: Generalized game theory's contribution to multi-agent modelling: addressing problems of social regulation, social order, and effective security. In: Dunin-Keplicz, B., Jankowski, A., Skowron, A., Szczuka, M. (eds.) Monitoring, Security and Rescue Techniques in Multiagent Systems, pp. 363–384. Springer Verlag, Berlin/London (2005)
5. Burns, T.R., Dietz T.: Revolution: an evolutionary perspective. Int. Soc. **16**(4), 531–555 (2001)
6. Burns, T.R., Flam H.: The Shaping of Social Organization: Social Rule System Theory with Applications. Sage Publications, London (1987, reprinted 1990)
7. Burns, T.R., Gomolińska, A.: Modeling social game systems by rule complexes. In: Polkowski, L., Skowron, A. (eds.) Rough Sets and Current Trends in Computing. Springer-Verlag, Berlin/Heidelberg (1998)
8. Burns, T.R., Gomolińska, A.: The theory of socially embedded games: the mathematics of social relationships, rule complexes, and action modalities. Qual. Quant. Int. J. Methodol. **34**(4), 379–406 (2000)
9. Burns, T.R., Gomolińska, A., Meeker, L.D.: The theory of socially embedded games: applications and extensions to open and closed games. Qual. Quant. Int. J. Methodol. **35**(1), 1–32 (2001)
10. Burns, T.R., Roszkowska, E.: Fuzzy games and equilibria: the perspective of the general theory of games on Nash and normative equilibria. In: Pal S.K., Polkowski L., Skowron A. (eds.) Rough-Neural Computing. Techniques for Computing with Words, pp. 435–470. Springer-Verlag, Berlin (2004)
11. Burns, T.R., Roszkowska E.: Multi-value decision-making and games: the perspective of generalized game theory on social and psychological complexity, contradiction, and equilibrium. In Shi Y. (ed.) Advances in Multiple Criteria Decision Making and Human Systems Management, pp. 75–107. IOS Press, Amsterdam (2007)
12. Burns, T.R., Roszkowska, E.: The social theory of choice: from Simon and Kahneman-Tversky to GGT modelling of socially contextualized decision situations. Optimum-Studia Ekonomiczne. **3**(39), 3–44 (2008)
13. Cassidy, J.: Economics: Which Way for Obama? Review of R.H. Thaler and C.R. Sunstein. New York Review of Books June 12, pp. 30–34 (2008)
14. Chapman, A.: Democratizing Technology. Risk, Responsibility and Regulation of Chemicals. Earthscan, London (2007)
15. Chen, C.T.: Extensions of the TOPSIS for group decision-making under fuzzy environment. Fuzzy Sets Syst. **114**(1), 1–9 (2000)
16. Chomsky, N.: Rules and Representation. Columbia University Press, New York (1980)
17. Chomsky, N.: Knowledge of Language: Its Nature, Origins, and Use. Praeger, New York (1986)
18. Dawes, R.M., Thaler, R.: Cooperation. J. or Econ. Perspect. **2**, 187–197 (1988)
19. Espeland, W., Stevens, M.: Commensuration as a social process. Annu. Rev. Sociol. **24**, 313–343 (1998)
20. Garfinkel, H.: Studies in Ethnomethodology. Prentice-Hall, Englewood Cliffs (1967)
21. Gomolińska, A.: Rule complexes for representing social actors and interactions. Studies Logic Gramm. Rhetor. **3**(16), 95–108 (1999)

22. Gomolińska, A.: Fundamental mathematical notions of the theory of socially embedded games: a granular computing perspective. In: Pal, S.K., Polkowski, L., Skowron, A. (eds.) Rough-Neuro Computing: Techniques for Computing with Words, pp. 411–434. Springer-Verlag, Berlin/London (2004)
23. Guo, M., Yang, J.B., Chin, K.S., Wang, H.W.: Evidential reasoning approach for multi-attribute decision analysis under both fuzzy and interval uncertainty. IEEE Tran. Fuzzy Syst. **17**(3), 683–697 (2009)
24. Hodgson, G.M.: The evolution of institutions: an agenda for future theoretical research. Const. Polit. Econ. **13**(2), 111–127 (2002)
25. Hwang, C.L., Yoon, K.: Multiple attribute decision making. Springer-Verlag, Berlin (1981)
26. Jahanshahloo, G.R., Lotfi, F.H., Izadikhah, M.: An algorithmic method to extend TOPSIS for decision-making problems with interval data. Appl. Math. Comput. **175**(2), 1375–1384 (2006a)
27. Jahanshahloo, G.R., Lotfi, F.H., Izadikhah, M.: Extension of the TOPSIS method for decision-making problems with fuzzy data. Appl. Math. Comput. **181**(2), 1544–1551 (2006b)
28. Jiang, J., Chen, Y., Wu Tang, D.W., Chen, Y.W.: TOPSIS with belief structure for group belief multiple criteria decision making. Int. J. Autom. Comput.
29. Jiang, J., Chen, Y., Chen, Y., Yang, K.: TOPSIS with fuzzy belief structure for group belief multiple criteria decision making. Expert Syst. Appl. **38**, 9400–9406 (2011)
30. Kahneman, D., Tversky, A.: Prospect theory: an analysis of decision under risk source. Econometrica. **47**(2), 263–291 (1979)
31. Knight, F.H.: Risk Uncertainty and Profit. Houghton Mifflin Company, Chicago (1921)
32. Linkov, I., Satterstrom, F.K., Kiker, G., Batchelor, C., Bridges, T., Ferguson, E.: From comparative risk assessment to multi-criteria decision analysis and adaptive management: recent developments and applications. Environ. Int. **32**, 1072–1093 (2006)
33. March, J.G., Olsen, J.P.: Rediscovering Institutions. Free Press, New York (1989)
34. North, N.C.: Institutions, Institutional Change, and Economic Performance. Cambridge University Press, Cambridge (1990)
35. Ostrom, E.: Governing the Commons: The Evolution of Institutions for Collective Action. Cambridge University Press, Cambridge (1990)
36. Ostrom, E., Gardner, R., Walker, J.: Rules, Games, and Common-Pool Resources. University of Michigan Press, Ann Arbor (1994)
37. Pinker, S.: Rules of language. Science **253**, 530–534 (1991)
38. Powell, W.W., DiMaggio, P.J. (eds.): The New Institutionalism in Organizational Analysis. University Press, Chicago (1991)
39. Roeser, S., Hillerbrand, R., Sandin, P., Peterson, M. (eds.): Handbook of Risk Theory Epistemology, Decision Theory, Ethics, and Social Implications of Risk. Springer, Dordrecht (2012)
40. Scharpf, F.W.: Games Real Actors Play: Actor-Centered Institutionalism in Policy Research. Westview Press, Boulder (1997)
41. Schmid, M., Wuketits, F.M. (eds.): Evolutionary Theory in the Social Sciences. Reidel, Dordrecht (1987)
42. Scott, W.R.: Institutions and Organizations. Sage Publications, London (1995)
43. Simon, H.A.: A behavioral model of rational choice. Q. J. Econ. **59**, 99–118 (1955)
44. Thaler, R.H., Sunstein C.R.: Nudge: Improving Decisions about Health, Wealth, and Happiness. Yale University Press, New Haven (2008)
45. Tversky, A., Kahneman, D.: Judgment under uncertainty: heuristics and biases. Science **185**, 1124–1131 (1974)
46. Tversky, A., Kahneman, D.: The framing of decisions and the psychology of choice. Science **211**, 453–458 (1981)
47. Wittgenstein, L.: Philosophical Investigations. Blackwell, Oxford (1953)
48. von Neumann, J., Morgenstern, O.: Theory of Games and Economic Behavior, 3rd edn. Princeston University Press, Princeton (1953)
49. Yang, J.B., Sen, P.: A general multi-level evaluation process for hybrid MADM with uncertainty. IEEE Trans. Syst. Man Cybern. Part A: Syst. Humans **24**(10), 1458–1473 (1994)

50. Yang, J.B., Singh, M.G.: An evidential reasoning approach for multiple attribute decision making with uncertainty. IEEE Trans. Syst. Man Cybern. Part A: Syst. Humans **24**(1), 1–18 (1994)
51. Yang, J.B., Xu, D.L.: On the evidential reasoning algorithm for multiattribute decision analysis under uncertainty. IEEE Trans. Syst. Man Cybern. Part A: Syst. Humans **32**(3), 289–304 (2002)
52. Yang, J.B., Wang, Y.M., Chin, K.S., Xu, D.L.: The evidential reasoning approach for MCDA under both probabilistic and fuzzy uncertainties. Eur. J. Oper. Res. **171**(1), 309–343 (2006)

Supporting Ill-Structured Negotiation Problems

Ewa Roszkowska, Jakub Brzostowski and Tomasz Wachowicz

Abstract The negotiation is a complex decision-making process in which two or more parties talk with one another in afford to resolve their opposing interests. It can be divided into consecutive stages, namely: pre-negotiation phase involving structuring the problem and the analysis of preferences, the intention phase involving the iterative exchange of offers and counter-offers, and the postoptimization phase aiming at the improvement of the agreement obtained in the intention phase. In this chapter, we focus on the analysis of negotiators′ preferences in ill-structured negotiation problems. We employ the modified FTOPSIS approach and the AHP method for determining the negotiation offers′ scoring system, which allows for the easy evaluation of both the incoming offers as well as the packages under preparation. The imprecision and vagueness of the packages and options′ descriptions is modeled by the fuzzy triangular numbers. The Analytic Hierarchy Process is used to derive the negotiation issue weights instead of directly assigning such values to the issues (a classic approach). The FTOPSIS method is used to build the final scoring system allowing for the evaluation of any potential negotiation package. The whole process of negotiation supported by the approach we proposed is illustrated with an numerical example.

E. Roszkowska (✉)
Faculty of Economy and Management, University of Bialystok, Warszawska 63, 15-062 Bialystok, Poland
e-mail: erosz@o2.pl

J. Brzostowski
University of Leipzig Faculty of Economics and Management Information Systems Institute, Grimmaische Str. 12 D, 04-109 Leipzig, Germany
e-mail: brzostowski@wifa.unileipzig.de

T. Wachowicz
Department of Operations Research, University of Economics in Katowice, 1 Maja 50, 40-287 Katowice, Poland
e-mail: tomasz.wachowicz@ue.katowice.pl

P. Guo and W. Pedrycz (eds.), *Human-Centric Decision-Making Models for Social Sciences*, Studies in Computational Intelligence 502,
DOI: 10.1007/978-3-642-39307-5_14,

Keywords negotiation support · preference elicitation · negotiation offers′ scoring system · Fuzzy TOPSIS · AHP · post-negotiation optimization

1 Introduction

Negotiation is an everyday activity that most people undertake to influence others and to achieve personal objectives [16]. It may be perceived as an iterative process of exchanging offers and messages between the interested parties that is conducted until a satisfying agreement is reached. In bilateral negotiations the agreement will be reached until an offer is satisfying for both parties, i.e. the negotiator and his counterpart. Therefore we may say that a negotiation is a non-individual decision making process, in which two (or more) parties decide how to divide the limited resources [4, 7, 27].

Decision context is crucial for negotiations. Since early 1980s a new discipline called negotiation analysis [18] has been being developed, aiming at formalizing the negotiation process and applying supportive decision making and game-theoretic methods and tools. The significant part of negotiation analysis is devoted to the problem of defining, evaluating and building the negotiation template, which is usually elaborated in the pre-negotiation preparation phase. The negotiation template specifies the negotiation space by defining the negotiation issues and their acceptable resolution levels (options), and it constitutes the framework for further decision analysis conducted by the negotiators themselves and/or the third party that supports the negotiation process. Since the negotiation template very often involves numerous and usually conflicting issues, multiple criteria decision making (MCDM) methods are applied to help negotiators with analyzing and eliciting their preferences, which allows to span the scoring system over the whole negotiation space. The simple additive weighting, SAW [13], is most commonly suggested for template analysis since it is considered to be a straightforward method easy to understand by decision makers (here, the negotiators). This approach is very often applied in negotiation support systems or electronic negotiation systems, such as Inspire [14], Negoisst [24] or NegoCalc [29]. There are also other MCDM methods used in negotiation support, such as AHP [21], even swaps [26] or ELECTRE-TRI [30].

Most methods and models that have been currently applied in negotiation software support tools operate with a well-defined negotiation problem. They require a precise definition of the negotiation template, i.e. a clear specification of negotiation issues as well as an accurate representation of their resolution levels. The latter need to be predefined by means of real numbers or linguistic labels with the real number equivalents. These systems also require the users to precisely define their preferences, so that the standard MCDM procedures can be applied to provide the user with their own negotiation offers′ scoring system. The scoring system is used later for supporting the negotiators in making their decision on rejecting or accepting the negotiation offers during the actual negotiation phase. It may be also used for visualizing the negotiation progress by means of negotiation history and negotiation dance graphs [19], and then in post-negotiation phase to conduct post-optimization analysis, which may help the negotiators to find improvements for the agreement negotiated. Unfortunately, the vast majority of negotiation processes are too complex and involve too many uncertain or risky factors that it is nearly impossible to describe them by means of crisp values without being accused of oversimplification. Therefore there is a strong need to develop methods and techniques that could handle the imprecise and fuzzy information occurring in the process of template construction and evaluation.

In this chapter we present an effective application of the fuzzy approach to support an ill-structured negotiation problem. We develop a negotiation model based on the analytic hierarchy process (AHP) and a technique for ordering performance by similarity to the ideal solution (TOPSIS), to support negotiation processes where the vagueness and subjectivity in problem definition are handled by means of triangular fuzzy numbers. The AHP is used to analyze the structure of the negotiation problem and to determine the importance weights of the negotiations issues. The fuzzy TOPSIS method is used to obtain the final ranking of the negotiation packages. Although there are many studies in the literature that use fuzzy TOPSIS as well as AHP to solve different MCDM problems, here we propose some modifications of those techniques to make them useful in and applicable to the negotiation context. Combining the fuzzy TOPSIS and AHP for the analysis of negotiators' preferences as well as applying of this approach to handling ill-structured negotiation problems are the major contributions of our chapter.

This approach is employed for the following reasons: the basic concepts of TOPSIS are rational and understandable; the computation processes are straightforward and take into account the negotiation space of each issue; the concepts of the ideal and anti-ideal solutions are easy to interpret and to represent in a simple mathematical form; the modified FTOPSIS procedure makes it possible to expand the negotiation template by introducing new options after the preference elicitation has been conducted (within the actual negotiation phase); and the importance weights are determined by subjective judgments using the AHP technique based on pairwise comparison procedures instead of being simply assigned. The full Fuzzy AHP or AHP technique is practically usable only if the number of criteria and alternatives is sufficiently low, usually not higher than 9. Therefore, to avoid a possibly large number of pairwise comparisons, the fuzzy TOPSIS is employed to obtain the final ranking of packages. It eliminates the negotiator's workload in the preference elicitation stage, makes it easier and allows to build the offers′ scoring system in a shorter time.

Our chapter consists of four more sections. In Sect. 2 we discuss the negotiation process and identify its elements that may be formalized and supported by means of the proposed approach. Then in Sect. 3 we introduce the fuzzy TOPSIS model for the evaluation of the negotiation template. In Sect. 4 we show how these scoring systems may be used by a third party (or a negotiation support system) to improve the negotiated agreement by applying a straightforward notions of fair solutions derived from game theory. In Sect. 5 an example of using a fuzzy TOPSIS model of negotiators' decisions in supporting a bilateral negotiation is presented, showing also the possibility of using the negotiators' scoring system to visualize the negotiation

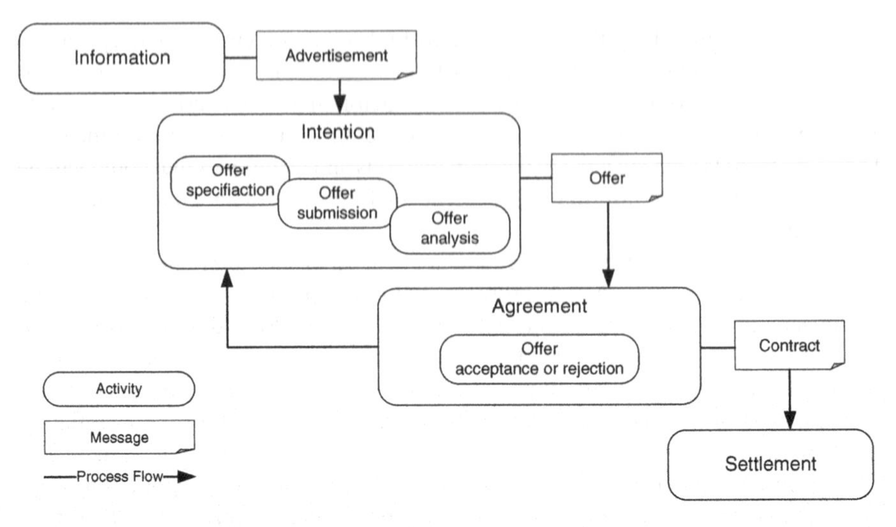

Fig. 1 Negotiation in Montreal Taxonomy. *Source* [25]

progress and concession scale, as well as the suggestion for the improvement of the agreement that was negotiated.

2 Negotiation Process and its Elements

To have a wider perspective on negotiation support it is worthwhile to look at the negotiation process and its elements in detail, i.e. to identify the negotiation phases, sub-processes, negotiators' activities, and the dataflow. The Media Reference Model [23], adopted later by Stroebel and Weinhardt [25] for electronic negotiation context, identifies four major phases of interaction with accompanying sub-phases. In this form it is known as the Montreal Taxonomy (Fig. 1).

The information phase consist of all the preparation activities the negotiators undertake to identify the problem and the counterparts and to elaborate scenarios for direct interactions with other parties. The intention phase consists of the exchange and evaluation of offers. In the agreement phase the parties decide whether to accept or reject the offer of their counterpart. Finally in the settlement phase of the negotiation the parties confirm the contract and execute it.

Montreal Taxonomy was verified and extended by Jertila and Schoop [11] to include the specificity of bilateral negotiations. In this form it distinguishes between private and public information (Fig. 2).

Here the negotiation process starts with the information phase, in which each party tries to recognize the negotiation problem individually. Then they contact each other and declare their willingness to negotiate. If both decided to negotiate, they pre-negotiate the organization of the negotiation process, i.e. what will be the meta-

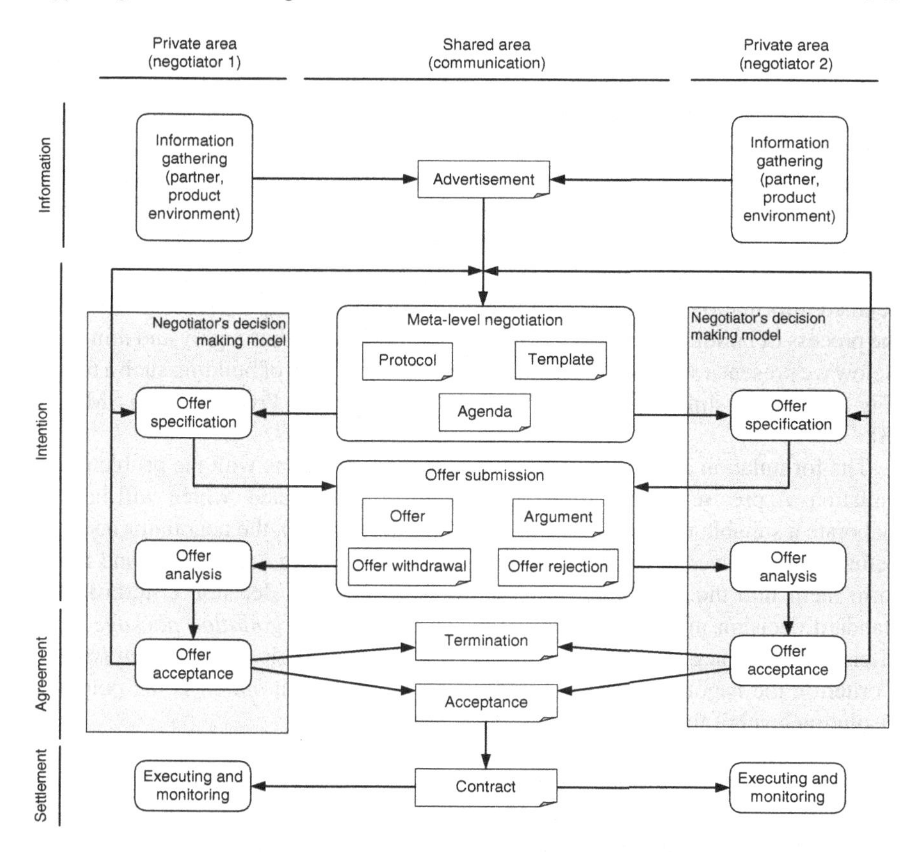

Fig. 2 Bilateral extension of Montreal Taxonomy. *Source* On the basis of [11]

level of the forthcoming negotiation. They agree on the negotiation protocol and agenda and specify the negotiation issues and options (negotiation space) building this way a negotiation template. This stage opens the intention phase. Then the parties start the process of exchanging offers (the actual negotiation). This process binds the intention phase with the forthcoming agreement phase, since the parties iteratively propose and receive the offers, conduct a detailed analysis of their pros and cons and decide whether to accept them or reject or even to terminate the negotiation. This analysis is based on the decision making model applied in the negotiation support tool or used individually by the parties. The important issue here is that the information exchanged during the offer submission may influence the fundamental descriptors of the meta-level negotiation. Hence, if the parties realize that the negotiation situation or the problem itself changes, they may also change the formal protocol, agenda or template and restart the negotiation process again. If one of the offers submitted is accepted, the process moves to the settlement phase, where the negotiators jointly build the final negotiation contract and then, individually monitor the process of its execution.

As we see, the strategic element in the intention and agreement phases of negotiation is the decision making model offered to the parties by the negotiation support system or the third party facilitating the process. Its parameters are initiated by the data developed as a result of the meta-level negotiation activity and may be changed by the feedback which the parties receive during the offer submission stage. The model itself provides the negotiator with adequate information about the quality of the offers submitted that is sufficient for the negotiator to decide whether an offer could be accepted. To do it well, the model has to take into consideration the detailed information about the focal negotiator's structure of preferences. Therefore the process of building such a model ought to be conducted thoroughly and minutely. Below we present a simple schema presenting the main steps of building such a model (Fig. 3), which is similar to other MCDM algorithms such as PrOACT [9] or SMART [3].

The formulation of the negotiator's decision model begins with the problem formulation. A precise definition of the problem must be stated, which will help to elaborate a suitable negotiation template. In the second step, the negotiator needs to define the objectives to be achieved during the forthcoming negotiations and transform them into the negotiation issues (the equivalents of decision criteria in the standard decision making problem). We can say that *a negotiation package* is an offer, which the negotiator may send to or receive from their opponent, *an issue* is a criterion the negotiator uses to evaluate the offers and an *option* is the potential resolution level of the criterion.

Then (Step 3) a feasible negotiation space is defined, which may be specified by means of a decision matrix if the negotiation problem is discrete, or may be more vaguely described by means of feasible ranges of the resolution levels for each negotiation issue defined in the previous step. These three steps of the algorithm allow for an objective definition of the negotiation template, while the next three operate with the subjective preferences of a negotiator to fine-tune the parameters of the model and build the scoring system of the negotiation offers that precisely reflects the negotiator's individual evaluations.

Step 4 requires the implementation of a selected MCDM method, that will be used for preparing the scoring system. In this paper we will apply the fuzzy TOPSIS modified algorithm together with the AHP single-level procedure; the latter one used only to estimate the weights of negotiation issues. Any other MCDM model can be applied here, but it naturally requires prior modifications and adequate adaptation of the scoring algorithm to the negotiation context and requirements.

In Step 5 we apply the TOPSIS scoring algorithm to evaluate the negotiation space and the predefined template (for details see Sect. 3). If the negotiator is not satisfied with the global scores, he can modify the model parameters to rescore the system and adjust it best to his preference structure (Step 5f, optional). A scored template allows to build the final ranking of all salient packages (Step 6). The negotiators are now able to identify groups of offers of acceptable and similar quality that can be used in the successive negotiation phases as alternative concessions.

Note that we assume that the specification of the negotiation space, as well as the negotiator's preferences, cannot be done in a precise and thorough way (which is

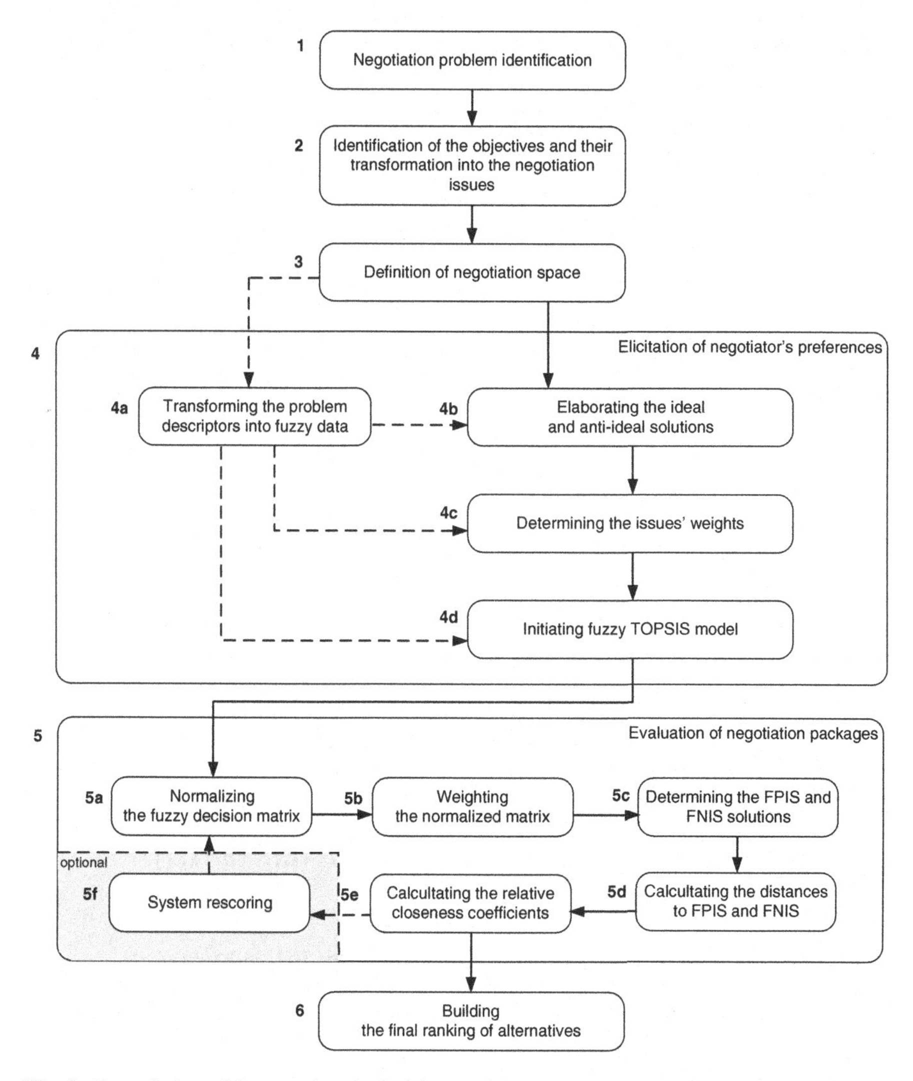

Fig. 3 Formulation of the negotiator's decision model

typical for very many negotiation situations involving uncertainty and risk). Therefore we allow the negotiators to express their preferences using imprecise and vague evaluations. They may operate directly with triangular fuzzy numbers (TFNs) for quantitative issues or express their preferences verbally using linguistic labels that are transformed later into their numerical equivalents. The specificity of fuzzy numbers allows also to transform other types of data into the TFNs. If there are some issues, for which the negotiator is able to precisely define their resolution levels, we may transform the crisp values into TFNs using simple transformation formulas (see [1]).

3 Negotiation Decision Making Model Based on Fuzzy TOPSIS

3.1 Triangular Fuzzy Numbers (TFN)

As mentioned before, our approach applies the fuzzy TOPSIS (FTOPSIS) operating with Fuzzy Triangular Numbers (TFNs). The reason for using a TFN is that it is intuitively easy for the decision-makers to interpret. It provides an effective way for formulating decision problems where the available information is subjective and imprecise or it is represented by linguistic values. Technically, a fuzzy number is characterized by an interval of real numbers with a grade of membership between 0 and 1. The membership function of a TFN is expressed in the following way:

$$\mu_{\widehat{A}}(x) = \begin{cases} 0 \ \textit{for} \quad x < a \\ \frac{x-a}{b-a} \ \textit{for} \ \ a \le x \le b \\ \frac{c-x}{c-b} \ \textit{for} \ \ b \le x \le c \\ 0 \ \textit{for} \ \ x > c \end{cases} \tag{1}$$

A TFN, denoted by $\widehat{A} = (a, b, c)$, is defined by three real numbers indicating the smallest possible value (a), the most promising value (b), and the largest possible value (c) of this TFN. A TNF $\widehat{A} = (a, b, c)$, is a non-negative fuzzy number, if (and only if) $a \ge 0$. Below we list major TFN operations used later in TOPSIS procedure. Let $\widehat{A}_1 = (a_1, b_1, c_1)$ and $\widehat{A}_2 = (a_2, b_2, c_2)$ be two positive triangular fuzzy numbers, then [15]:

- multiplication of TFN by a real number $k : k \otimes \widehat{A}_1 = (ka_1, kb_1, kc_1)$,
- the fuzzy inverse: $\left(\widehat{A}_1\right)^{-1} \cong \left(\frac{1}{c_1}, \frac{1}{b_1}, \frac{1}{a_1}\right)$
- max of TFN: $\max(\widehat{A}_1, \widehat{A}_2) \cong (\max(a_1, a_2),\ \max(b_1, b_2),\ \max(c_1, c_2))$,
- min of TFN: $\min(\widehat{A}_1, \widehat{A}_2) \cong (\min(a_1, a_2),\ \min(b_1, b_2),\ \min(c_1, c_2))$,
- the vertex distance: $d(\widehat{A}_1, \widehat{A}_2) = \sqrt{\frac{1}{3}\left((a_1 - a_2)^2 + (b_1 - b_2)^2 + (c_1 - c_2)^2\right)}$.

The linguistic variable is a variable whose values are words or sentences in a natural or artificial language [31]. The notion of a linguistic variable provides means to approximate a characterization of phenomena, which are too complex or too ill-defined to be described in conventional, crisp quantitative terms. These linguistic variables should have quantitative equivalents. One possibility is to define these equivalents in the form of positive triangular fuzzy numbers. An example of such a linguistic variable is given in Table 1.

In the linguistic scale shown in Table 1 there are three major categories describing the option ratings: Poor (P), Fair (F) and Good (G). Four more categories describe the intermediate ratings: Very poor (VP), Medium Poor (MP), Medium Good (MG) and Very Good (VG). Each of these categories has the TFN equivalents that are based on the consecutive odd numbers ranging form 0 to 10 including endpoints. Therefore

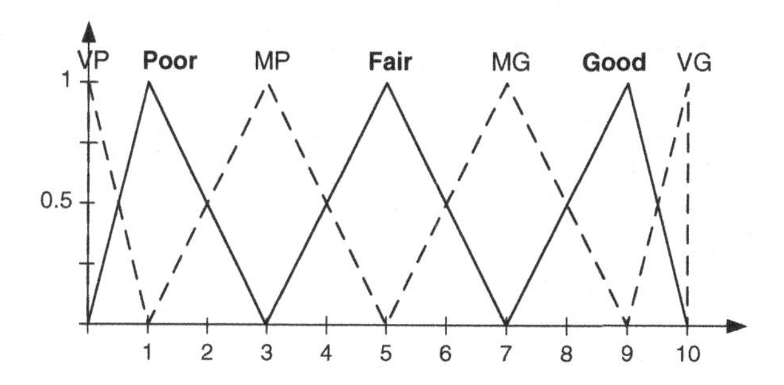

Fig. 4 Major (*bold*) and intermediate (*dashed*) categories of the linguistic scale based on TFNs

the TFN equivalent of (P) is (0, 1, 3), and for (F) we obtain (3, 5, 7). The intermediate levels begin with (0, 0, 1) for (VP) and end with (9, 9, 10) for (VG)—see Fig. 4.

3.2 TOPSIS

Technique for Order Preference by Similarity to an Ideal Solution (TOPSIS) has been developed by Hwang and Yoon [10] for solving the MCDM problems. The basic principle of TOPSIS is that the chosen alternative should have the "shortest distance" to the positive ideal solution (PIS) and the "longest distance" to the negative ideal solution (NIS), where PIS is the solution maximizing the benefit criteria and minimizes the cost ones, while NIS is the solution maximizing the cost criteria and minimizes the benefit ones. In the classical TOPSIS algorithm, the ratings and the weights of the criteria are defined precisely, so the application of this technique is limited to the well-structured decision (negotiation) problems only. However, since human preferences are often vague and cannot be expressed by exact and precise numerical values, the application of fuzzy concepts in negotiation support seems more relevant. Following the principles of FTOPSIS procedure [1] we need first to convert the decision matrix into a fuzzy decision matrix and then operate on these

Table 1 Linguistic variables for the ratings

Linguistic variables	Fuzzy triangular numbers
Very Poor (VP)	(0, 0, 1)
Poor (P)	(0, 1, 3)
Medium Poor (MP)	(1, 3, 5)
Fair (F)	(3, 5, 7)
Medium Good (MG)	(5, 7, 9)
Good (G)	(7, 9, 10)
Very Good (VG)	(9, 10, 10)

Source [1]

fuzzy evaluations (construct a normalized and weighted fuzzy decision matrix, apply vertex fuzzy distance notion form measuring the distances etc.).

We propose some modifications of the FTOPSIS procedure to fit it better to the support of the negotiation process (for details see [20]). Taking into account the fact that during the negotiation process the negotiator can introduce new alternatives (outside of the predefined template) we propose to define subjectively the ideal and anti-ideal solutions in the form of aspiration and reservation packages, which expands the initial negotiation space defined on the basis of TOPSIS-based max and min solutions. This will make our negotiation problem stable, which means that the new offer will not change the scoring system obtained before this offer was introduced into the predefined set of alternatives, nor will it result in rank reversal which can appear in the classic TOPSIS procedure [2, 6]. To obtain criteria weights the AHP method may be used, the pairwise comparisons of which would allow the negotiator to define their preferences in a natural and intuitive way. This technique can be also easily integrated with FTOPSIS procedure.

The process of preparing the negotiator's decision model (see Fig. 3) may be described as follows:

- **Step 1. Definition of the negotiation problem**
  The negotiator conducts a thorough analysis of the problem. He thinks of potential ways of solving it and considers how the negotiation with the potential counterpart may solve this problem.
- **Step 2. Identification of the objectives and their transformation into the negotiation issues.**
  The negotiator thinks of the major objectives connected with the forthcoming negotiation. These objectives are the evaluation criteria of the potential negotiation contract. The relevant negotiation issues are elaborated on the basis of these evaluation criteria and associated measurement scales are assigned to each of the issues.
  Let us denote by $Z = \{Z_1, Z_2, ..., Z_n\}$ the set of n issues. The set Z can be divided into two sets: I—a subset of benefit issues (the higher value the more preferable) and J—a subset of cost issues (the lower value the more preferable).
- **Step 3. Definition of the negotiation space.**
  Let D_j denote the negotiation space with respect to the jth issue. It is defined in the form of a set of feasible options and bounded by the lowest acceptable target value (reservation limit) $r_j \in D_j$ and an aspiration value $a_j \in D_j$, where $j = 1, 2, ..., n$. These values give the maximum limit of demands as well as the minimum limit of concessions and define the negotiation space for each issue.
- **Step 4a. Transforming the problem descriptors into fuzzy data.**
  All packages are measured with regard to every issue using a related measurement scale. We assume that in the preliminary step of the negotiation modeling the negotiators may choose the way of describing the resolution levels of the issues. These evaluations can be based on different types of data (numerical values, linguistic or mixed values) and subjective judgments.

A sufficiently representative and manageable set of packages is generated in the form of a finite set of negotiation packages $P = \{P_1, P_2, ..., P_m\}$, that are predefined as examples of the potential negotiation contract. The definition of the set P is a fairly important issue, since it puts frames on the future negotiation analysis. Each package can be represented as $P_i = [x_{i1}, ..., x_{in}]$– where $x_{ij} \in D_j$. The standard TOPSIS method requires the x_{ij} to be defined by means of precise values. Here we assume that if some of the criteria (or all of them) are uncertain or imprecise or have subjective characteristics, then the triangular fuzzy number (TFN) transformations are used to express the negotiator's assessments. All input data have to be represented in the form of TFNs $\widehat{x} = (a, b, c)$. The crisp values $x \in \Re$ are transformed into TFNs by the formula $\widehat{x} = (x, x, x)$.
Now every package P_i is represented by a vector $\hat{P}_i = \left[\widehat{x}_{i1}, \widehat{x}_{i2}, ..., \widehat{x}_{in}\right]$, where $\widehat{x}_{ij}$ is a TFN representation of the jth issue's option in the ith package.

- **Step 4b. Elaborating the ideal and anti-ideal solutions.**
In this step we construct two packages P_I and P_{AI}, that represent the aspiration and reservations levels of a negotiator [20]. The reservation package, BATNA,[1] has the form: $P_{AI} = [\widehat{x}_{P_{AI}1}, ..., \widehat{x}_{P_{AI}n}]$ where $\widehat{x}_{P_{AI}j}$ is a TFN representation of the reservation level for the jth issue and $P_I = [\widehat{x}_{P_I1}, ..., \widehat{x}_{P_In}]$ where $\widehat{x}_{P_Ij}$ is a TFN representation of the aspiration level of the jth issue. Let us denote by $\widehat{x}_{P_Ij} = (a^j_{P_I}, b^j_{P_I}, c^j_{P_I})$ and $\widehat{x}_{P_{AI}j} = (a^j_{P_{AI}}, b^j_{P_{AI}}, c^j_{P_{AI}})$ (any other option is described by a TFN as follows $\widehat{x}_{ij} = (a_{ij}, b_{ij}, c_{ij})$) the fuzzy triangular numbers that represent the resolution levels for the jth issue in the packages P_I, P_{AI}, respectively. If the criteria are defined numerically, we can take $\widehat{x}_{P_Ij} = (x^+_j, x^+_j, x^+_j)$, $\widehat{x}_{P_{AI}j} = (x^-_j, x^-_j, x^-_j)$, where x^+_j, x^-_j are extreme values such that

$$x^-_j \leq \max_i x_{ij} \quad \text{for} \quad i = 1, 2, ..., m; \; j = 1, 2, ..., n \tag{2}$$

$$x^+_j \geq \max_i x_{ij} \quad \text{for} \quad i = 1, 2, ..., m; \; j = 1, 2, ..., n \tag{3}$$

where x_{ij} is a value of the jth issue's option in the ith package.
Then

$$\widehat{x}_{P_Ij} = \begin{cases} (x^+_j, x^+_j, x^+_j), \text{ if} j \text{ is a benefit criterion} \\ (x^-_j, x^-_j, x^-_j), \text{ if } j \text{ is a cost criterion} \end{cases} \tag{4}$$

and

$$\widehat{x}_{P_{AI}j} = \begin{cases} (x^-_j, x^-_j, x^-_j), \text{ if } j \text{ is a benefit criterion} \\ (x^+_j, x^+_j, x^+_j), \text{ if} j \text{ is a cost criterion} \end{cases} \tag{5}$$

[1] Best Alternative to Negotiated Agreement (see [4])

Table 2 Summarization of the experiments

Importance intensity	Definition
1	Equal importance
3	Moderate importance
5	Strong importance
7	Very strong importance
9	Extreme importance
2, 4, 6, 8	Can be used to express intermediate values

Source [22]

are the TFNs that represent the resolution levels for the jth issue in the packages P_I , P_{AI}, respectively. For the criteria represented by linguistic variables we can use the numerical equivalents of extreme linguistic values such as: Very good (VG) or Very poor (VP).

- **Step 4c. Determining the issue weights.**
  We use the well-known Analytic Hierarchy Process (AHP) proposed by Saaty [22] for determining issue weights. When applying AHP, the negotiator's preferences are elicited by means of a pairwise comparison of the issues. Saaty [22] has recommended a 9-level verbal scale with the equivalent numerical evaluations to describe the evaluation of the given pair. In the negotiation context, if two criteria (issues) are of equal importance, the value of 1 is assigned to the pair being compared, whereas the value of 9 indicates the absolute importance of one criterion over the other (see Table 2).
  The judgments are put into the matrix $A = \left[a_{ij}\right]$, which contains the pairwise comparison elements,

$$a_{ij} = \frac{w_i}{w_j} \tag{6}$$

where w_i and w_j are the relative importances of criteria i and j, respectively, their reciprocals, $a_{ji} = \frac{1}{a_{ij}}$ and $a_{ii} = 1$.
The weights of the criteria can be calculated by averaging over the normalized columns using the following formula:

$$w_i = \frac{1}{n}\sum_{j=1}^{n}\left(\frac{a_{ij}}{\sum_{i=1}^{n} a_{ij}}\right) \tag{7}$$

Saaty established also a Consistency Index (CI) of the square matrix A. This measure can be used to verify the extent to which the judgments supplied are consistent.

$$CI = \frac{(\lambda_{\max} - n)}{n - 1} \tag{8}$$

where $\lambda_{\max}$ is the highest eigenvalue of the matrix A.

To decide whether the *CI* is acceptable or not, the Random Consistency Index (RI) is provided, which is the average *CI* of randomly generated reciprocal matrices of dimension n [22]. The degree of inconsistency of the square matrix A can be measured by the ratio of *CI* to*RI*, which is called the Consistency Ratio (*CR*).

$$CR = \frac{CI}{RI}.100\% \quad (9)$$

We can conclude that the matrix is sufficiently consistent and accept the matrix when $CR \leq 10\%$. Otherwise, it can be concluded that the inconsistency is too large and unacceptable, so that decision makers must revise their judgments.

- **Step 4d. Initiating the fuzzy TOPSIS model.**
  The negotiation context dependent factors discussed in the two previous steps are now transformed into the TOPSIS algorithm parameters. Thus we obtain the vector of criteria weights derived from the evaluation of the issues:

$$w = [w_1, w_2, \ldots, w_n], \quad (10)$$

as well as the ideal and anti-ideal solutions defined by the aspiration and reservation packages P_I and P_{AI}, and the fuzzy decision matrix representing the set of predefined negotiation offers

$$\hat{X}_{P \cup \{P_I, P_{AI}\}} = [\widehat{x}_{ij}], \quad (11)$$

for $i = 1, \ldots, m+2$, $j = 1, \ldots, n$.

- **Step 5. Evaluation of negotiation packages.**
  The scoring points for negotiation packages are determined using fuzzy arithmetic (see [1]). We follow the modified FTOPSIS algorithm:

  – *Step 5a. Construction of the normalized decision matrix.* Normalization allows for inter-criteria comparisons, since it reduces the empirical ranges of feasible options to the unified common range. The normalized fuzzy decision matrix $Z = \left[\hat{z}_{ij}\right]$ can be expressed by the following formula:

$$\widehat{z}_{ij} = \left(\frac{a_{ij}}{c_j^{\max}}, \frac{b_{ij}}{c_j^{\max}}, \frac{c_{ij}}{c_j^{\max}}\right), \text{ where } j \in I, j = 1, 2..., n \quad (12)$$

$$\widehat{z}_{ij} = \left(\frac{a_j^{\min}}{c_{ij}}, \frac{a_j^{\min}}{b_{ij}}, \frac{a_j^{\min}}{a_{ij}}\right), \text{ where } j \in J, j = 1, 2..., n \quad (13)$$

  where $a_j^{\min} = \min_i a_{ij}$, $c_j^{\max} = \max_i c_{ij}$ and the values a_{ij}, b_{ij}, c_{ij} are the descriptors of the fuzzy option $\widehat{x}_{ij} = (a_{ij}, b_{ij}, c_{ij})$.
  Note that if all options in the package being normalized belong to the negotiation space, the following relations are satisfied: $a_j^{\min} = a_{P_{AI}}^j$ and $c_j^{\max} = c_{P_I}^j$.

- *Step 5b. Construction of the weighted normalized decision matrix.* In the weighted normalized fuzzy decision matrix $V = \left[\widehat{r}_{ij}\right]$ the criteria importance is taken into consideration:

$$\widehat{r}_{ij} = w_j \otimes \widehat{z}_{ij} \text{ for } i = 1, \ldots, m+2;\ j = 1, \ldots, n. \tag{14}$$

- *Step 5c. Determination of the fuzzy positive ideal solution (FPIS) and the fuzzy negative ideal solution (FNIS).* The fuzzy ideal solution FPIS (A^+) and the fuzzy anti-ideal solution FNIS (A^-) have the form:

$$A^+ = \left(\widehat{v}_1^+, \widehat{v}_2^+, \ldots, \widehat{v}_n^+\right) = \left(\max_i \widehat{r}_{i1}, \max_i \widehat{r}_{i2}, \ldots, \max_i \widehat{r}_{in}\right) \tag{15}$$

$$A^- = \left(\widehat{v}_1^-, \widehat{v}_2^-, \ldots, \widehat{v}_n^-\right) = \left(\min_i \widehat{r}_{i1}, \min_i \widehat{r}_{i2}, \ldots, \min_i \widehat{r}_{in}\right) \tag{16}$$

- *Step 5d. Calculation of the distances of each alternative from FPIS and FNIS, respectively.*

$$d_i^+ = \sum_{j=1}^{n} d(\widehat{r}_{ij}, \widehat{v}_j^+),\quad i = 1, 2, \ldots, m+2 \tag{17}$$

$$d_i^- = \sum_{j=1}^{n} d(\widehat{r}_{ij} \widehat{v}_j^-),\quad i = 1, 2, \ldots, m+2 \tag{18}$$

  where $d(\hat{A}, \hat{B})$ is the vertex distance between two triangular numbers $\hat{A}$, $\hat{B}$.
- *Step 5e. Calculation of the relative closeness to the FPIS.* For each alternative the closeness coefficient is determined, that aggregates the values of d_i^+ and d_i^- into a scalar criterion. For the alternative i we compute:

$$CC_i = \frac{d_i^-}{d_i^- + d_i^+} \tag{19}$$

  where $0 \leq CC_i \leq 1,\quad i = 1, 2, \ldots, m+2.$

- **Step 6. Ranking all alternatives according to descending** CC_i.
  Let us denote by $C = \{CC_i, P_i \in P\}$ the set of closeness coefficients, the differences of which $\Delta CC_{i/k} = CC_i - CC_k$ can be interpreted as a cardinal measures of concessions made by the negotiator when moving from offer i to k $(i, k = 1, 2, \ldots, m)$. The negotiator's decision problem can be thus formally described as the nine-tuple:

$$(Z, P, I, J, w, P_I, P_{AI}, \hat{X}_{P \cup \{P_I, P_{AI}\}}, C). \tag{20}$$

Having defined the model in this form, the negotiator is able to conduct the intention phase, i.e. specify their offers, analyze the quality of their counterpart's proposals, measure the scale of concessions made by their counterpart in the successive stages of the offer submission phase etc.

Let t_r be the time variable denoting the negotiation round within the offer submission phase ($r = 1, ..., T$). In the round t_1 one party makes a proposal, i.e. submits an offer, which their counterpart may accept or reject. Its acceptance is equivalent to an *agreement* and the negotiation ends successfully. Its rejection constitutes a *disagreement* and the negotiation proceeds to the round t_2 , in which the counterpart makes their proposal (that the focal negotiator may accept or reject). The process continues as long as one of the offers will be accepted or rejected without any counteroffer. The negotiations are thus the paths of offers and counteroffers concluded with an agreement or disagreement. These paths may be visualized using the negotiation history and negotiation dance graphs that allow negotiators to track negotiation progress and consider the current negotiation offer in a wider perspective of previous proposals.

4 The Post-Agreement Improvements

After completing the intention phase the parties may conclude with an agreement. In this case they move to the settlement phase and build the final form of the negotiation contract. Having known the negotiators' scoring system the third party (mediator, negotiation support system etc.) may analyze the efficiency of the negotiated agreement and search for its possible improvements. In this situation the negotiation can be regarded as a cooperative game, for which bargaining solutions may be applied [28]. The first approach to solve the negotiation problem in this way was proposed by Nash [17]. Let us assume that the preferences of both parties are formally described by utility functions f_s , f_b:

$$f_s : P \rightarrow [0, 1] \quad f_b : P \rightarrow [0, 1]$$

These functions assign to each feasible alternative (a negotiation package) a potential level of satisfaction. In our particular application context the value of CC_i is assigned to the i-th negotiation package ($f_s(P_i) = CC_i^s \; f_b(P_i) = CC_i^b$). Let U be the set of all possible score profiles for both parties (CC_i^s, CC_i^b) where CC_i^s $\left(CC_i^b\right)$ is the closeness coefficient for the Seller (Buyer), respectively. By d we denote the disagreement point $d = (d_s, d_b) = (CC_p^s, CC_p^b)$ in the set of profiles, which is the outcome obtained when no agreement is met. In the particular application context this point corresponds to the solution obtained in the previous phase of negotiation, that is, to the pth package. Nash assumed a set of axioms that the negotiation solution $\delta(U, d)$ should satisfy:

1. The solution $\delta(U, d)$ should **be Pareto efficient** meaning that there does not exist a pair of score profiles $(CC_i^s, CC_i^b) \in U$ such that $(CC_i^s, CC_i^b) \prec \delta(U, d)$.

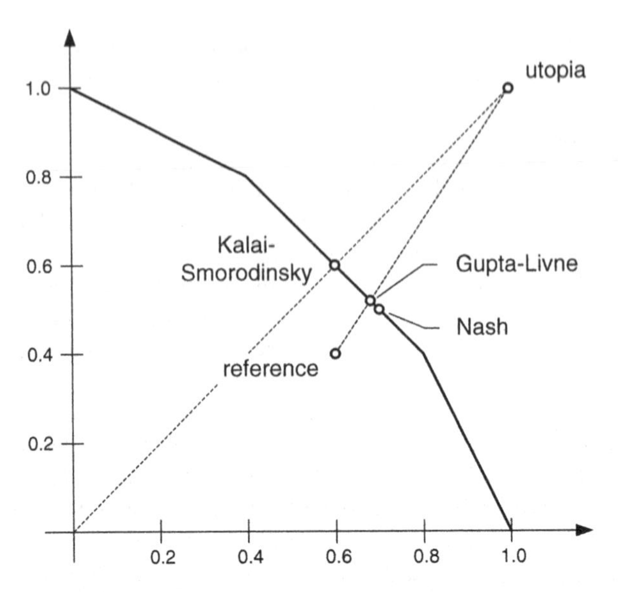

Fig. 5 Examples of Nash, Kalai-Smorodinsky and Gupta-Livine bargaining solutions. *Source* [5]

$$(CC_i^s, CC_i^b) \in U$$

2. **Symmetry** should be satisfied, meaning that if we consider a situation for which $d_s = d_b$ and $(CC_i^s, CC_i^b) \in U \Leftrightarrow (CC_i^b, CC_i^s) \in U$ then $\delta_1(U, d) = \delta_2(U, d)$.
3. **Invariance of the solution under equivalent scoring functions representations**. If the bargaining problem (U, d) is transformed into (U', d') by taking: $CC_i^{s\prime} = \alpha_s CC_i^s + \beta_s, \quad CC_i^{b\prime} = \alpha_b CC_i^b + \beta_b$ and $d_s{}' = \alpha_s d_s + \beta_s, \quad d_b{}' = \alpha_b d_b + \beta_b$ where $\alpha_s > 0$ $\alpha_b > 0$ then $\delta_s(U', d') = \alpha_s \delta_s(U', d') + \beta_s, \ \delta_b(U', d') = \alpha_b \delta_b(U', d') + \beta_b$.
4. **Independence of irrelevant alternatives (packages).** For two problems (U, d) and (U', d) with $U \subset U'$ and $\delta(U', d) \in U$, we obtain $\delta(U, d) = \delta(U', d)$.

Nash proved that there exists a unique bargaining solution satisfying these axioms and which has the following form (see also Fig. 5):

$$P_k = \arg \max_{P_i \in P} (f_s(P_i) - d_s)(f_b(P_i) - d_b) \tag{21}$$

Another popular bargaining solution is the Kalai-Smorodinsky solution [12], which is obtained when the fourth axiom (in the Nash solution) is replaced by the monotonicity axiom. Visually, the Kalai-Smorodinsky solution can be presented as the point of intersection of the Pareto frontier and the line connecting disagreement point with utopia (in the space of score profiles of both parties). Utopia can be determined by taking the pair of scores of the ideal points of both parties or the pair of aspiration levels of both parties (see Fig. 5). The Gupta-Livne solution [8] is deter-

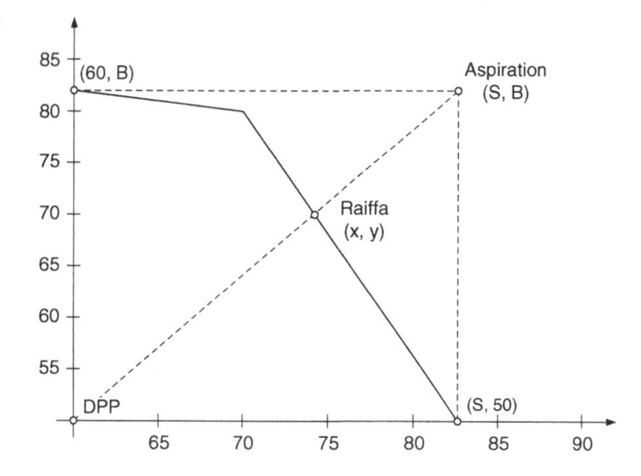

Fig. 6 The Raiffa pay-off pair determined by the arbitration scheme. *Source* [19]

mined in analogous way to the Kalai-Smorodinsky solution by taking the point of intersection of the efficient frontier with the line segment joining some intermediary negotiation solution (reference) with the global utopia point.

Before the Kalai-Smorodinsky solution was proved to satisfy the axioms mentioned Howard Raiffa (Raiffa 1952) had proposed an arbitration scheme consistent with this solution. For the sake of illustration we will use a different scale for the explanation of this scheme. Let us assume that during the negotiation the negotiators can gain pay-offs (scores) from the following intervals:

$$D_s = [60, 82], \quad D_b = [50, 82.5]$$

When the negotiators leave the negotiation table without an agreement their pay-offs are consistent with the lowest values $DPP = (60, 50)$ that correspond to the disagreement point. The aspiration levels of the negotiators are the maximal values of pay-offs they can potentially reach: $(S, B) = (82, 82.5)$. The region of feasible pairs of pay-offs is bounded by the Pareto frontier (see Fig. 6). All bargaining solutions constitute a pair of pay-offs located on this frontier. The pair of pay-offs determined by the aspiration levels corresponds to the utopia point which cannot be reached since it is located outside the set of feasible pairs of pay-offs. The Raiffa pay-off pair is the efficient pay-off pair on the line segment connecting the disagreement (DPP) and the aspiration pay-off pair (S,B).

As mentioned above, Kalai and Smorodinsky proposed axiomatizing this solution and proved the Raiffa characterization theorem stating that the Raiffa method is the only efficient, unbiased, scale invariant and individually monotone method (the axioms describing the Kalai-Smorodinsky solution).

The Raiffa method is an arbitration scheme meaning that the negotiation solution is dictated by a trusted third party. However, in the case of a protocol based on iterative exchange of offers and counter-offers we do not require the parties to use the help of a third party in the intention phase of the negotiation process. The help of such a party

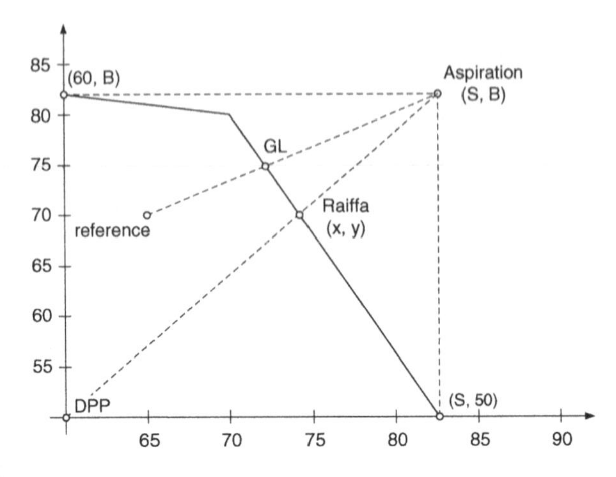

Fig. 7 Illustration of the Gupta-Livne solution (GL) as a possible improvement of the agreement

may be useful in the post-optimization phase which is non-mandatory, which means that both parties have to agree to participate in this phase. The additional phase of the agreement improvement may be useful since the parties very often do not reach a Pareto efficient outcome. The post-optimization phase assumes that the agreement obtained by the parties so far cannot be made worse; it can only be improved at least for one party if not for both.

The post-negotiation phase involves the disclosure of preferences to the third party (human mediator or software support system), which can determine the Pareto frontier in the space of score profiles in order to suggest an improved agreement. The Raiffa-Kalai-Smorodinsky solution can be adjusted to fit the problem of post-optimization. Let us assume that in the actual negotiation phase the negotiators agreed on an alternative and now they enter the post-optimization phase. Figure 7 shows that we do not need to consider the whole set of profiles. We know that the improved solution will dominate the reference (the current agreement). Therefore, we need only to consider a set of packages for which the score profiles dominate the reference.

If we consider only the set of profiles which dominate the reference (see Fig. 8) then we obtain an analogous situation to the original arbitration scheme. As mentioned before the Gupta- Livne solution is located on the line segment connecting the reference with aspiration level. As illustrated on Fig. 7 this solution resembles the Raiffa-Smorodinsky-Kalai solution except that it considers the global aspiration level (Aspiration1 in Fig. 8) taken from the full set of profiles.

We propose to change the aspiration level to make it consistent with the reduced set of profiles (Aspiration2 in Fig. 8). The reduction of aspiration levels from Aspiration 1 to Aspiration 2 is appropriate here since the set of profiles is reduced to the relevant profiles only and therefore the same should happen with the aspiration level. If the aspiration level (Aspiration1) is changed into the level Aspiration2 the GL1 solution is transformed into Raiffa 2 (Fig. 8).

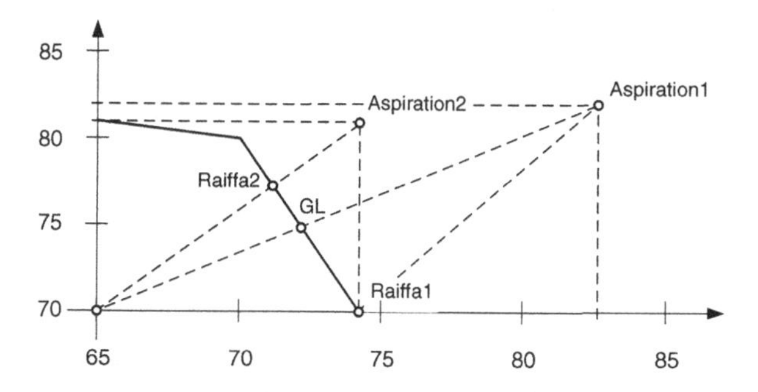

Fig. 8 The modification of aspiration levels after reducing the set of scores' profiles

Formally, the procedure of improvement has the following form. Let us consider a finite set $\boldsymbol{P}$ of feasible alternatives. The third party has at its disposal the set P and the scores assigned to all alternatives from the viewpoint of both negotiating parties:

$$CC_i^s = f_s(P_i), \quad CC_i^b = f_b(P_i) \quad i \in \{1, ..., n\}.$$

The pairs of scores for all alternatives considered are formed as follows:

$$U = \{(CC_i^s, CC_i^b), \quad i \in \{1, ...n\}\}.$$

In the next stage the Pareto front of the set U is determined in the form of the following set:

$$P = \{u \in U | \neg \exists v \in U | v \prec u\}.$$

where $\prec$ is the relation indicating that the profiles are in Pareto order. In the next step the reference (the current agreement) in the space of score profiles $u_r = (CC_r^s, CC_r^b)$ and the utopia (or aspiration level) u_u are connected with a straight line in the space of score profiles. The profiles located on this line are of the following form:

$$R = \{v \in U | v = u_r + t \cdot (u_u - u_r), t \in [0, 1]\}.$$

If the aspiration is used as utopia it can be determined in the following way:

$$u_u = (\max_{i \in \{1,..,n\}} CC_i^s, \max_{i \in \{1,..,n\}} CC_i^b).$$

For each package $P_k \in P$ (belonging to the Pareto front) the distance from the line R is computed: $d_k = d(P_k, R)$. By calculating the shortest distance $d_j = \min\{d_k = d(P_k, R) | P_k \in P\}$ we can find the efficient package which constitutes the approximation of Gupta-Livne (or the Kalai-Smorodinsky) solution. Since instead of a full set of efficient packages we have at our disposal a discrete set, the bargaining solutions are approximated. Therefore the package from the Pareto front P for which the profile score is nearest to the line R is selected as the agreement improvement.

5 Scoring Negotiation Offers with Modified Fuzzy TOPSIS: An Example

Let us now consider a buyer-seller negotiation that allows us to show how the proposed formalization and the FTOPSIS-based model can support the process of bilateral negotiation. The parties are negotiating the conditions of the potential business contract. Three issues are discussed: Z_1 – unitary price (EUR), Z_2 – payment conditions (days), Z_3 – warranty policy. The issues Z_1,Z_2 are represented directly by TFNs, although the initial definition of the negotiation's key issue (price) is precise and the price options are mapped to TFNs before the construction of the decision matrix. The issue Z_3 is of qualitative nature and can be defined as a combination of options describing the warranty period and place where the possible repairs will be done. Therefore Z_3's options are subjectively pre-evaluated by negotiators using linguistic variables, which are also encoded by TFNs and the corresponding linguistic scale presented in Table 1. The negotiation spaces defined by the crisp values for Z_1, Z_2 and the linguistic ones for Z_3 are the following:

- Price (EUR): $\langle 56, 68 \rangle$ for both parties;
- Payment (weeks): $\langle 1, 8 \rangle$ for both parties,
- Returns: ⟨Very Poor (VP), Very Good (VG)⟩ for both parties.

Z_1 is the benefit (cost) issue for the Seller (Buyer), Z_2 is cost (benefit) for the Seller (Buyer). Z_3 is assessed by both of them as benefit.

To build the initial set P the negotiators define the salient options for each issue. Let us assume they are given as:

- Price: 58, 60, 62, 64, 66, 68.
- Payment: (1, 2, 3), (2, 3, 4), (3, 4, 5), (4, 5, 6), (5, 6, 7), (6, 7, 8).
- Returns: '2 years at seller's repair centre (2S)', '2 years at buyer's office (2B)', '1 year at seller's repair centre (1S)', '4 years at buyer's office (4B)'. The buyer evaluates linguistically the options in the following way: 2S—Fair, 2B—Medium Good, 1S—Medium Poor, 4B—Good. The evaluations of the seller are: 2S—Fair, 2B—Medium Poor, 1S—Medium Good, 4B—Poor.

These options allow for building $6 \times 6 \times 4 = 144$ various packages that will constitute the set $\boldsymbol{P}$. Suppose that the matrix for subjective preference weight comparisons is:

- $A_S = \begin{bmatrix} 1 & 8 & 9 \\ 1/8 & 1 & 2 \\ 1/9 & 1/2 & 1 \end{bmatrix}$ for the Seller and $A_B = \begin{bmatrix} 1 & 6 & 8 \\ 1/6 & 1 & 2 \\ 1/8 & 1/2 & 1 \end{bmatrix}$ for the Buyer.

The weights vector determined by AHP is $w_S = [0.80,\ 0.12,\ 0.08]$ with $CR = 3,5\,\%$ and for the Buyer $w_B = [0.77,\ 0.15,\ 0.08]$ with $CR = 1,9\,\%$.

We assume that the negotiators have defined the ideal and anti-ideal packages and their TFN representations based on their aspiration and reservation levels (Table 3). The options VG and VP of issue Z_3 stand for Very Good and Very Poor and are represented by their TFNs as (9, 10, 10) and (0, 0, 1).

The negotiation offers can be obtained in the form of combinations of options drawn from different issues, located between aspiration and reservation levels with respect to different criteria. However, during the negotiation process only finitely many packages appear as Buyer or Seller offers.

Without loss of generality we can assume that the negotiations have been started by Buyer and the negotiation process took 10 rounds with offers $P_1, \ldots, P_{10}$ (see Table 4). According to the linguistic scale presented in Table 1, the linguistic variable P (Poor) is represented by the TFN equal to (0, 1, 3), MP (Medium Poor)—by (1, 3, 5), F (Fair)—by (3, 5, 7), MG (Medium Good)—by (5, 7, 9) and G (Good)—by (7, 9, 9).

The Buyer's and Seller's scoring systems for the selected set of packages with Ideal and Anti-Ideal package (Table 3) obtained by the FTOPSIS procedure (formulas (10)–(19)) are presented in Table 5. Observe that in the set of ten packages considered four: P_3, P_8,P_9,P_{10} are not Pareto optimal. Moreover, in every round of negotiation each party was making concessions. However, some concessions were viewed by the counterparts as being worse than the previously offered packages (see Table 5). The Buyer's second proposal (package P_3) was given a lower score than package P_1. Similarly, packages P_8 and P_{10} were scored by the Buyer lower that package P_6. The difference between scoring points of packages P_6 and P_4 is very small from the Buyer's point of view, so they can be treated as alternative packages. But from the Seller's perspective these packages differ to a high extent. That means that small concessions on the part of the Buyer may result in quite a high benefit for the Seller. The negotiation history depicted respectively for the Buyer and the Seller is presented in Figs. 9 and 10. They visualize the negotiation progress in each negotiator's scoring spaces, showing the scale of concessions made by both players from the perspective of Buyer and Seller, respectively. We may observe how differently the parties see the negotiation process. For instance, from the Buyer's perspective, all his offers were true concessions. The successive points that represent the concession path depict a monotonously decreasing line, which means that each consecutive offer was worse to him that the one proposed in the previous round. We can make an interesting

Table 3 The ideal and anti-ideal packages based on the negotiation spaces

package	Z_1 Price		Z_2 payment		Z_3 returns	
	Buyer	Seller	Buyer	Seller	Buyer	Seller
Ideal	(56, 56, 56)	(68, 68, 68)	(8, 8, 8)	(1, 1, 1)	VG	VG
Anty-ideal	(68, 68, 68)	(56, 56, 56)	(1, 1, 1)	(8, 8, 8)	VP	VP

Table 4 The negotiation offers and co-offers

Round	Offering party	Package	Z_1 price buyer/seller	Z_2 payment buyer/seller	Z_3 Returns Seller	Buyer
$r = 1$	Buyer	P_1	(60, 60, 60)	(4, 5, 6)	F	F
$r = 2$	Seller	P_2	(68, 68, 68)	(3, 4, 5)	P	MG
$r = 3$	Buyer	P_3	(58, 58, 58)	(3, 4, 5)	MG	MP
$r = 4$	Seller	P_4	(66, 66, 66)	(6, 7, 8)	MG	MP
$r = 5$	Buyer	P_5	(58, 58, 58)	(1, 2, 3)	F	F
$r = 6$	Seller	P_6	(64, 64, 64)	(4, 5, 6)	F	F
$r = 7$	Buyer	P_7	(60, 60, 60)	(1, 2, 3)	MP	MG
$r = 8$	Seller	P_8	(62, 62, 62)	(3, 4, 5)	MG	MP
$r = 9$	Buyer	P_9	(62, 62, 62)	(2, 3, 4)	MP	MG
$r = 10$	Seller	P_{10}	(64, 64, 64)	(2, 3, 4)	P	G

observation: the same concession path analyzed from the Seller's point of view is interpreted differently. The Seller sees the Buyer's second offer (P_3) to be worse than his counterpart's first proposal (P_1). Similarly, the Seller can claim that his whole strategy was build of true concessions (his own concession path decreases – see Fig. 10), while the Buyer will consider the offer P_8 to be worse than P_6.

As long as the parties operate with their individual history graphs they are unable to correctly recognize the true intentions of each other. If they decide to disclose their preferences to the third party (e.g. mediator or negotiation support system) the negotiation dance graph may be drawn, which shows the parties' concessions path in their joint evaluation space (Fig. 11).

Now they exactly know which offers proposed as concessions have been interpreted as true concessions by their counterparts. If we look at the Buyer's concession path we can see that he indeed made a concession but on his counterpart he made the impression of withdrawing. As the result they both lose (P_1 dominates P_3). They

Table 5 The Buyer-Seller scoring system

Package	Seller CC_S	Rank	Buyer CC_B	Rank
P_1	0.2966	9	0.5739	1
P_2	0.5602	1	0.3229	10
P_3	0.2931	10	0.5485	2
P_4	0.5390	2	0.4448	7
P_5	0.3773	8	0.4907	3
P_6	0.4416	3	0.4449	6
P_7	0.4013	6	0.4639	4
P_8	0.4377	4	0.4109	9
P_9	0.3818	7	0.4471	5
P_{10}	0.4138	5	0.4221	8

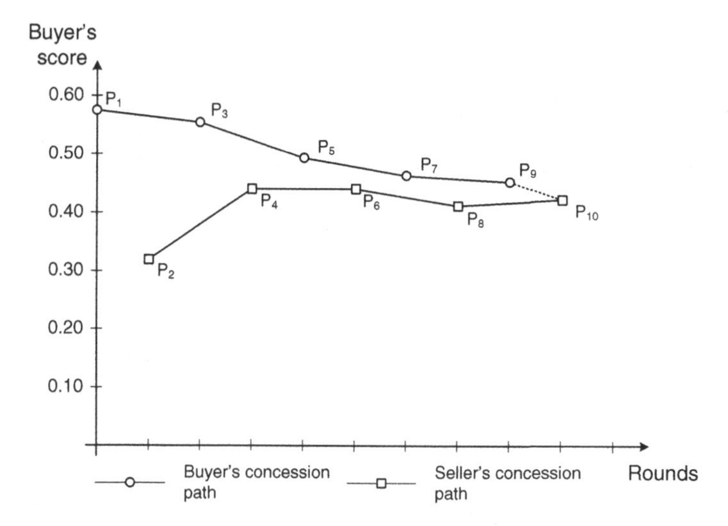

Fig. 9 Negotiation history graph for the Buyer

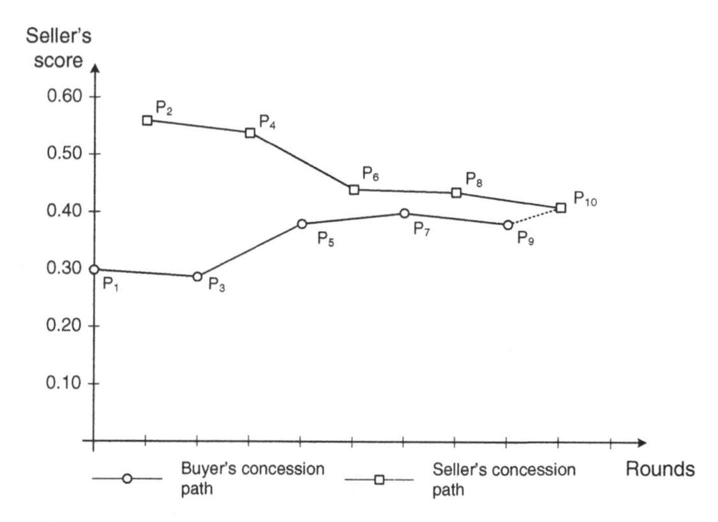

Fig. 10 Negotiation history graph for the Seller

also know that the compromise they negotiated is not Pareto efficient, and there are some other packages (negotiated before) that dominate the contract. They will see a strong need to improve the negotiated agreement.

Figure 12 illustrates the score profiles for a large set of packages. The points of Pareto front are marked with boxes. The approximations of Nash solution and the Raiffa-Kalai-Smorodinsky solution are computed and illustrated in Fig. 12 as well. Boxes correspond to non-dominated packages, while circles to dominated packages.

These solutions can be applied in a situation where the arbitration scheme is used instead of the iterative protocol. Let us assume that an agreement described by the package P_{26} was reached in the intention phase of the negotiation process. From the full set of packages, those dominating P_{26} have been selected as shown in Table 6.

Since all bargaining solutions are located on the Pareto frontier, first we derive this frontier (non-dominated packages) to derive the approximation of bargaining solution in the next step. We can see from Table 6 that the profile of scores of the package P_{117} (0.4920, 0.49) is not dominated by any other profile since the buyer's scores of other packages are all lower than 0.4920. If we consider the package P_{118} (0.4481, 0.54) we can see that some packages have the seller's score higher than 0.54, namely: P_{114}, P_{120}, P_{137}, P_{141} and P_{142}. However, the buyer's scores of these packages are all lower than 0.4481 which results in the conclusion that P_{118} is the next non-dominated package. This procedure of pairwise comparisons has to be performed for all packages to find the efficient ones. Column 8 in Table 6 contains the efficient packages.

For the derivation of the approximations of Kalai-Smorodinsky solution the distances of efficient points from the line segment between the reference (0.3793, 0.49) and the aspiration (0.492, 0.61) have been computed (column 9 of Table 6). From Table 6 we can see that for the package P_{141} the distance is minimal (0.005935), meaning that this package is the approximation of Kalai-Smorodinsky solution. The approximation of Nash solution is obtained as the package for which the product of both parties' scores is maximal taking into account the reference point (0.004032, the last column of Table 6 contains the products for all efficient packages). From Table 6 we can see that the package P_{143} maximizes the product of scores.

The same set of packages is illustrated in Fig. 13 together with the approximations of Nash and the Kalai-Smorodinsky solutions (P_{141}) which overlap in this case

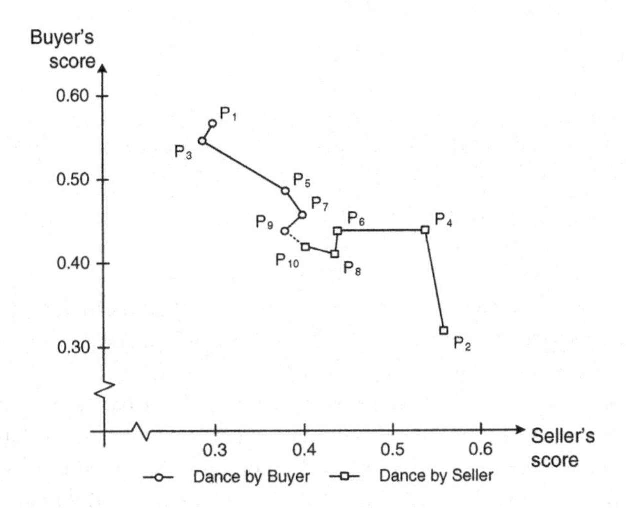

Fig. 11 Negotiation dance graph

Table 6 The packages considered during the procedure of agreement improvement

P_i	Price	Time of payment	Warranty conditions		Buyer's score	Seller's score	Pareto optimal	Distance from reference line	Product of parties' scores
			Buyer	**Seller**					
P_{26}	(60, 60, 60)	(1, 2, 3)	(1, 3, 5)	(5, 7, 9)	0.3793	0.4900	Agreement		
P_{86}	(64, 64, 64)	(4, 5, 6)	(1, 3, 5)	(5, 7, 9)	0.4039	0.4900			
P_{92}	(64, 64, 64)	(5, 6, 7)	(0, 1, 3)	(7, 9, 10)	0.4189	0.5200			
P_{96}	(64, 64, 64)	(6, 7, 8)	(0, 1, 3)	(7,9,10)	0.4694	0.5100			
P_{107}	(66, 66, 66)	(3, 4, 5)	(5, 7, 9)	(1, 3, 5)	0.3807	0.4900			
P_{109}	(66, 66, 66)	(4, 5, 6)	(3, 5, 7)	(3, 5, 7)	0.3891	0.5100			
P_{113}	(66, 66, 66)	(5, 6, 7)	(3, 5, 7)	(3, 5, 7)	0.4416	0.5000			
P_{114}	(66, 66, 66)	(5, 6, 7)	(1, 3, 5)	(5,7,9)	0.3975	0.5500			
P_{117}	(66, 66, 66)	(6, 7, 8)	(3, 5, 7)	(3, 5, 7)	**0.4920**	0.4900	TRUE	0.082151	0.000000
P_{118}	(66, 66, 66)	(6, 7, 8)	(1, 3, 5)	(5, 7, 9)	0.4481	0.5400	TRUE	0.015921	0.003440
P_{120}	(66, 66, 66)	(6, 7, 8)	(0, 1, 3)	(7, 9, 10)	0.4108	0.5800	TRUE	0.038652	0.002835
P_{137}	(68, 68, 68)	(5, 6, 7)	(3, 5, 7)	(3, 5, 7)	0.3861	0.5700			
P_{139}	(68, 68, 68)	(5, 6, 7)	(5, 7, 9)	(1, 3, 5)	0.4304	0.5300			
P_{141}	(68, 68, 68)	(6, 7, 8)	(3, 5, 7)	(3, 5, 7)	0.4369	0.5600	TRUE (**KS**)	**0.005935**	**0.004032**
P_{142}	(68, 68, 68)	(6, 7, 8)	(1, 3, 5)	(5, 7, 9)	0.3930	**0.6100**	TRUE	0.072164	0.001644
P_{143}	(68, 68, 68)	(6, 7, 8)	(5, 7, 9)	(1, 3, 5)	0.4807	0.5200	TRUE	0.053376	0.003042

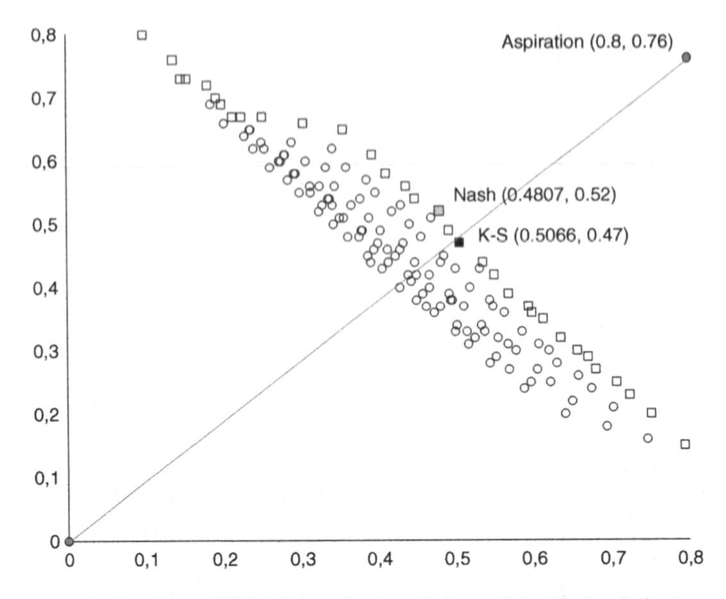

Fig. 12 The approximations of Kalai-Smorodinsky solution and the Nash solution

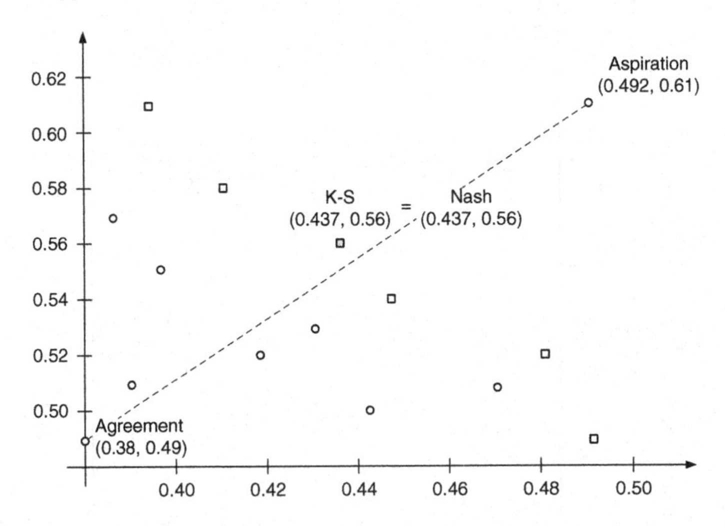

Fig. 13 The illustration of packages dominating the interim agreement P_{26}

(grey rectangle). The reference (agreement obtained before the post-optimization phase) was represented by (0.3793, 0.49) in the space of profiles that corresponds to the package P_{26}. After the reduction of the set of profiles the aspiration level was computed to be: (0.492, 0.61). As the improvement over the package P_{26} we recommend to use the approximation of Kalai-Smorodinsky solution which in our case is the package P_{141}.

6 Conclusions

The traditional approach to structuring the negotiation problem, analyzing negotiators' preferences and conducting the intention phase of negotiation assumes the precise resolution levels of negotiation issues. However, the negotiators' abilities to understand the precise values describing the resolution levels of some issues for some negotiators may be limited. Moreover, package evaluations may originate from various sources and may have various nature (crisp value, fuzzy, linguistic or mixed). Therefore in our paper we proposed a complete Negotiation Decision Making model for the support of ill-structured negotiation problems, that is based on the AHP and FTOPSIS techniques. The FTOPSIS allows for analyzing the negotiation problem using the representation of these different kinds of data in the form of triangular fuzzy numbers that can be further aggregated to derive the scores of any feasible package. The mathematical description of offers involves the translation of vague expressions into fuzzy numbers that can be easily used when performing arithmetic operations on such evaluations. Particularly, the description of option values by natural language seems relevant to ill-structured negotiations. The representation of values by verbal terms requires to form numerically expressed equivalents of imprecise and vague judgments.

On the other hand, a full scoring system able to assign a synthetic score to any feasible package is an extremely useful tool supporting the actual negotiation phase. That means that the negotiator can evaluate any package at any time of the intention negotiation phase in terms of its overall performance. Moreover, the proposed scoring system formation is resistant to the introduction of new packages and does not lead to the ranking reversal, which means that the scoring system is stable. The FTOPSIS method involving the computation of distance from the ideal and anti-ideal solutions is straightforward and computationally very efficient even when fuzzy numbers are used in the computations. Moreover, the AHP method is employed to elicit the importances of the issues instead of the usual direct assignment. Conducting the actual negotiation phase with an iterative exchange of offers and counter-offers allows to attain an agreement that can be further improved in the post-agreement phase using the concepts of Nash and Kalai-Smorodinsky bargaining solutions.

The proposed procedure will be validated empirically, namely different users will test the procedure by solving sample negotiation problems. The users will evaluate the approach in terms of different criteria such as: ease of use or transparency of the approach.

Acknowledgments This research was supported by a grant from the Polish National Science Center (DEC-2011/03/B/HS4/03857).

References

1. Chen, C.T.: Extension of the TOPSIS for group decision-making under fuzzy environment. Fuzzy Sets Syst. **114**, 1–9 (2000)
2. Chen, Y., Kilgour, D.M., Hipel, K.W.: An extreme-distance approach to multiple criteria ranking. Math. Comput. Model. **53**, 646–658 (2011)
3. Edwards, W.: Social utilities. Eng. Econ. Summer Symp. Series **6**, 119–129 (1971)
4. Fisher, R., Ury, W.: Getting to Yes. Penguin Books, New York (1996)
5. Faratin P.: Automated Service Negotiation between Autonomous Computational Agents Ph.D. Dissertation. University of London. Queen Mary College, Department of Electronic Engineering (2000)
6. García-Cascales, S.M., Lamata, M.T.: On rank reversal and TOPSIS method. Math. Comput. Model. **56**, 123–132 (2012)
7. Gimpel, H.: Preferences in Negotiations. The Attachement Effect. Springer, Berlin (2007)
8. Gupta, S., Livne, Z.A.: Resolving a conflict situation with a reference outcome: an axiomatic model. Manage. Sci. **34**(11), 1303–1314 (1988)
9. Hammond, J.S., Keeney, R.L., Raiffa, H.: Even Swaps: A Rational Method for Making Trade-offs. Harv. Bus. Rev. **76**, 137–149 (1998)
10. Hwang, C.L., Yoon, K.: Multiple Attribute Decision Making: Methods and Applications. Springer, Berlin (1981)
11. Jertila, A., Schoop, M.: Electronic contracts in negotiation support systems: Challenges, design and implementation. In: Proceedings of the 7th International IEEE Conference on E-Commerce Technology (CEC 2005), Lost Alamitos. IEEE Computer Society, pp. 396–399 (2005)
12. Kalai, E., Smorodinsky, M.: Other solutions to Nash's bargaining problem. Econometrica **43**(3), 513–518 (1975)
13. Keeney, R.L., Raiffa, H.: Decisions with Multiple Objectives. Wiley, New York (1976)
14. Kersten, G.E., Noronha, S.J.: WWW-based negotiation support: design. Implement. Use Decis. Support Syst. **25**(2), 135–154 (1999)
15. Kwang, H., Lee.: First Course on Fuzzy Theory and Applications, Springer, Berlin (2005)
16. Lewicki, R.J., Saunders, D.M., Minton, J.W.: Negotiation. Readings Exercises, and Cases. Irwin McGraw-Hill, New York (1999)
17. Nash, J.: The bargaining problem. Econometrica **18**(2), 155–162 (1950)
18. Raiffa, H.: The Art and Science of Negotiation. The Belknap Press of Harvard University Press, Cambridge (1982)
19. Raiffa, H., Richardson, J., Metcalfe, D.: Negotiation Analysis. The Science and Art of Collaborative Decision Making. The Balknap Press of Harvard University Press, Cambridge (2002)
20. Roszkowska, E., Wachowicz, T.: Negotiation Support with Fuzzy TOPSIS. In: Teixeira de Almeida, A., Costa Morais, D., de Franca Dantas Daher, S. (eds.) Group Decision and Negotiations 2012. Proceedings. Editoria Universitaria, Federal University of Pernambuco, Recife pp. 161–174 (2012)
21. Salo, A., Hamalainen, R.P.: Multicriteria decision analysis in group decision processes. In: Kilgour, D.M., Eden, C. (eds.) Handbook of Group Decision and Negotiation, pp. 269–283. Springer, New York (2010)
22. Saaty, T.L.: The Analytic Hierarchy Process. N.Y., McGraw Hill, New York (1980)
23. Schmid, B.F.: Elektronische maerkte—merkmale, organization und potentiale. In: Hermanns, A., Sauter, M. (eds.) Management Handbuch Electronic Commerce, pp. 31–48. Verlag Franz Vahlen, Munchen (1999)
24. Schoop, M., Jertila, A., List, T.: Negoisst: a negotiation support system for electronic business-to business negotiations in ecommerce. Data Knowl. Eng. **47**, 371–401 (2003)
25. Stroebel, M., Weinhardt, C.: The montreal taxonomy for electronic negotiations. Group Decis. Negotiat. **12**(2), 143–164 (2003)
26. Thiessen, E.M., Soberg, A.: Smartsettle described with the montreal taxonomy. Group Decision and Negotiation **12**, 165–170 (2003)

27. Thompson, L.: The Mind and Heart of the Negotiator. Prentice Hall, Upper Saddle River(1998)
28. Thompson W.: Cooperative models of bargaining. In: Aumann R.J., Hart S., (ed.) Handbook of Game Theory with Economic Applications, pp. 1237–1283. Tom II Elsevier, Amsterdam (1994)
29. Wachowicz, T.: NegoCalc: Spreadsheet Based Negotiation Support Tool with Even-Swap Analysis. In: Climaco, J., Kersten, G.E., Costa, J.P. (eds.) Group Decision and Negotiation 2008: Proceedings—Full Papers, pp. 323–329. INESC, Coimbra (2008)
30. Wachowicz, T.: Decision support in software supported negotiations. J. Bus. Econ. Manage. **11**(4), 576–597 (2010)
31. Zadeh, LA.: The concept of a linguistic variable and its application to approximate reasoning: Part 1. Inform. Sci. **8**, 199–249 (1975)

Personalised Property Investment Risk Analysis Model in the Real Estate Industry

Nur Atiqah Rochin Demong, Jie Lu and Farookh Khadeer Hussain

Abstract Property investment in the real estate industry entails high cost and high risk, but provides high yield for return on investment. Risk factors in the real estate industry are mostly uncertain and change dynamically with the surrounding developments. There are many existing risk analysis tools or techniques that help investors to find better solutions. Most techniques available refer to expert's opinions in ranking and weighting the risk factors. As a result, they create misinterpretation and varying judgments from the experts. In addition, investment purposes differ between investors for both commercial and residential properties. There is therefore a need for personalisation elements to enable investors to interact with the analysis. This chapter presents a personalised risk analysis model that enables investors to analyse the risk of their property investments and make correct decisions. The model has three main components: investor, decision support technologies, and the data. Real world data from the Australian real estate industry is used to validate the proposed model.

Keywords Risk analysis · Personalisation · Decision under uncertainty · Real estate industry · Property investment · Heuristic

---

N. A. R. Demong · J. Lu (✉) · F. K. Hussain
Decision Systems and e-Service Intelligence (DeSI) Lab, Quantum Computation and Intelligent Systems (QCIS) Centre, School of Software, Faculty of Engineering and Information Technology, University of Technology Sydney (UTS), PO Box 123, Broadway, NSW, Australia
e-mail: jie.lu@uts.edu.au

N. A. R. Demong
e-mail: ndemong@it.uts.edu.au; rochin@puncakalam.uitm.edu.my

F. K. Hussain
e-mail: farookh.hussain@uts.edu.au

N. A. R. Demong
Center of Applied Management Studies, Faculty of Business Management, Universiti Teknologi MARA, Puncak Alam Campus, 42300 Puncak Alam, Selangor, Malaysia

P. Guo and W. Pedrycz (eds.), *Human-Centric Decision-Making Models for Social Sciences*, Studies in Computational Intelligence 502,
DOI: 10.1007/978-3-642-39307-5_15,

1 Introduction

Risk analysis is a popular and useful method and tool that enables investors to make decisions on their investments. Property investment involves various risk factors including policy, environment, management, technical issues, schedule, contractual issues, location and finance. The current methods used to solve these issues are Delphi, brainstorming, fault tree analysis and strengths, weaknesses, opportunities and threats analysis [1]. Risk analysis consists of three stages: risk identification, risk estimation and risk assessment. Risk identification is commonly used to minimise the risk of the real estate losses [2]. Minimising the risk for property investment in the real estate industry is important as it involves high costs that lead to high risk, if not assessed properly.

Property investment risk analysis in the real estate industry involves decision making under uncertainty. In the real world, there are many risks and opportunities that can be measured as qualitative or quantitative factors that are subjective to different investors. Some examples of risk factors in the real estate industry include government policies, political risk, social risk, regulatory risk and contract risk. The uncertainty of the risk factor of property investment in the real estate industry will affect the risk analysis results. Existing risk analysis methods do not take into account the personalisation criteria for decision making and most refers to expert's judgment to rank and weight the risk factors. This is due to investors not having enough information about which property to invest in, given a set of constraints and goals.

Moreover, experts in the field know more about the surrounding environment of the property for investment. Main issue related to the application of experts or professional judgment is that their judgments were not aligned and create misinterpretations. The expert's interest must be aligned with the investors to gain trust and achieve the investor's goals and requirements. Therefore, personalisation criterion is needed for property investment risk analysis to achieve an investor's goals and requirements by using decision support technology. The decision support tool or technology helps to provide knowledge and process the input from both investors and data stored in the system.

The application of decision support technology to provide explicit knowledge to investors for property investment risk analysis will be more accurate and trustworthy when compared to an approach that refers to expert's judgment or opinion. However, the knowledge transferred from the system to the investors is fully dependent on data availability and completeness. Again, it depends on the investors' understanding and their level of knowledge to achieve a better result for their investment. Knowledge management such as learner's knowledge, learning material knowledge and learning process knowledge is used to enable personalisation [3].

Therefore, it is important to have a personalisation model that will provide guidelines for the investor to achieve their goals, mitigate risk and, at the same time, gain benefits from the investment made. The decision made for property investment risk analysis must achieve investor's goals and match with their limitations/constraints.

Property investment in the real estate industry normally focuses on individual or single user requirements either for residential or commercial types of investments.

Personalisation of risk factor ranking and measurement will be based on individual requirements. Personalisation, as a significant capability to maximize the effectiveness of decision support systems, provides useful information to support individuals' decision making processes. Individual investors, as the decision makers, interact with the system through personalisation. The personalisation technique proposed is important to ensure the comprehensive feasibility studies of the risk factors are parallel with investor's limitations/constraints. The significance of the proposed model is the user's ability to interact and be involved in setting their limitations and requirements. Investors rank and weight the risk factor based on their requirements to achieve an optimal decision. This model can be applied to other applications that involve an unstructured decision making process.

2 Risk Analysis in the Real Estate Industry

This section discusses the concepts of property investment risk analysis, decision under uncertainty, and personalisation. These concepts relate mainly to property investment risk analysis as applied in the real estate industry.

2.1 Property Investment Risk Analysis

It is universally agreed that real estate property investment creates big profits, compared to other types of investment such as cash and fixed interest investments, bonds and superannuation [4]. However, the risk and cost factors involved are also very high, compared to other types of investment [5–10]. Investment in the real estate industry incurs a slow liquidity and long term investment. Scientific investment decision making processes are needed to carefully analyse and ensure investment decisions are correct and effective. The feasibility studies of investment projects and property have been researched and various methods have been proposed and applied to measure risk analysis with the alternatives given [11, 12].

Comprehensive risk analysis is needed to help investors to make profits and, at the same time, achieve their goals, since the investment decisions to be made by the investors are complex and risky. The investment decisions are complex because of the uncertain risk factors in risk analysis in the real estate industry. Since uncertainties are involved in the factor determination over a period of time, it is very important to understand how sensitive the factors are to the variation of the investor's personalisation. Sensitivity analysis can be used to see the variation of the result if there is variation in the factors personalized by the investors that will affect the real estate property sold in a certain period of time.

Generally, risk analysis in the real estate industry is affected by many factors such as financial and interest rates, schedules, contracts, policies, and location. Property investment in the real estate industry is specifically affected by micro analysis of property features such as: price; size; number of bathrooms, bedrooms, number car spaces and/or garages; internal property features, eg. alarm, polished timber floors. In most cases, the property value will increase over time as development of the surrounding environment of the respective property will contribute to the price of the property.

Investor's goals and limitations have an influence on the measurement of risk factor ranking and weight, and the possible outcomes, or probabilities, will differ from each other. An example is the goal is to invest in either residential or commercial property. If the property investment is intended for commercial, the investor might plan to rent the property for profit. However, if the investment is for residential property, then the ranking and the weight of the risk factors affected would be different. This is due to the ability to meet the mortgage repayment would be affected by the constraint of investor's capability. The investor's character, for example, whether they are a risk taker or not, also will influence the results of property investment risk analysis. To include this factor in the property investment risk analysis, a personalisation criterion is needed.

A great deal of literature and researches have been undertaken related to investment in large real estate projects, and various techniques have been deployed to measure risk analysis. The most popular technique to rank and weight risk factors for risk analysis in the real estate industry is an analytical hierarchical process (AHP), developed by Saaty in 1980. However, no risk analysis for property investment to date includes personalisation criteria for individual investment. According to the analysis of current researchers, there has not been any attempt to include the personalisation element to rank and weight the risk factor for property investment risk analysis. This book chapter aims to redress this lack, and discusses in detail the personalisation model proposed.

2.2 Decision Under Uncertainty

The uncertainty of the risk factor in the real estate industry led to a high cost and a high risk for the risk analysis. Most existing technique for the risk analysis in the real estate industry refers to experts in the field to rank and weight the risk factors especially for the novice investor. This technique creates misinterpretation and different judgments from different experts in the field which led to inaccurate risk measurement.

Dynamic risk analysis in the real estate industry has always dealt with the uncertainty factors [13]. The uncertainty factors create high risk as it involves the high cost of investment in the real estate industry [14–16]. The uncertainty of the risk factor for investment risk analysis in the real estate industry includes financial risk, economic risk, location risk, scheduled risk, technical risk, policy risk, contractual risk and others. For example, the financial risk refers to the uncertainty of profits which originates from the process of financing, money allocation and transfer, interest payment as financial aspects of a project [1]. Another example, the economic risk includes economic risk, regional development risk, market supply and demand risk and inflation risk which is uncertain [17].

The investors need to have in depth knowledge in the field to decide the best investment and to reduce the risk of loss [18, 19]. Hence, knowledge management of dynamic risk analysis is important to help investors make better decision and reliable. Risk measurement to rank and weigh the risk factors is a major step to consider. Existing risk analysis techniques are still lack of sufficient and comprehensive evaluation for investors to make a good decision [20]. Tools to rank and weight the uncertain risk factors for dynamic risk analysis, such as an analytical hierarchy process (AHP) and Delphi methods have existed for some time. These techniques referred to expert's opinion in ranking and weighting the risk factors for risk analysis [9]. Thus, it creates misinterpretation and different judgments from the experts or professionals in the fields.

Moreover, these tools have significant shortcomings for settings personalized by constraining the investor's goals and objectives that change dynamically. In addition, different user will have different requirements and goals of the investment. Some requirements are simple and seem straight forward, while for others, are complex and require more analysis to making the decisions. Therefore, there is a need for a new method that is more reliable and trusted as compared to referring to the experts in the field. Expert opinion was unable to incorporate more empirical evidence that contribute to project failure [21].

Besides, other issues related to the application of expert opinion include lack of reliable reference [22], expert opinion may change [21] and it depends on the expert's level of experience in the field [23]. As a result, a more systematic approach to accumulating and reporting evidence that can provide in-depth knowledge and can solve the problems from different user's requirements.

2.3 Personalisation

The element of personalisation has been researched for many years and many personalisation algorithms have been investigated [24]. In this chapter, the term personalisation refers to "the mapping and satisfying of a user's/business's goal in a specific context with a service's/business's goal in its respective context" [25]. Personalisation is motivated by the recognition that a user has needs, and meeting them successfully is likely to lead to a satisfying relationship with what they require. Personalisation involves a process of gathering user-information during interaction with the user, which is then used to provide appropriate assistance or services, tailor-made to the user's needs [25–28].

The technology of *personalized service* has been applied to many different fields. Modern personalized service can provide pertinent service for different users so that their specific demands can be met. In an Internet field the technology of personalized service can improve the quality of a web service and the efficiency of users' access [29, 30]. Personalisation criteria help to solve several issues because the vast majority of queries to search engines are short and ambiguous, and different users may have completely different information needs and goals when using precisely the same query [24].

Personalisation enables users to fully utilize their constraints, such as budget or capital, and time constraints, for investment in the real estate industry that are characterized as long term planning. There is a gap in the literature of a need for the justification of risk factor weight and ranking that is based on historical data driven to decision support using knowledge discovery and investors' personalisation for real estate investment risk analysis. The application of data mining technology to find patterns of data helps to provide accurate and valid information for users to understand, analyse, and use as knowledge to enable them to make better decision. Personalisation helps to achieve a user's goals and requirements by making recommendations automatically, based on data available for analysis. Personalisation allows data to be delivered and matched with the user's requirements and interest to fully utilize the user's constraints.

Personalisation is a great advantage that enables users to fulfil their requirements with given constraints. For example, in property investment risk analysis, investors can set the rank and weight of the risk factors of the property features that align with their limitations and goals. It is, therefore, a more effective way of meeting the objective of property investment and achieves better results. In addition, no other existing approaches in real estate property investment analysis take into account the investors personalisation and applied the multi-dimensional analysis of the factors that will affect the decision making process. There are many substantial studies related to the application of AHP and Analytic Network Process for risk analysis in different fields, including analysis of investment in the real estate industry. However, limited study has been undertaken for property investment that does not include the personalisation criteria for risk analysis in the real estate industry, based on knowledge discovery from data mining processes. This book chapter proposes a new personalisation model for risk analysis in the real estate industry that meets the investor's requirements to achieve their goals and objectives.

3 Personalized Property Investment Risk Analysis Model in the Real Estate Industry

This section presents the personalized property investment risk analysis model. The components, main activities and personalisation session involved are also discussed in detail.

3.1 Personalized Property Investment Risk Analysis Model

A personalisation model for property investment risk analysis in the real estate industry is needed to fully utilize the investor's limitation and at the same time fulfilling their goals and requirements. Different investors will have different goals and may have completely different information needs. There has been an impressive development of methods for risk analysis in the real estate industry that focus more on real estate projects, rather than individual investment analysis specifically for property investment. The personalisation model proposed integrates the knowledge discovery approach with the investor's personalisation to enable effective decision making to deal with the property investment risk analysis and meet investor's requirements. A personalisation model for property investment risk analysis in the real estate industry is depicted in Fig. 1.

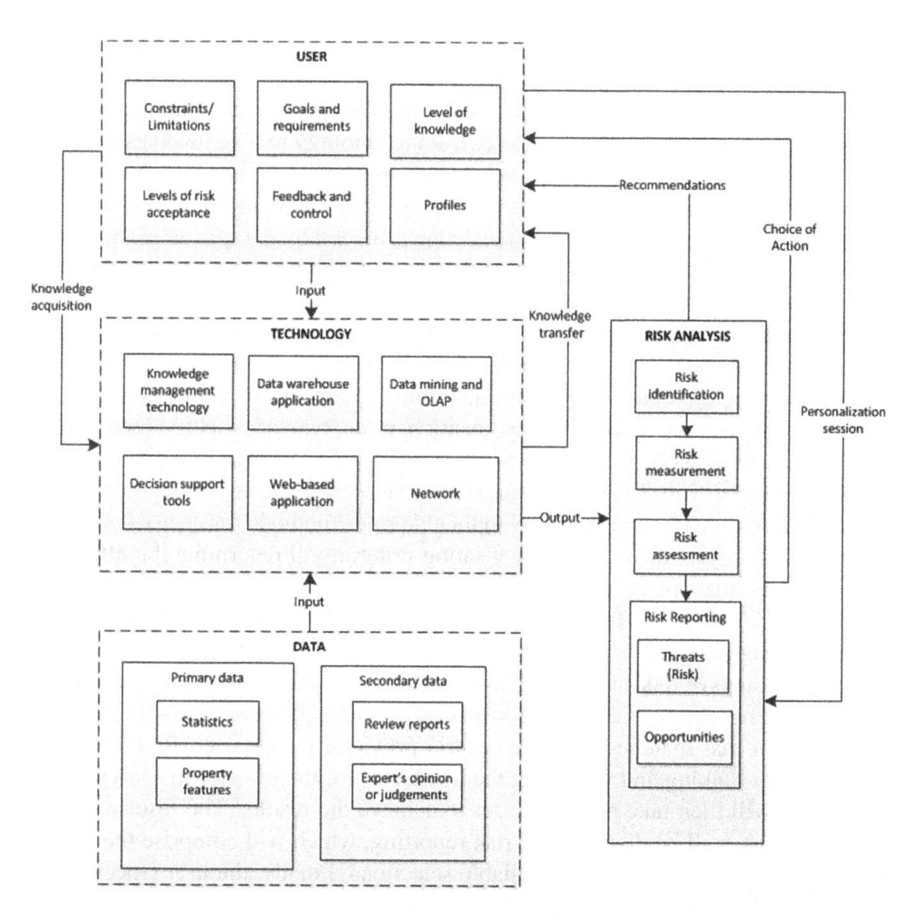

Fig. 1 The personalized property investment risk analysis model for the real estate industry

As shown in Fig. 1, there are three components included in the proposed personalisation model:

(1) User;
(2) Technology; and
(3) Data.

Each of these components has its own functions and is interconnected with each of the other components to support decision making for property investment risk analysis. Based on these three components, the user and data will generate input for the technology to process them and produce the output for the risk analysis.

The user will provide the input for the system that applies the decision support technology to process and match with the available data in the system. By providing input to the system, the user actually acquires knowledge that will help them to make decisions for the property investment risk analysis. The decision support technology, or tools, processes the input and data available in the system. The data stored in the system will be processed by this technology as output in the form of information for the user to exploit as explicit knowledge. Based on the explicit knowledge, the user, as investors, will have experience based on a heuristic approach to exploit the tacit knowledge. The knowledge transfer from technology to user involves different types of decision support tools, a web-based application, a network, knowledge management technology, data warehouse application and data mining, and online analytical processing (OLAP). The knowledge provided by the system to the user is based on the data, or input, available stored as either primary or secondary data.

The personalisation model will provide the pattern of data to determine the factors that contribute to analysis of buying or selling of real estate property, and raise the questions of what, why, and when? For example, the data driven approach will explain what factors contribute to the short time frame for the property sold. Is it because of the features of the property, location, price, type of property, type of sale, sale result, size of property for a certain period of time, or what real estate agency handles the transactions? This model presents a data driven system and a process from data to patterns, and from patterns to applicable rules/methods for decision support. The what-if analysis through personalisation criteria will determine the effects of any pattern changes made to the risk factor measurement that match the investor's requirements. The output from the technological components will be transferred to the user through the risk analysis process.

The three steps of risk analysis consist of risk identification, risk measurement and risk assessment. The risk identification will be classified based on the data available in the system that matches with the criteria provided by the user. Risk measurement involves ranking and weighting the risk factor for the investor to analyse. Risk assessment will then take place in order to achieve the results. The results, as recommendations, will be displayed as risk reporting, which will comprise the threats (risks) and opportunities for the available selections. Finally, the user (specifically, the investor) will run the personalisation session with the rank and the weight of the risk factor until it meets their requirements and they are satisfied with the results.

3.2 Components of the Personalized Property Investment Risk Analysis Model

The proposed personalized model for property investment risk analysis in the real estate industry has three main components: user, data and technology. The user is an independent component and consists of six main sub-components: constraints or limitations, goals and requirements, level of knowledge for the decision made, feedback and control, levels of risk acceptance, and their profiles. The constraints or limitations refer to capital or budget available for the investment and the time period for return on investment. A personalisation criterion helps to fully utilize their constraints and achieve better results. The user goals and requirements are all different and personalisation is needed to meet their goals.

The level of the investor's knowledge will also affect the result of the decision made, so what-if analysis through personalisation will help to provide them with valuable knowledge to make the right choices based on data available in the system. The explicit knowledge transfer from the system to the user as tacit knowledge will help the investor to provide feedback and maintain control of the what-if analysis to enable them to understand and assess the decision making process. Moreover, the level of risk acceptance will also impact the input given for the personalisation process as different investors will have different levels of risk acceptance. Different investors will have different profiles because they have different priorities and different risk tolerances that will reflect the risk analysis. Different investors will also define their goals and explicit knowledge transfer from the system in different ways. Each of these components is related and will impact the level of risk measurement. For example, an investor profile is a reflection of an investor's goals and objectives.

The technology component is dependent on the user and data for processing and consists of six main sub-components: knowledge management technology, data mining and OLAP technologies, data warehouse application, decision support tools, web-based application, and network. All of these technologies are integrated to process the input from users and match their inquiries with available data stored in the system. The technology refers to the tools used to process the input from the user, and stored primary or secondary data for the analysis.

The knowledge management technology applied in the personalisation model is a vital sub-component because it will support new strategies, processes, methods and techniques to better disseminate and apply the best knowledge at anytime and in anyplace. The web application to support decision making will be used to disseminate the knowledge. Data warehouse application, specifically the data mining technology, is applied to mine the data and provide hidden knowledge for investors to analyse. For example, data mining operations include link analysis (association), predictive modelling (regression), database segmentation (clustering) and deviation detection (visualization and statistics) [31].

All these techniques will provide an output or explicit knowledge for investors to personalize the risk factor analysis. OLAP, which applies the multi-dimensional data model, will enable investors to analyse data with more than two dimensions. The network is important to ensure that the data travel is input for the technology to be processed and transferred to the user. The personalisation session integrates all six main components to provide the best result by using the investor's limitation to meet their requirements. The technology collects, gathers and prepares data for analysis to build predictive models and make recommendations for investors to analyse.

The data can be categorized into two types, namely primary data and secondary data. Primary data includes statistics and property features, while secondary data includes review reports and expert's opinions. It is important to ensure that the data is valid and of a high dimension for accurate results and analysis. Features of data that provides accurate analysis should be up-to-date, standardized, integrated and include historical information for better analysis. As with the user, the data component is independent and gathered from different types of sources. Accuracy of prediction is dependent on the data available and stored in the system. The data itself must be correct, valid and integrated.

3.3 Main Activities of the Personalized Property Investment Risk Analysis Model

Five main activities involved with the personalisation session are: input, process, output, feedback and control. Two main sources of input for the personalisation model include input from the user and input from data collected and stored in the system. The input gathered from the user as knowledge acquisition for the system is comprised of the six sub-components of the user, discussed earlier. Input from data storage will be used to match the user's criteria. The data collected and stored in the system uses a web-based application. The input gathered from users and data stored in the system provides the technology to process and produce the output.

The technology is then applied to process the input and match with the data available in the system to produce the output to be given to the user as risk reporting. Processing of data and input from users is dependent on the technology applied to solve the problem. For example, data mining technology uses the data to build a data mining model and produce a hidden pattern of data as an output for users, while the OLAP technology will produce aggregate information. The output of the system as explicit knowledge will be displayed in the form of recommendations and choice of actions to the user as tacit knowledge, as shown in Fig. 2.

The user will then provide feedback through a personalisation session for the choice of action presented by the system as the output of risk reporting. The what-if analysis will be applied to personalize the rank and weight of the risk factors that meet with the investor's requirements. The user will have control of the degree level of the risk factor in order to achieve user goals, with available budget as a limitation.

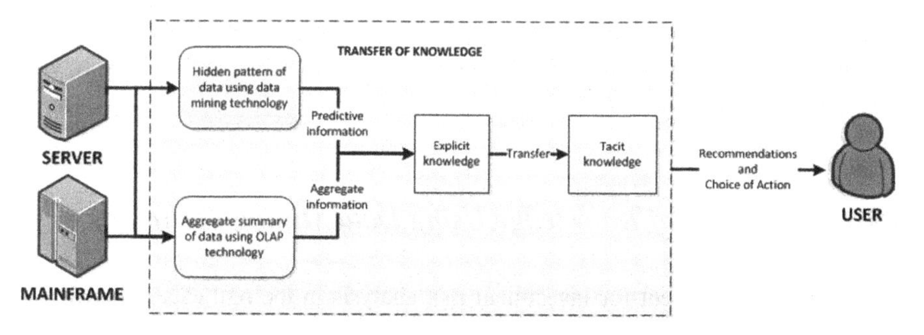

Fig. 2 The transfer of explicit knowledge to the user as tacit knowledge

3.4 Personalisation Session for Risk Analysis

Personalisation helps to meet the user's goals and fully utilize their constraints. Their profiles and preferences will be matched with data stored in the system. The personalisation session deals with the justification of risk factor weight and ranking, which is based on historical data driven by decision support using the knowledge discovery approach. The investor's personalised recommendations and choice of action are provided, fulfilling their requirements. The personalisation session starts with the user identifying their goals and limitations for processing by the technology, as shown in Fig. 3.

The system will then process the input by matching the user's criteria with available data stored in the system. The system produces the output as recommendations and choice of action to be analysed by the user, who will identify the risk factor for the risk analysis. The user will provide a degree level of the risk factor by rank and weight of each risk factor, which is based on the percentage; different users will definitely provide different measurements. Based on the input given by the user, the system will process the data by creating a data mining model and producing the risk report, which consists of the threats or risk, and opportunity. The predictive modelling of the data mining category—for example, the decision table technique—is

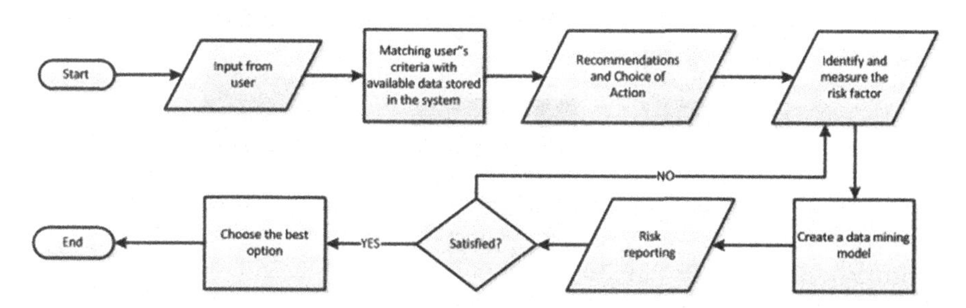

Fig. 3 Data flow of the personalisation session for property investment risk analysis in the real estate industry

applied to discover the pattern of data to support the decision making process. After that, the user will analyse the results displayed and, if satisfied with the analysis, will finally make a decision.

3.5 Development of Knowledge Using Data Mining Technology

Knowledge management for investment risk analysis in the real estate industry is an important field that needs to be focused on as it involves with high cost and high risk. An investor needs to know the vital process involved as it is dealing with different kinds of information either structured, semi-structured and unstructured. The information gathered from the risk analysis must be reliable and can be trusted. The application of the deterministic approach for disseminating the knowledge by discovering the hidden pattern of data using the data mining technique is more reliable as it refers to valid data stored in the data warehouse. This section will explain in detail the development of knowledge management for investment risk analysis.

It is important to think carefully about how to gain knowledge for investment risk analysis. The decision support tools and technologies could help to minimize the risk impacts since invested in the real estate industry involve with high capital. There are several questions that need to be asked at the initial step. Figure 4 depicts the several

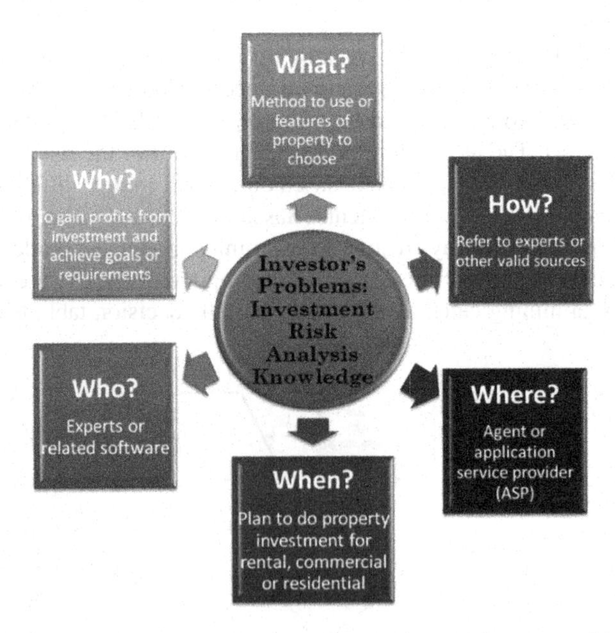

Fig. 4 Several questions that investors will face when dealing with investment risk analysis in the real estate industry

questions that the investors would face when dealing with investment risk analysis in the real estate industry.

As shown in Fig. 4, the 'Wh' questions namely what, where, when, why, how and who are the common types of issues that the decision maker would deal with when trying to figure out an idea of how to start with the analysis. Example of questions that would be asked when dealing with the investment risk analysis as follows:

- What method to use?
- What features of property to choose for the investment such as how many bedrooms, bathroom and car park?
- How to run the investment risk analysis?
- Why you need to have knowledge in the field?
- Who to refer?
- When you need to run the investment risk analysis?
- Where to get the knowledge about the investment risk analysis?

In order to answer these questions, the investor must have in-depth knowledge to make a decision as an investment in the real estate industry incurred with high cost and high risk. Based on the several question highlights above, it is very important to have the correct method, tools and technologies to handle this situation. A reliable and accurate decision support tool and technology is needed in determining the rank and weight of risk factors for investment in the real estate industry. The application of the deterministic approach as proposed in this paper helps the investor to gain knowledge and answer the questions highlighted.

3.6 Investment Risk Analysis Knowledge Management Development

The development of knowledge management for investment risk analysis proposed in this chapter is focused on knowledge embedded in individual specifically the investor. The transfer of knowledge here refers to the explicit knowledge generated by the system transferred to the investor as tacit knowledge. The tacit knowledge transferred as an experience for the investor to understand and evaluate the risk factors in the decision making process. The development of knowledge management by using heuristic through deterministic approach for investment risk analysis in the real estate industry consists of three different parts as illustrated in Fig. 5.

The first element is the investor as the decision maker that will have problems and need an in-depth knowledge to solve their problems. Second, the deterministic approach that processes the data based on investor's requirements and provides knowledge by showing hidden patterns of data using data mining techniques. Third, the application of heuristic approach in which the investor's will gain knowledge through deterministic approach to solve their problems that meets their requirements.

The process begins with the investor will set their query and requirements to the decision support technology. The user specifically the investor needs to define their

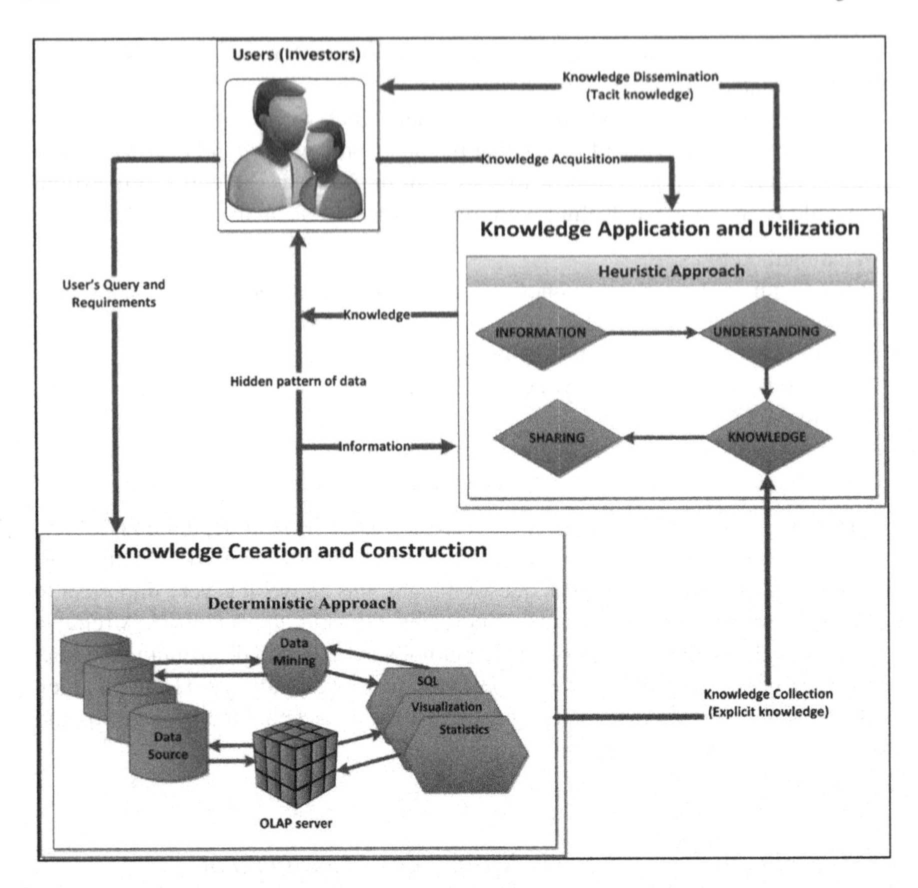

Fig. 5 Development of knowledge management by using heuristic through deterministic approach for investment risk analysis in the real estate industry

goals, limitations and requirements as a query to the system that would be processed using deterministic approach. Next, the result will be generated by the system that matches with investor's requirement. The deterministic approach will process the input gather from the investor to prepare the knowledge related to investor's requirements and limitations. By using deterministic approach, the transfer from explicit knowledge to tacit knowledge refers to the experience that the investor will get to analyze the risk analysis for investment. Data warehouse end user application such as data mining, online analytical processing (OLAP), structured query language (SQL), visualization and statistics will create and construct the knowledge as explicit knowledge. For instance, the data mining techniques such as association and prediction techniques will try to find the hidden pattern of data and make forecasts of house price as knowledge collection. Based on these results, the investor will receive, understand and analyze this information heuristically as tacit knowledge. The heuristic approach refers to the knowledge acquisition, application and utilization in making decision. The investor will rank and weight the risk factor analysis based on hidden knowledge generated by the system and map it with their requirements for the investment risk analysis.

The evolution of knowledge management applications for stage 5 (future age) important activities is to support business intelligence [32]. The paper move towards this milestone in which the development of knowledge management proposed comprehend with the personalization technique. The application of the deterministic approach solved the problem of fulfilling the investor demand better than referring to the experts in the field. The main contributions of this chapter are (1) it proposes a new technique to produce explicit knowledge through deterministic approach through personalization model; (2) it proposes the acquisition of tacit knowledge through heuristic approach; (3) it proposes a novel knowledge management development for investment risk analysis in the real estate industry by using heuristic through deterministic approach; (4) risk measurement of ranking and weighting the risk factors personalized by investors that meets with their requirements. The application of decision support technology for deterministic approach helps to speed up the transfer of knowledge hence faster decision can be made. Moreover, the personalization technique applied helps the investor to achieve their goals, within their limitations and fulfill their requirements.

4 Experiments

This section describes the application of the decision table technique to demonstrate how the personalisation criterion affects both of the results, and the data mining techniques, for knowledge discovery. Two data mining techniques, namely the clustering technique and the forecasting technique, were chosen to discover the hidden pattern of data.

4.1 Decision Table Technique

An example of a decision table, as depicted in Table 1, is created to examine the combination of inputs, which produce different results by using this technique. The user's limitation will be the conditions for the decision table and each condition has a different number of values. For example, the cost of investment can be defined as three values, known as low (L), medium (M) or high (H); the level of risk acceptance also can be defined as three values, known as low (L), medium (M) or high (H) and the objective of the investment can be defined as two values, known as residential (R) or commercial (C). The combination of these conditions will generate a number of rules with respective actions. For each combination, the action can be single or multiple. The investor profiles or constraints will be linked with their goals, objectives and strategy for investment. For example, as shown in Table 1, the conditions, or

Table 1 A decision table showing a scenario of recommendations for a property investment based on a user's constraints and requirements

Conditions (User's constraints/limitation/ profiles)	Rules																		
Cost of investment	L	M	H	L	M	H	L	M	H	L	M	H	L	M	H	L	M	H	...
Level of risk acceptance	L	L	L	M	M	M	H	H	H	L	L	L	M	M	M	H	H	H	...
Objective of investment	R	R	R	R	R	R	R	R	R	C	C	C	C	C	C	C	C	C	...
⋮	⋮	⋮	⋮	⋮	⋮	⋮	⋮	⋮	⋮	⋮	⋮	⋮	⋮	⋮	⋮	⋮	⋮	⋮	⋮
Actions																			
Highly recommended					X		X						X						...
Moderately recommended	X			X				X		X	X	X		X		X	X	X	...
Low recommended		X	X			X			X						X				...
⋮	⋮	⋮	⋮	⋮	⋮	⋮	⋮	⋮	⋮	⋮	⋮	⋮	⋮	⋮	⋮	⋮	⋮	⋮	⋮

Legend:
Cost of investment: *L* Low, *M* Medium, *H* High
Level of risk acceptance: *L* Low, *M* Medium, *H* High
Objective of investment: *R* Residential, *C* Commercial

user's limitations, include the cost of the investment, level of risk acceptance and objective of investment. The combination of these constraints will produce different recommendations for the user to choose.

The decision table is one of the best techniques to model complicated logic as personalisation because property investment risk analysis in the real estate industry has many limitations. Each action in the decision table corresponds to associate conditions, as shown in Table 1. The what-if analysis is described through the decision table to identify the risk level of the risk analysis factor that will affect the analysis results. The recommendations and choice of actions given are a guideline for the user to understand and discover the alternatives that meet their goals and requirements. Based on this scenario using the decision table technique, the investors achieve better results if they employ a personalisation session for risk measurement to achieve their goals.

4.2 Data Mining Techniques for Knowledge Discovery

This section explains in detail how the application of data mining techniques helps to discover hidden patterns of data. Two data mining techniques have been chosen for the analysis, namely the clustering technique to group or cluster similar records, and the predictive technique for forecasting. The resulting reports derived from an experimental data which consist of 619 rows of data selected for the analysis with four attributes as shown in Table 2. Data was collected from the Australian Property

Table 2 Sample data used for the analysis collected from the Australian Property Monitor domain database

Property type	Land size (Sqm)	Year	Rental
Commercial	250	2005	750
Commercial	5312	2006	132800
Commercial	80	2006	31200
House	342	2004	350
House	334	2004	430
House	567	2005	650
Industrial	2801	2012	60
Industrial	48	2012	24800
Other residential	929	2012	420
Semi	171	2006	360
Terrace	134	2012	485
Terrace	108	2012	500
Townhouse	149	2012	700
Unit	464	2003	225
Unit	930	2003	230
Unit	976	2003	190
Unit	1882	2003	220
Villa	811	2012	380
⋮	⋮	⋮	⋮

Monitor domain database and analysed using MS SQL Server 2008, integrated with Microsoft SQL Data Mining Add-ins.

The sample of property data located in the Eastlakes suburb of Sydney in New South Wales was selected for the experiments. Four attributes have been chosen for the experiments, namely the type of property, land size, year and rental rates.

As shown in Fig. 6, the most common type of property available for rental at the Eastlakes suburb is dominated by unit, followed by house, commercial, terrace, industrial, villa and semi-detached. This information will give investors an idea of what type of property they should focus on if they intend to invest in commercial or business real estate.

4.2.1 Clustering Technique

A clustering technique organizes data by abstracting underlying structures, either as a grouping of individuals, or as a hierarchy of groups. The representation can then be investigated to see if the data group is in accordance with preconceived ideas, or to suggest new experiments. Cluster analysis groups data objects into clusters so that objects belonging to the same cluster are similar, while those belonging to different ones are dissimilar [33]. Based on sample data collected for the analysis,

Variables	Values	Probability
Property Type	Unit	80 %
Rental	4,524 - 19,775	25 %
Land Size _Sqm_	3,697 - 12,084	25 %
Rental	19,776 - 72,364	25 %
Land Size _Sqm_	12,085 - 41,004	25 %
Year	2012	21 %
Year	2006	20 %
Year	2004	15 %
Land Size _Sqm_	18 - 3,696	12 %
Year	2011	11 %
Property Type	House	10 %
Rental	60 - 4,523	8 %
Year	2009	8 %
Year	2007	8 %
Year	2005	7 %
Property Type	Commercial	6 %
Year	2008	5 %
Year	2010	5 %
Year	2003	2 %
Property Type	Terrace	1 %
Property Type	Industrial	1 %
Property Type	Villa	1 %
Property Type	Semi	1 %

Fig. 6 Summary of data selected based on Table 2

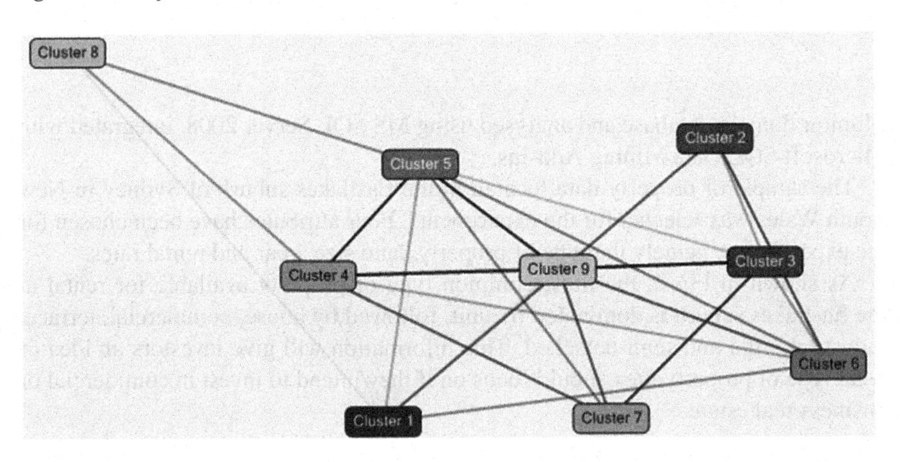

Fig. 7 Clustering technique based on selected Eastlakes property data

nine clusters have been created and the associations between clusters are linked using lines as shown in Fig. 7.

The associations between clusters provide knowledge about the most influential features of property for rental that the investors need to choose when planning a rental property investment.

Variables	States	Population (All)	Cluster 1	Cluster 3	Cluster 2	Cluster 5	Cluster 6	Cluster 4	Cluster 7	Cluster 8	Cluster 9
Size		173	39	28	27	20	18	13	12	9	7
Land Size _Sqm_	Mean	3,697.00	907.3	897.87	745.02	401.64	1,288.20	370.34	6,316.00	1,563.78	63,156.00
Land Size _Sqm_	Deviation	12,435.90	353.43	262.69	299.54	250.15	975.49	216.33	6,615.72	2,176.20	
Property Type	Unit	139	99 %	97 %	100 %	45 %	98 %	24 %	87 %	0 %	100 %
Property Type	House	18	1 %	0 %	0 %	44 %	2 %	61 %	0 %	0 %	0 %
Property Type	Commercial	11	0 %	0 %	0 %	3 %	0 %	6 %	4 %	100 %	0 %
Property Type	Terrace	2	0 %	1 %	0 %	2 %	0 %	8 %	0 %	0 %	0 %
Property Type	Industrial	1	0 %	0 %	0 %	0 %	0 %	0 %	9 %	0 %	0 %
Property Type	Villa	1	0 %	2 %	0 %	0 %	0 %	1 %	0 %	0 %	0 %
Property Type	Semi	1	0 %	0 %	0 %	6 %	0 %	0 %	0 %	0 %	0 %
Property Type	...	...	...	...	...	...	...	...	...	...	...
Rental	Mean	4,523.00	230.32	387.59	216.5	305	246.12	390.85	483.98	80,738.36	580
Rental	Deviation	22,613.70	25.77	50.76	28.38	112.83	64.86	200.48	197.38	63,461.45	50
Year	2012	36	0 %	55 %	0 %	11 %	0 %	24 %	82 %	11 %	57 %
Year	2006	34	48 %	0 %	17 %	24 %	0 %	6 %	0 %	67 %	0 %
Year	2004	26	1 %	0 %	64 %	11 %	15 %	5 %	0 %	0 %	0 %
Year	2011	19	0 %	30 %	0 %	27 %	4 %	18 %	0 %	0 %	29 %
Year	2009	13	5 %	4 %	1 %	6 %	46 %	14 %	0 %	0 %	0 %
Year	2007	13	23 %	0 %	1 %	6 %	4 %	12 %	0 %	11 %	0 %
Year	2005	12	22 %	0 %	1 %	0 %	5 %	11 %	8 %	0 %	0 %
Year	2008	8	1 %	1 %	12 %	2 %	14 %	1 %	0 %	11 %	0 %
Year	...	...	...	...	...	...	...	...	...	...	...

Fig. 8 Cluster profiles based on Fig. 5

Figure 8 illustrates the cluster's profile, in which cluster 1 dominates the population, followed by cluster 3, 2, 5, 6, 4, 7, 8 and 9, respectively. The rental rate mean for cluster 1 is 230 dollars per week, with a standard deviation 25.77. Based on this result, the investor can calculate the mortgage instalment if they plan to invest in property for rental and need to organise a loan for the capital. The investor should utilize this information as knowledge and guidelines in order to buy the best property for investment, if their objective is to invest in property for rent.

4.2.2 Predictive Technique for Forecasting

The predictive technique is used for forecasting based on time series and historical data available to generate further analysis. The forecast data generated by the system will provide more valuable information and knowledge to the investor. Out of 619 rows of data, 239 rows are characterized as 'unit' type of property that has been selected to forecast the rental rate. Figure 9 depicts the five year forecast of rental rate predicted up to 2017 for investors to analyse, based on selected data shown in Table 2.

As shown in Fig. 9, the historical information appears to the left of the vertical line (straight line), which represents the data that the algorithm uses to create the model, while predicted information appears to the right of the vertical line (dotted line) and represents the forecast that the model makes. The forecasting value will help investors make better decisions based on their limitations and goals or objectives.

Based on the experiment and the results shown, it is important to consider the time series for investment and for forecasting, depending on investor's requirements. By using an expert survey method, there is no customization on the time frame that fits with decision makers' requirements. Based on the historical data and valid data available in the database, investor's confidence is increased by using the output generated by the system.

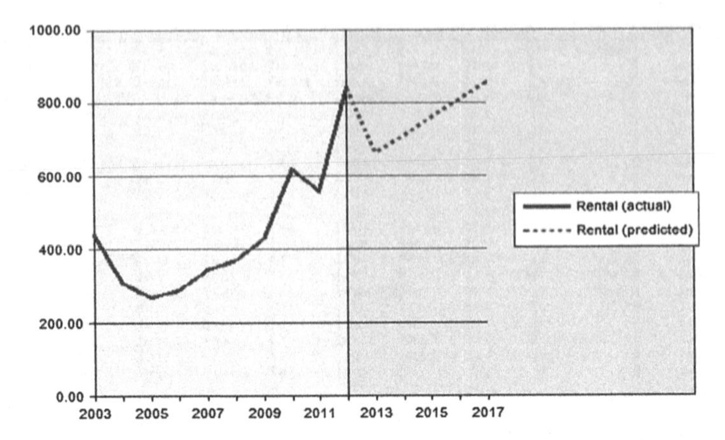

Fig. 9 Rental rate prediction until 2017, based on data selected in Table 2

Origin of the knowledge is coming from the data stored in the system and created using data mining technique to discover and find the hidden pattern of data that would be useful for the investors as the heuristic approach. The application of the heuristic through deterministic approach to disseminate knowledge and its implications provide reliable and accurate information. Investor as the decision maker need to have knowledge in the field to speed up the decision making process. Heuristic commonly utilized in the process of decision making to help users obtaining relevant ideas, experience and gain knowledge in managing problems they dealt with. In addition, the hidden pattern of data discovered will enables the investor to make efficient decisions with the assistance of heuristic approach to personalize the criteria based on their requirements. The technology of data mining could be used for daily practice of analysing property for investment, and those in the field of analysing uncertain factors for decision making process. The proposed model can be easily understood by the investor thus providing a practical assessment tool for decision making about investment risk analysis.

5 Conclusions

This chapter has detailed a personalisation model for property investment risk analysis in the real estate industry. The main objective of proposing this model is to improve the measurement of risk factors that align with investor's goals and limitations. Personalisation is an important element that needs to be considered when dealing with property investment risk analysis in the real estate industry. The main consideration in any real estate investment analysis is that it is a risk analysis with uncertainty factors. The proposed personalisation model provides a guideline for investors to achieve their goals, because the ability to accept different levels of risk varies significantly from one investor to another. The investor must prepare and specify valid, accurate information of their limitations and requirements before proceeding with the property investment risk analysis. Different goals and limitations have different values for risk factor rank and weight for property investment risk analysis in the real estate industry. It is important to have a computer system in place for personalized property investment risk analysis that achieves investor's goals, taking into account their limitations.

As knowledge, the proposed personalisation model for property investment risk analysis introduces a new perspective to investors to measure the risk analysis factor. This model will also be used as a decision support tool in property or real estate investment risk analysis. Further research on personalisation algorithms is needed to evaluate more clearly, systematically and mathematically how effectively the personalisation model may be applied to property investment risk analysis.

References

1. Yu, J., Xuan, H.: Study of a practical method for real estate investment risk decision making. International Conference on Management and Service Science, pp. 1–4 (2010)
2. Chen, L.: Research on the risk identification and evaluation of real estate development. 2nd International Conference on Information Science and Engineering, pp. 3087–3090 (2010)
3. Khribi, M. K., Jemni, M., Nasraoui, O.: Recommendations for e-learning personalisation based on web usage mining techniques and information retrieval. Eight International Conference on Advanced Learning Technologies, pp. 241–245 (2008)
4. Hui, S., Fan, Z.Q., Shi, Y.: Study of impact of real estate development and management risk on economic benefit. 16th International Conference on Industrial Engineering and, Engineering Management, pp. 1299–1303 (2009)
5. Yu, S.W., Wang, J.P., Guo, N.: Application of project portfolio management in the real estate corporations. 16th International Conference on Industrial Engineering and, Engineering Management, pp. 1225–1228 (2009)
6. Ren, H., Yang, X.: Risk measurement of real estate portfolio investment based on CVaR model. International Conference on Management and Service Science, pp. 1–4 (2009)
7. Gao, J., Wang, Z.: Research on real estate supply chain risk identification and precaution using Scenario analysis method. 16th International Conference on Industrial Engineering and, Engineering Management, pp. 1279–1284 (2009)
8. Rocha, K., Salles, L., Garcia, F.A.A., Sardinha, J.A., Teixeira, J.P.: Real estate and real options—a case study. Emerg. Markets Rev. **8**, 67–79 (2007)
9. Sun, Y., Huang, R., Chen, D., Li, H.: Fuzzy set-based risk evaluation model for real estate projects. Tsinghua Sci. Technol. **13**, 158–164 (2008)
10. Li, K.: Empirical study on Chinese real estate investment transmission via interest rate adjustment. International Conference on Business Management and Electronic, Information, pp. 271–274 (2011)
11. Wang, Y.M., Liu, D.D.: The financial risk and precaution of real estate enterprises. International conference of Information Technology, Computer Engineering and Management Sciences, pp. 96–99 (2011)
12. Cai, M., Li, Y.: Decision making under uncertainty and its application. International Conference on E-Business and E-Government, pp. 1795–1798 (2010)
13. Lv, F., Gao, Y.: Study of multiple criteria decision making for real estate investment environment based on AHP. 2nd IEEE International Conference on Emergency Management and Management Sciences (ICEMMS), pp. 89–92 (2011)

14. Liu, L., Zhao, E., Liu, Y.: Research into the risk analysis and decision-making of real estate projects. International Conference on Wireless Communications, Networking and Mobile Computing, WiCom 2007, pp. 4610–4613 (2007)
15. Wu, C., Guo, Y., Wang, D.: Study on capital risk assessment model of Real Estate enterprises based on support vector machines and fuzzy integral. Control and Decision Conference, pp. 2317–2320 (2008)
16. Tang, D.Z., Li, L.H.: Real estate investment decision-making based on analytic network process. International Conference on Business Intelligence and Financial, Engineering, pp. 544–547 (2009)
17. Zhi, Q.M., Qing, B.M.: The research on risk evaluation of real estate development project based on RBF neural network. Second International Conference on Information and Computing Science, vol. 2, pp. 273–276 (2009)
18. Yuan, X.E., Yuan, Z.Z.: Innovation project investment risk evaluation model. International Conference on Future BioMedical Information, Engineering, pp. 409–412 (2009)
19. Wang, X.: Model of investment risk prediction based on neural network and data mining technique for construction project. International Symposium on Computational Intelligence and Design, vol. 1, pp. 373–378 (2008)
20. Zhou, S.J., Li, Y.C., Zhang, Z.D.: Self-adaptive ant algorithm for solving real estate portfolio optimization. International Symposium on Computational Intelligence and Design, vol. 1, pp. 241–244 (2008)
21. Kitchenham, B., Budgen, D., Brereton, P., Turner, M., Charters, S., Linkman, S.: Large-scale software engineering questions—expert opinion or empirical evidence?. Software, IET, (1:5), pp. 161–171 (2007)
22. Zhou, J., Zhu, Y.Q., Tang, W.Q.: Approach of expert opinion acquisition based on cloud model and evidence theory. International Conference on Management and Service Science, pp. 1–4 (2009)
23. Modi, S., Yingzi, L., Long, C., Guosheng, Y., Lizhi, L., Zhang, W.W.J.: A socially inspired framework for human state inference using expert opinion integration. IEEE/ASME Transactions on Mechatronics, (16:5), pp. 874–878 (2011)
24. Dou, Z.C., Song, R.H.S., Wen, J.R., Yuan, X.J.: Evaluating the effectiveness of personalized web search. IEEE Trans. Knowl. Data Eng. **21**, 1178–1190 (2009)
25. Zhang, P., Zeng, Z.: A framework for personalized service website based on TAM. International Conference on Service Systems and Service Management, vol. 2, pp. 1598–1603 (2006)
26. Zheng, H.: Construction of a personalized service oriented learning resource management system framework. First International Workshop on Knowledge Discovery and Data Mining, pp. 422–425 (2008)
27. Soujanya, M., Kumar, S.: Personalized IVR system in contact center. International Conference On Electronics and Information Engineering (ICEIE), vol. 1, pp. 451–457 (2010)
28. Kao, S.C., Tseng, Y.F., Lee, T.Z.: The design of personalized knowledge integration platform using digitized information resources. 40th International Conference on, Computers and Industrial Engineering, pp. 1–6 (2010)
29. Zhen, H., Yongchun, H., Yanquan, Z., Cong, W., Chunxiao, F.: Research on personalized information service on mobile networks based on mining user's interest. IEEE International Conference on Industrial Informatics, pp. 1052–1056 (2006)
30. Xiong, Y.N., Geng, L.X.: Personalized intelligent hotel recommendation system for online reservation-a perspective of product and user characteristics. International Conference on Management and Service Science (MASS), pp. 1–5 (2010)
31. Connolly, T.M., Begg, C.E.: Database systems: a practical approach to design, implementation and management, 4th edn. Addison-Wesley, Reading (2005)
32. Wang, T., Xu, Z., Gao, G.: Development and strategy of knowledge management for e-services. International Conference on Service Systems and, Service Management, October, pp. 30–35 (2006)
33. Andritsos, P., Tsaparas, P.: Categorical data clustering, encyclopedia of machine learning 2010, pp. 154–159. http://www.springerlink.com/content/k11383n8ur119m35 (2010). Accessed 14 June 2012

The Logic and Ontology of Assessment of Conditions in Older People

Patrik Eklund

Abstract In this paper we present some views on ontologies and assessments, and the relation between logic and guidelines within municipal decision-making in elderly care. Logic is seen, on the one hand, as carrier of information, and, on the other hand, as including mechanisms for inference as underlying decision-making. The ontology and logic for the framework is based on a non-classical typing system where uncertainty is canonically developed in a category theory framework involving term monads both composed with other monads, and as viewed over other categories than just the category of sets. The main question is where uncertainty actually resides, so that they are canonically retrieved rather than amalgamated in ad hoc approaches.

Keywords Assessment · Diagnosis · Home care · Smart homes · Database management.

1 Introduction

Elderly care includes personnel with various skills and expertise, e.g., social workers, nurses, gerontologists, therapists, psychologists and physicians, general practitioners, neurologists and geriatricians. It should be noted that the home care staff in its vast majority consists of a selected mix of social workers and nurses and thus social care overweighs the health care. In residential and nursing homes though, social and health care should be in balance, while in hospital wards the provision of health/medical care is the pivotal aim. In a population of 100,000 citizens we typically have 2–3 geriatricians and up to 100 GPs, but GPs spend only fractions of their time for geriatric/gerontological problems. In home care we would, on the other hand, have hundreds of social workers. This clearly calls for management of Social

P. Eklund (✉)
Department of Computing Science, Umeå University, Umeå, Sweden
e-mail: peklund@cs.umu.se

P. Guo and W. Pedrycz (eds.), *Human-Centric Decision-Making Models for Social Sciences*, Studies in Computational Intelligence 502,
DOI: 10.1007/978-3-642-39307-5_16,

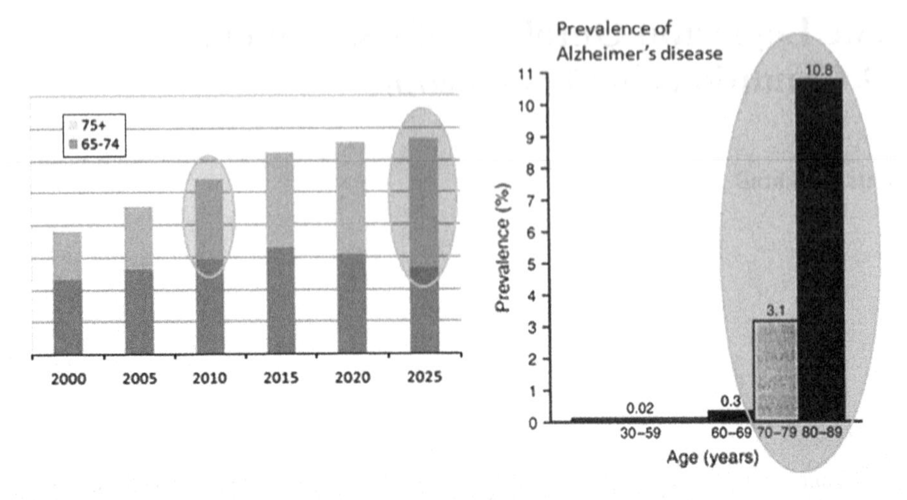

Fig. 1 Growth of 75+ population and its consequences for Alzheimers disease prevalence

Records, but also for the social workers and nurses to be prepared for information gathering into the Social Record and its related repository data. In larger municipalities and cities, the care processes and organization of workflow must be hierarchical and geographical, including area managers and team leaders. Service requirements per customer in home care vary between one visit per week to three visits per day. A social workers visit is up to 1 h, whereas a nursing visit should be down to 30 min. A team of 6–10 social workers and nurses should thus be able to manage up to 50 customers, or even more.

Demographic change means in particular an increase of dementia. In these scenarios focus must be not only on the 65+ population, but also on the 75+ population which in fact accelerates faster than the 65+ population. The prevalence of Alzheimers disease increases rapidly in the 75+ population (Fig. 1).

Faster growth of the 75+ population means an increase in dementia and non-cognitive/behavioral syndromes, which in turn requires not only more services because of growth, but also new types of services because of the changing condition spectra.

The objective of ontology and assessment is to establish municipal and regional best practices for strategic planning and management of ageing. This is achieved by developing accurate socio-economic modelling tools based on rigorous design of information and processes. Older person conditions and related monitoring of information is crucial. Further, information must be appropriately structured so that ontology supports e.g. interoperability. The demographic model will enable the analysis and prediction of demographic change, and the socio-economic model, based on ageing information and process design, is sensitive and specific in particular concerning variables related to demographic change. This approach to socio-economic modelling based strategic planning is both customer-centric with respect to information and process design as well as care-centric with respect to care management. Socio-economic modelling of the social welfare effect due to demographic change is therefore utmost important, on the one hand, for municipality resource planning and objective decision making, and on the other hand, for enabling required accuracy of business models as used by public and private actors in the social sector.

The minimal set of assessment scales is usually some ADL scales combined with suitable cognitive scales, like MMSE [10]. Combination scales, like the CDR for ADL/DEMENTIA, are also widely used. Non-cognitive signs are captured, e.g., by the neuro-psychiatric inventory NPI [4], the Cohen-Mansfield agitation inventory CMAI [3], and BEHAVE-AD [13] for assessment of behavious. NPI is particularly useful in home care. Depression is usually captured in its own right, where e.g the geriatric depression scale GDS [15] is widely used in home care. Depression is known to accelerate cognitive decline. Nutrition scales are important, like also scales for social conditions, and so on and so forth. The selection of assessment scales to be used is of utmost importance and must be optimized with respect to professional resources available.

Figure 2 illustrates the minimal set of assessment scales, or rather, the set of subsets of assessment scales The selection of assessment scales to be used is of utmost importance and must be optimized with respect to professional resources available in the particular service field where the OAD gerontechnological platform is to be installed and used.

Accurate monitoring of assessment scale further supports dementia differential diagnosis [5], and these are based on consensus guidelines as provided e.g. by DSM-IV [1] and NINCDS-ARDRA. Early detection of dementia is important e.g. to achieve desirable effects of pharmacologic treatment.

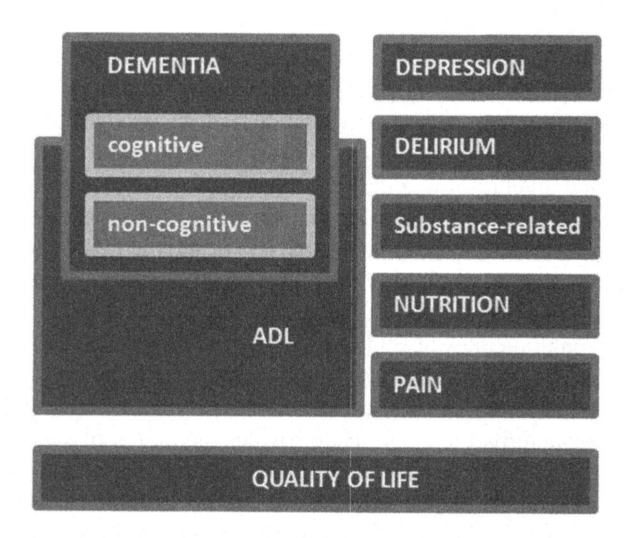

Fig. 2 Subareas of assessments—The OAD (Observe-Assess-Decide) framework

This paper proposes to use a typing framework for assessment scales to support their integration into decision-making processes, including pathways for regional and municipal decision and policy making.

2 Signatures and Type Constructors

In [8] we proposed a three-level arrangement of signatures, where the basic signature Σ is on level one, and Σ' is on level three. On level two we have the superseding type signature as a one-sorted signature $\mathtt{S}_\Sigma = (\{\mathtt{type}\}, \mathtt{Q})$, where $\mathtt{Q}$ is a set of *type constructors* satisfying

(i) $\mathtt{s} : \to \mathtt{type}$ **is in** $\mathtt{Q}$ **for all** $\mathtt{s} \in S$
(ii) **there is a** $\Rrightarrow : \mathtt{type} \times \mathtt{type} \to \mathtt{type}$ **in** $\mathtt{Q}$.

If $\mathtt{Q}$ does not contain any other type constructors, apart from those given by (i) and (ii), we say that $\mathtt{S}_\Sigma$ is a (Σ-)*superseding simple type signature.*

For any Σ-superseding type signature $\mathtt{S}_\Sigma$ we have the type term monad $\mathbf{T}_{\mathtt{S}_\Sigma}$, over a category $\mathtt{Var}$ of variables, so that $\mathsf{T}_{\mathtt{S}_\Sigma} X$, $X \in \mathrm{Ob}(\mathtt{Var})$, contains all type terms.

The signature $\Sigma' = (S', \Omega')$ on level three is based on $S' = \mathsf{T}_{\mathtt{S}_\Sigma} \emptyset$. See [8] for detail on the term constructions.

3 Assessment of Gerontological Conditions

It is typical to distinguish between information and knowledge, but the meaning of *information* and *knowledge* in a particular context is usually not explained. We adopt a quite strict logical view on information and knowledge, and we therefore need a few remarks on our take on logic. A logic has its signature with sorts (types) and operators, and algebras providing the meaning of the signature. Terms are formally constructed [8] using operators in the signature. Algebras must be carefully introduced. Substitutions and assignments have to be handled very carefully. Signatures and terms are then basis for providing representation of *information*. Sentences have terms as ingredients, and conglomerates of sentences can be formally treated. Sentences and conglomerates of sentences is what we would mean by *knowledge*. Entailment is the relation between these conglomerates representing what we already know, and sentences representing knowledge we are trying to arrive at. Satisfaction as the semantic counterpart to entailment provides the notion of valid conclusions. Axioms represent what we assume at start, and inference rules support chains of entailments. Concerning *logic* we must be aware of the distinction between logic as in 'logic as foundation for mathematics' and logic as in 'mathematics as foundation for logic'. Our take on logic is focused on 'category theory as foundations for logic';

and this logic we call *substitution logic*, which is an extension of developments presented in [9].

Gerontological conditions and circumstances is the about information and knowledge and in this paper we focus on the information part. We therefore provide some examples on how gerontological data can be properly typed so as to open up possibilities to invoke appropriate constant values and operators.

Before we start to develop our examples, let us make the reader aware of the apparent non-uniqueness of these representations. The medical term 'dementia' may serve as a good example. Medically speaking, dementia is a syndrome rather than a disease, but it can be encoded as a diagnosis. There are different types of dementia, like Alzheimer's disease and vascular dementia, and Alzheimer's disease in turn can be diagnosed as connected with other diseases. Encoding dementia hierarchically in this can be done e.g. by the World Health Organization ICD-10 standard, or using DSM-IV encoding, or a combination of both. ICD-10 and DSM-IV.

Hierarchies of sets, or sets of sets, sets of sets of sets, and so on, can be modelled by the 'powerset' type constructor $\mathsf{P} :\rightarrow$ **`type`** on level two, i.e., intuitively thinking that the algebra P would be the ordinary powerset functor over the category `Set` of sets and functions. However, more structure can be added to powerset functors, e.g., by allowing 'double powerset' type constructors, or considering powersets with structure. The typical example in [8] considered the distinction between fruit basket and fruit plate, where a plate intuitively is more of of a set or structured set, but a basket is may be more of sets of sets.

The outcome of an assessment scale is usually a number, like in the case of MMSE, which has a range 0–30, with single digit values indicating a severe dementia, and values around 20 are more appear mostly in a mild dementia. Note that an Alzheimer's disease can be diagnosed at early or late stages, so the severity and progrediation of that particular dementia is not included as more specific information in the diagnosis code. The content of the MMSE test builds up from questions related e.g. to orientation and language, but the final MMSE value obviously hides the underlying specific information. A loss of points in orientation is not distinguished from corresponding loss of points in language aspects, which means that these two apparently different neurological aspects are hidden in "MMSE=22". Most of the 8 point loss may be attributed of orientation or language, and this is knowledge may be additionally useful e.g. when providing decisions about further examinations. It is quite common in non-professional discussions to see an identification between dementia and Alzheimer's disease, and the latter is indeed the most common form of dementia, representing something like two thirds, or more, of all dementia diseases. Difficulties an impairments in speech, like in aphasia, can be caused by circumstances related to vascular diseases. Indeed the Hackinski scale indicates vascular dementia if an older person has had a previous stroke, is medicated for hypertension, and shows signs of depression. Differentiating e.g. between Alzheimer's disease, dementia with Lewy bodies and vascular dementia is then important, e.g. since inhibitor drugs may be used for Alzheimer's disease and dementia with Lewy body but have no effect for vascular dementia.

Now we inevitably come to the discussion on side-effects of drugs, and wonder how those cholinesterase inhibitors may do good or harm for vascular dementia patients, if prescribed for a dementia patient under the assumption the specific diagnosis really is Alzheimer's disease and not vascular dementia. Before we continue, note here that even the autopsy diagnosis of Alzheimer's disease is not always clear, so the clinical diagnosis is obviously even more difficult. Accuracy in diagnosis is indeed not very high. In [2] there is a result that cholinesterase inhibitors reduce falls in Parkinson's disease. Recall that many other drugs like sedatives, in particular benzodiazepines, affect balance and increase falls in older persons. Parkinson's disease sometimes combines with dementia, and there is a diagnosis for Parkinson's disease with dementia, where dementia is related to Alzheimer's disease, but there is no single diagnosis code for "Parkinson's disease with Alzheimer's disease". Adding to this complicated picture, we should mention "vascular parkinsonism", which is also recognized and named, but not encoded as a particular diagnosis. An older person may be prescribed to use cholinesterase inhibitors because of Parkinson's disease which appears to be a Parkinson's disease with dementia. An inhibitor drug may have good effects on preventing falls. There are, however, other studies, e.g. [11], showing bad affects of psychotropic medications, including sedatives and cholinesterase inhibiting drugs, on falls risk. The medical literature with all its studies clearly provide mostly lots of good and useful recommendations, but the example above shows that sometimes there are contradictions. These contradictions cannot be overcome with some smart mathematical encoding of rules, but a more strict encoding about "what is what" certainly improves the understanding of underlying information and knowledge structures.

We need also distinguish between *Observation* and *Assessment*. Roughly speaking, Assessments assess situations which are described and represented by data, which in turn represent values in an Observation.

Even if assessment scales often come with an outcome index, we should not oversimplify the possible meaning of "assessment of assessments" as an outmost representation of the overall assessment of an older person's gerontological condition. This overall condition we could call a *Gerontium*, and we expect it to present all relevant information about the older person's gerontological condition, and not to hide any specific information behind some oversimplified index. The structure for this Gerontium must embrace the structures for Observation and Assessment.

A Gerontium is obviously not just a sort or an operator at some level of a signature. A Gerontium *is* the specific signature embracing the structures for Observation and Assessment. Even more so, a Gerontium is all three levels of signatures, including the relevant sorts and operators on all three levels. We may therefore introduce the notation

$$Gerontium = (\Sigma, \mathtt{S}_\Sigma, \Sigma', X_\Sigma, X_{\mathtt{S}_\Sigma}, X_{\Sigma'})$$

where $X_\Sigma = \{X_\mathtt{s}\}_{\mathtt{s} \in S}$ is in Ob($\mathtt{Set}_S$), $X_{\mathtt{S}_\Sigma}$ is in Ob($\mathtt{Set}$), and $X_{\Sigma'} = \{X_{\mathtt{s}'}\}_{\mathtt{s}' \in S'}$ is in Ob($\mathtt{Set}_{S'}$)

A gerontological condition can be initially specified as

$$\mathbf{GerontologicalCondition} :\rightarrow \mathbf{type}$$

on level two. This enables to have specific person data structures in form of a term and non-aggregated e.g. as

$$\mathbf{HighFallRisk}, \mathbf{DementiaWithLewyBodies} \Rrightarrow \mathbf{GerontologicalCondition}$$

on level two. On level three, we may then declare variables x and y e.g. according to x :: **HighFallRisk** and y :: **DementiaWithLewyBodies**. In this fashion we may continue to build up type declarations including assessment scales, and considering scales and scale data being part of sets, sets of sets, and so on, and further being specific about structured sets when ever they inevitably have to be invoked.

On level one we assume to have standard types, e.g. like nat, bool and string, which are implemented in respective programming languages or environments. These implementations are obviously algebras of these types and signatures at large. For instance, strings and string operations in C++ in general and languages within .NET cannot be assumed to be perfectly comparably, in particular when we would need to manipulate byte and bit representations related to strings. Within .NET, however, languages like VB and C# are expected to handle strings equivalently since these *core language* share the same language for *machine code*. Note then how we expect to have rather precise meanings e.g. of the algebras $\mathfrak{A}_{C++}(\mathbf{string})$ and $\mathfrak{A}_{C\#}(\mathbf{string})$. In the same way we interpret the arrangement of algebras for sorts and operators in *Gerontium*, so that e.g. $\mathfrak{A}_{Gerontium}(\mathbf{GerontologicalCondition})$ is specific for the end-user, or for a particular use in a older population regional repository.

In summary, part of the typing on level two can be as follows.

$$\begin{aligned}
\Rrightarrow &: \mathbf{type} \times \mathbf{type} \rightarrow \mathbf{type}\\
\boxtimes &: \mathbf{type} \times \mathbf{type} \rightarrow \mathbf{type}\\
\mathbf{nat}, \mathbf{bool} &: \rightarrow \mathbf{type}\\
\mathbf{bool3}, \mathbf{bool4}, \ldots, \mathbf{bool}n &: \rightarrow \mathbf{type}\\
\mathbf{bool}\natural, \mathbf{bool3}\natural, \mathbf{bool4}\natural, \ldots, \mathbf{bool}n\natural &: \rightarrow \mathbf{type}\\
\mathbf{boolStandardAssessmentScale} &: \rightarrow \mathbf{type}\\
\mathbf{Observation} &: \rightarrow \mathbf{type}\\
\mathbf{GerontologicalCondition} &: \rightarrow \mathbf{type}\\
\mathbf{Dementia}, \mathbf{ADL}, \mathbf{Depression}, \mathbf{Nutrition} &: \rightarrow \mathbf{type}\\
\mathbf{CognitiveDementia}, \mathbf{Non\text{-}CognitiveDementia} &: \rightarrow \mathbf{type}\\
\mathbf{GlobDetS}, \mathbf{Hackinski} &: \rightarrow \mathbf{type}\\
\mathsf{P}, \mathbf{Assessment}, \mathbf{AssessmentScale} &: \mathbf{type} \rightarrow \mathbf{type}
\end{aligned}$$

and on level three

$$\mathbf{DementiaDiagnosis} : \to \mathbf{Assessment}(\mathbf{Dementia})$$
$$\mathbf{CognitiveImpairment}, \mathbf{FallRiskAssessment} : \to \mathsf{P}(\mathbf{AssessmentScale})$$
$$\mathbf{MMSE}, \mathbf{GDS}, \mathbf{NPI} : \to \mathbf{AssessmentScale}$$
$$\mathbf{StandardAssessmentScale} : \to (\mathbf{boolStandardAssessmentScale}^n \Rrightarrow \mathbf{nat})$$
$$\mathbf{FES\text{-}I} : \to (\mathbf{bool4}\natural^{16} \Rrightarrow \mathbf{nat})$$

Obviously, the above are technical examples of typing in the signature *Gerontium*, and we have e.g. not included any typing for drugs. The ATC (Anatomic, Therapeutic, Chemical) encoding of drugs, however, is a good example which leaves only little room for various implementations, since the ATC code is already very strict and well-defined. Further, given that older person conditions are recognized and monitored quite differently depending on the purpose of and motivation for underlying needs of decision-making, e.g., as related to care levels, the structure and exemplification above is indicative only for pragmatic use. The structure is theoretically solid in general, but must be adapted as required to specific needs.

On level three, given a sort $\mathtt{s}$ we have to be careful to distinguish between $\mathtt{s}^n$ as appearing before $\to$ and $\Rrightarrow$, where in the first case 'product' is $\times$ and in the latter case $\boxtimes$. Clearly, neither of $\Rrightarrow$ and $\boxtimes$ can be assumed to be associative. Indeed, the algebra of $\Rrightarrow$ is a hom functor. However, $\boxtimes$ can be viewed as being associative, if we intuitively think of its algebra as being the categorical product.

Example 1 The algebra of **FES-I**, adopted by a certain organization org, is then a mapping $\mathfrak{A}_{org}(\mathbf{FES\text{-}I}) : \mathfrak{A}_{org}(\mathtt{bool4}\natural)^{16} \to \mathfrak{A}_{org}(nat)$, so that with org e.g. being THL (National Institute for Health and Welfare) in Finland, the mapping is specifically $\mathfrak{A}_{THL}(\mathtt{FES{-}I}) : \mathfrak{A}_{THL}(\mathtt{bool4}\natural)^{16} \to \mathfrak{A}_{THL}(nat)$ which uses, $\mathfrak{A}_{THL}(\mathtt{bool4}\natural) = \{\natural, 1, 2, 3, 4\}$ and $\mathfrak{A}_{THL}(nat) = \{\natural\} \cup \mathbb{N}$, so that whenever the number of missing answers, i.e. answer values being $\natural$, is less of equal to 4, then we divide the sum of all answer values with the number of answers and multiply it by 16, the total number of questions. If the number of missing answers exceed 4, then $\mathfrak{A}_{THL}(\mathtt{bool4}\natural)$ will return the $\natural$. THL says in their recommendations "then it cannot be computed", and our formalization translates that to the FES-I index being "missing". It should be remarked that, even if THL is a national authority, $\mathfrak{A}_{THL}(\mathtt{FES{-}I})$ is only a recommendation, so there is no legacy that disables any municipality in Finland to adopt their own specific $\mathfrak{A}_{municipality}(\mathbf{FES\text{-}I})$, if they, for any reason, would wish to do so. The situation is the same basically in all EU member states, and in countries and regions all over the world.

Clearly, more operators need to be included also for transformations between these types and operators. In particular, transformations like $\varphi : \mathsf{P}(\mathbf{Assessment}) \to \mathtt{Gerontium}$. Note also how terms like $t :: \mathsf{P}(\mathbf{Gerontium})$ could be seen as a population or set of older person's for which a municipality or regional decision-making body or group providing optimizations within care planning e.g. in home and residential care environments.

This example can then be extended further to work over selections of underlying categories with $\mathtt{Set}(\mathfrak{A})$ as the prime example for such an underlying category. Here

$\mathfrak{A}$ can be a lattice, quantale, or even a Kleene algebra, representing uncertainties. For more detail, see [8].

4 Conclusions and Future Work

This paper is related to on-going research and development in ICT support in the area of *Active Healthy Ageing*. Underlying ontologies are shown to be of utmost important, and a strict typing of information is a key factor in implementation of information management systems. We have focused on signatures and terms, and future work embraces decision-making involving selections of sentence functors and all the way through inference mechanisms to substitutions logics in its full strength as a support for ontology and decision-support with ageing.

References

1. American Psychiatric Association: Diagnostic and statistical manual of mental disorders, 4th edn. (DSM-IV-TR). Text Revisions, American Psychiatric Association (2000)
2. Chung, K.A., Lobb, B.M., Nutt, J.G., Horak, F.B.: Effects of a central cholinesterase inhibitor on reducing falls in Parkinson disease. Neurology **14**, 1263–1269 (2010)
3. Cohen-Mansfield, J., Marx, M., Rosenthal, A.: A description of agitation in a nursing home. J. Gerontol. **44**, M77–M84 (1989)
4. Cummings, J.L., Mega, M., Gray, K., Rosenberg-Thompson, S., Carusi, D.A., Gornbein, J.: The neuropsychiatric inventory: comprehensive assessment of psychopathology in dementia. Neurology **44**, 2308–2314 (1994)
5. Eklund, P.: Assessment scales and consensus guidelines encoded in formal logic, 19th IAGG World Congress of Gerontology and Geriatrics, Paris. J. Nutr. Health Aging **13**(Suppl 1), S558–S559 (2009)
6. Eklund, P.: signatures for assessment, diagnosis and decision-making in ageing, E. Hüllermeier, R. Kruse, and F. Hoffmann (Eds.): IPMU: Part II, CCIS, vol. 81, pp. 271–279 . Springer, Berlin (2010)
7. Eklund, P. Galán, M.A. Helgesson, R. Kortelainen, J.: Paradigms for many-sorted non-classical substitutions. 41st IEEE international symposium on multiple-valued logic (ISMVL 2011), 318–321 (2011).
8. Eklund, P. Galán, M.A. Helgesson, R. Kortelainen, J.: Fuzzy terms (submitted)
9. Eklund, P., Helgesson, R.: Monadic extensions of institutions. Fuzzy Sets Syst. **161**, 2354–2368 (2010)
10. Folstein, M., Folstein, S., McHugh, P.: Mini Mental State: a practical method for grading the cognitive state on patients for the clinician, J. Psychiatr. Res. **12**, 189–198 (1975)
11. Kim, D.H., Brown, R.T., Ding, E.L., Kiel, D.P., Berry, S.D.: Dementia medications and risk of falls syncope, and related adverse events meta-analysis of randomized controlled trials, J. Am. Geriatr. Soc. **59**, 1019–1031 (2011)
12. Morris, J.C.: The clinical dementia rating (CDR): current version and scoring rules. Neurology **43**, 2412–2414 (1993)
13. Reisberg, B. Borenstein, J. Salob, S.P. Ferris, S.H. Franssen, E. Georgotas, A.: Behavioral symptoms in Alzheimer's disease: phenomenology and treatment, J. Clin. Psychiatr. **48**(Suppl), 9–15 (1987)

14. Selbaek, G., Kirkevold, Ø., Engedal, K.: The prevalence of psychiatric symptoms and behavioural disturbances and the use of psychotropic drugs in Norwegian nursing homes. Int. J. Geriatr. Psychiatr. **22**(9), 843–849 (2007)
15. Yesavage, J.: Development and validation of geriatric depression screening scale: a preliminary report. J. Psychiatr. **17**, 37–49 (1983)

Decision Making on Energy Options: A Case Study

V. Jain, D. Datta and A. Deshpande

Abstract Major decisions are made without advance knowledge of their consequences including decision on energy options. In spite of the best efforts initiated in the development of renewable energy resources, it is too early to visualize that the ever-increasing gap between supply and demand of energy, for peaceful purposes, should be bridged in the near future. A mix of low carbon sources, including nuclear energy and renewable energy, while limiting greenhouse gases is considered a viable solution with less/no computations. In this chapter, a brief write up on Kahneman and Tversky's Prospect Theory is presented. Authors believe that the perception of gain, loss and risk are intrinsically fuzzy due to limited or no information about the future scenario. Computing with words, a facet of Restriction—Centered Theory of Reasoning and Computation (RCC) proposed by Prof. Lotfi A. Zadeh, could therefore be a useful armamentarium in decision making under risk and uncertainty. The case study, describing decision-making for the energy prospects (options) in India under risk and uncertainty, is presented by using prospect theory, type-1 and type-2 fuzzy relational calculus- a subset of Computing with Words. A commentary on safety of nuclear plants in India is an integral part of the chapter.

V. Jain (✉)
Department of Mathematics, Central University of Rajasthan, Kishangarh, Rajasthan, India
e-mail: vidyottama.jain@curaj.ac.in

D. Datta
Computational Radiation Physics Section, Health Physics Division, Bhabha Atomic Research Centre, Mumbai, India
e-mail: ddatta@barc.gov.in

A. Deshpande
Berkeley Initiative in Soft Computing (BISC)-Special Interest Group (SIG)—Environment Management Systems (EMS), College of Engineering Pune, Pune, India
e-mail: ashok_deshpande@hotmail.com

P. Guo and W. Pedrycz (eds.), *Human-Centric Decision-Making Models for Social Sciences*, Studies in Computational Intelligence 502,
DOI: 10.1007/978-3-642-39307-5_17,

Keywords Decision making · Prospect theory · Cumulative prospect theory · Type-1 and Type-2 fuzzy relations · Renewable and nonrenewable energy resources · Computing with words · Reference point

1 Introduction

Exponential population growth, rapid urbanization and industrialization resulted into increased energy needs and environmental degradation in almost all countries. Man has been exploiting the use of natural resources such as: water, minerals, oil · gas, and coal -the precious gift available from the Nature *termed as a real estate* for the welfare of mankind since time immemorial. Realizing the importance of these depleting resources, there have been concerted efforts on conducting research in the development of, preferably environmental friendly, non-conventional/alternate energy resources as the viable energy options. Some countries have made substantial financial allocation for application-oriented investigations on a variety of energy issues. It is too early to state that only renewable energy will not be able to fulfill the increasing energy needs of countries like China, India and so on. In practicality, the need of the hour is the energy mix.

In real life, most decisions are made without advance knowledge of their consequences with some degree of risk or uncertainty. The first and predominant economic theory for decision making under risk, formulated axiomatically by Von Neumann and Morgenstern [11], is expected utility theory. Further, experimental evidence shows that people violate the axioms of Von Neumann and Morgenstern [11]. Allais paradox is the most perfect and celebrated violation of expected utility theory. Kahneman and Tversky [7, 13] proposed a leading behavioral model of decision-making under risk and uncertainty based on decision utility, named as prospect theory, which accommodates Allais paradox and the violation of Von Neumann and Morgenstern axioms [11]. Recently, Guo [4] has proposed a thought provoking "one shot theory" to understand different behavior of decision makers and to find the best solution based on his/her attitude. This theory is typically for those situations where a decision is made only once.

It is noticeable that our discussion about decision making under risk is centered on decision utility (refers to the weight of an outcome in a decision or wanting) rather than expectation utility (represents to its hedonic quality or liking).

The chapter is organized as follows. Section 2 presents a summary of possible approaches in decision making under risk and uncertainty with focus on Prospect Theory. Contents of Sect. 3 are focused on Computing with Words (CWW) methodology with a few limitations of prospect theory perceived by the authors. In Sect. 4, a case study describing decision-making on India's viable energy option is detailed, using Cumulative Prospect Theory (CPT), type-1 and type-2 fuzzy relational calculus. We propose a computational scheme to demonstrate that nuclear energy will be the viable option, especially, for India to fulfill energy needs till 2030. Further, a brief write up on safety issues of the identified energy option for India is presented in Sect. 5. Concluding remarks and future directions for further research work are presented in Sect. 6.

2 Decision Making Under Risk and Uncertainty: Prospect Theory

Decision making theory is a holy grail of numerous studies in management science, economics and other areas. It comprises a broad diversity of approaches to modeling behavior of a decision maker realized under various information frameworks. In essence, the solution to the decision-making problem is defined by a preference framework and a type of decision-relevant information. In its turn preference and decision-relevant information frameworks are closely related. One of the approaches to formally describe preferences on the base of decision-relevant information is the use of a utility function. A utility function is a quantitative representation of preferences of a decision-maker (DM) [1].

Expected utility based on utility functions gained greater currency in economic theorizing when Von Neumann and Morgenstern [11] articulated a set of axioms that are necessary and sufficient to allow one to represent preferences by expected utility maximization. But very soon, these axioms were called into question. The Allais Paradox is perhaps the starkest and most celebrated violation of expected utility theory.

Further, Tversky and Kahneman have demonstrated in numerous highly controlled experiments that most people systematically violate all of the basic axioms of expected utility theory in their actual decision making behavior at least some times. In response to their findings, Tversky and Kahneman proposed a theory of choice, based on psychophysical model, which accurately describes how people go about making their decisions. The Original Prospect Theory (OPT), suggested by Kahneman and Tversky [7] in 1979, is based on non-linear transformation of outcome and probabilities, which allow describing psychological aspects of decision-making. The OPT developed for simple prospects with monetary outcomes and stated probabilities has three major characteristics:

Reference point dependence: An individual views consequences (monetary or other) in terms of changes from the reference point, which is usually that individual's status quo.
Diminishing sensitivity: The values of the outcomes for both positive and negative consequences of the choice have the diminishing returns characteristic. That means limit values of gains and losses decrease with an increase of their absolute values.
Loss aversion: Losses loom larger than gains, which mean people prefer "not to bear losses".

OPT predicts that people go through two distinct stages while making decisions. In the first phase, decision makers are predicted to edit a complicated decision into a simpler prospect, usually specified in terms of gains or losses. In the second phase, the decision makers evaluate each of the edited prospects available to them and choose the prospect of the highest value between the edited prospects. This evaluation is expressed in terms of two scales π and v. The first scale π associates with each probability p a decision weight $\pi(p)$ which shows the impact of p on the overall value of the prospect. The second scale, v, assigns to each outcome a number $v(x)$ that gives the subjective value of that outcome x. Therefore, the evaluation function for a prospect (x_i, p_i) is given by

$$V(x_i,\ p_i) = \sum v(x_i) * \pi(p_i)$$

where p_i is perceived probability of outcome x_i, $\pi\ (p_i)$ is the probability weighting function and $v\ (x_i)$ is value function.

The value function $v(x_i)$, depicted in Fig. 1, is defined on deviations from a reference point, concave for gains and convex for losses and steeper for losses than for gains.

Here, the probability weighting function is a monotonic function defined over (0,1). Consequently, the weighting function does not always satisfy stochastic dominance.

Also, in their experiments Kahneman and Tversky observed that the interplay of over weighting of small probabilities and concavity-convexity of the value function leads to the so-called fourfold pattern of risk attitudes: risk-averse for high probability gains and low probability losses; risk-seeking for low probability gains and high probability losses.

In brief, OPT encounters two problems:

1. Weighting function does not always satisfy stochastic dominance, and
2. OPT cannot be applied to prospects with a large number of outcomes.

These problems can be resolved by the rank dependent model or cumulative functional first proposed by Quiggin [12] for decision under risk. On the basis of rank dependent model, Tversky and Kahneman [13] proposed cumulative representation of prospect theory, which applies rank dependent model separately to gains and losses. Also, this cumulative prospect theory can be applied to uncertain as well as risky prospects with any number of outcomes.

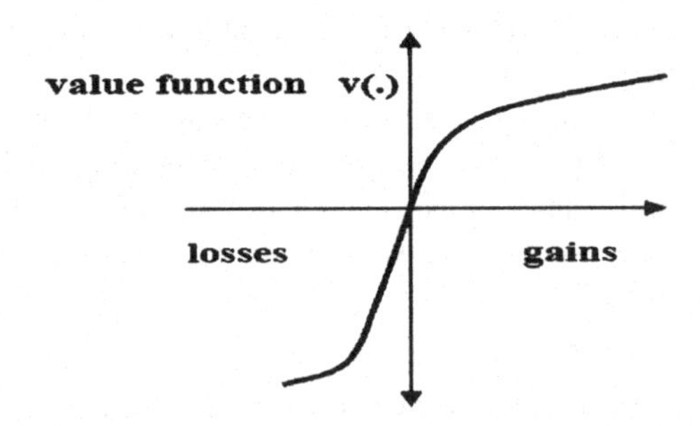

Fig. 1 Value function v as a function of gains and losses

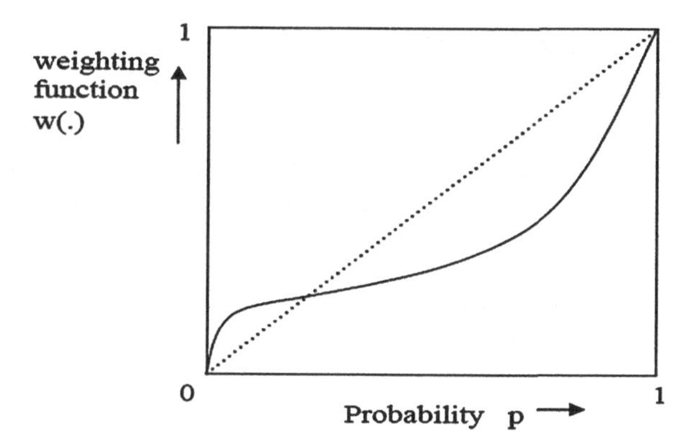

Fig. 2 Weighting function w for gains as a function of the probability p of a chance event

Following Tversky and Kahneman [13], the value function can be parameterized as a power function

$$v(x) = \begin{cases} x^{\alpha}, x \geq 0 \\ -\lambda(-x)^{\beta}, x < 0 \end{cases}$$

where α, β measure the curvature of the value function for gains and losses, respectively, and λ is the coefficient of loss aversion. This value function for gains and losses is increasingly concave and convex respectively for α, β <1. The weighting function, defined by Tversky and Kahneman [13], is an inverse-S-shaped weighting function. It is concave near 0 and convex near 1 as presented in the Fig. 2. It is very clearly explaining the fourfold pattern of risk attitudes as the low are overweighted (leading to risk seeking for gains and risk aversion for losses) and high probabilities are underweighted the weighting function (leading to risk seeking for losses and risk aversion for gains). It also satisfies Allais paradox. Therefore, this modified inverse-S-shaped weighting is more consistent with a range of empirical findings.

Following Lattimore et al. [10], the weighting function can be parameterized in the following form

$$w(p) = \left(\frac{\delta p^{\gamma}}{\delta p^{\gamma} + (1-p)^{\gamma}}\right)$$

It assumes that the relation between w and p is linear in a log-odds metric. Here δ measures the elevation of the weighting function and γ measures its degree of curvature.

3 Decision Making Under Risk and Uncertainty: Computing with Words

While making decision, the decision maker always thinks about gains and losses, however, we realize that his/her brain never perceive concave or steep convex functions as suggested in PT. In addition, the estimation of the functions suggested in the PT models [7, 13] is almost impossible to achieve in practice, especially energy scenario. The authors believe that, in real life, most of the decisions are taken by the domain knowledge experts with recourse to their perceptions.

3.1 Computing with Words (CWW) Methodology

Computing with Words (CWW) [15–17] offers an important capability to compute with information described in natural language. This opens the door to a wide-ranging enlargement of the role of natural languages in scientific theories and engineering systems.

The importance of Computing with Words derives from the fact that much of human knowledge is based on perceptions.
Levels of Complexity in CWW:

Level 0 CWW: Dana is 25; Tandy is 3 years older, than Dana Tandy is $(25+3)$ years old.
Level 1 CWW: Dana is young; Tandy is a few years Older than Dana; Tandy is $(\text{young}+\text{few})$ years old.
Level 2 CWW: Most Swedes are tall; Most tall Swedes are blond; Most^2 Swedes are blond.

Here, Level 1 CWW and Level 2 CWW require *precisiation* of meaning, which is much simpler in Level 1 CWW than in Level 2 CWW.

The fuzzy relation formalism based on type-1 and type-2 fuzzy sets - level 1 CWW approach [8, 9, 15–17] is used in the following section to rank the conventional energy options.

4 Ranking of Conventional Energy Options: Case Study

In this section, applications of Prospect theory and of variants of Computing with Words methodology are presented in deciding viable conventional energy options in Indian Scenario.

Let us assume that the projected total energy needs of India in the year 2030 are 1,000,000 MWe [5]. Assume 70 % power generation using non-conventional energy resources. Therefore, the balance energy needs of 300,000 MWe will have to be met from conventional energy resources. Keeping in view the ground realities that the country has already installed power plants of around 160,000 MWe. Thus, the additional electrical energy requirements in 2030 will be approximately 140,000 MWe.

In real life, domain expert's knowledgebase plays central role in most of the policy issues. Projection of future energy needs is invariabilty based on some computational method and this is the first level of uncertainty involved in decision analysis. The computational framework in this section is, therefore, based on energy expert's tacit knowledge and their logical assumptions. The discussion with energy experts was centered around on availability of energy resources, subjective probability in expected gains and losses (while using prospect theory) and the possbility of association between gains and loss while using fuzzy relational calculus via computing with words. The numerical values reported in this document are, therefore, not probability measures but are subjective measures.

Formalism I (Prospect Theory)

The collation of the information with proposed break up is as follows.

Generated power through hydro energy is 30,000 MWe (gain) on the basis of water availability with probability, Say, 0.98. The probability of its failure i.e. risk due to accident (man made/ natural disaster) is, say, 0.02. It is difficult to quantify the loss of human life due to natural/ man made disasters; but in an economic analysis/decision making, we are left with no other option but to convert these losses into cost units; furthermore assuming that the cost units are transformed into energy equivalent, and is 8,000 MWe (loss)—a pessimistic scenario. In this case, low value of availability is assumed with the reasoning that even if the neighbor nation, Nepal, has entered into bi-lateral agreement till 2030, however, the construction phase of such water storages (dam) would take at least 10–15 years—more than the plan period. In addition to this, there has been growing concern on seismic hazards!

Computed need for the thermal power could be a maximum of 10,000 MWe (gain). The assumed production capacity is very less but the probability of availability can be considered as 0.5 because at one side coal, oil and gas reserves are depleting, while on the other side India could be in a position to import good quality coal for their future thermal power plants. Coal with low ash content is being imported from Indonesia and Australia. As the coal reserves are depleting, these countries might rethink on their export policy. Coal washing does help to somewhat in reducing the problem. The health risks due to releases of SO_2, NO_x, Particulate Matter (PM), and other harmful gases from these plants is a common scene in India. Therefore, the probability of risk due to accidental releases of hazardous gases, chemicals, and high ash is assumed to be, say, 0.50. Assume that the cost can be converted in economic units and then could be transformed into energy equivalent as 1,000 MWe (loss).

Argue that India has developed nuclear fuel—say Thorium in adequate quantity to meet future needs for another five decades. India has also entered into an agreement for the supply of nuclear fuel such as uranium for the sustained nuclear power generation for peaceful purpose, using advanced nuclear technologies. Thus, maximum expected nuclear power could be 100,000 MWe with the probability, say, 0.8. Its availability is high due to the exploration for the fuel and the assurance given by another country under the bilateral agreement, say for 20 years. Assuming high probability value signifies risk adverse attitude of a common man as he has almost no realization of other even involuntary and voluntary risks. Therefore, the probability for risk (loss) could be assumed as 0.2 which indicates that thermal power plants are prone to high risk than nuclear power plants—a debatable issue, world over! Assuming that the energy equivalent, to the loss of life and other effects due to nuclear radiation, could be predicted as 10,000 MWe (loss). Typical decision tree for nuclear energy resource is presented in Fig. 3.

Consider x_1 is the generated power (outcome as gain) depending on the availability and cost of the source with probability p_1 and x_2 is failure of power (outcome as loss) with the probability p_2. In order to decide the most viable prospect under risk, the computations are performed for the estimation of the evaluation function for 30 % and 50 % using following expression with the above given assumptions:

$$V(f) = v(x_1)\,\pi(p_1) + v(x_2)\,\pi(p_2)$$

where $v(x_1)$, $\pi(p_1)$, $v(x_2)$ and $\pi(p_2)$ will be calculated as explained in Sect. 2.

The cumulative prospect theory parameters, $\alpha, \beta, \lambda, \delta$ and γ, can all be estimated for individuals using simple choices tasks on computer. Although the typically measured values of these parameters suggest an S-shaped value function $0 < \alpha, \beta < 1$with loss aversion ($\lambda > 1$) and an inverse S-shaped weighting function ($0 < \gamma < 1$) that crosses the identity line below 0.5 ($0 < \delta < 1$), there is considerable heterogeneity between individuals in these measured parameters. For instance, in a sample of ten psychology graduate students evaluating gambles involving only the possibility of gains, Gonzalez and Wu [3, 14] obtained measures of α ranging from 0.23 to 0.68, δ ranging from 0.21 to 1.51, and γ ranging from 0.15 to 0.89. Also, Tversky and Kahneman [13] estimated median values of $\alpha = \beta = 0.88$ and $\lambda = 2.25$ among their samples of college students.

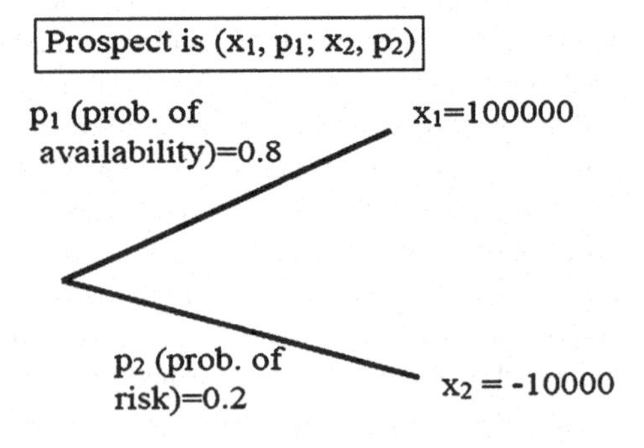

Fig. 3 Typical decision tree for nuclear power

Table 1 30 % energy is from conventional sources

Energy options	x_1 (MWe)	p_1	x_2 (MWe)	p_2	V(f) (MWe)
Hydro	30,000	0.95	−8,000	0.02	6920
Thermal	10,000	0.5	−1,000	0.5	1080
Nuclear	1,00,000	0.8	−10,000	0.2	13850

Conduct an opinion poll amongst different well informed stakeholders including the decision makers in order to estimate these parameters based on their tacit knowledge. This might result into somewhat similar probability values.

The values of these parameters, used in this illustration, are $\alpha = \beta = 0.88$, $\lambda = 2.25$, $\delta = 0.86$ and $\gamma = 0.52$ [10, 13].

Therefore, two cases are illustrated as follows:

30 % energy of the total energy needs (300,000 MWe) is from conventional sources as described in Table 1 after considering that the existed installed capacity is 160,000 MWe.

50 % energy of the total energy needs (500,000 MWe) is from conventional sources as described in Table 2 after considering that the existed installed capacity is 270,000 MWe.

Finally, it can be seen that Country *A* should go for nuclear power as a viable option.

Formalism II (Computing with Words (CW) Methodology)

In this methodology, the collation of the logically assumed information is as follows.

Power generation (gain) through hydro energy is **medium** (as the water resources are diminishing due to environmental degradation) on the basis of water availability. The probability of its failure i.e. risk due to accident (man made/natural disaster) is also assumed to be **medium** by the domain expert group.

Computed energy generation using coal and oil/and gas termed as thermal power is **very low**. Also its risk of failure is **very low**.

Consider Country A is in the final phase of its development of nuclear fuel and also in agreement with some other country for supply of nuclear fuel for the sustained

Table 2 50 % energy is from conventional sources

Energy options	x_1 (MWe)	p_1	x_2 (MWe)	p_2	V(f) (MWe)
Hydro	60,000	0.95	−13000	0.02	12930
Thermal	20,000	0.5	−2700	0.5	1700
Nuclear	1,50,000	0.8	−20000	0.2	18880

power generation. Therefore, power generation through nuclear option is **very high**. Somehow, nuclear power plants have been over criticized because of one single unfortunate accident at Chernobyl plant. Keeping in view the pubic sentiments alone, risk of its failure is assumed as **very high**.

A word of caution: It is hoped that the concerned organizations will take cognizance of diligently carried out environment impact assessment studies and the public debate about the feasibility these plant locations. Only intellectuals, with clean image who can work selflessly with the local community, should be involved in the process of site location for any plant.

Type-I Fuzzy Relational Calculus

Consider Fuzzy set A for generated power wherein outcome is as gain, X, which depends upon the availability and overall cost / MWe. Fuzzy set B refers to the overall energy requirement, Y, while set C expresses the failure of power (outcome as loss), Z, termed equivalent. The membership functions for X, Y and Z are considered as linear increasing functions.

Note: Understanding of meaning is a prerequisite to precisiation of meaning. Precisiation of meaning is a prerequisite to computation; the expressions given below justify the concept of precisiation proposed by Professor Zadeh in his seminal work on CW methodology. It is of interest to note that the concept of precisiation, in the sense in which it is used in CW, does not exist within linguistics or computational linguistics. Expressions stated below explain the concept of precisiation.

The mathematical functions of membership functions for gains, total energy requirement, and losses (expressed as energy equivalent) are μ_A(x), μ_B(y) and μ_C(z) respectively and given as:

$$\mu_A(x) = \begin{cases} 0, & 0 \le x \le 10000 \\ \frac{(x-a)}{(b-a)}, & 10000 \le x \le 110000 \\ 1, & x \ge 110000 \end{cases}$$

$$\mu_B(y) = \begin{cases} 0, & 0 \le y \le 110000 \\ \frac{(y-a)}{(b-a)}, & 110000 \le y \le 160000 \\ 1, & y \ge 160000 \end{cases}$$

and

$$\mu_C(z) = \begin{cases} 0, & 0 \le z \le 1000 \\ \frac{(z-a)}{(b-a)}, & 1000 \le z \le 14000 \\ 1, & z \ge 14000 \end{cases}$$

We have considered total energy requirement as approx. 140,000 MWe, Y can be considered as [140,000, 150,000]; where $y_1 = 140{,}000$ MWe and $y_2 = 150{,}000$ MWe.

Nuclear Energy Option

In case of the nuclear option, consider *high energy gain* and *high risk,* defined by the user as follows

Gain Loss

Consider a fuzzy set $\underset{\sim}{A}^N$ (near optimum-gain) with $x_1 = 90{,}000$ MWe and $x_2 = 100{,}000$ MWe, and a set $\underset{\sim}{C}^N$ (expected-loss) with $z_1 = 10{,}000$ MWe and $z_2 = 11{,}000$ MWe, based on user's input.

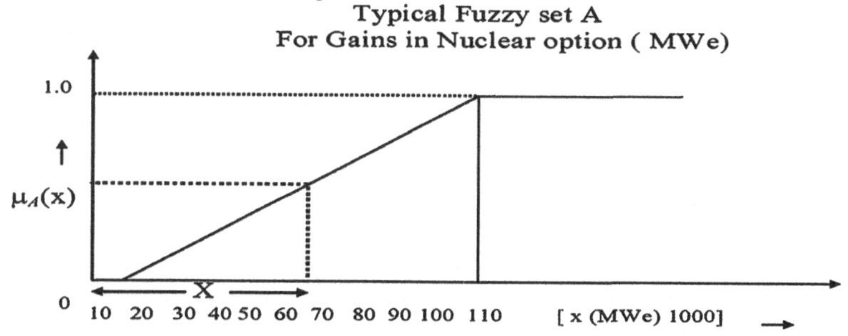

Membership function of fuzzy set A:

$$\mu_A(x) = \begin{cases} 0, & 0 \le x \le 10,000 \\ \frac{(x-10000)}{100000}, & 10{,}000 \le x \le 110{,}000 \\ 1, & x \ge 110{,}000 \end{cases}$$

The fuzzy Cartesian product between $\underset{\sim}{A}^N$ and $\underset{\sim}{B}^N$ and $\underset{\sim}{B}^N$ and $\underset{\sim}{C}^N$ could be worked out as given below:

$$\underset{\sim}{R}^N = \begin{matrix} & y_1 & y_2 \\ x_1 & 0.6 & 0.8 \\ x_2 & 0.6 & 0.8 \end{matrix}$$

$$\underset{\sim}{S}^N = \begin{matrix} & z_1 & z_2 \\ y_1 & 0.6 & 0.6 \\ y_2 & 0.69 & 0.77 \end{matrix}$$

Since fuzzy relation $\underset{\sim}{R}^N$ is defined from X to Y and fuzzy relation $\underset{\sim}{S}^N$ is defined from Y to Z, then fuzzy max-min composition between $\underset{\sim}{R}^N$ and $\underset{\sim}{S}^N$ results into the following fuzzy relation matrix $\underset{\sim}{T}^N$ as

$$\underset{\sim}{T}^N = \begin{array}{c} \\ x_1 \\ x_2 \end{array} \begin{array}{c} z_1 \quad z_2 \\ \begin{bmatrix} 0.69 & 0.77 \\ 0.69 & 0.77 \end{bmatrix} \end{array}$$

Hydro Energy Option

In case of the Hydro energy option, the user defines medium energy gain and medium risk as X (50,000, 60,000) and Z (8,000, 9,000). Following similar computational procedure presented in nuclear option, the final matrix after using the compositional rule of inference or max–min composition works out to be:

$$\underset{\sim}{T}^{Hy} = \begin{array}{c} \\ x_1 \\ x_2 \end{array} \begin{array}{c} z_1 \quad z_2 \\ \begin{bmatrix} 0.4 & 0.4 \\ 0.5 & 0.5 \end{bmatrix} \end{array}$$

Thermal Energy Option

In case of the thermal energy option, very low energy gain and very low risk are defined by the user as follows X (13,000, 18,000) and Z (1,500, 2,200). Since fuzzy relation $\underset{\sim}{R}^{Th}$ is defined from X to Y and fuzzy relation $\underset{\sim}{S}^{Th}$ is defined fuzzy relation matrix $\underset{\sim}{T}^{Th}$ as

$$\underset{\sim}{T}^{Th} = \begin{array}{c} \\ x_1 \\ x_2 \end{array} \begin{array}{c} z_1 \quad z_2 \\ \begin{bmatrix} 0.03 & 0.03 \\ 0.08 & 0.08 \end{bmatrix} \end{array}$$

Now, defuzzify all the fuzzy relation matrices $\underset{\sim}{T}^N$, $\underset{\sim}{T}^{Hy}$ and $\underset{\sim}{T}^{Th}$ by considering $\alpha = 0.08$, therefore, we have

$$T^N = \begin{array}{c} \\ x_1 \\ x_2 \end{array} \begin{array}{c} z_1 \quad z_2 \\ \begin{bmatrix} 1 & 1 \\ 1 & 1 \end{bmatrix} \end{array}, \quad T^{Hy} = \begin{array}{c} \\ x_1 \\ x_2 \end{array} \begin{array}{c} z_1 \quad z_2 \\ \begin{bmatrix} 1 & 1 \\ 1 & 1 \end{bmatrix} \end{array}, \quad T^{Th} = \begin{array}{c} \\ x_1 \\ x_2 \end{array} \begin{array}{c} z_1 \quad z_2 \\ \begin{bmatrix} 0 & 0 \\ 0 & 1 \end{bmatrix} \end{array}$$

It is prudent to calculate the ratio as that could alone help in deciding the viable option.
Calculate ratio of loss and gain
For thermal:

$$\frac{z_2}{x_2} = \frac{2200}{18000} = 0.122$$

For hydro:

$$\frac{z_1}{x_1} = 0.16, \quad \frac{z_1}{x_2} = 0.133, \quad \frac{z_2}{x_1} = 0.18, \quad \frac{z_2}{x_2} = 0.15$$

For Nuclear:

$$\frac{z_1}{x_1} = 0.11, \quad \frac{z_1}{x_2} = 0.1, \quad \frac{z_2}{x_1} = 0.122, \quad \frac{z_2}{x_2} = 0.11$$

We can draw useful conclusion form the computational procedure that the ration of loss and gain is less for Nuclear energy option. Therefore, Nuclear Energy will be the viable option or prospect.

Type-II Fuzzy Relational Calculus

Nuclear Energy Option

Consider a fuzzy set $\tilde{A}_N$ (near optimum-gain) with $x_1 = 90{,}000\,\text{MWe}$ and $x_2 = 100{,}000\,\text{MWe}$, and a set $\tilde{C}_N$ (expected-loss) with $z_1 = 10{,}000\,\text{MWe}$ and $z_2 = 11{,}000\,\text{MWe}$, based on expert's input.

The fuzzy Cartesian product between $\tilde{A}_N$ and $\tilde{B}_N$ and, $\tilde{B}_N$ and $\tilde{C}_N$ could be worked out as given below:

$$\tilde{R}_N = \begin{array}{c} \\ x_1 \\ x_2 \end{array} \begin{array}{c} y_1 \; y_2 \\ \left|\begin{array}{cc} 0.6 & 0.8 \\ 0.6 & 0.8 \end{array}\right| \end{array} \text{ and } \tilde{S}_N = \begin{array}{c} \\ y_1 \\ y_2 \end{array} \begin{array}{c} z_1 \; z_2 \\ \left|\begin{array}{cc} 0.6 & 0.6 \\ 0.69 & 0.77 \end{array}\right| \end{array}$$

Let us consider the type-2 fuzzy relations by adding some uncertainty to the type-1 fuzzy relations.

$$\underset{\approx}{R}_N = \begin{array}{c} \\ x_1 \\ x_2 \end{array} \begin{array}{c} \qquad y_1 \qquad\qquad\qquad y_2 \\ \left|\begin{array}{cc} \frac{0.25}{0.45} + \frac{1}{0.6} + \frac{0.5}{0.7} & \frac{0.33}{0.6} + \frac{1}{0.8} + \frac{0.5}{0.9} \\ \frac{0.25}{0.45} + \frac{1}{0.6} + \frac{0.5}{0.7} & \frac{0.33}{0.6} + \frac{1}{0.8} + \frac{0.5}{0.9} \end{array}\right| \end{array}$$

and

$$\underset{\approx}{S}_N = \begin{array}{c} \\ y_1 \\ y_2 \end{array} \begin{array}{c} \qquad z_1 \qquad\qquad\qquad z_2 \\ \left|\begin{array}{cc} \frac{0.25}{0.45} + \frac{1}{0.6} + \frac{0.5}{0.7} & \frac{0.25}{0.45} + \frac{1}{0.6} + \frac{0.5}{0.7} \\ \frac{0.53}{0.55} + \frac{1}{0.69} + \frac{0.63}{0.8} & \frac{0.23}{0.6} + \frac{1}{0.77} + \frac{0.87}{0.8} \end{array}\right| \end{array}$$

Since fuzzy relation $\underset{\approx}{R}_N$ is defined from X to Y and fuzzy relation $\underset{\approx}{S}_N$ is defined from Y to Z, then fuzzy max-min composition between $\underset{\approx}{R}_N$ and $\underset{\approx}{S}_N$ results into the following fuzzy relation matrix $\underset{\approx}{T}_N$ as

$$\underset{\approx}{T_N} = \begin{matrix} x_1 \\ x_2 \end{matrix} \left| \begin{matrix} \overbrace{\frac{0.25}{0.55} + \frac{0.53}{0.6} + \frac{1}{0.69} + \frac{0.5}{0.7} + \frac{0.63}{0.8}}^{z_1} & \overbrace{\frac{0.33}{0.6} + \frac{0.33}{0.7} + \frac{1}{0.77} + \frac{0.87}{0.8}}^{z_2} \\ \frac{0.25}{0.55} + \frac{0.53}{0.6} + \frac{1}{0.69} + \frac{0.5}{0.7} + \frac{0.63}{0.8} & \frac{0.33}{0.6} + \frac{0.33}{0.7} + \frac{1}{0.77} + \frac{0.87}{0.8} \end{matrix} \right|$$

Hydro energy option

In case of the Hydro energy option, the user defines medium energy gain and medium risk as $X = (50{,}000, 60{,}000)$ and $Z = (8{,}000, 9{,}000)$. Following similar computational procedure presented for nuclear option, the final matrix after using the compositional rule of inference or fuzzy max-min composition works out to be:

$$\underset{\approx}{T_{Hy}} = \begin{matrix} x_1 \\ x_2 \end{matrix} \left| \begin{matrix} \overbrace{\frac{0.5}{0.2} + \frac{1}{0.4} + \frac{0.25}{0.53} + \frac{0.25}{0.6}}^{z_1} & \overbrace{\frac{0.5}{0.2} + \frac{1}{0.4} + \frac{0.25}{0.53} + \frac{0.25}{0.6}}^{z_2} \\ \frac{0.3}{0.2} + \frac{0.7}{0.4} + \frac{1}{0.5} + \frac{0.5}{0.53} + \frac{0.5}{0.7} & \frac{0.3}{0.2} + \frac{0.25}{045} + \frac{1}{0.5} + \frac{0.5}{0.6} + \frac{0.5}{0.62} + \frac{0.5}{0.7} \end{matrix} \right|$$

Thermal energy option

In case of the thermal energy option, *very low* energy gain and *very low* risk are defined by the user as follows $X = (13{,}000; 18{,}000)$ and $Z = (1{,}500; 2{,}200)$. Since fuzzy relation $\underset{\approx}{R_{Th}}$ is defined from X to Y and fuzzy relation $\underset{\approx}{S_{Th}}$ is defined from Y to Z, then fuzzy max-min composition between $\underset{\approx}{R_{Th}}$ and $\underset{\approx}{S_{Th}}$ results into the following fuzzy relation matrix $\underset{\approx}{T_{Th}}$ as

$$\underset{\approx}{T_{Th}} = \begin{matrix} x_1 \\ x_2 \end{matrix} \left| \begin{matrix} \overbrace{\frac{0.4}{0.01} + \frac{0.4}{0.02} + \frac{1}{0.03} + \frac{0.5}{0.04} + \frac{0.5}{0.05}}^{z_1} & \overbrace{\frac{0.4}{0.01} + \frac{0.4}{0.02} + \frac{1}{0.03} + \frac{0.5}{0.04} + \frac{0.5}{0.05}}^{z_2} \\ \frac{0.4}{0.05} + \frac{0.5}{0.07} + \frac{1}{0.08} + \frac{0.3}{0.09} + \frac{0.3}{0.1} & \frac{0.4}{0.05} + \frac{0.5}{0.07} + \frac{1}{0.08} + \frac{0.3}{0.09} + \frac{0.3}{0.1} \end{matrix} \right|$$

By finding centroids, $\underset{\approx}{T_N}$, $\underset{\approx}{T_{Hy}}$ and $\underset{\approx}{T_{Th}}$ can be converted to the following

$$\tilde{T}_N = \begin{matrix} & z_1 & z_2 \\ x_1 & 0.687 & 0.749 \\ x_2 & 0.687 & 0.749 \end{matrix}, \quad \tilde{T}_{Hy} = \begin{matrix} & z_1 & z_2 \\ x_1 & 0.391 & 0.391 \\ x_2 & 0.485 & 0.535 \end{matrix}$$

and

$$\tilde{T}_{Th} = \begin{matrix} & z_1 & z_2 \\ x_1 & 0.031 & 0.031 \\ x_2 & 0.0768 & 0.0768 \end{matrix}$$

It can be seen that fuzzy relation matrix $\tilde{T}$ shows the relationship between energy gain and energy loss with respect to energy options in fuzzy scenario. Components of matrix $\underset{\sim}{T}$ are called membership values, which are expressing degrees of strength of the relation between loss and gain on the unit interval [0,1].

In case of thermal energy option, these membership values are comparatively very low. Therefore, the possibility of relationship of loss and gain for thermal energy option is less. This clearly shows that in 2030, thermal energy option can be ignored. In case of the other options i.e. Hydro and Nuclear Energy options, the membership value of relationship is high. Therefore, the possibility of that relation is more. Further, it is visible that the highest membership value or the highest possibility is **0.77**, which is the nuclear option.

Mathematically, after ignoring thermal energy option, by considering $\alpha = 0.485$, fuzzy relation matrices $\overset{\approx}{T}_N$ and $\overset{\approx}{T}_{Hy}$ can be defuzzified as follows:

$$\tilde{T}_N = \begin{matrix} & z_1 \; z_2 \\ \begin{matrix} x_1 \\ x_2 \end{matrix} & \begin{vmatrix} 1 & 1 \\ 1 & 1 \end{vmatrix} \end{matrix} \quad \text{and} \quad \tilde{T}_{Hy} = \begin{matrix} & z_1 \; z_2 \\ \begin{matrix} x_1 \\ x_2 \end{matrix} & \begin{vmatrix} 0 & 0 \\ 1 & 1 \end{vmatrix} \end{matrix}$$

To decide the viable option, calculate the ratio of loss and gain:

For the Nuclear energy option

$$\frac{z_1}{x_1} = 0.11, \quad \frac{z_1}{x_2} = 0.1, \quad \frac{z_2}{x_1} = 0.122, \quad \frac{z_2}{x_2} = 0.11$$

For Hydro energy option

$$\frac{z_1}{x_2} = 0.133, \quad \frac{z_2}{x_2} = 0.15$$

We can draw useful conclusion from the computational procedure that the ration of loss and gain is minimum in case of Nuclear energy option. It could be revealed from the computations that the final ranking of energy options is nuclear, i.e., Nuclear energy, using either type-1 [2] or type-2 fuzzy relational algorithm, ranks first. Hence, it can be finally concluded through fuzzy relation approach that Nuclear energy is the topmost viable option or prospect in order to meet the growing energy needs of India till 2030.

5 A Word on Safety Provisions in Nuclear Plants in India

Enhanced safety provisions are built in the nuclear reactors to avoid any accident. In nuclear reactors, energy is generated by fission of fissile (Uranium or Plutonium) nuclei in a continuous chain reaction. In addition to energy, the nuclear fission produces fission products that are radioactive. Nuclear power plants use fissile materials to produce energy in the form of heat, which is converted to electricity by conventional generating plant. Radioactive materials are produced as a by-product of this process. Whilst radioactive materials can have beneficial uses, such as in cancer therapy, they are generally harmful to health. Their use, and the process by which they are produced, must be strictly regulated to ensure nuclear safety. Apart from the management of fuel, nuclear safety particularly covers the design, construction, operation and decommissioning of all nuclear installations. It is important to understand why Fukoshuima type disaster is most unlikely in India before we discuss on nuclear safety.

Fukushima Daiichi accident caused due to earthquake followed by Tsunami is a series of equipment failures, nuclear meltdowns and release of radioactive materials. All the six reactors are light water, boiling water reactor. Many of the internal components and fuel assembly cladding being made from Zircalloy, cooling the reactor was essential because at temperature above 500o C in presence of steam zircalloy undergoes an exothermic reaction, zircalloy oxidizes and free hydrogen got produced. The reaction between the zirconium cladding and the fuel lowered the melting point of the fuel and melting of core got speed up. Inadequate cooling caused the release of radioactivity in the environment. Keeping in view of this disaster, it is obvious to put a question about the safety of nuclear plants during its design and operation under the normal as well as accident condition. It is also true that location wise (from the earthquake and tsunami pruned areas) nuclear plants should be safe first. It is known fact from the latest version of the earthquake resistant design code of India that four levels of seismicity have been assigned in terms of zone factors and zonings are ordered from 2 (extreme least) to 9 (extreme most). According to IS code the effective peak horizontal ground acceleration for zone 1 is 0.36 (36 % of gravity) also known as zone factor. Similarly the zone factor for zone 2 is 0.24, for zone 3 is 0.16 and for zone 4 is 0.1. Indian nuclear power plants are situated in Zone 2 and 3 except Narora plant in Uttar Pradesh, which is situated in Zone 4. **Japan's nuclear plants are in Zones 7, 8 and 9.**

The basic objective of nuclear safety, as a concept, is to protect the public, workers in the nuclear industry and the environment from radiological risks. In order to ensure safe operation of nuclear reactor, three basic safety functions have to be achieved in a sustained manner and these are:

Control of fission reaction:
The reactors are controlled by controlling the population of neutrons by use of neutron absorbers like boron and cadmium. The shutdown of the reactor takes place by inserting the control rods into the core and shutdown systems are designed to be fail-safe. Hence in case of power failure the rods drop due to gravity or the liquid poison is injected due to accumulator gas pressure.

Cooling of the reactor core:
During normal operation, heat is generated in the core due to nuclear fission. Even when the reactor is in shutdown state a small amount of heat is generated due to the decay of fission products (decay heat). The intensity of decay heat reduces slowly with time. The reactor therefore needs cooling continuously in all states. Reliable cooling normally takes place by two or more coolant circuits that help in removal of heat in case of failure of one circuit. Improvement is further achieved by the coolant

pumps provided with backup power supply from diesel generators and battery banks, which supply power during grid failure. In addition to this all the reactors are also provided with emergency core cooling system, which is independent from normal cooling systems. The emergency core cooling system ensures cooling of the core even if there is a leak in the coolant circuit.

Containment of the radioactive fission products:
Radioactive material is produced in the core of the reactor when fission occurs. Most of these fission products remain within the fuel itself under normal circumstances. However, to prevent their release to the environment under transient or accident conditions, at least three successive barriers are provided. The first barrier is the fuel clad within which the fuel is enclosed. The second barrier is the leak tight coolant circuit. The third barrier is the containment building around the coolant system. In some of the reactors, a secondary containment is provided for further protection.

So, in summary, safety in design of nuclear power reactor is provided by a graded approach called as "Defense in Depth". Safety in operation is achieved by a stringent administrative control with rigorous training and especially high skilled manpower. Indian nuclear plants are equipped with all the fore referred safety features.

6 Concluding Remaks

It is a deep-seated tradition in science to employ the conceptual structure of bivalent logic and probability theory as a basis for formulation of definitions and concepts. What is widely unrecognized is that, in reality, most concepts are based on perception of domain experts/individual. Fuzzy logic via computing with words could, therefore, be useful in perception-based modeling.

Perceptions of the domain experts remain as the backbone for CW methodology used in any decision-making problems. In this paper, authors infer that nuclear energy ranks as the first viable option in order to meet the energy needs of the country till 2030. In a real life scenario, the decision makers may not invariably carry out extensive social surveys for the estimation of several parameters required in Cumulative Prospect Theory/ Prospect Theory. The CW based techniques used in this paper offers a reasonable solution with less computational complexities as compare to the prospect theory.

Future energy needs of India are so large that unlike some other countries, no one or two energy resources will be adequate to bridge the gap between supply and demand. Therefore, the country is left with no other option than to decide on the proportions of these renewable and nonrenweable energy sources in order to meet ever increasing future energy needs. The case study presented herein is, therefore, a need based applied research. The data on future energy needs have been freely used from the literature published by the Government of India. The policy issues are invariably based on the expert's perception which is considered via the restricted centered theory of reasoning and computation (RCC) [17, 15].

Realising the complexity of energy policy issues reported in this sequel, application of other methods such as: Fuzzy Multi criteria Decision Making (FCDM), Fuzzy Analytical Hierarchical Procedure (FAHP) and alike will be attempted in our continued research efforts.

Acknowledgments We are deeply indebted and would like to express our immense gratitude towards Prof. Lotfi A. Zadeh, the father of fuzzy logic for his motivation and all helpful and insightful suggestions. Also, the authors thank the esteemed referees for their valuable comments and suggestions.

References

1. Aliev, R., Pedrycz, W., Fazlollahi, B., Huseynov, O.H., Alizadeh, A.V., Guirimov, B.G.: Fuzzy logic-based generalized decision theory with imperfect information. Inf. Sci. **189**, 18–42 (2012)
2. Deshpande, A., Jain, V.: Computing with words on energy options?—towards decision under risk and uncertainty. Int. J. Nucl. Knowl. Manag. **5**(2), 219–232 (2011)
3. Gonzalez, R., Wu, G.: On the shape of the probability weighting function. Cogn. Psychol. **38**, 129–166 (1999)
4. Guo, P.: One-shot decision theory. IEEE Trans. Syst. Man Cybern. Part A Syst. Hum. **4**(5), 917–926 (2011)
5. http://www.energias-renovables.com/articulo/grid-connected-policy-framework
6. Jain, V., Deshpande, A. : Prospect theory on energy options?—towards decision making under risk. In: 2nd International Conference on Reliability, Safety & Hazard, pp. 112–117 (2010)
7. Kahneman, D., Tversky, A.: Prospect theory: an analysis of decision under risk. Econometrica **4**, 263–291 (1979)
8. Karnik, N.N., Mendel, J.M.: An introduction to type-2 fuzzy logic systems. USC Report, http://sipi.usc.edu/_mendel/report (1998)
9. Karnik, N.N., Mendel, J.M.: Operations on type-2 fuzzy sets. Fuzzy Sets Syst. **122**, 327–348 (2001)
10. Lattimore, P.K., Baker, J.R., Witte, A.D.: The influence of probability on risky choice—a parametric examination. J. Econ. Behav. Organ. **17**(3), 377–400 (1992)
11. Von Neumann, J., Morgenstern, O.: Theory of games and economic behavior. Princeton Univ. Press, Princeton (1944)
12. Quiggin, J.: A theory of anticipated utility. J. Econ. Behav. Org. **3**(4), 323–343 (1982)
13. Tversky, A., Kahneman, D.: Advances in prospect theory: cumulative representation of uncertainty. J. Risk Uncertain. **5**(4), 297–323 (1992)
14. Wu, G., Gonzalez, R.: Curvature of the probability weighting function. Manag. Sci. **42**, 1676–1690 (1996)
15. Zadeh, L.A.: Computing with words—why? and how? WORLDCOMP 2010. Las Vegas, NV, USA (2010)
16. Zadeh, L.A.: Fuzzy sets. Inf. Control **8**(3), 338–353 (1965)
17. Zadeh, L.A.: Toward a perception based theory of probabilistic reasoning with imprecise probabilities. J. Stat. Planning Infer. **105**, 233–264 (2002)

Printed in the United States
By Bookmasters